FLUID FLOW PHENOMENA

FLUID MECHANICS AND ITS APPLICATIONS
Volume 55

Series Editor: R. MOREAU
MADYLAM
Ecole Nationale Supérieure d'Hydraulique de Grenoble
Boîte Postale 95
38402 Saint Martin d'Hères Cedex, France

Aims and Scope of the Series

The purpose of this series is to focus on subjects in which fluid mechanics plays a fundamental role.

As well as the more traditional applications of aeronautics, hydraulics, heat and mass transfer etc., books will be published dealing with topics which are currently in a state of rapid development, such as turbulence, suspensions and multiphase fluids, super and hypersonic flows and numerical modelling techniques.

It is a widely held view that it is the interdisciplinary subjects that will receive intense scientific attention, bringing them to the forefront of technological advancement. Fluids have the ability to transport matter and its properties as well as transmit force, therefore fluid mechanics is a subject that is particulary open to cross fertilisation with other sciences and disciplines of engineering. The subject of fluid mechanics will be highly relevant in domains such as chemical, metallurgical, biological and ecological engineering. This series is particularly open to such new multidisciplinary domains.

The median level of presentation is the first year graduate student. Some texts are monographs defining the current state of a field; others are accessible to final year undergraduates; but essentially the emphasis is on readability and clarity.

For a list of related mechanics titles, see final pages.

Fluid Flow Phenomena

A Numerical Toolkit

Edited by

PAOLO ORLANDI

Department of Mechanics and Aeronautics,
University of Rome 'La Sapienza', Italy

KLUWER ACADEMIC PUBLISHERS
DORDRECHT / BOSTON / LONDON

A C.I.P. Catalogue record for this book is available from the Library of Congress.

Additional material to this book can be downloaded from http://extras.springer.com.

ISBN 0-7923-6095-8

Published by Kluwer Academic Publishers,
P.O. Box 17, 3300 AA Dordrecht, The Netherlands.

Sold and distributed in North, Central and South America
by Kluwer Academic Publishers,
101 Philip Drive, Norwell, MA 02061, U.S.A.

In all other countries, sold and distributed
by Kluwer Academic Publishers,
P.O. Box 322, 3300 AH Dordrecht, The Netherlands.

Reprinted 2001

Printed on acid-free paper

Fluid Flow Phenomena

A Numerical Toolkit

Edited by

PAOLO ORLANDI

Department of Mechanics and Aeronautics,
University of Rome 'La Sapienza', Italy

KLUWER ACADEMIC PUBLISHERS
DORDRECHT / BOSTON / LONDON

A C.I.P. Catalogue record for this book is available from the Library of Congress.

Additional material to this book can be downloaded from http://extras.springer.com.

ISBN 0-7923-6095-8

Published by Kluwer Academic Publishers,
P.O. Box 17, 3300 AA Dordrecht, The Netherlands.

Sold and distributed in North, Central and South America
by Kluwer Academic Publishers,
101 Philip Drive, Norwell, MA 02061, U.S.A.

In all other countries, sold and distributed
by Kluwer Academic Publishers,
P.O. Box 322, 3300 AH Dordrecht, The Netherlands.

Reprinted 2001

Printed on acid-free paper

to Gino and Roberto

Table of Contents

Preface and Acknowledgments

I would like to dedicate this book to Prof. Luigi (Gino) Crocco. He completely changed my way of doing research, with him I passed from the dark to the light. I was really fascinated to see Luigi, a true scientist, in the last period of his life struggling with turbulence and checking on his pocket calculator the results I was producing with computers. He taught me to always double check all results. He passed away while he was writing a paper on the last research we did together on EDQNM closures for anisotropic sheared turbulence. I never had the capability to finish that paper.

In addition, I would like to point out that I increased my understanding of fluid mechanics with the help of computers and that I benefited from visits to the Mechanical Engineering Department at Stanford. At that prestigious university, I realized that they reached a high level of interchange of ideas between students, professors and visitors, and this was the reason for their exceptional level of achievement. By comparing my professional life at the University of Roma, and that at Stanford I was delighted when I was invited to initiate a collaboration with the group at the Mechanical Engineering Department of Stanford. In the United States I began to learn how to use the computer to study flow physics. For this I would like to thank Parviz Moin and John Kim, since, when they were at NASA Ames, they supported me writing a code to perform simulations of turbulence in complex geometries. From that period initiated all the material of this book. The numerical method for complex flows was not efficient; I remember those frustrating days when the iterative schemes to solve the "pressure" very often did not converge. The problem was too difficult for me, and so I was dreaming of dealing with direct methods. This is the reason why I decided to move to flows in cylindrical and Cartesian coordinates.

I had success in this new circumstance, and the main reason for this success is related to my encounter with Roberto Verzicco. Everybody knows that professors are always looking for good students. I was lucky; we enjoyed using numerics to understand vortex dynamics. Roberto was supposed to be the coauthor of this book, but, for beaurocratic reasons, he decided to spend his time becoming a professor, a position that he deserved for his hard work during these past years. To him I express my particular thanks, but I would like to acknowledge also the work I did together with the few other students I had: Marco Briscolini, Paola Cavazza, Massimiliano Fatica, Domenico Manna, Ever Fadlun, Roberta Serravall, Martin Vega, and David Ebstein.

In my life I collaborated with several scientist more famous than me; for instance, Emil Hopfinger, Gert Jan van Heijst and Bob Antonia. It is difficult to acknowledge all the scientists I met at CTR (Center of Turbulence Research), but particular thanks go to Karim Shariff for the scientific discussion we had during my periods of work at the building 202A1 at NASA Ames. At CTR I met Javier Jimenez and discussions with him sometimes depressed me; I was always losing, but as a result I was learning something new. During my visits at NASA Ames I remember the lunch discussions, where the most brilliant fellow was Alan Wray.

I devoted a large part of the book to geophysical flows; I learned this topic interacting with George Carnevale. I would like to thank him for his visits to Roma that every time brought the American atmosphere full of scientific discussions.

My research was supported by Italian money, in large part from the funds from the Ministero della Universita' e della Ricerca Scientifica e Tecnologica. I used the grant money to travel and to buy workstations; these machines changed my life allowing me to perform simulations without arguing with the administrators of the computer centers in Italy who never understood the requirements of scholars performing flow simulations. Particular thanks go to the administrator of NASA who allowed me to take the TURB3D package and with it to perform flow visualizations, a very important tool in the study of flow physics. Without Piergiorgio Esposito it would not have been possible to use this package on our IBM workstations.

Finally I would like in advance to thanks the users of the codes furnished with this book. I am sure they had a lot of patience on trying to understand the codes confiding on comments that in several places were not appropriate to that particular code. I am also sure that they found some bugs. For the latter I know that they can understand that the elimination of all the bugs is a very difficult task. For the former instead I am asking their comprehension; the reason is due to the fact that really I am enjoying to perform the simulations and I prefer this to spend my time on writing comments on codes. Any way in the case this book will have such a success to have a second edition I am promising that in the new edition the codes will have better comments. With these thanks I would like to encourage the users of the codes to send me their comments and their suggestions to improve the use of this toolkit.

1 Introductory comments and summary

1.1 Introduction

This book for the Kluwer series in Fluid Mechanics and its applications, concerns the simulation of the incompressible Navier-Stokes equations for laminar and turbulent flows. In this monograph the variety of numerical methods that has been developed in the last decades is not described, the book is limited only to explain and to employ the finite difference method that, to my point of view, is the simplest to understand. This method, in fact, is based on the concept of the Taylor series expansion, a concept comprehensible even by high school students; the basic ideas of the method can easily be understood by scholars which do not have a very deep mathematical background. Scholars interested to increase their knowledge of the different numerical methods to solve the incompressible Navier-Stokes equations can find a large number of books as those by Roache (1972), Peyret & Taylor (183), Hirsch (1991), Andersson *et al.*(1984), Fletcher (1987), Canuto *et al.* (1987) and several others. This book intends to be different from the previous ones, since the aim is also to furnish a large number of source codes which permit to play with the Navier-Stokes equations and to understand the complex physics related to fluid mechanics. My belief is that a numerical simulation is a useful tool to understand the complexity of the flows which often is difficult to derive from laboratory experiments. With this belief in mind, however, I would like to point out that the numerics cannot, completely, replace the information from experiments. At last everyone recognizes that the numerics together with the experiments is the only way to get the perfect knowledge of the flows, to reach the condition of a good flow control and thus to obtain benefits which could improve the efficiency of the products based on flow technology.

This book, then, can be very useful to scholars doing laboratory experiments. They usually spend a large amount of time in understanding new experimental techniques and they do not have extra time to study the large variety of numerical methods; furthermore they cannot spend more time in transferring one of the methods into a computer language. This procedure is annoying, since a lot of time is necessary to debug and test the codes and it is well known that a bug left in the code is very difficult and almost impossible to find. The experimentalists in these days need workstations or personal computers in the laboratories, so that they can use this material to have a complementary tool to understand flow behaviours and physical mechanisms difficult to detect from the experiments. In this case the experimentalists have to address their experiments to those flows that at low Re number maintain the complexity of the phenomena at higher Re. By means of numerical simulations, for example, insights into the vorticity field can be obtained and since a passive scalar can be introduced in numerical simulations the numerical flow visualizations can be validated by the real flow visualizations and then the differences between the vorticity field and the flow markers are understood.

This book can be used by graduate as well as undergraduate students while reading books on theoretical fluid mechanics; they learn how to simulate on their personal computers the dynamics of flow fields. This exercise will give them a better way to understand the theory. As an example I would like to mention the vorticity generation at solid walls that plays a fundamental role in several practical applications. The theory explains the generation only in the first stage but cannot be used to understand the successive flow evolution. This book together and in straight connection with the Lugt (1983) book on vortex dynamics that has the quality to describe the physics neglecting the complex mathematical formalisms could give to graduate and undergraduate students the bases to reach a deep comprehension of the complexity of vortex dynamics.

Someone can argue that a large number of commercial codes are sold by software companies to solve the Navier-Stokes equations; however, these products are directed to industries with the aim to give solutions for practical applications and in particular at high Reynolds numbers. This software requires turbulence models based on Reynolds averages. The defects of these turbulence models is well known and this is the reason why this subject is neglected in this book. However, two chapters on Large Eddy Simulations (LES) have been inserted since in these days these models have the possibility to have in the near future more universal models for practical applications. To reach this goal the way is long and difficult, and stronger efforts are necessary to reach the conditions to have reliable sub-grid models for LES. On this topic, still there is no clear idea whether the sub-grid model depends on the truncation errors of the numerical method; this is the reason why this topic has been vaguely attached. However, as mentioned in the book, finite-differences will always be used in LES for practical applications, thus these chapters can be a useful starting point to analyze the lacks and the qualities of LES.

On the contrary the direct simulation of the Navier-Stokes equations is a simple task, its limitations are related to the memory and speed of the available computers. The requirement is to resolve all the scales of the flows and to see whether or not this condition is satisfied; the grid refinement check is mandatory. Recently consensus has been reached that finite-differences are satisfactory for the direct simulation of turbulent flows, therefore a part of the book is devoted to the simulation of turbulent flows and in particular to wall turbulent flows.

This book, as mentioned above, can be used to enjoy the vortex dynamics, a very interesting subject that attracted the interests of several scientists in the past starting from Leonardo da Vinci who discovered the flow visualization technique. Since it is very difficult for today scholars to have Leonardo's skill in drawing, this book furnishes a tool to create the data-bases which by a good graphic package can make everyone a Leonardo sitting on the Arno river and, as Leonardo did, he can reproduce the fascinating evolution of vortical structures.

1.2 Summary of the chapters

A brief summary of the content of each chapter follows.

The solution of the Burgers equation is considered to give the first insights on finite differences. The use of a non-uniform grid is described for the Burgers equation, since non-uniform grids are necessary to save computational time in solving thin shear layers. The solution of the Burgers equation has also been used to introduce the time advancement discretization scheme.

As a second step, the simulation of two-dimensional flows is considered; in this case the vorticity streamfunction formulation is introduced together with the concept of global conservation in the discrete space. In the inviscid limit the vorticity and velocity invariants must be theoretically conserved, on the contrary in the "discrete world" the complexity of the numerical scheme is related to the number of conserved invariants. In the major part of the simulations it is mandatory to conserve in 2D energy and enstrophy, and this is very simple in the ω, ψ formulation. On the contrary, it is shown that the Navier-Stokes equations in primitive variables require a more elaborate discretization of the nonlinear terms to conserve the same quantities as in the ω, ψ formulation. It has also been shown, however, that it is more important to conserve energy than enstrophy to have satisfactory and accurate results. This constraint is very simple to achieve in primitive variables when staggered velocities are used, and this scheme is very important in 3D when the theory shows that only the energy is conserved. To prove the efficiency of the method the evolution of compact patches of vorticity is considered, e.g. Lamb's dipolar vortices which, in the inviscid limit, preserve also the linear relationship between ω and ψ in a frame moving with the vortical structure. The formation of more complicated vortical structures, when the viscosity amalgamates the patches, has also been addressed. In the two-dimensional case the generation of vorticity at solid boundaries is described and this phenomenon is considered by simulating the impact of dipoles on no-slip walls. Uniform and non-uniform grids are used and the accuracy and the saving of computational time are analyzed by means of non-uniform grids.

The simulation of these coherent 2D structures is important in geophysical applications; in this case it is interesting to consider simulations with the β effect and/or the presence of bottom topography. In geophysical flows radiative boundary conditions are necessary and the implementation of these conditions is explained. Radiative boundary conditions are very important e.g. when Rossby waves are generated in a β plane; this has been shown by a monopolar structure that in the β plane moves in the north-west direction.

To simulate the evolution of compact vorticity patches in many applications the use of fine grids is required. In the solution of the elliptic equation, in incompressible flows, very fine grids produce a very slow convergence when iterative schemes are chosen. The direct inversion overcomes these lacks of convergence, but this choice limits the simulations to cases where non-uniform grids occur only in one direction. The use of a multi-grid method is also shortly addressed. This method is very important in the presence of general curvilinear coordinates when direct method cannot be used. The direct inversion of the elliptic equations is easily obtained by FFTs in the directions with uniform grids. This method is used for the evaluation of the streamfunction and of the "pressure" in primitive variables. The use of the FFT has been explained and since efficient FFT routines are machine dependent, subroutines to solve the elliptic equations for the most common computers, such as CRAY and IBM, are given. For those who do not want to be computer dependent, subroutines using the Temperton routines are also given. In this case the codes can easily be transported from one to an other computer.

Axisymmetric flows are the intermediate condition between two-dimensional and three-dimensional flows, where vortex stretching plays a role. This mechanism makes the Navier-Stokes equations interesting. In this case the motion of free vortex rings at high Reynolds number has been efficiently simulated since the equations can be written in a form which permits to use, as in the two-dimensional case, discretizations for the nonlinear terms which conserve second order global invariants. The collision between vortex rings has been considered and this is a useful test for aeroacoustics problems. On the other hand collisions of vortex rings with solid walls show similarities and differences with the two-dimensional case. In the axisymmetric case the simulations require a higher resolution for the effect of vortex stretching. The axisymmetric equations have been also used to solve the time evolution of an axisymmetric jet. In the axisymmetric case simulations have been performed by primitive variables.

The three-dimensional equations have been solved in Cartesian and in cylindrical coordinates. In the first case at least two periodic directions have been used and this is useful in turbulent flows. Homogeneous directions, in fact, reduce the cost of the numerical simulations to have a reasonable number of samples to evaluate the turbulence statistics. This code can be used for the simulation of turbulent channel flows with a variety of boundary conditions for longitudinal and transversal velocities. This is an interesting and useful game to play to understand the wall drag and to have insights on drag control. The simulations at low Reynolds number with different boundary conditions permit to understand the role of the wall vortical structures to generate drag and turbulent kinetic energy.

Finite-differences have the further advantage to adapt this code to mixing layers by changing the coordinate transformation in the vertical directions and the longitudinal and transversal velocities boundary conditions at the upper and lower boundaries. Of course, also the initial velocity profiles must be changed accordingly. A similar procedure is not feasible by means of spectral methods. A version of this code with three periodic directions permits the direct simulation of isotropic turbulence and hence can be used to verify the large number of theories available for isotropic turbulence. Also in this case the conservation of the total energy in the inviscid limit has been verified independently of the grid size. The user thus can use the code to evaluate low and high order turbulent statistics. Finally this version can be used to study the vortex dynamics of vortical structures of different kind and shape.

Since in several laboratories or universities there are plenty of different codes in Cartesian coordinates, the previous section has a reduced number of users. On the contrary, to my knowledge, in few places are available codes performing simulations in cylindrical coordinates, thus a very efficient finite-difference scheme is presented for flows in cylindrical coordinates. This method has been applied to study the growth of azimuthal instabilities of free vortex rings at high Reynolds numbers. The method has also been applied to simulate the interaction of vortex rings with solid walls to understand the mechanism driving the faster growth of the azimuthal instabilities in the secondary ring. The introduction of a passive scalar produced the visualization of the ejection of a new vortex ring far from the wall as was observed in laboratories. The limitation of the grid resolution restricted the simulation at low values of the Schmidt number. This accurate numerical scheme in cylindrical coordinates was applied to a large variety of flows. The simulation of turbulent rotating and non-rotating pipe flows has been also addressed. In this case a post-processing code is given to calculate the helicity density and to show the utility of this quantity. The code with inlet and outlet conditions permits to simulate the evolution of space developing round and coaxial jets, the basic flows necessary to study problems related to combustion and to aeroacoustics.

Two chapters have been devoted to the use of LES for turbulent channel flow and for stratified turbulence. The necessity to test the qualities of finite-differences methods to solve turbulent flows by the introduction of a sub-grid model is very important because finite difference methods are very efficient solving turbulent flows in complex geometries.

1.3 Conclusions

From this outline it clearly comes out that the reader and the utilizer of the codes has a large number of different choices. Since the codes are written in a standard and not a sophisticated Fortran language, the reader can combine parts of the codes, insert new parts and thus analyze flows I did not have time to study or flows I did not imagine. I hope to give to the reader tools which will put him in a condition to better understand the complexity of fluid dynamics without spending his time in developing and coding numerical methods.

2 Finite differences

2.1 Introduction

The Finite Difference (FD) method is the simplest numerical method to solve Partial Differential Equations (PDE) in regions bounded by simple as well as by complex geometries. It is based on the assumption that the continuous physical space is replaced by a space subdivided into a grid where the unknowns are located at certain location of the cell. The grid suitable to FD is a structured or regular grid, where each point is numbered consecutively and is identified by a set of indices, e.g. (i,j,k). This passage from a continuous to a discrete space leads to the substitution of the differential operators by algebraic operators. When the decision has been taken to move into the discrete space, each further equation derived by applying differential operators to the original equations should be obtained by applying the algebraic operators.

The basic operators in the Navier-Stokes equations are the *div*, *grad* and *curl* operators. From these operators it follows for instance that the Laplacian is defined as $\nabla^2 = div\, grad$. The use of differential operators instead of algebraic operators on deriving the pressure equation produces an inconsistency as Peyret & Taylor (1983) show at Pg.156 of their book. To avoid any inconvenience in this book the discretized expressions of higher order derivatives are constructed by subsequent applications of first derivative operators. FD are based on the concept that the discretized operators should be the most compact possible to reduce the truncation errors independently of the chosen accuracy. These statements are the building blocks of the methods here applied since this book describes only simulations with staggered variables. This belief in 3D simulations by primitive variables does not have any inconvenience for flows in simple geometries. On the other hand in complex geometries an enormous increase of memory occupancy to store a large number of metric quantities occurs. In this chapter the difference operators that permits the discretization of the Navier-Stokes equations in Cartesian and in Cylindrical coordinates are described. In incompressible Navier-Stokes equations parabolic and elliptic equations appear hence in this introductory chapter the method of solution for parabolic and elliptic equations are described. The description is limited to the methods used in this book; for a more general treatment it is necessary to consult the large number of books dedicated to numerical fluid dynamics as Roache (1972), Peyret & Taylor (1983), Fletcher (1987) and Ferziger & Peric (1996) to cite only some of them.

2.2 Difference operators

2.2.1 Space discretization

The operators necessary to discretize a set of differential equations are meaningless without giving the associated truncation error that defines the accuracy of the operator. The Taylor-series expansion permits in a very simple way to obtain finite differences expressions for derivatives of any order and to calculate the truncation errors. This Taylor-expansion links the value of a quantity at a point $i \pm n$ to that at the point i and to its derivatives. The Taylor expansion is given by

$$u_{i\pm n} = u_i \pm n\Delta x \frac{\partial u}{\partial x}|_i + \frac{(n\Delta x)^2}{2!}\frac{\partial^2 u}{\partial x^2}|_i \pm \frac{(n\Delta x)^3}{3!}\frac{\partial^3 u}{\partial x^3}|_i + O(\Delta x)^4 \qquad 2.1$$

The discrete expression of any order n derivative, can be derived by Eq.(2.1). This equation furthermore is used to define interpolation and filtering operators. The interpolation operators are often necessary to evaluate a quantity in points where it is not defined. Usually the linear interpolation is second order accurate and it is sufficient. The following interpolation operators can be introduced

$$\overline{\alpha}^x = \frac{1}{2}[\alpha(x+\frac{\Delta x}{2})+\alpha(x-\frac{\Delta x}{2})] = \alpha_i + O(\Delta x)^2 \qquad 2.2$$

$$\overline{\alpha}^{x^\pm} = \frac{1}{2}[\alpha(x\pm\Delta x)+\alpha(x)] = \alpha_{i\pm\frac{1}{2}} + O(\Delta x)^2 \qquad 2.3$$

where $x^\pm$ indicates the position displaced $\frac{\Delta x}{2}$ on the right or on the left of the x position

$$\overline{\alpha}^{2x} = \frac{1}{2}[\alpha(x+\Delta x)+\alpha(x-\Delta x)] = \alpha_i + O(\Delta x)^2 \qquad 2.4$$

A filtering procedure has also a large use, in particular in the solution of the Navier-Stokes equations for filtered variables, (Large Eddy Simulations) described in Chapters 12 and 13. It is expressed by

$$\tilde{\alpha}^x = \frac{1}{2}[\overline{\alpha}^{x+} + \overline{\alpha}^{x-}] \qquad 2.5$$

where the value at the x position is expressed as a function of the value of the quantity at the same position and of those on the sides. A smoother filter is obtained by taking wider filters

$$\tilde{\alpha}^{xx} = \frac{1}{2}[\overline{\alpha}^{2x}(x+\Delta x)+\overline{\alpha}^{2x}(x-\Delta x)] \qquad 2.6$$

where the value is related to a greater number of values on the sides.

The operator related to the first central derivative is

$$\alpha_x = \frac{1}{\Delta x}[\alpha(x+\frac{\Delta x}{2})-\alpha(x-\frac{\Delta x}{2})] = \frac{\partial \alpha}{\partial x}|_i + O(\Delta x)^2 \qquad 2.7$$

It can also be evaluated at the points $x \pm \frac{\Delta x}{2}$ and the expression is

$$\alpha_{x\pm} = \pm\frac{1}{\Delta x}[\pm\alpha(x\pm\Delta x)-\alpha(x)] = \frac{\partial \alpha}{\partial x}|_{i\pm\frac{1}{2}} + O(\Delta x)^2 \qquad 2.8$$

The second derivatives can be constructed by the previous expression as

$$\alpha_{xx} = \frac{1}{\Delta x}[\alpha_{x^+} - \alpha_{x^-}] \qquad 2.9$$

leading to the usual expression

$$\alpha_{xx} = \frac{1}{\Delta x^2}[\alpha(x+\Delta x)-2\alpha(x)-\alpha(x-\Delta x)] = \frac{\partial^2 \alpha}{\partial x^2}|_i + O(\Delta x)^2 \qquad 2.10$$

The derivative of a product can be defined as

$$(\alpha\beta)_x = \overline{\alpha}^x\beta_x + \overline{\beta}^x\alpha_x \qquad 2.11$$

$$\overline{(\alpha\beta)_x}^x = \overline{\alpha}^{2x}\overline{\beta_x}^x + \overline{\beta}^{2x}\overline{\alpha_x}^x \qquad 2.12$$

These expressions are used to construct an algebra based on discrete operators. The extension of these operators to 2D and 3D is straightforward and is left to the readers.

2.2.2 Non-uniform grids

In a large number of practical applications high gradients are localised in certain regions; to reduce the computational load the physical space should account for this occurrence. The present book deals with viscous incompressible flows hence the resolution of discontinuities as those occurring in compressible flows are not discussed. The concept of upwinding of any order is completely neglected, although it is often used in simulations of incompressible flows (Leonard 1984, Rai & Moin 1991). My belief is that for a direct simulation of the incompressible Navier-Stokes equations, the grid should be related to the physics of the flow. In unsteady flows the large gradients move with the flow and then the optimal choice should be to have grids adapting to these movements. Attempts have been made to develop methods with moving grids as described by Ferziger & Peric (1996) from Pg. 321 to Pg. 327. Still this technique is too cumbersome to be applied to 3D flows.

In the present book the methods permit to have a grid refinement mainly in one direction and the refinement is *a priori* chosen by considerations related to experimental observations or by theoretical arguments. In several cases it is sufficient to have a refinement in certain regions and a stretching in others. These highly non-uniform grids, in a very simple and efficient way, can be generated by introducing the concept of a computational and a physical grid. This is a fundamental concept that has been often used to perform FD simulations in complex geometries. For a detailed and complete description of these techniques it is important to read the Levi Civita (1997) book for a theoretical comprehension and the book of Thomson *et al.* (1985) for the recent applications of the tensorial calculus to fluid dynamics. A coordinate transformation allows to map the physical grid into a computational grid where the algebraic operators are defined; as a consequence, the operators apply to the variables and to the coordinates. This procedure is useful to clarify the previous assertion: in the computational space it is mandatory to operate by discrete rather than by differential operators. In one dimension the truncation errors are easily evaluated to find the transformation which brings the greatest reduction.

The physical grid x is expressed in terms of the computational grid ξ as $x = f(\xi)$, leading to the following expression for the first derivative

$$\frac{\partial u}{\partial x} = \frac{d\xi}{dx}\frac{\partial u}{\partial \xi} \tag{2.13}$$

using the relationship in the previous paragraph it results

$$\overline{u}^{x} = \frac{1}{\overline{x}^{\xi}}\overline{u}^{\xi} + O(\Delta\xi)^2 \tag{2.14}$$

Although the accuracy is second order in the computational space, it can be shown that it is $O(\delta x)$ in the physical space.

The second derivative in the continuous space can be expressed in two equivalent ways; the first one as

$$\frac{\partial^2 u}{\partial x^2} = \frac{d}{d\xi}\left(\frac{d\xi}{dx}\frac{\partial u}{\partial \xi}\right)\frac{d\xi}{dx} \tag{2.15}$$

the second one, where each derivative has been explicited, leads to

$$\frac{\partial^2 u}{\partial x^2} = \frac{\frac{\partial^2 u}{\partial \xi^2}\left(\frac{d\xi}{dx}\right)^2 - \frac{\partial u}{\partial \xi}\frac{d^2 x}{d\xi^2}}{\left(\frac{dx}{d\xi}\right)^3} \qquad 2.16$$

Immediately it turns out that even in this very simple 1D case the second expression is more cumbersome requiring a greater number of operations and even a large memory occupancy to store the metric quantities.

As a first remark, it is important to point out that it is convenient to look for a smooth transformation, to avoid discontinuities that could effect the solution. An analytical transformation fulfills this request, but it leads to several options. For instance, in presence of an analytical transformation, someone could be tempted to evaluate analytically the metric of the transformation and to use this to evaluate the discrete derivatives in Eqs. (2.14, 15, 16). This is a dangerous choice especially in presence of enhanced grid clustering. The evaluation of the truncation errors, by Taylors series expansion, shows that error cancelation occur when in Eq.(2.15) each term is evaluated by the operators in the previous paragraph. Cunsolo & Orlandi (1978) gave the analytical expression of the truncation errors for the second derivatives in Eq.(2.15) and in Eq.(2.16). Someone could be reluctant to believe this statement; to convince him that this is indeed true, the code *secdml.f* used to obtain Fig.2.1, is in the file CODESCH2/SECD of the diskette.

In the code Eq.(2.15) is expressed as

$$\frac{\partial^2 u}{\partial x^2} = \frac{1}{(\Delta\xi)^2}\left(\frac{1}{x_{\xi+}}u_{\xi+} - \frac{1}{x_{\xi-}}u_{\xi-}\right)\frac{1}{x_\xi} + O(\Delta\xi)^2 \qquad 2.17$$

and Eq.(2.16) as

$$\frac{\partial^2 u}{\partial x^2} = \left(\frac{(x_\xi)^2 u_{\xi\xi} - u_\xi x_{\xi\xi}}{(x_\xi)^3}\right) + O(\Delta\xi)^2 \qquad 2.18$$

To prove that Eq.(2.15) is more accurate than Eq.(2.16) one could evaluate the cumbersome expressions of the truncation errors but it is more rapid to write a short code which evaluates the second derivative by Eq.(2.17) and (2.18) and the differences between the analytical and the numerical results. The calculation of the second derivatives is performed only at the grid points inside the domain and not at the boundary. The treatment of the boundaries is described in section 2.2.4. The results depend on the coordinate transformation used. In the code *secdml.f* for u has been used the initial velocity profile for a time developing mixing layer simulation:

$$u = -1 + \frac{tgh\alpha_u(y+1)}{tgh\alpha_u} \qquad for \qquad y < 0.$$

$$u = +1 + \frac{tgh\alpha_u(y-1)}{tgh\alpha_u} \qquad for \qquad 0 < y. \qquad 2.19$$

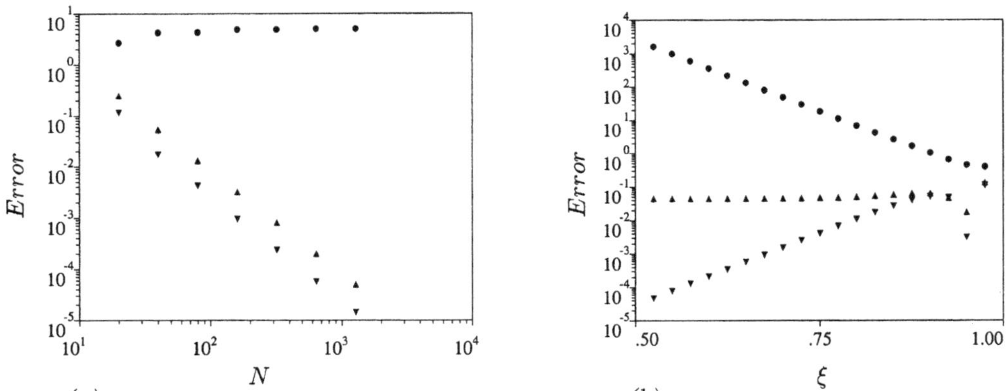

(a) (b)

Fig. 2.1 a) Percentual error versus the grid size, b) Percentual error versus ξ ; ($\triangledown$ by Eq.(2.17) with numerical metrics), ($\triangle$ by Eq.(2.17) with analytical metrics), ($\bullet$ by Eq.(2.16) with numerical metrics).

The coordinate transformation in section 2.2.3c has been used. In Fig.2.1a the relative error at a distance $y = 0.05$ from the centerline is plotted versus the number of grid points; it shows an accuracy close to the second order. Fig.2.1 has been obtained by setting $\alpha_u = 2$ and the stretching parameter of the coordinate transformation $\alpha_c = 6$. By 41 points in the whole domain the profile of the relative error distribution shows in Fig.2.1b that Eq.(2.15) gives errors smaller than those by Eq.(2.16). When the metrics are evaluated numerically instead of analytically the errors in the region of clustering are reduced. The reader can play with the code and investigate the dependence on the parameter α_u and thus he will be convinced that Eq.(2.15) gives the best accuracy. In finite difference, transformations which are a combination of *tanh* are very convenient and leave the freedom to find that one fulfilling the necessity of the flow physics. An example of the most useful coordinate transformations is given in the next section together with suggestions of the more appropriate flows. The computational coordinate varies in the range $0 < \xi < 1$.

2.2.3 Coordinate transformations

2.2.3a Two walls clustering

For the simulation of a flow between two parallel walls as in a plane channel, a coordinate transformation, clustering the points near the walls $x = \pm 1$ in order to resolve the thin vorticity layers, is given by

$$x = \frac{tgh(a(\xi - 0.5)}{tgh(a/2))} \qquad 2.20$$

2.2.3b Uniform in the center, stretched in the far field

In problems related with shear flows, such as plane mixing layers, it is desirable to have an uniform and fine grid in a central layer of thickness x_c. This refined region is important to represent accurately the roll up of the thin vorticity layer and the successive pairing of the vorticity patches. Usually the interest is to simulate free mixing layers with a negligible influence of the external boundaries. Outside the central region the velocity gradients are small; a stretched grid permits to locate the external boundaries at $x = \pm x_F$ as far as possible from $x = 0$. If the total number of grid points is n_F; and by imposing that the central region, with constant grid size, is described by n_C points, the transformation for the central region is

$$x = \eta x_C / \eta_C = b\eta \qquad for \qquad |\eta| < \eta_c \qquad 2.21$$

where $\eta = \xi - 0.5$. The transformation in the external part is

$$x = \frac{\eta}{|\eta|} x_F + \frac{tgh(\alpha_c(\eta - 0.5\frac{\eta}{|\eta|}))}{tgh\alpha_c(0.5 - \eta_C)}(x_F - x_C) \qquad 2.22$$

the stretching parameter α_c can be evaluated by imposing that the derivative is continuous at $x = x_C$. α_c is evaluated solving the following transcendental equation

$$\frac{\alpha_c}{sinh(\alpha_c \zeta_F) cosh(\alpha_c \zeta_F)}(x_F - x_C) = b \qquad 2.23$$

by a Newton-Rapson iterative scheme, where $\zeta_F = 0.5 - \eta_C$.

2.2.3c Clustered in the center

This transformation is also useful for the simulation of thin shear layers. It differs from the previous one because of the grid spacing continuously growing from the center to the external boundaries. The previous one, however, is more appropriate because it gives a constant grid spacing in a wide region where the unstable shear layer rolls up and later on these patches merge. The transformation from the computational grid $0 < \xi < 1$ to the physical coordinate $-h < x < h$ is

$$x = h(-1 + \frac{tgh2a\xi}{tgha}) \qquad for \qquad 0 < \xi < 0.5$$

$$x = h(1 + \frac{tgh2a(\xi - 1)}{tgha}) \qquad for \qquad 0.5 < \xi < 1. \qquad 2.24$$

2.2.3d Clustered around two inner regions

This expression could be used in the simulation of a wake within two free-slip boundaries. In this case it has been assumed that the same number of grind points are required to describe the internal and the external region of the wake. The transformation clustering the grid around $\xi = 0.25$ and $\xi = 0.75$ is

$$x = -0.5(\frac{arctgha(0.25 - \xi)}{arctgh0.25a} + 1.) \qquad for \qquad 0 < \xi < 0.5$$

$$x = +0.5(\frac{arctgha(\xi - 0.75)}{arctgh0.25a} + 1.) \qquad for \qquad 0.5 < \xi < 1. \qquad 2.25$$

2.2.3e Clustered near two walls and at the center

This transformation could be used to map the vertical direction for the flow past a backward facing step or to map the radial coordinate to simulate the evolution of a round jet inside an enclosure. If H is the vertical dimension and h the step height, the transformation is

$$x = \frac{H-h}{2}\left(\frac{-tgha(0.25-\xi)}{tgh0.25a}+1\right) \qquad for \qquad 0<\xi<0.5$$

$$x = H-h+h\left(\frac{tgha(\xi-0.75)}{tgh0.25a}+1\right) \qquad for \qquad 0.5<\xi<1. \qquad 2.26$$

By this transformation the boundary layers of the inlet velocity profile and those forming at the upper and lower walls are described by a satisfactory number of points. The central clustering is necessary to describe the shear layer originating at the separation point.

2.2.3f Clustered in one region and stretched elsewhere

This transformation is appropriate to map the streamwise direction for the flow past a backward facing step. The clustering here allows to describe the velocity gradients near the reattachment point. The stretching, on the other hand, locates the outlet boundary far away from the reattachment position to let the flow to recover the fully developed condition. An enhanced coordinate stretching is necessary for laminar flows if a parabolic velocity profile is assigned at the outlet. However, this is not a satisfactory condition in transitional or turbulent conditions where radiative boundary conditions are more appropriate. For these flows the outlet could be located sufficiently close to the reattachment point and the number of points in the horizontal direction is thus reduced. This is an important requirement in three-dimensional simulations where a good resolution in the spanwise direction is also required to resolve the longitudinal vortical structures.

In this transformation the the number N_R of points within the recirculating region is preassigned. The physical dimension X_R together with the location of the outlet boundary X_O must also be chosen. The coordinate transformation for $x = X/X_O$, fixing X_R in correspondence to $\xi_R = N_R/(N_F)$ is

$$x_1(\xi) = x_R \frac{tgh\alpha_1\xi}{tgh\xi_R\alpha_1}$$

The value of $x_1(1)$ is necessary for a further transformation

$$x_2(\xi) = \frac{1.}{x_1(1)} + \left(1 - \frac{1.}{x_1(1)}\right)\frac{tgh\alpha_2(\xi-1.)}{tgh(\xi_R-1)\alpha_2} \qquad 2.27$$

The coordinate transformation in the whole span is $x(\xi) = x_1(\xi)x_2(\xi)$. There are two parameters α_1 and α_2 controlling the grid point clustering and stretching. These are evaluated empirically.

The different transformations can be evaluated by the code *trascor.f* given in the file CODESCH2/CORD. The data required for each transformation are in the file *trascor.d*. In Fig.2.2a-f the transformation are shown.

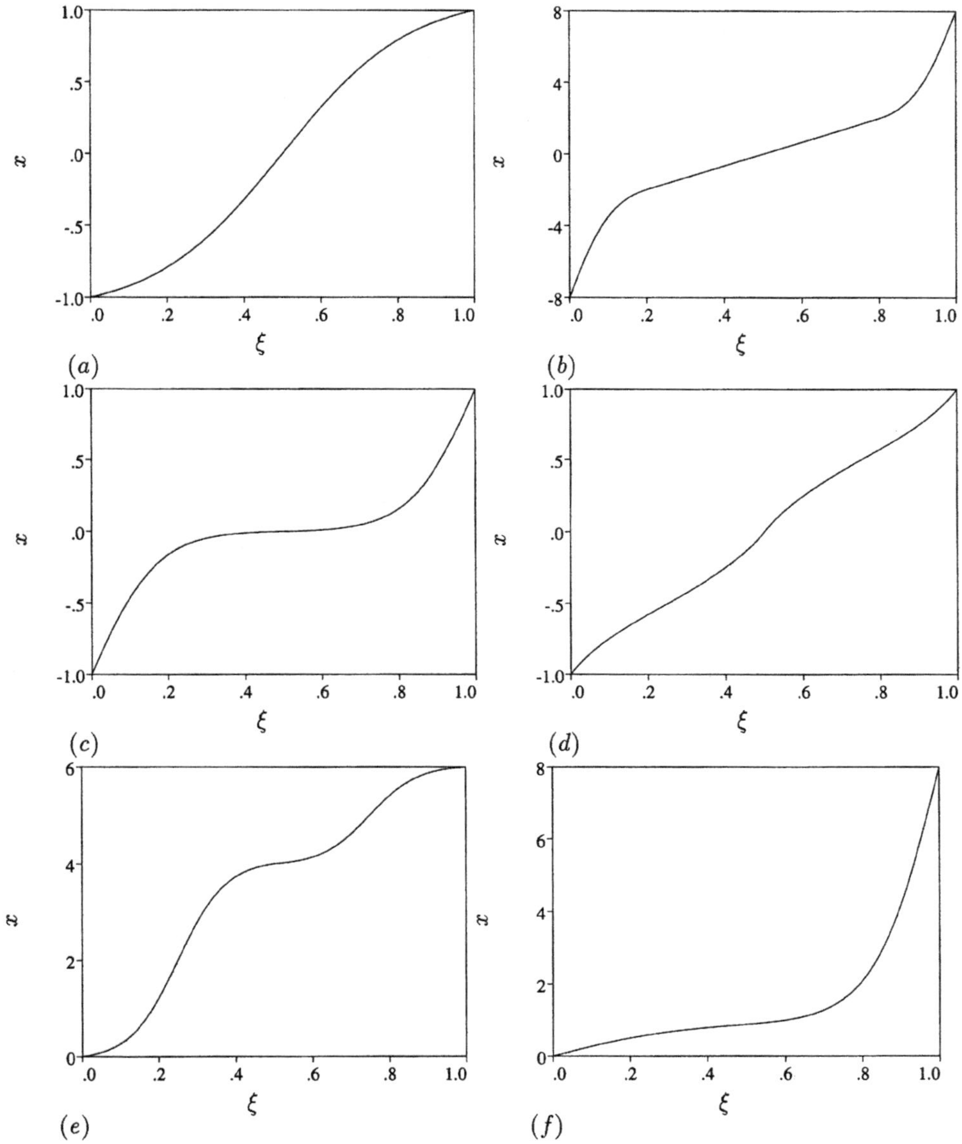

Fig. 2.2 Grid transformation for non-uniform coordinates.

2.2.4 Non-uniform grids boundary conditions

The Navier-Stokes equations in primitive variables and in the vorticity streamfunction formulation require boundary conditions near free-slip or no-slip boundaries. The free-slip boundaries are simple on the other hand, no-slip boundaries require attention. In Roache's (1972) book several expressions are given for the wall vorticity boundary and it was asserted that Woods' (1954) expression was satisfactory. This expression has a second order accuracy; Roache (1972) reported expressions with higher accuracy which requires unknowns at a greater distance from the wall. I do not recommend these expressions from a physical consideration: it is contradictory that quantities at a certain distance from the wall, not linked to the vorticity at $w+1$ instead are linked to the wall vorticity. However the merit of this expression is based on the consideration that if the wall vorticity is strictly connected to the adjacent vorticity, eventual errors on the discretization of the right hand side affect in a less measure the calculation. Thus an efficient boundary conditions for a no-slip wall is obtained by the vorticity gradient at $w+\frac{1}{2}$

$$\frac{\partial \omega}{\partial y} = \frac{\partial^2 u}{\partial y^2} - \frac{\partial^2 v}{\partial y \partial x} \qquad 2.28$$

The calculation of the second term on the right hand side does not need any approximation, in fact the velocity v is null at impermeable walls and in the case of a transpiring wall the profile of the inlet velocity is given. On the other hand the discretization of the first term is crucial because it requires approximations. This is the leading term that generates the wall vorticity in viscous flows and eventual inaccuracies could affect the entire solution. This second derivative appears also in the Navier-Stokes equations in primitive variables; the following discretization permits to understand how the no-slip boundary conditions are assigned in all the codes described in this book. Since this expression appears in the two formulations; the results obtained by the primitive variables and by the ω, ψ formulations are identical.

The Taylor series in a non-uniform grid allows to relate $\frac{\partial^2 u}{\partial y^2}$ at $w+\frac{1}{2}$ to the quantities $u_{w+\frac{3}{2}}$, $u_{w+\frac{1}{2}}$ and u_w. The quantities at w and at $w+\frac{3}{2}$ are linked to u_w and its derivatives by

$$u_w = u_{w+\frac{1}{2}} - \Delta y_1 \frac{\partial u}{\partial y}|_{w+\frac{1}{2}} + \frac{\Delta y_1^2}{2}\frac{\partial^2 u}{\partial y^2}|_{w+\frac{1}{2}} + +O(\Delta y)^3 \qquad 2.29$$

$$u_{w+\frac{3}{2}} = u_{w+\frac{1}{2}} + \Delta y_2 \frac{\partial u}{\partial y}|_{w+\frac{1}{2}} + \frac{\Delta y_2^2}{2}\frac{\partial^2 u}{\partial y^2}|_{w+\frac{1}{2}} + +O(\Delta y)^3 \qquad 2.30$$

by eliminating $\frac{\partial u}{\partial y}|_{w+\frac{1}{2}}$ the second derivative at $w+\frac{1}{2}$ is

$$\frac{\partial^2 u}{\partial y^2}|_{w+\frac{1}{2}} = \frac{u_{w+\frac{3}{2}}2}{\Delta y_2(\Delta y_2+\Delta y_1)} + \frac{u_w 2}{\Delta y_1(\Delta y_2+\Delta y_1)} - \frac{u_{w+\frac{1}{2}}2}{\Delta y_2 \Delta y_1} + O(\Delta y) \qquad 2.31$$

In the codes the metric quantities at the node and at the middle position in the cell are used, hence it is convenient to express Eq.(2.31) in terms of these quantities. To understand how to substitute the increments by the metrics it is useful to remind that the second derivative at the internal points evaluated at the mid points is

$$\frac{\partial^2 u}{\partial y^2}|_{i+\frac{1}{2}} = \frac{1}{\Delta\xi^2}\frac{1}{y_\xi|_{i+\frac{1}{2}}}\left[\frac{u_{i+\frac{3}{2}} - u_{i+\frac{1}{2}}}{y_\xi|_{i+1}} - \frac{u_{i+\frac{1}{2}} - u_{i-\frac{1}{2}}}{y_\xi|_i}\right] + O(\Delta\xi^2) \qquad 2.32$$

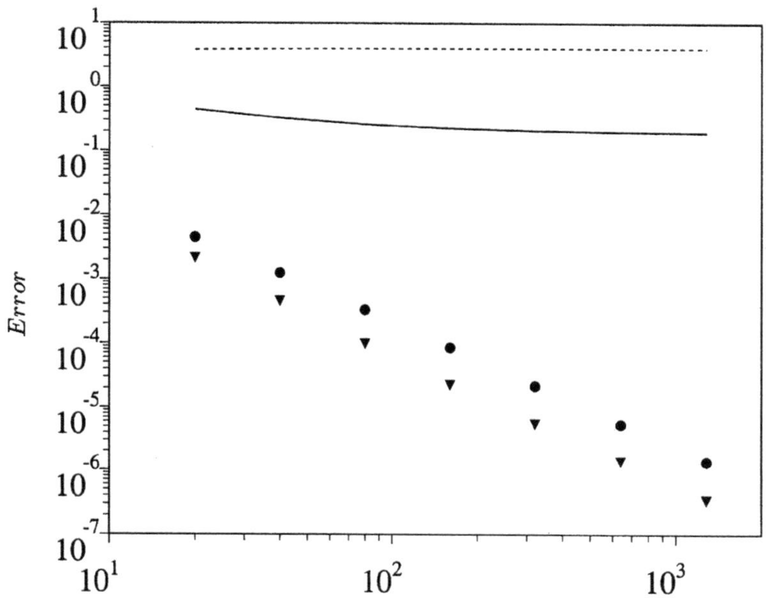

Fig. 2.3 Percentual error versus the number N of points, at $y = 0.05$ $\triangledown$ numerical metrics • analytical metrics; first half grid point close $y = 0.0$ —— numerical metrics - - - - analytical metrics.

and that the expression expliciting the physical coordinate is

$$\frac{\partial^2 u}{\partial y^2}|_{i+\frac{1}{2}} = \frac{1}{y_{i+1} - y_i}\left[\frac{u_{i+\frac{3}{2}} - u_{i+\frac{1}{2}}}{y_{i+\frac{3}{2}} - y_i + \frac{1}{2}} - \frac{u_{i+\frac{1}{2}} - u_{i-\frac{1}{2}}}{y_{i+\frac{1}{2}} - y_i - \frac{1}{2}}\right] + O(\Delta\xi^2) \qquad 2.33$$

Δy_2 and Δy_1 expressed by the metrics at the points near the boundary when substituted in Eq.(2.31) leads to

$$\frac{\partial^2 u}{\partial y^2}|_{w+\frac{1}{2}} = \frac{1}{\Delta\xi^2}\left[\frac{u_{w+\frac{3}{2}}2}{y_\xi|_{w+1}(\frac{y_\xi|_w}{2} + y_\xi|_{w+1})} + \frac{u_w 4}{y_\xi|_w(\frac{y_\xi|_w}{2} + y_\xi|_{w+1})} - \frac{u_{w+\frac{1}{2}}4}{y_\xi|_w y_\xi|_{w+1}}\right] + O(\Delta y) \qquad 2.34$$

This expression has a satisfactory accuracy and can be empirically proved by a small code varying the number of grid points. The code *secdbls.f* in the file CODESCH/SECDBL evaluates the second derivative of the function $u = y^{1/7}$ in the full domain that has been mapped by a coordinate transformation clustering the grid points near $y = 0$. This function resembles a turbulent boundary layer profile. In Fig.2.3 the accuracy for a point located at a distance 0.05 from the wall and at the first grid point close to $y = 0$ are shown. As expected the accuracy near the boundary is reduced.

2.3 Parabolic equations in time

The spatial discretization of a system of parabolic partial differential equations, as the Navier–Stokes equations, leads to a system of ordinary differential equations for each grid point. Linear and nonlinear terms appearing in this system can be treated differently in the time integration . The simplest way is to treat both terms explicitly even if, as described in several textbooks, severe restrictions for the time step size occur. For example, Ferziger and Perič (1996, p.135) considered an explicit method to solve the linear equation

$$\frac{\partial \phi}{\partial t} + u\frac{\partial \phi}{\partial x} = \frac{1}{Re}\frac{\partial^2 \phi}{\partial x^2} \qquad 2.35$$

which serves as a model problem for more complicated equations. With all the terms explicitly computed the discretized form is:

$$\frac{\phi_i^{n+1} - \phi_i^n}{\Delta t} + u\frac{\phi_i^n - \phi_{i-1}^n}{\Delta x} = \frac{\phi_{i+1}^n - 2\phi_i^n + \phi_{i-1}^n}{\Delta x^2}. + O(\Delta t) \qquad 2.36$$

This scheme is referred to as explicit forward Euler and gives a straightforward procedure to advance in time. A serious drawback of this scheme is its instability. A standard stability analysis leads to satisfy the following inequalities to prevent the solution from diverging:

$$\frac{u\Delta t}{\Delta x} < 1, \qquad \frac{\Delta t}{\Delta x^2 Re} < \frac{1}{2}. \qquad 2.37$$

The first relation is referred to as the Courant, Friedrichs and Lewy (CFL) condition and is due to the explicit treatment of the convective term of equation (2.35). In this simple example this condition could be bypassed by computing implicitly the convective term (i.e. by evaluating the second term of equation (2.36) at the time level $n+1$) and this would yield an unconditionally stable scheme. In the Navier–Stokes equations, however, the convective terms are nonlinear, since the unknowns appear also as multiplicative coefficients of the derivative and the implicit treatment is unpractical. The linearization of the nonlinear terms permits the implicit treatment; however, by neglecting iterations this reduces the accuracy of the scheme.

The second inequality of 2.2.3 is the viscous stability condition, usually more restrictive than the previous one. This is particularly true for low Reynolds number flows and in the presence of fine spatial discretizations (for instance near walls or in any region where the grid points are clustered). In addition it turns out that while the CFL condition remains unchanged when two– or three–dimensional flows are considered, the viscous stability condition becomes more restrictive since the right hand side reads $1/(2n)$ with n the number of spatial dimensions of the problem. On the other hand, the viscous term in the Navier–Stokes equations is linear and an implicit treatment does not require any particular effort, except that to invert tridiagonal matrices.

In the following of this section, an implicit scheme for the viscous term and a couple of explicit schemes for the convective term are described. When choosing these schemes one has to do a compromise since as the time accuracy of a scheme increases the number of arrays to store increases as well. The storage requirements for a computation might become soon too large for the available computer. Reduced storage requirement is a very important property of a scheme in the direct simulation of turbulent flows since every dummy array is very large. When using modern supercomputers the reduced memory allocation might not be a mandatory requirement; nevertheless the queue system penalizes the large storage simulations, resulting in a more expensive and long run. On the other hand, three dimensional simulations can be run also on desk workstations or personal computers; in this case an optimization of the memory requirement is desirable owing to the limited resources of these computers.

2.3.1 Crank–Nicholson implicit scheme

Let us consider Eq.(2.35) without the advective term

$$\frac{\partial \phi}{\partial t} = \frac{1}{Re}\frac{\partial^2 \phi}{\partial x^2}. \qquad 2.38$$

By evaluating Eq.(2.38) at the time level $n + \frac{1}{2}$ it follows

$$\phi_i^{n+1} = \phi_i^n + \frac{\Delta t}{2Re}[\delta^2 \phi_i^{n+1} + \delta^2 \phi_i^n], \qquad 2.39$$

where the symbol δ^2 indicates the discrete spatial derivative operator. This scheme is second–order accurate in time and is unconditionally stable meaning that time steps Δt of arbitrary size could be used. However, even if stability does not restrict the Δt, accuracy considerations do. In fact the error $\mathcal{O}(\Delta t^2)$ implies that large time steps result in inaccurate solutions.

The effective solution of Eq.(2.39) is explained later on, here it is important to mention that the procedure depends on the spatial discretization of the term $\delta^2 \phi_i^{n+1}$ and on the dimensions of the problems. Instead of using Eq.(2.39) directly it could be convenient to introduce the quantity $\Delta\phi_i = \phi_i^{n+1} - \phi_i^n$ and to write Eq.(2.39) as

$$\Delta\phi_i - \frac{\Delta t}{2Re}\delta^2 \Delta\phi_i = \frac{\Delta t}{Re}\delta^2 \phi_i^n, \qquad 2.40$$

This passage has been done since in the whole book the Navier-Stokes equations will be recasted in this form, that, as will be shown, later on in the ω, ψ formulation reduces the time accuracy of the boundary conditions. For the one–dimensional case, by a second–order–accurate discretization for the spatial derivatives, Eq.(2.40) becomes

$$-\frac{\Delta t}{2Re\Delta x^2}\Delta\phi_{i+1} + \left(1 + \frac{\Delta t}{Re\Delta x}\right)\Delta\phi_i - \frac{\Delta t}{2Re\Delta x}\Delta\phi_{i-1} = R_i^n, \qquad 2.41$$

where Δx is the size of the spatial mesh and in R_i^n all the known quantities at the time level n are gathered. This equation, written for all the i–nodes and completed with the boundary conditions can be solved by inverting a tri–diagonal matrix. The procedure becomes more complicated in two or tree dimensions and when curvilinear coordinates are used.

2.3.2 Adams–Bashfort explicit

Coming back to Eq.(2.35), this time the diffusive term is neglected . In a more general way $f(\phi, t)$ indicates the advective term so that Eq.(2.35) is

$$\frac{d\phi}{dt} = f(\phi, t), \tag{2.42}$$

which represents a linear or non linear ordinary differential equation. Now, also the terms are evaluated at the time level $n + \frac{1}{2}$ and the application of the discretization leads to

$$\frac{\phi_i^{n+1} - \phi_i^n}{\Delta t} = f_i^{n+\frac{1}{2}} + O(\Delta t^2). \tag{2.43}$$

In an explicit scheme, however, different from the previous example, quantities at the time $n+1$ do not permit to evaluate $f_i^{n+\frac{1}{2}}$; therefore a different expression has to be adopted. Using the Taylor series, an extrapolation allows the evaluation of $f_i^{n+\frac{1}{2}}$ in terms of known quantities with the same accuracy as the time derivative (Δt^2). The Taylor series gives

$$f_i^{n+\frac{1}{2}} = \frac{1}{2}(3f_i^n - f_i^{n-1}) + O(\Delta t^2), \tag{2.44}$$

that when substituted in 2.2.9 readily yields ϕ_i^{n+1}.

It has to be noted that differently from the implicit Crank–Nicholson scheme this scheme requires an extra array to store the quantity f_i^{n-1} that increases the storage requirements. On the other hand it is very easy to code and the solution of equation 2.2.9 does not imply the inversion of a linear system. Being an explicit scheme, it is conditionally stable with the CFL condition $CFL < 1$. A drawback of this method is that it requires a constant time step during the integration since, in order to preserve the second order accuracy, a variable Δt would require a recalculation of the extrapolation coefficients each time the Δt is varied. This occurrence suggests that simulations at constant CFL rather than at constant Δt should be avoided when using the Adams-Bashfort scheme.

2.3.3 Low storage third-order Runge–Kutta Explicit

An efficient algorithm of integration for equation (2.42) was developed by A. Wray (1987) to satisfy the following conditions:

a)Only 2 memory spaces are allocated, and $f(\phi, t)$ initially occupies one of them.

b)$\phi(t)$ at the new time is approximated by a Taylor series expansion

c)Cyclic sequential operations are used; by starting at time n, which has known conditions ϕ_i^n and f_i^n at time $n+1$ ϕ_i^{n+1} and f_i^{n+1} are in two memory spaces.

If the quantities ϕ_i^n and f_i^n are known, it turns out from equation (2.42) that the only dimensionless function which can be formed is

$$f_i^n \frac{\Delta t_n}{\phi_i^n}, \tag{2.45}$$

thus any successive approximation to ϕ must be of the form

$$F\left[f_i^n \frac{\Delta t_n}{\phi_i^n}\right]\phi_i^n. \qquad 2.46$$

According to condition (b), however, F can only be a linear function of the argument implying

$$\phi_i' = \phi_i^n + a\Delta t_n f_i^n. \qquad 2.47$$

The first memory space is allocated to this quantity; the other is occupied by

$$\phi_i^A = \phi_i^n + A\Delta t_n f_i^n. \qquad 2.48$$

Calculating the quantity

$$f(\phi_i^A, t_n + A\Delta t_n) \equiv f_i^A \qquad 2.49$$

and writing this in the memory space which was occupied by ϕ_i^A, the same conditions as those at t_n are reached, but with a solution that has advanced in time. This sequence of operations is repeated to reach the desired order of accuracy. In the case of a third order accurate scheme the calculation proceeds as follows: consider the scheme below, where variables to the left of the symbol | occupy the first memory space, those to the right the second.

$$\begin{aligned}
&\phi_i^n \mid f_i^n \\
\phi_i' &\equiv \phi_i^n + a\Delta t_n f_i^n \mid \phi_i^n + A\Delta t_n f_i^n \equiv \phi_i^A \\
\phi_i'' &\equiv \phi_i' + b\Delta t_n f(\phi_i^A, t_n + A\Delta t_n) \mid \phi_i' + B\Delta t_n f(\phi_i^A, t_n + A\Delta t_n) \equiv \phi_i^B \\
\phi_i^{n+1} &\equiv \phi_i'' + c\Delta t_n f(\phi_i^B, t_n + (a+B)\Delta t_n) \mid \text{empty for the } 3^o \text{ order.}
\end{aligned} \qquad 2.50$$

The final step consists of substituting in the expression for ϕ_i^{n+1} all the terms given in 2.2.16 and comparing the result with those by the Taylor expansion. By matching the coefficients of the two relations a system of equation allows to determine the unknowns a, b, c, A and B. The coefficients a, b, c, A and B must satisfy the following equations:

$$\begin{aligned}
a + b + c &= 1 \\
(a+B)c + Ab &= \frac{1}{2} \\
(a+B)^2 c + A^2 b &= \frac{1}{3} \\
ABc &= \frac{1}{6}
\end{aligned} \qquad 2.51$$

this is a system of 4 equations in 5 unknowns with a one parameter family solution. A commonly used solution is obtained by setting $b = 0$ which gives

$$a = \frac{1}{4}, \qquad A = \frac{8}{15}, \qquad B = \frac{5}{12}, \qquad c = \frac{3}{4}. \qquad 2.52$$

2.3.4 Hybrid third-order Runge-Kutta/Crank-Nicholson

The coefficients given in 2.2.18 are not usable because most of the passages described in the previous section are skipped during the integration of an equation. In addition, the fluid dynamics equations generally have advective and diffusive terms and the combination of explicit and implicit schemes modifies the coefficients of the latter. To find the coefficients instead of Eq.(2.35) the more general equation

$$\frac{\partial \phi}{\partial t} = N(\phi) + L\phi \qquad 2.53$$

is considered. N is the nonlinear term operator and L represents the linear term. Similarly to the previous section the following steps are done

$$\begin{aligned} \phi' &= \phi^n + \gamma_1 \Delta t N^n + \rho_1 \Delta t N^{-''} + \alpha_1 \Delta t \frac{L\phi' + L\phi^n}{2}, \\ \phi'' &= \phi' + \gamma_2 \Delta t N' + \rho_2 \Delta t N^n + \alpha_2 \Delta t \frac{L\phi'' + L\phi'}{2}, \\ \phi^{n+1} &= \phi'' + \gamma_3 \Delta t N'' + \rho_3 \Delta t N' + \alpha_3 \Delta t \frac{L\phi^{n+1} + L\phi''}{2}. \end{aligned} \qquad 2.54$$

The term $N^{-''}$ is the counterpart of N'' but at the time level $n-1$. Since at $t = 0$, any previous information is not provided, it is desirable that the coefficient ρ_1 vanishes.

The linear term at each substep is evaluated by a Crank–Nicholson scheme and the coefficients α_i are determined by imposing that nonlinear and linear terms advance at the same time step fraction. This gives the relationship

$$\alpha_i = \rho_i + \gamma_i. \qquad 2.55$$

A summation of the three substeps of equation (2.54) (the linear terms are neglected since α_i is related to the other coefficients) gives

$$\phi^{n+1} = \phi^n + \Delta t[(\gamma_1 + \rho_2)N^n + (\gamma_2 + \rho_3)N' + \gamma_3 N'']. \qquad 2.56$$

A similar procedure applied to the equations (2.50) yields

$$\phi^{n+1} = \phi^n + a\Delta t f(\phi^n, t_n) + b\Delta t f(\phi^A, t_n + A\Delta t) + c\Delta t f(\phi^B, t_n + (a+B)\Delta t) \qquad 2.57$$

that compared to the previous one gives

$$\begin{aligned} a &= \gamma_1 + \rho_2, \\ b &= \gamma_2 + \rho_3, \\ c &= \gamma_3. \end{aligned} \qquad 2.58$$

Taking into account the expressions for ϕ^A and ϕ^B we have

$$\gamma_1 = A, \qquad \gamma_2 = B \qquad 2.59$$

and finally

$$\alpha_1 = \frac{8}{15}, \alpha_2 = \frac{2}{15}, \alpha_3 = \frac{1}{3}$$

$$\gamma_1 = \frac{8}{15}, \gamma_2 = \frac{5}{12}, \gamma_3 = \frac{3}{4}$$

$$\rho_1 = 0, \rho_2 = -\frac{17}{60}, \rho_3 = -\frac{5}{12} \qquad 2.60$$

These are the coefficients which are used through the book for the solution of the Burgers and the Navier-Stokes equations in two and three-dimensions.

It should be stressed that the third order accuracy is indeed achieved only if the linear term is neglected since the condition 2.2.22 reduces the accuracy to second order. The advantage of this scheme, however, is that the stability limit is increased up to $CFL \leq \sqrt{3}$ with respect to the second–order Adams–Bashfort scheme whose limit is $CFL \leq 1$. Although the stability is increased, the third order Runge–Kutta scheme needs three substeps to advance a time step Δt thus it requires three times more operations than the Adams–Bashfort. Nevertheless it is still convenient to use this Runge-Kutta scheme because it is very common during a simulation to have transients in which the magnitude of the velocity is large, followed by periods of quiet dynamics. During these transients the CFL number reaches peak values exceeding the stability limit and while the Runge–Kutta scheme can go through without diverging, the Adams–Bashfort blows up. As a matter of fact, numerical simulations using the Adams–Bashfort have to be run with a $CFL \leq 0.6$ while, with the help of the viscous terms, simulations with the Runge–Kutta scheme can run at $CFL = 2$, and even higher.

In addition, a further advantage of this scheme, and perhaps the most considerable, is the coefficient $\rho_1 = 0$. This is an interesting property because each time step is self–starting and no information from previous times is needed. This allows to perform the integration using always the maximum CFL, by adjusting at each step the size of Δt, without decreasing the accuracy of the scheme or introducing any interpolation procedure.

2.3.5 Third order Runge–Kutta vs Adams–Bashfort.

The comparison between the performances of the third order Runge-Kutta and the Adams-Bashfort schemes are checked by solving the Burgers equation. This equation is often used to test the numerical methods since, even if it seems quite simple, it retains the main characteristics of the Navier-Stokes equations, consisting of a nonlinear and a viscous term. The contribution of the pressure does not appear, but the correct treatment of the pressure in the incompressible Navier-Stokes equations is largely discussed in other chapters of this book. The Burgers' equation

$$\frac{\partial u}{\partial t} + \frac{1}{2}\frac{\partial uu}{\partial x} = \nu \frac{\partial^2 u}{\partial x^2}, \qquad 2.61$$

is described in detail in Chapter 3. It is discretized in space on a non-uniform grid which is mapped to the computational grid by the coordinate transformation in equation (2.24); fine resolution is thus achieved around $x = 0$ in a domain $-1 < x < 1$.

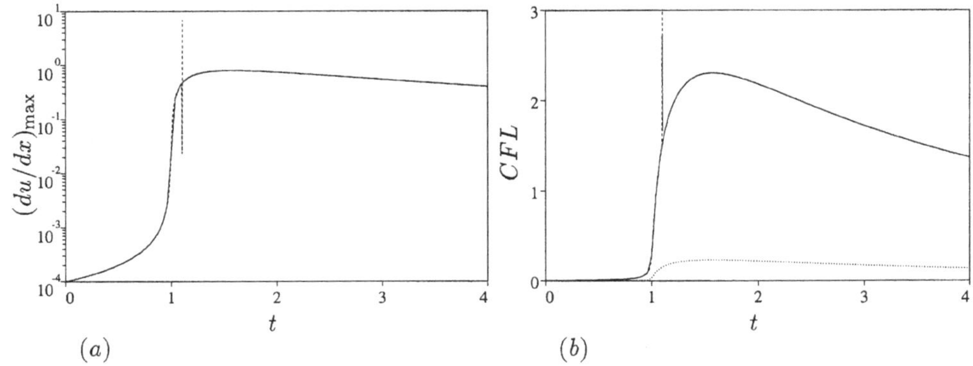

Fig. 2.4 (a) Time evolution of the maximum velocity derivative at $x = 0$ and (b) of the CFL number: ——— third order Runge–Kutta, - - - - Adams–Bashfort, ········ third order Runge–Kutta with $\Delta t = 3 \cdot 10^{-5}$.

The advancement in time uses an Adams–Bashfort or a third order Runge-Kutta explicit scheme for the nonlinear term while the viscous term employs a Crank–Nicholson implicit scheme. The initial condition is

$$u(x,0) = -sin(\pi x).$$

The solution is computed for $\nu = 10^{-4}$, which is small enough to produce around $x = 0$ a jump resembling a shock. With a value for $\alpha = 4$ in the coordinate transformation of section 2.2.3c the solution is obtained by 200 grid points.

The characteristic quantity for the Burgers equation is the spatial derivative of the velocity du/dx at $x = 0$. In figure 2.4 the time histories of $(du/dx)|_{max}$ is plotted for the two different schemes. Using a constant time step $\Delta t = 4 \cdot 10^{-4}$ the Runge–Kutta scheme is stable while the Adams–Bashfort scheme diverges. For comparison it has been also computed a solution by means of the Runge–Kutta scheme with the time step 10 times smaller ($\Delta t = 4 \cdot 10^{-5}$) and the results agree, within the precision of the plot, with those for the larger time step.

Figure 2.4a also shows that the velocity derivative increases three orders of magnitude in a short time, and then it decreases owing to viscous effects. Together with the velocity derivative also the peak CFL increases and its evolution is shown in figure 2.4b. For this one–dimensional example it is very interesting to notice that for the third order Runge–Kutta scheme the simulation passes through a transient during which the peak CFL reaches a value of about 2.3, without observing any numerical instability. In contrast, for the second order Adams–Bashfort the solution diverges a soon as the CFL approaches the stability limit of 1.

2.4. FFT for elliptic partial differential equations.

Dealing with incompressible Navier-Stokes equations in two- and three-dimensions the incompressibility condition is characterized by an elliptic equation. The pressure thus is calculated by solving an elliptic equation with the source term linked to the nonlinear terms. This equation transmits any local disturbance to the rest of the domain at an infinite speed. This physical phenomenon, for instance in space developing flows, explains why any eventual disturbance created by inaccurate boundary conditions propagates inside the field and even affects the inlet conditions. The same phenomenon occurs in a wind tunnel where a disturbance at the outflow affects the entire field. The solution of an elliptic equation in the primitive variables formulation permits to calculate a scalar quantity which is related to the pressure; this quantity projects a non-solenoidal velocity field in a solenoidal one.

In two-dimensional flows in the ω, ψ formulation the kinematic equation between vorticity and streamfunction is an elliptic equation. In a vorticity velocity formulation elliptic equations express the relation between the velocity and the vorticity field. The solution of these elliptic equations would require a large amount of computational time by inefficient numerical algorithms. In the past these equations were often solved by iterative schemes, which appeared very inefficient in particular in presence of a very fine grid. The slow convergence was related to the slower damping of the low wave numbers with respect to the high wave number components of the errors introduced by the first guess solution. The remedy for the annoying and sometimes frustrating convergence procedure is the use of direct solvers. The inversion of large band matrices by direct solvers such as LU decompositions are time consuming and require a large memory occupancy, that becomes prohibitive if the flow physics requires a very fine grid, necessary at high Reynolds numbers.

As a rule the direct solution is achieved by reducing the large band of the associated matrix to the equation to a set of tridiagonal matrices that are easily inverted. The Fourier series fulfills this requirement by reducing the size of the band of the matrix, in the direction of an uniform grid. By doing this operation the equation is transformed from the physical to the wave number space only in the directions where the Fourier transforms are applied. Depending on the boundary conditions, in the direction of uniform grid, FFT, *sin*FFT, or *cos*FFT can be applied. In the other direction tridiagonal solvers permit uniform and non-uniform grids. In the present book we describe the routines to solve elliptic equations for variables defined at the grid points (e.g. the stream function ψ) and for variables at the mid point of the cell, for instance the "pressure" ϕ in the primitive variable formulation. The efficiency of the routines is evaluated by comparing the CPU time with that by the routine *fishpack.f* developed at NCAR. This routine has been used to solve the 3D Navier-Stokes equations in cylindrical coordinates with non-uniform grids in the radial and axial directions. Very easily this method can be extended to flows in Cartesian coordinates to cluster the grid near walls and in certain positions along the streamwise direction.

For flows in general curvilinear coordinates in two or in three directions the associated matrix cannot be reduced to a tridiagonal one, thus iterative methods are mandatory. Multigrid methods speed-up the convergence, and thus the routines for general coordinates in two directions are addressed in this book.

2.4.1 Solution by FFT

If the elliptic equation is defined at the node of a cell the number of unknowns is equal to the number of grid points. For the Navier-Stokes equations this occurs mainly in the solution of the stream function, thus only the solutions in 2D are described. It has been assumed that the grid is uniform in the direction x. In this direction periodic,

Dirichlet and von Neuman boundary conditions can be assumed, and depending on this choice FFT, *sinFFT* and *cosFFT* are used. Along the y direction any kind of boundary condition and grid distribution are permitted. If a periodic condition is assumed in y as a consequence a constant grid is necessary. For flows with radiative boundary conditions along the $y = const$ boundary and with a non-uniform grid in x, the *sinFFT* in y schemes can be applied if a new variable transforms the non-homogeneous Dirichlet boundary condition in a homogeneous boundary. This new variable has been used to solve the stream function for the flows described in chapter 4. In all the different flows studied in chapter 4 tridiagonal solvers are applied along the x direction.

Two periodic directions permit to apply two successive FFT. This has been done to show that the solution is less efficient than that using one FFT and a tridiagonal solver.

The performances of the routines have been evaluated on a RISK 380 using the Temperton routines for the FFT. The Fortran source code of the Temperton routine is given in this book and thus it makes the codes portable on any kind of computers. A better efficiency of the elliptic solvers results by the machine FFT's such as those in the *ESSL* libraries of the IBM or in the CRAY's libraries. To give an idea of the speed-up of a code in the CRAY C90 at NASA Ames research center the code in chapter 8 on a 128^3 grid by the Temperton FFTs had a speed of 250 megaflops, by the CRAY's libraries the performance increased up to 500 megaflops.

2.4.2 Two periodic directions

The elliptic equation

$$\nabla^2 u = f \tag{2.62}$$

in the two-dimensional case becomes

$$\frac{\partial^2 u}{\partial x^2} + \frac{\partial^2 u}{\partial y^2} = f(x,y) \tag{2.63}$$

In a domain of size L_x and L_y the equation is solved by a finite difference scheme by subdividing L_x and L_y in equally spaced intervals of size $\Delta x = L_x/N_x$ and $\Delta y = L_y/N_y$. The discretized equation becomes

$$\frac{u_{i-1,j} - 2u_{i,j} + u_{i+1,j}}{\Delta x^2} + \frac{u_{i,j-1} - 2u_{i,j} + u_{i,j+1}}{\Delta y^2} = f_{i,j} \tag{2.64}$$

In the direction x the Fourier series can be applied

$$u_{i,j} = \sum_{-N_x/2}^{N_x/2-1} \hat{u}_{k,j} e^{\frac{\iota 2\pi k i}{N_x}} \tag{2.65}$$

where ι is the imaginary number. By substituting this expression in Eq.(2.64), we have

$$\sum_{-N_x/2}^{N_x/2-1} e^{\frac{\iota 2\pi k i}{N_x}} \left[\hat{u}_{k,j} \frac{e^{\frac{-\iota 2\pi k}{N_x}} - 2 + e^{\frac{+\iota 2\pi k}{N_x}}}{\Delta x^2} + \frac{\hat{u}_{k,j-1} - 2\hat{u}_{k,j} + \hat{u}_{k,j+1}}{\Delta y^2} - \hat{f}_{k,j} \right] = 0 \tag{2.66}$$

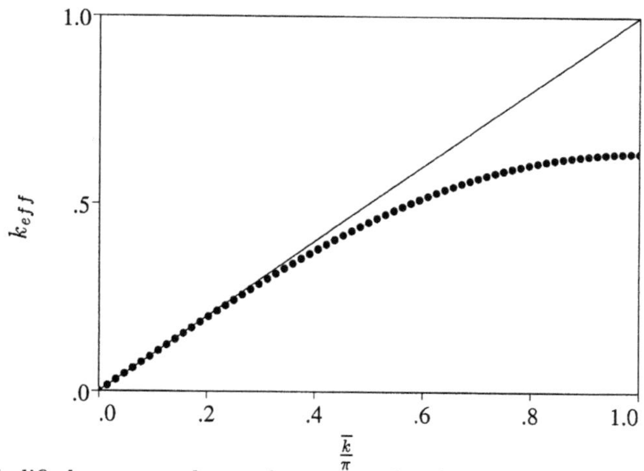

Fig. 2.5 Modified wave number and wave number for second derivative.

A modified wave number k' can be defined as

$$k'^2 = -\frac{2(\cos \overline{k} - 1)}{\Delta x^2} \qquad 2.67$$

with $\overline{k} = \frac{2\pi k}{N_x}$.

The modified wave number permits to evaluate the difference between the finite difference approximation and the spectral approximation of a second derivative. The spectral approximation of the second derivative is

$$\frac{\partial^2 u}{\partial x^2} = \sum_{-N_x/2}^{N_x/2-1} \left(-\frac{2\pi k}{L_x}\right)^2 \hat{u}(y)_k e^{\frac{\iota 2\pi k x}{L_x}} \qquad 2.68$$

$\hat{k} = \left(\frac{2\pi k}{L_x}\right) = \frac{2\pi\overline{k}}{\Delta x N_x}$ is the wave number. By plotting $k_{eff} = \frac{\hat{k}\Delta x}{\pi}$ and $k'_{eff} = \sqrt{k'^2 \Delta x^2}$ as function of $\overline{k}$, Fig.2.5 shows that the second order discretization of the second derivative represents well the low wave number and poorly the high wave numbers. The same procedure can be applied to the first derivative.

By introducing the modified wave number into Eq.(2.66)

$$\sum_{-N_x/2}^{N_x/2-1} \left[\left(\delta_y^2 - k'^2\right) \hat{u}_{k,j} - \hat{f}_{k,j}\right] e^{\frac{\iota 2\pi k i}{N_x}} = 0 \qquad 2.69$$

where δ_y^2 is the discrete operator of the second derivative in the y direction. There are N_x of these equation at each wave number $\overline{k}$ and these are solved by tridiagonal matrices. The subroutine related to the solution of an elliptic equation with periodic conditions in two directions is given in *pftpe.f.* The calculation proceeds as follows: as a first step N_y FFT are applied to $f_{i,j}$. Since $f_{i,j}$ is real the subroutine $fft99$ is used. The output is a complex array $\hat{f}_{k,j}$ with $-N_x/2 < k < N_x/2$. At each wave number $\overline{k}$ Eq.(2.69) is solved and the $\hat{u}_{k,j}$ matrix is obtained. The tridiagonal solver is different depending on the boundary conditions, in the case of periodical conditions the appropriate tridiagonal solver is given in the routine *tripv.f* This routine has the do loops ordered to run efficiently on vectorised computers as the CRAY machines. N_y inverse FFTs by the $fft99$ routine give the solution $u_{i,j}$ of the elliptic equation.

This solution is very efficient since the number of operations by the FFTs is proportional to $N_y N_x log(N_x)$. Someone could argue that in the y direction the Fourier transform could be also applied. In this case in the second sweep the $cfft99$ routine is used because of the complex $\hat{u}_{k,j}$ resulting from $fft99$. The FFT routines are machine dependent e.g. on the RISK IBM workstations the routines $SRCFT$ perform the same operations which use the Temperton FFTs. The same efficient routines are available on the CRAY machines. The machine dependent routines are usually much faster than the Temperton's which are compiled in each computer. To show the relative speed of the elliptic solvers with two periodic directions, the solution of Eq.(2.62) has been performed by assigning as source term $f = -(n^2 + m^2)\sin(2\pi x)\sin(2\pi y)$. The solution of Eq.(2.62) is $u = \sin(2\pi mx)\sin(2\pi ny)$ which permits a comparison between the numerical and the analytical solutions. In this contest the interest is on evaluating the efficiency of the routines rather than the accuracy. Since some of the routines necessitate a first call to evaluate arrays that recoursely are used to get the solution, the calculation has been performed 5 times to have a comparable efficiency between the different methods. In fact, the first call could take as much time as that necessary to get the solution in the successive call. A high resolution (a 512×512 grid) has been used since in this condition the direct solvers are often necessary. With this elevated number of points the *fishpack* package necessitates double precision; thus all the calculations were performed in double precision on the RISK 380. The *fishpack* CPU is 36.00 secs. The CPU necessary by two Temperton FFTs is 16 secs. The CPU by the real FFT and a tridiagonal solver is 12 secs. Finally the CPU by the ESSL FFTs routine is only 6 secs. The user interested to repeat the tests with different grids could use the programs in the directory CODESCH2/ELLIP. In this file there is the make file *capros.mak* linking different files which solve the elliptic equations in the following ways :

a) The file *pfishck.f* that uses the *fishpack* package in the directory NCARFFT.

b) The file *p2fft.f* using two Temperton FFTs in the periodic directions; the package for these FFTs is named *ffts.f* and is in the directory NCARFFT.

c) The file *p2fftess.f* using two ESSL FFTs and these are linked in the make file by calling the esslp2 library. This option should be discarded if RISK IBM machines are not available.

d) The file *pftpe.f* using one Temperton FFT and a tridiagonal solver.

e) The file *psncs.f* using the Temperton FFT for a cos transformation in the y direction and a tridiagonal solver in x with Dirichlet boundary.

f) The file *pcssn.f* using the Temperton FFT for a cos transformation in the x direction and a tridiagonal solver in y with Dirichlet boundary.

In the 3D case with two periodic directions and with the third non-homogeneous direction, a very good efficiency for the solution of the elliptic equation requires two consecutive FFTs in the homogeneous direction and a tridiagonal solver in the third direction.

In several 2D flows it is necessary to calculate the streamfunction in a region with one periodic direction and with one or two no-slip walls in the other. To have more resolution near the no-slip wall a non-uniform mesh by a coordinate transformation clusters the grid points near this wall. The equation is still the equation 2.63 but in this condition the operator in the y direction is that in Eq.(2.17) and the coefficients for the tridiagonal solver for $2 < j < N2$ are

$$AP(j) = \frac{1}{\Delta x_2^2 y_{x_2} y_{x_2^+}} AM(j) = \frac{1}{\Delta x_2^2 y_{x_2} y_{x_2^-}}, AC(J) = -(k'^2 + AP(j) + AM(j)) \quad 2.70$$

The coefficients of the boundary conditions are

$$AP(1) = AM(1) = AP(N_y + 1) = AM(N_y + 1) = 0, AC(1) = AC(N_y + 1) = 1 \quad 2.71$$

Also in this case the CPU required by the present routine for a 512×512 grid and for 5 equal calculations is 28 secs by the *fishpack* and that by the routine in the code *pftpe.f* is 13 secs. This computational time was obtained by the Temperton routine; a substantial saving can be reached by the ESSL routines, on the RISK IBM workstations.

2.4.3 sinFFT

This case can be applied to evaluate the streamfunction when in one direction, for instance the y direction, there are no-slip conditions and in the x direction there are two symmetric or two free-slip boundaries ($\omega = \psi = 0$). In the y direction uniform and non-uniform grids can be considered; on the other hand, in the x direction uniform grids are enforced. In this direction the *sin* FFT satisfies the boundary conditions. By the same notations used in the previous section the unknowns in the physical space are expressed as a function of variables in wave number space

$$u_{i,j} = \sum_{0}^{N_x - 1} \hat{u}_{k,j} \sin \frac{2\pi k x}{L_x} = \sum_{0}^{N_x - 1} \hat{u}_{k,j} \sin \frac{2\pi k i}{N_x} \quad 2.72$$

By introducing this expression in Eq.(2.64) we have the following set of algebraic equations

$$\sum_{0}^{N_x - 1} \sin \frac{2\pi k i}{N_x} \left[-\hat{u}_{k,j} \frac{2(\cos \overline{k} - 1)}{\Delta x^2} + \frac{\hat{u}_{k,j-1} - 2\hat{u}_{k,j} + \hat{u}_{k,j+1}}{\Delta y^2} - \hat{f}_{k,j} \right] = 0 \quad 2.73$$

The reduced wave number k'^2 is equal to that in the previous section. In the computers libraries there are routines based on the sin FFT transform. However, since these are not always available, it is interesting to show how the *sin* FFT can be performed by routines based on the FFT transform. The *cfft99* routine performs the following transformations

$$g_i = \sum_{k=0}^{N} \hat{g}_k e^{\frac{\iota 2\pi k i}{N}} \quad , \quad \hat{g}_k = \frac{1}{N} \sum_{i=0}^{N} g_i e^{\frac{-\iota 2\pi k i}{N}} \quad 2.74$$

where $\hat{g}_k$ and g_i are complex arrays. To go from the physical to the wave number space the real part of g_i enters on the $cfft99$ routine with $isign = -1$.

The $sinFFT$ transformation gives

$$f_i = \sum_{k=0}^{M} \hat{f}_k \sin\frac{\pi ki}{M} \quad , \quad \hat{f}_k = \frac{2}{M}\sum_{k=0}^{M} f_i \sin\frac{\pi ki}{M} \quad . \qquad 2.75$$

where f_i and $\hat{f}_k$ are real. By the identity $e^{\pm\iota x} = \cos x \pm \iota \sin x$ and by $\hat{g}_k = \hat{a}_k + \iota\hat{b}_k$ Eq.(2.74) becomes

$$\hat{a}_k + \iota\hat{b}_k = \frac{1}{N}\sum_{k=0}^{N} g_i \left(\cos\frac{2\pi ki}{N} - \iota\sin\frac{2\pi ki}{N}\right) \qquad 2.76$$

by equating the real and imaginary parts it follows

$$\hat{a}_k = \frac{1}{N}\sum_{k=0}^{N} g_i \cos\frac{2\pi ki}{N} \quad , \quad \hat{b}_k = -\frac{1}{N}\sum_{k=0}^{N} g_i \sin\frac{2\pi ki}{N} \quad . \qquad 2.77$$

$\hat{b}_k$ in Eq.(2.77) is equal to Eq.(2.75) if $N = 2M$ and it follows

$$\begin{aligned} g_i &= -f_i \qquad & i &= 0, ..., M \\ g_i &= 0 \qquad & i &= M+1, ..., 2M \\ \hat{f}_k &= 4\hat{b}_k & & \end{aligned} \qquad 2.78$$

On using different FFT routines it is necessary to see whether in the inverse FFT the output is normalised by $\frac{1}{N}$. Since in the Temperton routines the normalization factor is 1 in Eq.(2.78) $\hat{f}_k = \hat{b}_k 2/M$.

A similar procedure permits to evaluate the coefficients for the inverse FFT . By decomposing $\hat{g}_k$ into the real and imaginary part the expression for g_i in Eq.(2.74) becomes

$$g_i = \sum_{k=0}^{N} (\hat{a}_k + \iota\hat{b}_k)\left(\cos\frac{2\pi ki}{N} + \iota\sin\frac{2\pi ki}{N}\right) \qquad 2.79$$

Thus f_i in Eq.(2.75) and g_i in Eq.(2.79) coincides if

$$\begin{aligned} N &= 2M & & \\ \hat{a}_k &= 0 \qquad & k &= 0, ..., 2M \\ \hat{b}_k &= -\hat{f}_k \qquad & k &= 1, ..., M \\ \hat{b}_k &= 0 \qquad & k &= M+1, ..., 2M \\ f_i &= Re[g_i] \qquad & i &= 1, ..., N \end{aligned} \qquad 2.80$$

From the previous consideration it turns out that the *sinFFT* necessitates a number of points twice that for the usual FFT.

The solution of Eq.(2.73) proceeds as in the previous paragraph, by calling the *cfft99* routine with $isign = -1$ and $2N_x$ points to get $\hat{f}_{k,j}$. A tridiagonal solver with the coefficients in Eq.(2.70) and Eq.(2.71) gives $\hat{u}_{k,j}$. As a final step a new call of *cfft99* with $isign = 1$ gives $u_{i,j}$ at each grid point. As previously done the efficiency of this method has been compared with that for the *fishpack* by solving five times Eq.(2.62) in a 512×512 grid. The CPU for the *fishpack* package is 27 secs and that for the present scheme is 41 secs. A better efficiency can be reached by the *ESSL* routines. However the results shows that in this case the fishpack is superior and, as a further merit, it has a larger degree of freedom by allowing non-uniform grids in both directions.

2.4.4 cosFFT centered

The *cosFFT* routines can be used to solve equations with Neumann conditions at the boundaries. In the solution of the Navier-Stokes equations such an elliptic equation with a Neumann condition is the equation for the "pressure" Φ. Since this variable is defined at the center of the grid the method requires a *cosFFT* routine on a staggered grid. The comprehension of the sequences of operations in a staggered grid is facilitated if it is clear how to evaluate the *cosFFT* on a centered grid. This is the reason of briefly describing the *cosFFT* routine that does not greatly differ from the *sinFFT*.

The *cosFFT* is defined as

$$f_i = \frac{\hat{f}_0}{2} + \sum_{k=1}^{M} \hat{f}_k \cos\frac{\pi k i}{M} \quad , \quad \hat{f}_k = \frac{2}{M} \sum_{k=0}^{M} f_i \cos\frac{\pi k i}{M} \quad . \qquad 2.81$$

To perform the *cosFFT* by the standard FFT routines which are based on the transformation in Eq.(2.74), even in this case, a number of points $N = 2M$ must be used. To get the coefficients $\hat{f}_k$ in Eq.(2.81) from the $\hat{a}_k$ in Eq.(2.77), for the inverse FFT, that is for $isign = -1$, the following relationship holds

$$\begin{aligned} g_i &= f_i & i &= 0, ..., M \\ g_i &= 0 & i &= M+1..., 2M \\ \hat{f}_k &= 4\hat{a}_k & k &= 1, ..., N \\ \hat{f}_0 &= 2\hat{a}_k \end{aligned} \qquad 2.82$$

It is important to remember that the zero wave number has half amplitude.

To have the coefficients in the physical from those in wave number space, f_i in Eq.(2.81) must be equal to Eq.(2.79); then we have

$$\begin{aligned} N &= 2M \\ \hat{b}_k &= 0 & k &= 0, ..., 2M \\ \hat{a}_k &= -\hat{f}_k & k &= 0, ..., M \\ \hat{a}_k &= 0 & k &= M+1, ..., 2M \\ f_i &= Re[g_i] & i &= 1, ..., N \end{aligned} \qquad 2.83$$

As mentioned before particular care should be devoted to the different normalization factors in each FFT routine. In the solution of Eq.(2.62) by the *cosFFT* routine in one direction and Dirichlet conditions in the other direction, the CPU is the same as the CPU necessary to solve the equation with the *sinFFT*. Thus also in this case the *fishpack* package is more efficient.

2.4.5 cosFFT "staggered"

For the *cosFFT* as for the *sinFFT* the reduced wave number is still equal to that in the periodic case independently from the variable location. Before we explain the procedure it is convenient to recall that this equation occurs in the solution of the "pressure" ϕ in the solution of the Navier-Stokes equations. In Cartesian coordinates and in the 3D case the equation is

$$\frac{\partial^2 \phi}{\partial x^2} + \frac{\partial^2 \phi}{\partial y^2} + \frac{\partial^2 \phi}{\partial z^2} = q(x,y,z) \qquad 2.84$$

The right hand side is the divergence of the non-solenoidal field which has been assigned or has been calculated by solving a set of parabolic equations that differ from the Navier-Stokes equations by a term related to the real pressure. The source $q(x,y,z)$ does not have any spatial coherence. Thus in wave number space there is the same probability to have equal amplitudes at low or at high wave numbers. At this point it is important to understand the role of the boundary conditions on the solution of Eq.(2.84) by considering for instance what occurs in space developing mixing layers. In these flows periodicity in the streamwise direction z is assumed. At the inlet velocity profiles must be assigned and at the outlet radiative conditions are those reproducing the physics accurately. Free-slip or no-slip conditions apply at the horizontal $x-z$ planes. Eq.(2.84) is evaluated only at the interior of the computational domain and when it is discretized at the points near a boundary it happens that the normal velocity on the boundary is prescribed. The equation appears then as if it is discretized by assuming on the boundary $\frac{\partial \Phi}{\partial n} = 0$. These considerations hold in any kind of boundary in the cases studied in this book, except when periodicity is assumed. These arguments lead to the use of the *cosFFT* in the streamwise and in the normal directions. In this paragraph the step related to the *cosFFT* is explained and in section 8.7 the entire solution of Eq.(2.84) is described. Replacing $q(i+\frac{1}{2}, j+\frac{1}{2}, k+\frac{1}{2})$ by $f(i)$ or more realistically by setting $f_i = \hat{q}(i+\frac{1}{2}, j+\frac{1}{2}, k_z)$, if $q(i+\frac{1}{2}, j+\frac{1}{2}, k+\frac{1}{2})$ has been transformed in the spanwise direction through the $fft99$ routine. The *cosFFT* gives

$$f_i = \frac{\hat{f}_0}{2} + \sum_{k=1}^{M} \hat{f}_k \cos\frac{\pi k(i+\frac{1}{2})}{M} \quad , \quad \hat{f}_k = \frac{2}{M}\sum_{i=0}^{M} f_i \cos\frac{\pi k(i+\frac{1}{2})}{M} \quad . \qquad 2.85$$

f_i should be considered as an abbreviation for $\sum_{kz=1}^{N_z} \hat{q}_{k,j+\frac{1}{2},kz}$. Eq.(2.85) is necessary to go from the physical to the wave number space and it can be written as

$$\hat{f}_k = Re\left[\frac{2}{M}\sum_{k=0}^{M} f_i e^{\frac{-\iota\pi k(i+\frac{1}{2})}{M}}\right] \qquad 2.86$$

The following relationship should be satisfied to let Eq.(2.86) coincide with Eq.(2.74)

$$\begin{aligned}
N &= 2M \\
g_i &= f_i \qquad && i = 0, ..., M \\
g_i &= 0 \qquad && i = M+1..., 2M \\
\hat{f}_k &= Re\left[4\hat{g}_k e^{\iota\frac{k\pi}{2M}}\right] \\
= 4\left[Re(\hat{g}_k)\cos\frac{k\pi}{2M} \right. &+ \left. Im(\hat{g}_k)\sin\frac{k\pi}{2M}\right] \qquad && k = 1..., N \\
\hat{f}_0 &= 2Re(\hat{g}_0)
\end{aligned} \tag{2.87}$$

Once more we have to recall the FFT normalization. The complex quantity $\hat{q}_{k,j+\frac{1}{2},kz}$ derives from a call of $cfft99$ with $isign = -1$ where the $\hat{q}_{k,j+\frac{1}{2},kz}$ enters; it is the right hand side of the linear system of equations. A tridiagonal solver in the j direction gives the complex coefficients $\hat{\phi}_{k,j,kz}$. A first application of the $cfft99$ routine from the wave number to the physical space produces the $\hat{\phi}_{i+\frac{1}{2},j+\frac{1}{2},k_z} = \hat{f}_k$.

The coefficients of the staggered $cosFFT$ from the wave number to the physical space are calculated by setting f_i in Eq.(2.85) as

$$f_i = \frac{\hat{f}_0}{2} + \sum_{k=1}^{M} \hat{f}_k \left[\cos\frac{\pi k i}{M}\cos\frac{\pi k}{2M} - \sin\frac{\pi k i}{M}\sin\frac{\pi k}{2M}\right] \tag{2.88}$$

g_i in Eq.(2.74) can be written as

$$g_i = \sum_{k=0}^{N} (Re(\hat{g}_k) + \iota Im(\hat{g}_k))\left(\cos\frac{\iota 2\pi k i}{N} + \sin\frac{\iota 2\pi k i}{N}\right) \tag{2.89}$$

The quantities entering in the $cfft99$ to get the unknowns in physical space are

$$\begin{aligned}
N &= 2M \\
Re(\hat{g}_k) &= \hat{f}_k \cos\frac{k\pi}{2M} \qquad && k = 0, ..., M \\
Im(\hat{g}_k) &= \hat{f}_k \sin\frac{k\pi}{2M} \qquad && k = 0, ..., M \\
Re(\hat{g}_k) &= Im(\hat{g}_k) = 0 \qquad && k = M+1, ..., 2M \\
f_i &= Re[g_i] \qquad && i = 0, ..., N
\end{aligned} \tag{2.90}$$

The staggered $cosFFT$ is restricted to a uniform grid in the x direction.

Often in space developing flows and mainly for transitional flows it is necessary to cluster the grid near the inlet and in the vertical region where the velocity gradients are higher. With this grid distribution the staggered $cosFFT$ cannot any longer be applied and the ***fishpack*** package is used to find the solution at each wave number k_z, for $1 \leq k_z \leq N_z/2$ for the complex coefficients $\hat{\phi}_{i+\frac{1}{2},j+\frac{1}{2},k_z}$.

2.5 Multigrid methods for elliptic partial differential equations

As before mentioned, the Navier-Stokes equations for incompressible flows in general curvilinear coordinates to achieve a solenoidal velocity field require the solution of elliptic equations. Dealing with Cartesian or cylindrical coordinates, the FFT permits direct inversion of the matrix and the limitations of the grid size depend only on the computer available. On the contrary, dealing with general curvilinear coordinates in two directions, direct solvers are not available. The inversion is always possible by the direct inversion of the full matrix but this is possible only for coarse grids. In fact, for a grid of 100×100 points the direct inversion necessitates the storage of a matrix of size 10000×10000 and this is an enormous memory request. Band storage routines reduce the memory occupancy but still it is too large. Standard iterative methods, as point SOR or line SOR, do not work because the large wave numbers errors are dumped very slowly and the convergence to round-off precision is difficult.

Several methods for a faster convergence have been developed but in this book only the multigrid method is considered because of its efficiency. The equations for general curvilinear coordinates are considered these include also the case of non-uniform grids in two directions. The method describes the solution for a scalar quantity ϕ defined at the location i, j. This is the equation relating the streamfunction to the vorticity in chapter 5 where the 2D Navier-Stokes equations in general curvilinear coordinates are solved. A similar equations is considered when the primitive variables are solved in curvilinear coordinates and Φ, in this case, is the the "pressure" that projects the non-solenoidal velocity field in the solenoidal one. In this case Φ is defined at the center of the computational grid. A similar equation is encountered to evaluate the temperature field in heat conduction problems or the potential flow around bodies of complex structures. In 3D the equations have an extra second derivative in the third direction; this can be eliminated by using the FFT in this direction with uniform grid spacing. Thus an elliptic equation should be solved for each wave number. By small modifications, the method here explained for the variables at the nodes, could be adapted for equations with variables at the center of the cell.

2.5.1 Multigrid method

As it is shown in chapter 5 in general curvilinear coordinates the elliptic equation, in conservative form, is

$$\frac{1}{\sqrt{a}}\left(\frac{\delta}{\delta x^m}\alpha^{mn}\frac{\delta \Phi}{\delta x^n}\right) = \hat{Q} \quad ; \quad m,n = 1,2 \qquad 2.91$$

As a matrix Eq.(2.91) can be written:

$$\sum_{i=1}^{9} C_i \Phi_i = Q \qquad 2.92$$

where

$$\Phi_1 = \Phi_{i+1,j+1} \qquad \Phi_2 = \Phi_{i+1,j} \qquad \Phi_3 = \Phi_{i+1,j-1}$$

$$\Phi_4 = \Phi_{i,j-1} \qquad \Phi_5 = \Phi_{i-1,j-1} \qquad \Phi_6 = \Phi_{i-1,j}$$

$$\Phi_7 = \Phi_{i-1,j+1} \qquad \Phi_8 = \Phi_{i,j+1} \qquad \Phi_9 = \Phi_{i,j}$$

Eq.(2.91) is written at each point at the interior of the computational box, that is for $2 \leq i \leq n1-1$ and $2 \leq j \leq n2-1$. The coefficients are related to the metric quantities α^{mn}. At this point to explain the multigrid method it is not important to know how the quantities α^{mn} are defined by the derivatives of the physical coordinates with respect to the computational. It is important to know that at the smallest grid, the grid at which Eq.(2.91) is solved, it is necessary to evaluate the coefficients α^{1n} at the points $i+\frac{1}{2}, j$ (indicated as α_i^{1n}) and the coefficients α^{2n} at the points $i, j+\frac{1}{2}$ (indicated as α_j^{2n}). The discretization of Eq.(2.91) leads to the C_l coefficients, which are stored in one array $C_l(ij)$. In the same array are stored also the coefficients evaluated at the coarser grid. The index ij is then a function of the grid point location and of the grid level.

$$ij(ic, jc, l) = ic^l + (jc^l - 1) * n1g(l) + n12g(l-1) \qquad 2.93$$

The best way to understand how the coefficients are linked to the metric and how the indices of the $C_l(ij)$ and of α^{mn} vary is considering the Fortran language. If the variable nl indicates the grid level l (= 1 indicates the fine level), $mlev$ is the number of levels, and the number of grid points between one level and the other are $n1g(l-1) = 2n1g(l)$ and $n2g(l-1) = 2n2g(l)$. The part of the Fortran code where the coefficients $C_l(ij)$ are calculated is

```
igjg(i,j,l) = i + (j - 1) * n1g(l) + n12g(l - 1)
do100nl = 1, mlev
dx1q = float((n1g(nl) - 1) * *2)
dx2q = float((n2g(nl) - 1) * *2)
dxcd = float(n2g(nl) - 1) * float(n1g(nl) - 1)/4.
write(6,335)nl, n1g(nl), n2g(nl), dx1q, dx2q, dxcd
335format(3x,' nl =', i3, 2x, 2(2x, i3), 2x,' mesh', 3e12.4)
dojc = 2, n2g(nl) - 1
doic = 2, n1g(nl) - 1
ip = ic + 1
im = ic - 1
jm = jc - 1
jp = jc + 1
jci = (jc - 1) * 2 * *(nl - 1) + 1
ici = ic * 2 * *(nl - 1) - (nl - 2)
imi = (ic - 1) * 2 * *(nl - 1) - (nl - 2)
icj = (ic - 1) * 2 * *(nl - 1) + 1
jcj = jc * 2 * *(nl - 1) - (nl - 2)
jmj = (jc - 1) * 2 * *(nl - 1) - (nl - 2)
if(nl.eq.1)a22jcc = alfj(2, 2, ic, jc)
if(nl.gt.1)a22jcc = alfc(2, 2, icj, jcj)
if(nl.eq.1)a21jcc = alfj(2, 1, ic, jc)
if(nl.gt.1)a21jcc = alfc(2, 1, icj, jcj)
if(nl.eq.1)a22jcm = alfj(2, 2, ic, jm)
if(nl.gt.1)a22jcm = alfc(2, 2, icj, jmj)
if(nl.eq.1)a21jcm = alfj(2, 1, ic, jm)
```

```
if(nl.gt.1)a21jcm = alfc(2,1,icj,jmj)
if(nl.eq.1)a11icc = alfi(1,1,ic,jc)
if(nl.gt.1)a11icc = alfc(1,1,ici,jci)
if(nl.eq.1)a12icc = alfi(1,2,ic,jc)
if(nl.gt.1)a12icc = alfc(1,2,ici,jci)
if(nl.eq.1)a11imc = alfi(1,1,im,jc)
if(nl.gt.1)a11imc = alfc(1,1,imi,jci)
if(nl.eq.1)a12imc = alfi(1,2,im,jc)
if(nl.gt.1)a12imc = alfc(1,2,imi,jci)
if(nl.eq.1)ugccc = 1./gccc(ic,jc)
if(nl.gt.1)ugccc = 1./gccc(icj,jci)
ij = igjg(ic,jc,nl)
co(1,ij) = (a12icc + a21jcc) * dxcd * ugccc
co(2,ij) = (a11icc * dx1q + (+a21jcc - a21jcm) * dxcd) * ugccc
co(3,ij) = (-a12icc - a21jcm) * dxcd * ugccc
co(4,ij) = (a22jcm * dx2q + (-a12icc + a12imc) * dxcd) * ugccc
co(5,ij) = (+a12imc + a21jcm) * dxcd * ugccc
co(6,ij) = (a11imc * dx1q + (-a21jcc + a21jcm) * dxcd) * ugccc
co(7,ij) = (-a12imc - a21jcc) * dxcd * ugccc
co(8,ij) = (a22jcc * dx2q + (+a12icc - a12imc) * dxcd) * ugccc
co(9,ij) = -((a11icc + a11imc) * dx1q + (a22jcc + a22jcm) * dx2q) * ugccc
co(10,ij) = 1./ugccc
enddo
enddo
100continue
```

This part of the code is given in the routine MATSB in the file *psomcurmg.f* in the diskette in the file CODESCH5. This file contains the multigrid solver that is used to evaluate the streamfunction from the vorticity field. The calculation of the C_l is performed once at the beginning.

The multigrid method is applied to the equation

$$\sum_{l=1}^{9} C_l(ij)\Phi_l(ij) = Q(ij) \qquad 2.94$$

Starting from a initial distribution of Φ e.g. that at the previous time step, Eq.(2.94) is relaxed by an iterative procedure for q iterations. The other iterative procedures are explained later on and are given in the file *psomcurit.f* in CODESCH5. From the solution $\Phi_l^{1,q}$, where the superscript 1 stands for the first level, the residual $R^1(ij)$ is calculated in the routine $RESCA$ of *psomcurmg.f*

$$R^1(ij^f) = Q(ij^f) - \sum_{l=1}^{9} C_l(ij^f)\Phi_l^{1,q}(ij^f) \qquad 2.95$$

By means of the q iterations the small scale errors at wave numbers close to that of the fine grid have been dumped. The large scale errors remain and these are dumped at the coarser grids. The residual $R^1(ij)$ is transferred to the successive grid with size $\Delta x_{n+1} = 2\Delta x_n$. Since the $\Phi(ij)$ is defined at the modes, the residual at the coarse grid is transferred by the following relationship

$$\begin{aligned}\tilde{R}^2(ij^l)\sqrt{a}(ij^l) =&[2(R^1(imj^f) + R^1(ipj^f) + R^1(ijp^f) + R^1(ijm^f)) \\ &+R^1(ipjp^f) + R^1(imjp^f) + R^1(ipjm^f) + R^1(imjm^f) + 4R^1(ij^f)]/16.\end{aligned} \tag{2.96}$$

where $R^1 = R^1/\sqrt{a}$. The transfer from fine to large grids is performed in the subroutine $TFILA$; this operation is indicated as

$$\tilde{R}^2 = I_1^2 R^1 \tag{2.97}$$

The superscripts in Eq.(2.97) are related to the large (l) and fine (f) grid; $ij^l = (ic, jc, nl+1)$ is given by Eq.(2.93). ic and jc are the indices of the large grid which are related to those of the fine grid by

$$\begin{aligned} ij^f &= ij(2ic-1, 2jc-1, nl) \ , imj^f = ij^f - 1, \ \ , ipj^f = ij^f + 1, \\ ijp^f &= ij(2ic-1, 2jc, nl) \ , imjp^f = ijp^f - 1 \ , ijm^f = ij(2ic-1, 2jc-2, nl) \ , \\ ipjp^f &= ijp^f + 1 \ , imjp^f = ijp^f - 1 \ , imjm^f = jm^f - 1 \end{aligned} \tag{2.98}$$

By Eq. (2.96) it follows that $\sum_{ic=1}^{ng1(l)} \sum_{jc=1}^{ng2(l)} \tilde{R}^2(ij^l) = 0$. The residual at the coarse grid $\hat{\Phi}(ij^l)$ is evaluated by

$$\sum_{l=1}^{9} C_l(ij^l)\Phi_l(ij^l) = R(ij^l) \tag{2.99}$$

where a good approximation gives $R(ij^l) = \tilde{R}^2(ij^l)$. The calculation proceeds as before and after q iterations $\hat{\Phi}^{2,q}(ij^l)$ is obtained and from this

$$R^2(ij^l) = R(ij^l) - \sum_{l=1}^{9} C_2(ij^l)\Phi_l^{2,q}(ij^l) \tag{2.100}$$

It is important to point out that at each coarse grid the initial guess for the iterative solver must be $\Phi_l^{2,0}(ij^l) = 0$. Only by this assumption a good convergence for the multigrid algorithm is achieved.

The procedure is performed until the coarsest level indicated by ml is reached. At this grid the solution can be obtained by an iterative procedure or by a direct solver, since the small number of points permits the direct inversion within round-off error. The whole procedure to go from the fine to the coarsest grid is coded in the subroutine $CFILA$.

The solution $\hat{\Phi}(ij^m l)$ is transferred from the level ml to the finer level $ml-1$, and it is used as an initial guess for the iterative procedure

$$\sum_{l=1}^{9} C_l(ij^{ml-1})\Phi_l(ij^{ml-1}) = R(ij^{ml-1}) \tag{2.101}$$

where $R(ij^{ml-1})$ is the residual that was used in the iterative relaxation from fine to large grids. At each level $l \leq nl-1$ the solution to initiate the relaxation is

$$\hat{\Phi}(ij^{ml-1,0}) = \hat{\Phi}(ij^{ml-1,q}) + P_{ml}^{ml-1}\hat{\Phi}(ij^{ml,q}) \qquad 2.102$$

where P_{ml}^{ml-1} is the operator to transfer the solution from a coarse to a fine grid. This operation is described in the subroutine $TLAFI$ where the do loops are on the indices of the fine grid. Even here the Fortran code explains Eq.(2.102).

```
igjg(i,j,l) = i + (j - 1) * n1g(l) + n12g(l - 1)
dojc = 2, n2g(nl) - 1, 2
doic = 2, n1g(nl) - 1, 2
icjcl = igjg(ic, jc, nl)
ipjcl = icjcl + 1
imjcl = icjcl - 1
icjpl = igjg(ic, jc + 1, nl)
ipjpl = icjpl + 1
imjpl = icjpl - 1
icjml = igjg(ic, jc - 1, nl)
ipjml = icjpl + 1
imjml = icjpl - 1
icjcf = igjg(2 * ic - 1, 2 * jc - 1, nl - 1)
imjcf = icjcf - 1
ipjcf = icjcf + 1
icjmf = igjg(2 * ic - 1, 2 * jc - 2, nl - 1)
imjmf = icjmf - 1
ipjmf = icjmf + 1
icjpf = igjg(2 * ic - 1, 2 * jc, nl - 1)
imjpf = icjpf - 1
ipjpf = icjpf + 1
ph(icjcf) = ph(icjcl) + ph(icjcf)
ph(icjpf) = (ph(icjpl) + ph(icjcl)) * 0.5 + ph(icjpf)
ph(ipjpf) = (ph(ipjpl) + ph(ipjcl) + ph(icjcl) + ph(icjpl)) * 0.25 + ph(ipjpf)
ph(ipjcf) = (ph(ipjcl) + ph(icjcl)) * 0.5 + ph(ipjcf)
ph(ipjmf) = (ph(ipjcl) + ph(ipjml) + ph(icjml) + ph(icjcl)) * 0.25 + ph(ipjmf)
ph(icjmf) = (ph(icjcl) + ph(icjml)) * 0.5 + ph(icjmf)
ph(imjmf) = (ph(icjcl) + ph(icjml) + ph(imjml) + ph(imjcl)) * 0.25 + ph(imjmf)
ph(imjcf) = (ph(icjcl) + ph(imjcl)) * 0.5 + ph(imjcf)
ph(imjpf) = (ph(icjpl) + ph(icjcl) + ph(imjcl) + ph(imjpl)) * 0.25 + ph(imjpf)
enddo
enddo
```

By solving Eq.(2.101) for q iterations $\hat{\Phi}(ij^{ml,q})$ is obtained and from this by the same procedure the finest grid is reached. The solution $\hat{\Phi}(ij^{1,q})$ permits to evaluate the residue by Eq.(2.95). The maximum value of the residual at each multigrid cycle m is R^m_{max} from which the convergence coefficients $\epsilon = R^m_{max}/R^{m-1}_{max}$ are evaluated.

Different types of multigrid cycles are possible, the V types and the W types. Usually the W cycles are more efficient than the V types.

In the multigrid methods the relaxation procedure is a very important issue from which the efficiency depends. The iterative procedure is coded in the subroutine $ITSOLV$ in the file *psomcurit.f*. This routine has been inserted in a separated file to have the possibility to change the iterative scheme whose efficiency is computer-dependent. In the present case we chosen the classical point SOR scheme which is not very efficient on the CRAY computers. For these vectorised computers the check-board schemes with two sweeps are more efficient. In each sweep two do loops are opened. Even four colors check board iterative procedures can be used. Roache's (1972) book explains some of these iterative schemes.

The point SOR scheme is based on evaluating at each grid point $ij(ic, jc, l)$ the quantity $F(ij) = -\sum_{l=1}^{8} \mathcal{C}_l(ij)\Phi_l^{p-1}(ij) + Q(ij)$ from Eq.(2.94). Somewhere in the array $\Phi^{p-1}(ij)$ there is the current value in the computer memory; it can be at the new p or the old $p-1$ iteration depending whether it was or it was evaluated or not. From $F(ij)$ the value at the new iteration is yielded:

$$\Phi_l^p(ij) = \omega\frac{F(ij)}{\mathcal{C}_9} + (1-\omega)\Phi_l^{p-1}(ij) \qquad 2.103$$

The acceleration parameter ω is assigned, and as reported by Roache (1972) a good value is $\omega = 1.85$.

The efficiency of the multigrid method has not been quantified since this method is used only in chapter 5 to calculate the streamfunction. The reader wants to investigate in more detail this method could use the code described in chapter 5 by performing only one time step for different grids. Thus he can verify that the computer time increases linearly with the number of grid points.

A version of the multigrid method to solve the elliptic equation when Φ is defined at the cell center instead of the grid modes can be easily derived from that above described. This solver is necessary to get the "pressure" Φ in the primitive variables formulation for complex geometries. This is not furnished because the solution of the 3D Navier-Stokes equation in curvilinear coordinates in 2D has been neglected; however the reader who is interested can look at the system of equations to solve in Orlandi (1989) and in Choi *et al.*(1994).

2.6 Conclusions

In this chapter the basic ideas on the finite difference methods have been recalled, and in particular only the fundamentals have been described to solve the Navier-Stokes equations in 2D and 3D. This book is not devoted to whom would like to know the basic ideas for a computational fluid dynamics course. The main intent is to furnish a tool to understand the physics of fluid flow by numerical simulations. Those who are interested in the basic principle of computational fluid dynamics have in literature a large amount of books devoted to this topic; some of these have been cited in the list of references.

3 The Burgers equation

3.1 Physical considerations

The Burgers equation is a simple equation to understand the main properties of the Navier-Stokes equations. In this one-dimensional equation the pressure is neglected but the effects of the nonlinear and viscous terms remain, hence as in the Navier-Stokes equations a Reynolds number can be defined. This number expresses the ratio between the advective and the viscous contribution in a flow. The present book deals with flows at high Reynolds numbers where the nonlinear terms play a fundamental role, and the physics is more complicated than that when the viscous term dominates. The simulation of the flow evolution then necessitates the use of accurate and robust numerical methods. In 3D turbulent flows, where the number of degrees of freedom is greater than in high Re laminar flows, to get solutions it is necessary to introduce some sort of closure to account for the impossibility to resolve the small scales. Before applying any new idea about numerical methods to 3D flows, the good sense suggests to find the simplest equation to test these ideas. This consideration explains why the Burgers equation was often used to check new numerical methods or closure for turbulent flows. The equation is

$$\frac{\partial u}{\partial t} + \frac{1}{2}\frac{\partial uu}{\partial x} = +\nu \frac{\partial^2 u}{\partial x^2} \qquad |x| \leq 1 \qquad 3.1$$

Since a numerical method should account for the physics linked to the equation it is important to understand the role each term in Eq.(3.1) is playing. This effort is facilitated if the equations are written not only in the physical but also in the wave number space. To get the equation on the wave number space periodicity in the x direction is assumed and the initial velocity profile should be periodic. The simplest initial condition is to assign a single wave number

$$u(x,0) = -sin(\pi x) \qquad 3.2$$

The solution can be expressed by a truncated Fourier series

$$u(x,t) = \sum_{k=-N/2-1}^{N/2} u_k(t) e^{ik^* x} \qquad 3.3$$

with $k^* = k2\pi/L_x$. By substituting Eq.(3.3) into Eq.(3.1) the equation becomes

$$\frac{du_k}{dt} = -\frac{ik^*}{2} \sum_{p=-N}^{N} u_p u_{k-p} - \nu k^2 u_k \qquad 3.4$$

In physical space Eq.(3.1) shows that the effect of the viscous terms is a diffusion of u, that is the variable u has to migrate into the external space. This physical process occurs if the domain is large enough. On the other hand, for instance, if periodicity is limiting the space the physical process is affected by this geometrical limitation. From physical reasoning it turns out that if the equation is solved in a region completely filled by the initial condition, there is an unphysical diffusion of $u(x) > 0$ in the region of $u(x) < 0$ resulting in a greater decay of the peak velocity and hence of $\frac{\partial u}{\partial x}|_{x=0}$. On the contrary this does not occur when an external region encircles the space where the initial conditions are assigned. This reasoning can be checked by numerical simulations, and it suggests that 2D and 3D turbulence simulations have physical validity until the size of the energy containing eddies are smaller than the computational box.

Eq.(3.4) shows that viscosity reduces the amplitude of each wave number u_k in time. In fact, by neglecting the nonlinear terms, Eq.(3.4) becomes

$$\frac{du_k}{dt} = -\nu k^2 u_k \qquad 3.5$$

with solution

$$u_k(t) = u_k(0)e^{-2\nu k^2 t} \qquad 3.6$$

Eq.(3.6) shows that the high wave numbers components are dumped at a rate that increases with the wave number, and with ν . This physical consideration is important in numerical simulations; in fact when the resolution is not satisfactory to reproduce high gradients or equivalently the smallest scales, the truncation errors produce unphysical components with amplitudes at high wave numbers. Eq.(3.6) shows that these errors at high k decay fast. If the simulations are performed at a value of the viscosity at which the resolution is not satisfactory, during a transient, energy is transferred at small scales and it piles up at high wave numbers. This energy is dissipated very fast and at a certain time the form of the spectrum is close to that of a fully resolved simulation. Only from that instant the simulation has a physical meaning. This procedure is often used in the DNS of isotropic turbulence since a transient is necessary to generate the phases of the velocities that were random in the initial field . In other words in this transient the structures of turbulence are produced, and then the simulation cannot be fully resolved. In chapter 8 this procedure is described.

Eq.(3.4) shows that a convolution product represents the nonlinear terms in wave number space and thus it turns out that the amplitude at each wave number is affected by the amplitude of all the other wave numbers. The evolution equation for the energy $E_k = u_{-k}u_k$ is obtained by multiplying the equation for u_k by u_{-k} and that for u_{-k} by u_k, it follows

$$\frac{dE_k}{dt} = T(k) - \nu k^2 E_k \qquad 3.7$$

$T(k)$ expresses that the transfer of energy occurs by interactions among wavenumbers k, p, q such that $p + q + k = 0$. In one-dimensional and in three-dimensional space the energy goes from low to high wavenumbers. On the contrary in two-dimensional space energy is transferred from high to low wave numbers. For the Burgers equation, the transfer of energy to small wave numbers, in physical space is equivalent to the formation of very sharp gradients that are the steeper the smaller is ν.

To understand the physics of a non-resolved simulation that was described for the 3D case, it is useful to see what occurs for the Burgers equation. It has to be expected that in the region of high gradients, if the grid is not sufficiently small, oscillations grow which successively are dumped by viscosity. In the wave number representation this process is

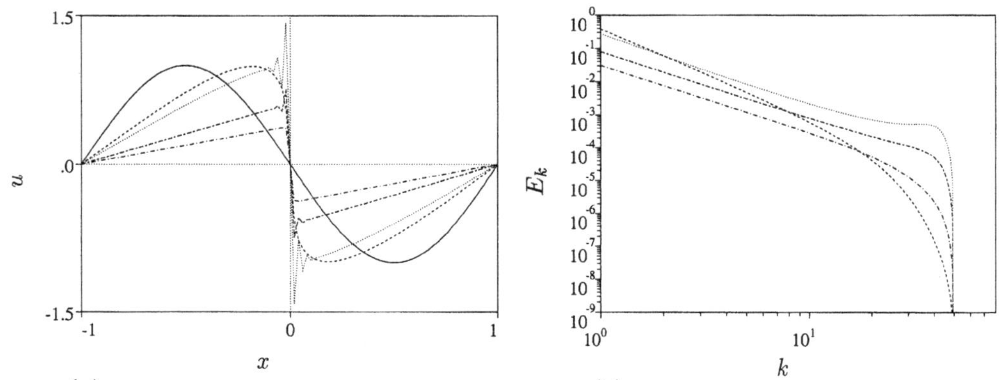

(a) (b)

Fig. 3.1 a) Velocity profiles , b) Energy spectrum ; —— $t = 0$, ---- $t = 1.$, ······· $t = 1.5$, —-— $t = 4.$, —·— $t = 7$.

shown by an energy pile-up at high wave numbers. Fig.3.1 indeed shows the two point of view by a simulation with 101 grid points equally distributed at $\nu = 10^2\pi$. At $t = 1.5$, when the transfer of energy to small scales is not completed, the insufficient resolution produces wiggles with high amplitude, localized around $x = 0$; these wiggles correspond to the bump in the energy spectrum. At $t = 7$ these unphysical energy components are not any longer observable, and Fig.3.1 shows that the spectrum and the velocity profile are smooth. Since the method conserves, in the inviscid limit, global energy, this unphysical transient does not bring the calculation to diverge, but the real physics is disrupted.

The evolution of the velocity gradient at $x = 0$ characterizes the physics of the Burgers equation and, as it was shown by Orlandi & Briscolini (1983), if it is scaled by $\pi\nu$, it has a self-similar behavior. Two simulations were performed at $\pi\nu = 10^{-2}$ and 10^{-3} with a fully resolved grid; the results in Fig.3.2 show that at the same value of πt the same peak value of the normalized velocity gradient is reached, even if one started from different values. A self-similar decay in time is also predicted. In the same figure a simulation, non-fully resolved, shows that the unphysical dissipation, during the transient, affects the whole simulation.

Local and total momentum conservation are the fundamental properties of the Navier-Stokes equations in the inviscid limit and these should be preserved for the set of the discretized equations. This property holds in the physical as well as in the wave number space. The preservation of these properties can be easily shown for the Burgers equation to facilitate the comprehension of the fulfillment in 3D. By recalling that $u_k = u_{-k}$ immediately it turns out, in Eq.(3.4), that $\iota \sum_{k=-N}^{N} k \sum_{p=-N}^{N} u_p u_{k-p} = 0$. From the physical point of view it means that the nonlinear terms redistribute the wave number amplitude among different wave numbers with a total null contribution. The same conclusion is obtained by integrating Eq.(3.1) in the whole space, for $\nu = 0$ and with periodic or with Dirichlet conditions.

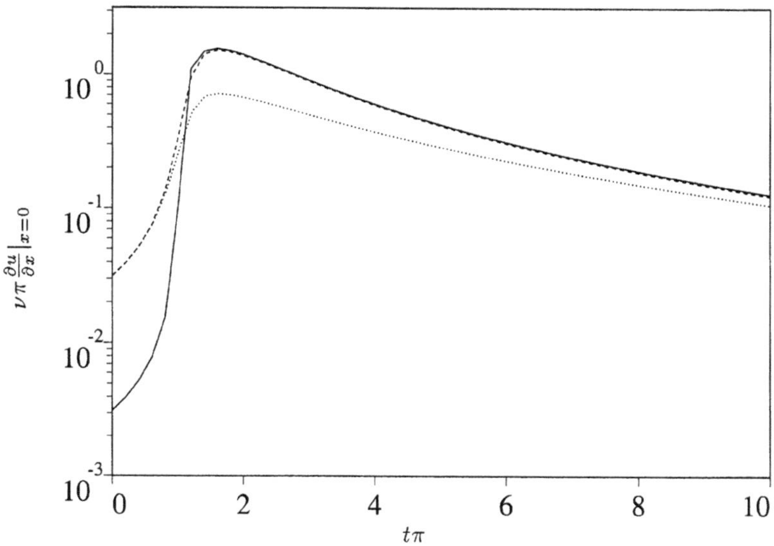

Fig. 3.2 time evolution of the velocity gradient at $x = 0$, —— $\pi\nu = 10^{-3}$, $n = 1600, \alpha = 3.$, ---- $\pi\nu = 10^{-2}$, $n = 800, \alpha = 3.$, ······· $\pi\nu = 10^{-2}$, $n = 100, \alpha = 0$.

The conservation of energy is a further property of the Burgers and Navier-Stokes equations. The expression of $T(k)$ shows that $\sum_{-N}^{N} T(k) = 0$, that is, that the energy is transferred by triadic interactions among different wave numbers, and hence for $\nu = 0$ the total energy is conserved. The energy equation in physical space, by defining $q = u^2/2$, is

$$\frac{\partial q}{\partial t} + \frac{2}{3}\frac{\partial uq}{\partial x} = +\nu\frac{\partial^2 q}{\partial x^2} - \nu\left(\frac{\partial u}{\partial x}\right)^2 \qquad 3.8$$

By integrating the second term in Eq.(3.8) it turns out that the total contribution of the energy advection is zero. On the right hand side appears the viscous energy diffusion, and the energy dissipation. For the Burgers as well as for 3D flows the integral of the turbulent diffusion is null for periodical conditions and hence the total energy decays in time because of the negative right hand side that is called the rate of energy dissipation.

From these physical considerations two rules hold: the first suggests that for incompressible viscous flows, without discontinuities for the presence of ν, it is convenient to write the transport equations of each variable, in the continuous as well as in the discrete form, such that global conservation of the total quantity and of the high moments occurs. The second rule forces to have a sufficient number of computational points across the sharp gradients to fully resolve them. Since the occurrence of these thin layers depends on ν, it turns out that the simulation of equations with a small ν requires more care.

3.2 Spatial discretization

In accordance with the previous considerations the Burgers equation evolves from the profile in Eq.(3.2) towards one with a gradient at $x = 0$ with the slope the steeper the smaller is ν. To get the solution by a reasonable number of grid points a non-uniform grid is necessary, and the transformation in subsection 2.4c is appropriate. The right hand side of Eq.(3.1) explicited by Eq.(2.15), and the convective term, discretized to maintain in the computational space the conservation properties, gives

$$\frac{du}{dt} + \frac{\Delta\xi}{x_\xi}(\overline{q}^\xi_\xi) = +\frac{\nu}{\Delta\xi}\frac{1}{x_\xi}\left(\frac{u_{\xi+}}{x_{\xi+}} - \frac{u_{\xi-}}{x_{\xi-}}\right) \qquad 3.9$$

The momentum in the computational space for $\nu = 0$ is conserved, in fact in the discretized form the total momentum is $U = \Sigma_i u_i \frac{x_\xi}{\Delta\xi}$. The summation of the second term of Eq.(3.9), by recalling Eq.(2.2), is null because of the cancelation of all the $q_{i+\frac{1}{2}}$ except those at the two extrema that, in accord to the boundary conditions, are equal. The conservation holds even if it is evaluated in half the domain. Eq.(3.9) can be written in the schematic form

$$\frac{du_i}{dt} = H_i + \nu L(u_i) \qquad 3.10$$

where H_i represents the nonlinear term and $L(u_i)$ is the linear operator $a_i^p u_{i+1} + a_i^c u_i + a_i^m u_{i-1}$.

3.3 Time discretization

In the solution of parabolic equations, and in particular in the presence of a very fine grid, in order to overcome the viscous stability restriction of the explicit schemes, it is convenient to treat implicitly the linear terms. The nonlinear terms, on the other hand, are treated explicitly in order to retain a high time accuracy; thus CFL is the only restriction to the time step, it is given by $\frac{U\Delta t}{\Delta x} < 1$.

Other choices can be done for the nonlinear terms and some are briefly mentioned by referring to the 3D flows which require them. For example the nonlinear terms can be linearized for their implicit treatment to allow the use of large Δt. More computational time is indeed requested; this effort is convenient only when the grid, in some region, is so small that the CFL restriction reduces the time step at a value so small, making the calculation almost impossible. This occurs, for instance, in simulations of flows in cylindrical coordinates where the azimuthal velocity near the center is not negligible (Aksevoll & Moin 1996 , Verzicco & Camussi 1997) or in complex geometries, near sharp angles, where large velocity gradients require a very high resolution (Choi *et al.* 1993). However, in the largest part of physical cases of interest, in particular in the simulation of unsteady flows, a restriction on the time step is required to have an accurate time evolution of the phenomenon, and this consideration suggests to retain the explicit treatment of the nonlinear terms. Moreover, in some applications, the physics requires very small time increments, much smaller than those given by the CFL restriction. When this happens it is worth, also for the linear terms, to use explicit schemes and then the computational time needed to invert the tridiagonal matrices is saved.

To understand the previous considerations the solution of the Burgers equation is very useful. To investigate the whole range of time integration schemes the Pade' formulation for the time derivative is applied. This compact formulation permits to go from one to another scheme by changing the values of two constants. The time derivative is

$$\frac{\partial u_i}{\partial t} = \frac{(1+\zeta)\Delta - \zeta\nabla}{1+\theta\Delta} u_i + (\theta - \zeta - \frac{1}{2})O(\Delta t^2) \qquad 3.11$$

Eq.(3.11) applied to Eq.(3.10) (by recalling that the denominator should be applied to the right hand side of Eq.(3.10)) gives

$$\Delta u_i^n - \frac{\zeta}{(1+\zeta)}\nabla u_i^n = \frac{\Delta t}{(1+\zeta)}\left[H(u_i^n) + \nu L(u_i^n) + \theta\left(\Delta H(u_i) + \nu L(\Delta u_i^n)\right)\right]$$

$$+(\theta - \zeta - 0.5)O(\Delta t^2) \qquad 3.12$$

Where $\Delta u_i^n = u_i^{n+1} - u_i^n$ and $\nabla u_i^n = u_i^n - u_i^{n-1}$. The increment of H can be expressed as

$$\frac{\Delta}{2}\frac{\partial u_i^2}{\partial x} = \frac{\Delta \xi}{x_\xi}(\overline{q}_\xi^\xi) \qquad 3.13$$

where $q = u^n \Delta u^n$. Eq.(3.12) can be easily coded. In the file *burgexim.f* in the directory CODESCH3/PADE there is a code that permits to check the stability conditions for the different numerical schemes. For $\theta = 0$ the set of different order explicit schemes are considered. On the other hand, for $\theta \neq 0$ the implicit schemes are treated.

3.4 Results

The simulations have been performed for $\nu\pi = 10^{-2}$ and on 101 non-equispaced grid points. In accordance with the viscous stability restrictions it has been found, from the simulation, that this enhanced grid clustering requires for the explicit time advancement schemes a extremely small Δt. The results for the explicit schemes are not shown since these are very accurate. On the other hand, the results of an implicit method are more interesting and are given for three different schemes: Euler implicit ($\theta = 1, \zeta = 0$), Crank Nicolson ($\theta = \frac{1}{2}, \zeta = 0$), and the three points backward ($\theta = 1, \zeta = 0.5$). The first derivative at $x = 0$ is the quantity most sensitive to the numerics and it has been plotted in Fig.3.3a-c

The results show that the three points backward is the most reliable scheme producing results independent of the time step. On the other hand, the Euler implicit scheme shows that the scheme has a poor accuracy at large Δt and that the Crank Nicolson generates oscillations around the exact solution. These oscillations never cause the divergence of the solution; however, if the oscillations are too large the solution is degraded.

The Padé formulation is onerous when it is applied to the 3D Navier-Stokes equations; in particular this is true for the linearization of the nonlinear terms, that, as mentioned before, is necessary only in a few flows. On the other hand, in unsteady simulations the physics requires accurate time solutions and it is better to have a reduced programming work, in order to eliminate the possibilities to make errors. In direct numerical simulations of turbulent flows two second order time accurate schemes were in large part used: the first one treats the linear terms by a Crank-Nicolson scheme and the nonlinear terms by the

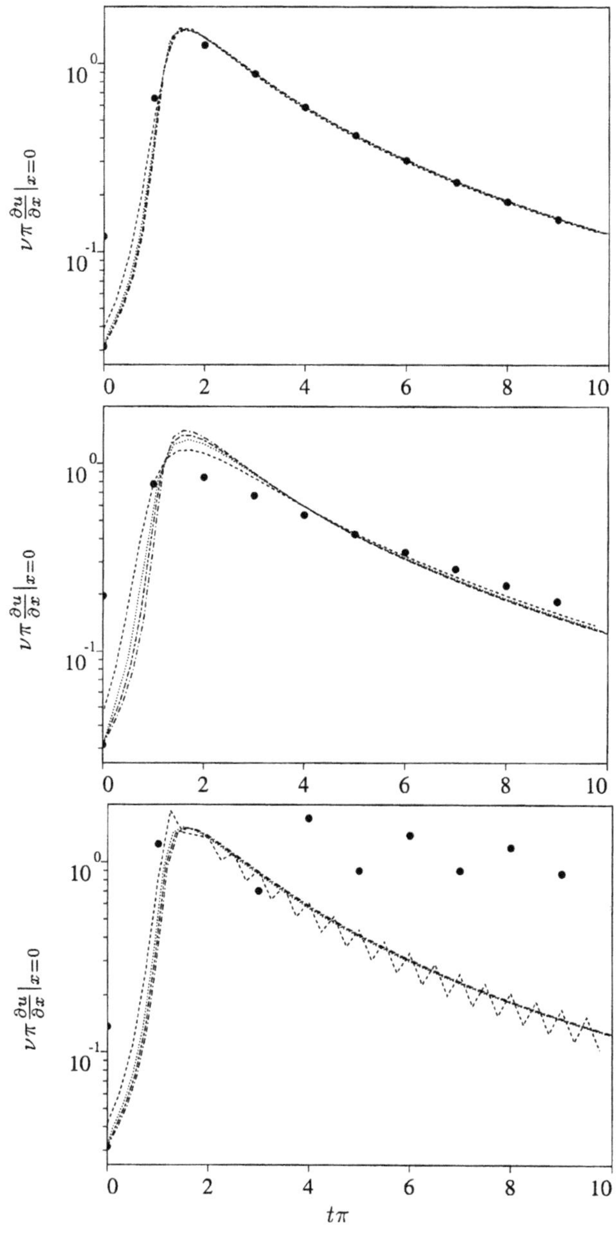

Fig. 3.3 time evolution of the velocity gradient at $x = 0$, ∘ $\pi\Delta t = 1.$, ---- $\pi\Delta t = .25$, ······· $\pi\Delta t = .1$ — - — $\pi\Delta t = .05$ —·— $\pi\Delta t = .01$ a) Three points backward, b) Euler implicit c) Crank Nicolson.

Adams-Bashfort scheme. The other one is a low storage third order Runge-Kutta scheme which splits each time step in three substeps.

These methods were described in Chapter 2, and a comparison between them was shown in section 2.3.5. In that section a description of the discretization was not given and it is explained briefly here. The relative code is given in the directory CODESCH3/RK and it is named *burgts.f*. This code is very simple and has a structure close to that adopted for the Navier-Stokes equations in 2D and 3D. This choice is important because those who are not experts in the Fortran language could be acquainted with the more complicate codes described in other chapters by starting from this simple example. The time integration schemes applied to the Burgers equation give

$$\frac{u_i^h - u_i^k}{\Delta t} - \left(a_i^p(u_{i+1}^h + u_{i+1}^k) + a_i^c(u_i^h + u_i^k) + a_i^m(u_{i-1}^h + u_{i-1}^k)\right)\nu\alpha^h$$

$$= \gamma^h H_i^k + \rho^h H_i^q \qquad 3.14$$

The two schemes differ only by assigning the appropriate values of the parameters. The Adams-Bashfort advances in one time step and the Runge-Kutta of 3rd order in three substeps; a do loop calls the different subroutines to evaluate each term of Eq.(3.14) the appropriate number of times. For the Adams-Bashfort the values of the parameters are

$$h = n+1, \quad k = n, \quad q = n-1, \quad \gamma^h = 1.5, \quad \rho^h = -0.5 \qquad 3.15$$

α in Eq.(3.14) is linked to the other two parameters: $\alpha^h = 0.5(\gamma^h + \rho^h)$.

For the third order Runge-Kutta scheme, for the first substep we have

$$h = n + \frac{1}{3}, \quad k = n, \quad q = n - \frac{1}{3}, \quad \gamma^h = \frac{4}{15}, \quad \rho^h = 0 \qquad 3.16$$

for the second

$$h = n + \frac{2}{3}, \quad k = n + \frac{1}{3}, \quad q = n, \quad \gamma^h = \frac{5}{12}, \quad \rho^h = -\frac{17}{60} \qquad 3.17$$

for the third

$$h = n + 1, \quad k = n + \frac{2}{3}, \quad q = n + \frac{1}{3}, \quad \gamma^h = \frac{3}{4}, \quad \rho^h = -\frac{5}{12} \qquad 3.18$$

Eq.(3.14) can be expressed in a more suitable form by introducing the quantity $\Delta u_i = u_i^h - u_i^k$, leading to

$$\left[(-a_i^p \Delta u_{i+1} + (1 - a_i^c)\Delta u_i - a_i^m \Delta u_{i-1}\right] \nu\alpha^h \Delta t$$

$$= \Delta t(\gamma^h H_i^k + \rho^h H_i^q) + 2\left[a_i^p u_{i+1}^k + a_i^c u_i^k + a_i^m u_{i-1}^k\right] \nu\alpha^h \Delta t \qquad 3.19$$

The results obtained by the code *burgts.f* were compared with those obtained by the three points backward scheme shown in Fig.3.3c. In Fig.2.4 it was shown by simulations at

$\pi\nu = 10^{-2}$ and 10^{-3} that the improvement of the Runge-Kutta scheme increases when the viscosity decreases. The time history of the gradient at $x = 0$ shows unphysical oscillations in Fig.3.4a-b at $\pi\nu = 10^{-2}$ by the Adams-Bashfort with $\pi\Delta t = 0.03$. These disappear when the CFL number is approximately 4 at $\pi\Delta t = 0.025$. The Runge-Kutta, on the other hand, gives good results by using $\pi\Delta t = 0.033$ with the CFL approximately 5. At this relative high value of ν it turns out that the Runge-Kutta with three substeps does not reduce the CPU time. The time histories at a smaller viscosity in Fig.3.4b were obtained by $\pi\Delta t = 0.003$ for the Adams-Bashfort and by $\pi\Delta t = 0.008$ for the Runge-Kutta. The Runge-Kutta even at this ν does not increase its efficiency. Those who are interested could prove that a further decrease of ν leads to a CPU saving by the Runge-Kutta as compared to the Adams-Bashfort.

The comparison between the coefficients for the first substep shows that the Runge-Kutta has the great advantage that it is possible to operate with a variable Δt without disrupting the time accuracy, in fact $\rho^h = 0$ does not need H_i^q at the first substep. This property permits to initiate the simulation without any assumption, because at $t = 0$ it is impossible to evaluate H_i^q. A further merit of the Runge-Kutta is that if there is the necessity to continue the simulation, that for any reason was interrupted, there is no necessity to save the H_i^q arrays. It is clear that in 3D simulations the restarting files are very large and any memory saving is an important requirement.

In several unsteady simulations large velocity gradients occur in the first part of the calculation; these gradients later on decrease for viscous effects. During the advancement of the solution the Δt can be increased in accord to this physical consideration and thus we have a CPU saving. The amplitude of the non-uniform step can be assigned by fixing the CFL; in the theoretical case for the Runge-Kutta scheme this is $\sqrt{3}$ and for the Adams-Bashfort it is < 1. Often (depending on the physical case) the theoretical limit for the Runge-Kutta can be increased because the peak CFL occurs not always at the same spatial location. From these observations it turns out that it is better to avoid the Adams-Bashfort with a non-uniform Δt since the necessity of H_i^q reduces the time accuracy. To maintain this accuracy an interpolation to get H_i^q increases the CPU. However it is also better to avoid, if possible, any sort of interpolations in solving the Navier-Stokes equations. The time history of the velocity gradients in Fig.3.4c obtained by applying the Runge-Kutta by fixing $CFL = 5$ and $CFL = 8$ were obtained respectively by 250 and 365 time steps. The results by $CFL = 5$ compare well with those obtained by the fully implicit three-points backward scheme with an uniform $\Delta t = 0.1$

3.5 Conclusions

From the Burgers equation it turns out that this simple equation has several physical properties in common with the Navier-Stokes equations. For those who do not have a deep background in fluid mechanics or in the properties of nonlinear systems of partial differential equations this chapter is useful. I am referring in this context to graduate or undergraduate students. From this chapter they got a first impression of the importance of the nonlinear terms in transferring energy from the large to the small scales. The comprehension of this mechanism has relevance in the comprehension of the effects of the vortex stretching term; that is the leading term producing a complex physics in 3D flows.

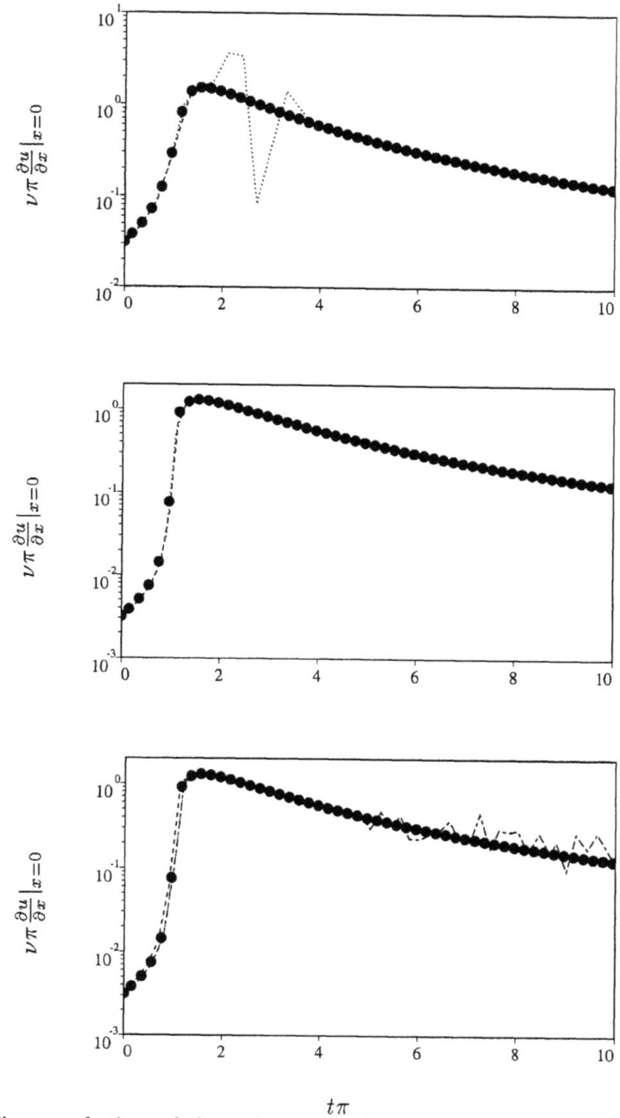

Fig. 3.4 Time evolution of the velocity gradient at $x = 0$, a) $\pi\nu = 10^{-2}$ o three-points backward $\pi\Delta t = .05.$, ---- Runge-Kutta $\pi\Delta t = .033$, ... Adams-Bashfort $\pi\Delta t = 0.03$; b) $\pi\nu = 10^{-3}$ o three-points backward $\pi\Delta t = .05.$, ---- Runge-Kutta $\pi\Delta t = .008$, ... Adams-Bashfort $\pi\Delta t = 0.003$; c) $\pi\nu = 10^{-3}$ o three-points backward $\pi\Delta t = .1.$, ---- Runge-Kutta $CFL = 8$, ... Runge-Kutta $CFL = 5$,

It is almost impossible to get solutions of the nonlinear system of equations in closed form, even in the one-dimensional case of the Burgers equation. Basdevant *et al.* (1986) showed that the computer is in any case necessary. From this remark it turns out that, the direct simulation in the physical space or in the wave number space is the simplest way to get the solution without having any deep mathematical background. In this chapter the most important rules to follow have been given for the Burgers equation.

In the spatial discretization it has been shown that a non-uniform grid spacing reduces the memory occupancy and the CPU time. With regard to the time integration it has been shown that the low storage Runge-Kutta scheme is very efficient and if used at constant CFL it is as fast and reliable as fully implicit schemes. These can use greater time increments but they require more programming work. In this book the interest is devoted to unsteady flows and therefore the Runge-Kutta is always employed. However, the users interested in reaching the steady state solution could employ fully implicit schemes since these schemes are very robust and in few time steps the solution is reached. In this case the time steps should be considered as iterations since the time evolution does not have a physical meaning.

This 1D equation permits us also to perform a comparison between the different methods available in literature to go from the continuous space to a discrete space. In this book the interest is to solve the equations in the physical space. Canuto *et al.* (1987) dedicated a book to spectral methods, listing all the different methods to solve the equation in the spectral space. These methods range from the full spectral to the pseudospectral and to the large variety of collocation methods. The comparison between a large part of these methods and the finite-differences was presented by Basdevant *et al.* (1986) for the Burgers equation.

4 Two-dimensional flows in Cartesian coordinates

4.1. Introduction

The Navier-Stokes equations in 2D can be expressed in primitive variables, in vorticity-streamfunction and in vorticity-velocity formulations. The second formulation has been largely employed in the past and the first simulations were related to geophysical flows (Arakawa 1956). Later one this formulation was also used to simulate 2D turbulence. These flows have a complex physics which requires fine resolution, thus it is important to chose the most appropriate formulation together with very efficient numerical methods. The necessity of a large number of grid points is the reason why the ω, ψ formulation, with only one parabolic and one elliptic equation, is the most convenient.

The vorticity-velocity formulation is equivalent to the ω, ψ formulation when velocities are staggered but it has the disadvantage solving two elliptic equations, also by direct methods, to require a large computational time. A tentative was done by Orlandi (1987) to develop a method based on the vorticity-velocity formulation without the solution of elliptic equations. The first order time accuracy, linked to the linearization, limited the method to the achievement of stationary solutions.

In flows with two periodic directions, as for example 2D turbulence, pseudo-spectral methods are very efficient and accurate as shown by Orszag (1971). These methods can be extended to flows with no-slip boundaries and, with a large effort, to more complicated boundaries. If the amount of work is too high and if it is difficult to prove that pseudo-spectral methods maintain the spectral accuracy, it could be worth while to use a standard finite difference second order accurate scheme in space and time. The simplicity of this method is such that, even without any good mathematical background, any scholar can, in a short time, understand the method and write a code. There are flows where the boundary conditions for the ω, ψ set of equations are difficult to formulate; this, for instance, occurs for flows in multiply connected domains and here the primitive variables have an advantage. Since these flows are not treated in this monograph, this chapter is devoted to the solution of the Navier-Stokes equations in the ω, ψ formulation.

4.2 The equations in vorticity-streamfunction formulation

As reported in several fluid dynamics books, for instance (Batchelor (1970) Pg.266 ,Phanton (1984) Pg.366) the vorticity equation is derived by applying the *curl* operator to the Navier-Stokes equations. This operation eliminates the pressure and a transport equation for the vorticity is obtained.

$$\omega = -\frac{\partial u_1}{\partial x_2} + \frac{\partial u_2}{\partial x_1} \qquad 4.1$$

The velocity components are related to the streamfunction ψ, by the relationship $\partial\psi/\partial x_2 = u_1$ and $\partial\psi/\partial x_1 = -u_2$. The equation describing the evolution of the vorticity is

$$\frac{\partial \omega}{\partial t} + J(\zeta, \psi) = +\frac{1}{Re}\nabla^2\omega. \qquad 4.2$$

where ζ in the most general case is the potential vorticity defined as $\zeta = \omega + \beta x_2 + h(x_1, x_2)$. This equation holds for engineering as well as for geophysical flows. For engineering applications $\zeta = \omega$; on the contrary, in geophysical applications, these equations are valid in the quasi-geostrofic approximation where β accounts for the variation of the earth rotation with the curvature and $h(x_1, x_2)$ accounts for topographic effects. Pedlosky (1986) book explains how this equation is derived from the full sets of Navier-Stokes equations and under which conditions it accurately reproduces physical flows. $J(\cdot, \cdot)$ is the Jacobian of the argument functions. The Reynolds number has been obtained by introducing a reference length L and a reference velocity U which depend on the flow. The Reynolds number is $Re = UL/\nu$. The vorticity and streamfunction are connected by the kinematic relationship

$$\nabla^2 \psi = -\omega \qquad 4.3$$

These equations must be solved with the appropriate boundary conditions.

4.3 Vorticity boundary conditions

4.3a Periodicity

If periodicity is imposed in one direction, boundary conditions are not necessary and the numerical method accounts for this assumption.

4.3b Free-slip

In case of symmetry or equivalently for a free-slip boundary $\omega = 0$ holds. This condition is valid only if the boundary is flat. A free-slip boundary also is a mathematical artifact; in fact in reality a solid boundary is no-slip. In laboratory experiments a condition close to that of a free-slip boundary can be obtained by an interface between two inmiscible fluids with a large density difference. Bernall *et al.* (1989) used this condition to reproduce the interaction of a vortex ring with a free-slip boundary. To be as close as possible to the ideal condition the fluid should not have any impurity and the interface should not deform.

A free-slip boundary can be assumed in the simulations when Re is so high that the boundary layer has a negligible thickness with respect to the size of the eddies. This assumption is often done in geophysical flows when solid boundaries are present. On the other hand, in engineering applications the assumption of free-slip boundaries reproduces flows in an unbounded space. In this case it is, however, important to locate the boundary as far as possible from the region where the vorticity is concentrated. A coordinate stretching fulfills this requirement without increasing the computational time and the memory occupancy. The large grid spacing near the boundary does not lead to inaccuracy since, there, the amount of vorticity is negligible.

4.3c No-slip

In the realistic case of a no-slip wall the boundary conditions for the vorticity must be derived from the assumption of zero velocity. In the past several numerical approximations were used to derive the wall vorticity from Eq.(4.3); they are described by Roache (1972) at Pg.140 and by Fletcher (1987) at Pg.374. All these approximations produced inaccurate solutions at high Reynolds numbers. A different choice leads to evaluate the boundary conditions for ω from the expression used in primitive variables with the staggered formulation. Dealing with the momentum equations the no-slip condition is imposed by setting, in the discretized expression of the second derivative of the tangential velocity at the first point near the wall, the wall velocity equal to zero This second derivative occurs in the expression of the normal vorticity flux. As mentioned in section 2.1.4 the boundary conditions for the vorticity equation are obtained by

$$\frac{\partial \omega}{\partial x_2}|_{\frac{w+1}{2}} = \left(\frac{\partial^2 u_2}{\partial x_1 \partial x_2} - \frac{\partial^2 u_1}{\partial x_2^2}\right)|_{\frac{w+1}{2}} \qquad 4.4$$

Since the boundary conditions in the pressure velocity formulation lead to very accurate solutions even for wall turbulent flows it is expected that Eq.(4.4) is appropriate to find accurate solutions at high *Re* in 2D flows. A check was done by Orlandi (1990) in the simulation of the impact of a vortex dipole with a wall at high Reynolds numbers, where *Re* was defined through by means of the circulation of the vortex patch. The values of *Re* used by Orlandi (1990) are greater than those of the streamwise vortices measured by Robinson (1991) in the direct simulations of a plane channel (Kim *et al.* (1987)) and of boundary layers (Spalart (1988)). Despite the high *Re* it is important to remind that in 2D the vortex dynamics is simpler due the absence of vortex stretching. The reason why Eq.(4.4) works well, is related to the fact that the wall vorticity is related to the vorticity inside the field. A similar argument was used by Woods (1954) to derive an expression for the wall vorticity. The reason why the expressions without the vorticity gradient did not work satisfactorily at high *Re* is linked to the contamination of the solution by the truncation errors which is unavoidable discretizing Eq.(4.3) at the wall. Non-uniform grids necessary to represent the thin and strong vorticity layers near no-slip walls increase these errors and it is difficult to have satisfactory results.

4.3d Outlet

In several applications it is necessary to impose outlet boundary conditions. This for instance occurs in simulations of wakes or mixing layers. Often, in the past, the assumption of a zero vorticity gradient was used, and the results were satisfactory only if the boundary was far enough from the inlet. This assumption, however, is not physical since it presumes that the vorticity is convected out of the domain with an infinite velocity. In fact the vorticity transport equation in the inviscid case, setting the velocity $u_1 = C$ and $u_2 = 0$ is

$$\frac{\partial \omega}{\partial t} + C\frac{\partial \omega}{\partial x_1} = 0 \qquad 4.5$$

Thus to have a finite rate of vorticity change if $\frac{\partial \omega}{\partial x_1} = 0$, C should be infinite. On the other hand if $\frac{\partial \omega}{\partial x_1} = 0$ and $C \neq \infty$ the rate of velocity change is zero and this condition does not reproduce reality. Eq.(4.5), instead, can be used as a first approximation to let any vorticity disturbance or even any vorticity to radiate out from the region of calculation. If $C = 0$ Eq.(4.5) implies $\frac{\partial \omega}{\partial t} = 0$. This condition was often used to solve, for instance, the flow past a backward facing step, by assuming a fully developed parabolic velocity profile. To conserve the total mass the outlet flow rate was set equal to that at the inlet. The assumptions $C = 0$ and $C = \infty$ to be realistic imply that the boundary is far enough from the inlet. Recently, in engineering applications, it was observed that the vorticity generated inside the field can exit smoothly from the domain if the velocity C is set equal to the mean flow rate at the inlet (Pauley *et al.* (1992)).

This approximation does not properly work in geophysical flows. For example in simulations with propagating Rosby waves, there is no inlet mean flow and furthermore the propagation speed of each wave is different. Orlansky (1972) suggested to use Eq.(4.5) and he also suggested to evaluate at each point, on the boundary, the propagation speed. The calculation is possible by the knowledge of the flow at the nearest grid point and at the previous time step. The restriction is that the velocity C is positive and limited, that is, it is assumed that signals can only propagate outwards and that their maximum velocity corresponds to $CFL = 1$. The final form for the radiation velocity at the boundary is

$$\begin{aligned} C =&\Delta x_1/\Delta t, && if - \omega_t/\omega_{x_1} > \Delta x_1/\Delta t, \\ C =&- \omega_t/\omega_{x_1} && if \ \ 0 < -\omega_t/\omega_{x_1} < \Delta x_1/\Delta t, \\ C =&0 && if - \omega_t/\omega_{x_1} < 0. \end{aligned} \tag{4.6}$$

4.4 Streamfunction boundary conditions

For no-slip and free-slip walls the streamfunction has assigned values. For instance for the flow within two parallel plates ψ has on one side a zero value and on the other side the value of the total flow rate entering in the computational region.

For boundaries with radiative boundary conditions equation (4.3.2) with ω replaced by ψ holds. Even for this variable in applications with an inlet flow the convection velocity C is set equal to that for the vorticity. On the contrary, when waves are radiating out, C is evaluated, as well as ω at each grid point on the boundaries. The velocity C for ψ could be different from that for ω.

In the solution of the elliptic equation for the streamfunction it could be useful to have Dirichlet boundary condition on the walls where the radiative conditions hold. This convenience is related to the possibility to employ the FFT routines as it was shown in chapter 2. To get a homogeneous condition a new variable ϕ can be introduced, linked to ψ as follows. Let us to assume, for example, that along the boundaries at $x_2 = 0$ and at $x_2 = 1$ respectively $\psi(x_1, 0) = g(x_1)$ and $\psi(x_1, 1) = f(x_1)$; the new variable is

$$\phi(x_1, x_2) = \psi(x_1, x_2) + x_2 \left[g(x_1) - f(x_1)\right] - g(x_1) \tag{4.7}$$

This gets zero on the boundary at $x_2 = 0$ and at $x_2 = 1$, and thus in this direction the $\sin FFT$ can be used. In the direction x_1 any value of ϕ is allowed because a tridiagonal solver is used. By this transformation, instead of Eq.(4.3), the equation

$$\nabla^2\phi = -\omega - x_2 \frac{\partial^2 (g-f)}{\partial x_1^2} - x_2 \frac{\partial^2 g}{\partial x_1^2} - \tag{4.8}$$

is solved which has boundary conditions

$$\begin{aligned} x_2 = 0,&\quad 0 < x_1 < 1, && \phi = 0,, \\ x_2 = 1,&\quad 0 < x_1 < 1, && \phi = 0,, \\ x_1 = 0,&\quad 0 < x_2 < 1, && \phi = \psi(0, x_2) + x_2(g(0) - f(0)) - g(0),, \\ x_1 = 1,&\quad 0 < x_2 < 1, && \phi = \psi(1, x_2) + x_2(g(1) - f(1)) - g(1),, \end{aligned} \tag{4.9}$$

4.5 Physical aspects of nonlinear terms

In 2D flows, the Euler equations, that is, the Navier-Stokes equations in the inviscid limit have the property to conserve the total energy $E = \int\int (u_1^2 + u_2^2)/2 dx_1 dx_2$ as well as all the integrals of the vorticity powers ω^n. It is easy to prove that in the ω, ψ formulation for periodic, free-slip and no-slip flows the total energy can also be expressed as $E = \int\int \psi\omega dx_1 dx_2$. The skew symmetry of the vorticity field is also conserved. These properties hold in continuous space, that is, with an infinite number of degrees of freedom. A numerical method can be thought of as a tool limiting the number of degrees of freedom, and this number is related to the computer memory. Even if the number of degrees of freedom is reduced, the numerical method should maintain as much as possible the properties of the equations in the continuous space. The ω, ψ formulation has the mass conservation built-in, since the streamfunction enforces the fulfillment of the continuity equation. To conserve the circulation $\Gamma = \int\int \omega dx_1 dx_2$ and the higher order ω invariants, on the other hand, a particular care for the discretization of the nonlinear terms is necessary. These physical observations in the past were partially neglected in engineering calculations where the viscosity was relatively high. The conservative form for the circulation was often used, but not for the high order invariants. A confirmation of this statement can be found in Roache (1972) and Fletcher (1987), devoted to engineering applications, in which the importance of global conservation properties of high order invariants in the inviscid limit are barely mentioned. In geophysical applications, the interest, instead, is on simulations as close as possible to the inviscid limit; there, at least, the global conservation properties for the second order invariants were considered. The numerics leading to these conservation properties is not very cumbersome, hence it could be beneficial to get solutions, at high Re, for engineering calculations.

In the early development of CFD in aeronautics there was more demand for compressible flows since in those periods, the research was devoted, and still is, to the design of aircrafts at the highest possible speed. At high Mac numbers discontinuities in the flow occur; thus a large number of considerations were done, based on physical ground, to develop accurate numerical methods to represent the shocks. This approach drove Fletcher (1987 pg.280) and still is driving a large number of scholars (e.g. Tafti (1996)) to extend upwind schemes to incompressible flows. My personal point of view is that the physics associated with compressible flows is different from that of incompressible flows, and hence it is convenient to use the numerical methods developed for geophysical flows.

From these arguments it turns out that the discretization of the nonlinear terms is, perhaps, the most delicate part of the solution of the Navier-Stokes equations. If it is not adequate instabilities lead the calculation to diverge. This phenomenon takes its origin from the misrepresentation of all scales with wavelength smaller than 2 grid intervals. In fact, using N_1 points in the x_1 direction and N_2 points in the x_2 direction, the streamfunction ψ, for periodic flows in i and with no-slip or free-slip conditions in j, is expressed by the series

$$\psi_{i,j} = \sum_{l=0}^{N1/2} \sum_{k=0}^{N2-1} \left(a_{k,l} \cos \frac{2\pi i l}{N_1} + b_{k,l} \sin \frac{2\pi i l}{N_1} \right) \sin \frac{\pi k j}{N_2} \qquad 4.10$$

The smallest wavelength represented in the x_1 direction is $\lambda_1 = 2\Delta_{x_1}$ and in the x_2

direction is $\lambda_2 = 2\Delta_{x_2}(N_2 - 1)/N_2 \simeq 2\Delta_{x_2}$. The nonlinear terms in Eq.(4.2), in the wave number space, are expressed by a convolution product. From the interaction between two components, e.g. (l_1, k_1) and (l_2, k_2), the wave-numbers $(l_1 + l_2, k_1 + k_2)$, $(l_1 + l_2, k_1 - k_2)$, $(l_1 - l_2, k_1 + k_2)$ and $(l_1 - l_2, k_1 - k_2)$ are affected. Because of the limited number of modes, in Eq.(4.2) not all the wave-numbers can be represented; for example, $l_1 + l_2 = N_1 - p > N_1/2$ coincides with the smaller wavenumber p. In fact

$$\cos\frac{2\pi i(l_1 + l_2)}{N1} = \cos\frac{2\pi i(N_1 - p)}{N_1} = \cos\frac{2\pi i p}{N_1}$$

$$\sin\frac{2\pi i(l_1 + l_2)}{N_1} = \sin\frac{2\pi i(I - p)}{N_1} = -\sin\frac{2\pi i p}{N_1} \qquad 4.11$$

and by this mechanism the energy is transferred into the wave p instead of into the short wave $N_1 - p$. The same phenomenon occurs in the x_2 direction. Phillips (1959), by this artificial example, showed that the misrepresentation is the cause of the nonlinear instability and that it cannot be removed by decreasing the time step.

Arakawa (1966) proved that, to avoid the nonlinear instability, the sufficient condition is to conserve at least one of the two second order invariants energy or enstrophy. In the same paper it was also proved that for 2D incompressible inviscid flows, the mean wave number must be conserved. The mean wavenumber is conserved if both energy and enstrophy are conserved; in fact from the expression of energy and enstrophy

$$E = \frac{1}{2}\overline{(\nabla\psi)^2} \qquad \Omega = \frac{1}{2}\overline{(\nabla^2\psi)} \qquad 4.12$$

and from Eq.(4.10), expanding ψ by orthogonal harmonic functions, at each wave number, E and Ω are related to each other through

$$\Omega_l = \frac{1}{2}\overline{(\nabla^2\psi_l)^2} = \frac{1}{2}\overline{(k_l^2\psi_l)^2} = k_l^2 E_l \qquad 4.13$$

and so the mean wave number

$$K = \left(\frac{\sum_l k_l^2 E_l}{\sum_l E_l}\right)^{\frac{1}{2}} \qquad 4.14$$

is conserved if the total energy $E = \sum_l E_l$ and the total enstrophy $\Omega = \sum_l k_l^2 E_l$ are conserved. This shows that no systematic one-way cascade of energy into shorter waves occurs. Moreover it turns out that, in 2D, there is no scaling of the energy density. The solution in the inviscid case depends on the initial energy spectra. The mean wave number conservation, by considering only three waves for sake of brevity, permits the energy to go from an intermediate wave k_M to a smaller k_L and a higher k_S or to go from low and high waves into the intermediate ones. If the viscosity produces an energy dissipation at the high waves k_S the only possible mechanism thus is that energy moves from the intermediate to high and low waves. At the small scales the energy is dissipated faster than at large scales, hence in time the energy tends to accumulate at low wave numbers.

4.6 Arakawa's scheme

Arakawa (1966) suggested that the conservation of each of the quadratic invariants is sufficient to avoid computational instabilities; however, to get numerical simulations as close as possible to the ideal in the continuous space, it is desirable to conserve both the quadratic invariants. Moreover, the conservation of only one invariant does not fulfill the Jacobian property that $J(\omega,\psi) = J(-\psi,\omega)$. Since in fully spectral or in pseudo-spectral dealiased simulations the total energy and enstrophy are conserved, finite differences with the same conservation properties should produce results close to those obtained by the spectral methods. In 2D simulations, with all the scales fully resolved, are more feasible for the fine resolution achievable with the present computers. 2D flows are difficult to reproduce in laboratories and therefore these ideal flows are interesting only for the comprehension of the physics of two-dimensional flows. If this is true, the simulations should be considered real experiments and the resolution is requested to capture the smallest vorticity structures. In fully resolved flows the accuracy does not play a substantial role as, it is supposed to have in LES (Large Eddy Simulations) of 3D flows. In fact in LES the subgrid energy could be comparable to that of second order truncation errors, and therefore the more accurate pseudospectral methods could be desirable. In chapter 12 it is shown, however that staggered velocity second order numerical schemes give reasonable results for LES.

From these considerations my opinion is that the second order Arakawa scheme has all the ingredients to use finite differences to study the physics of several 2D flows, as it is shown in this chapter. The Arakawa scheme can straightforwardly be applied to the equations in general curvilinear coordinates and this is an advantages of finite differences as compared to spectral methods.

To show the global conservation properties, Arakawa evaluated the Jacobian $J(\omega,\psi)$ as a linear combination of three terms

$$J(\omega,\psi) = \alpha_1 J_1(\omega,\psi) + \alpha_2 J_2(\omega,\psi) + \alpha_3 J_3(\omega,\psi) \qquad 4.15$$

where $\sum \alpha_i = 1$, and the three expressions for the Jacobians are

$$\begin{aligned}
J_1(\omega,\psi) &= \frac{1}{4\Delta x_1 \Delta x_2}[(\omega_{i+1,j}-\omega_{i-1,j})(\psi_{i,j+1}-\psi_{i,j-1}) \\
&\qquad - (\omega_{i,j+1}-\omega_{i,j-1})(\psi_{i+1,j}-\psi_{i-1,j})] \\
J_2(\omega,\psi) &= \frac{1}{4\Delta x_1 \Delta x_2}[\omega_{i+1,j}(\psi_{i+1,j+1}-\psi_{i+1,j-1}) - \omega_{i-1,j}(\psi_{i-1,j+1}-\psi_{i-1,j-1}) \\
&\qquad - (\omega_{i,j+1}(\psi_{i+1,j+1}-\psi_{i-1,j+1}) + \omega_{i,j-1}(\psi_{i+1,j-1}-\psi_{i-1,j-1}))] \\
J_3(\omega,\psi) &= \frac{1}{4\Delta x_1 \Delta x_2}[(\omega_{i+1,j+1}-\omega_{i-1,j-1})(\psi_{i-1,j+1}-\psi_{i+1,j-1}) \\
&\qquad - (\omega_{i-1,j+1}-\omega_{i+1,j-1})(\psi_{i+1,j+1}-\psi_{i-1,j-1}]
\end{aligned} \qquad 4.16$$

The long and perhaps tedious proof of the conservation properties of these expressions are given in detail by Arakawa (1966); here it is important to point out that each term or their combinations fulfill conservation requirements of different properties. Tab.1 summarizes these properties

$J(\omega,\psi)$	J_1	J_3	J_2	$\frac{J_1+J_3}{2}$	$\frac{J_3+J_2}{2}$	$\frac{J_1+J_2}{2}$	$\frac{J_1+J_2+J_3}{3}$
skew symmetry	X				X		X
Enstrophy cons.			X	X			X
Energy cons.		X				X	X

TABLE 1. Properties of the typical Jacobian

Arakawa showed that the accuracy of such a scheme is second order; he gave an expression to get a fourth order accuracy by using additional grid points. Since in this chapter the interest is to show that second order accuracy is good in a large number of applications, the expression of the fourth order scheme is not reported here. A further reason to avoid the fourth order is that, while for periodic flows the scheme can be used in each point of the computational domain, on the contrary near solid boundaries the high order accuracy is not maintained. The overall accuracy, therefore, is reduced and a simple second order scheme is convenient. In chapter 6 devoted to 2D turbulence within a biperiodic domain the fourth order Arakawa scheme is reported.

Arakawa proved that the properties of the scheme hold in any orthogonal curvilinear grid. It is simple to show that these properties hold in a general curvilinear coordinate system, as it is reported in chapter 5.

4.7 Nonlinear terms in primitive variables

The Navier-Stokes equations in a general curvilinear coordinate system have more terms related to the metrics as has been shown by Orlandi (1989). In this formulation is difficult to find a scheme which at least conserves the energy since interpolations are required not only for the variables but also for the metrics. Even for a regular coordinate system with only non-uniform grids it is difficult to find a scheme conserving the second order invariants similar to that for the ω, ψ formulation. Only in the case of a Cartesian coordinate system, with an equispaced grid, it is possible to find a numerical scheme, that for staggered primitive variables retains the conservation properties for energy and enstrophy in 2D. In this section this scheme is given to show its complexity.

The dimensionless Navier-Stokes equations in primitive variables are

$$\frac{\partial u_i}{\partial t} + \frac{\partial u_i u_j}{\partial x_j} = \frac{\partial u_i}{\partial t} + N_i = -\frac{\partial p}{\partial x_i} + \frac{1}{Re}\nabla^2 u_i. \qquad 4.17$$

These equations usually are solved in 3D where energy conservation is the requirement for the discretized convection terms. In 2D the development of a numerical scheme for primitive variables is an exercise to get some practice that later on helps in 3D.

An useful exercise consists on finding whether there is the possibility to build a scheme, analogous to the Arakawa formulation, conserving energy, enstrophy and skew symmetry.

To obtain similar expressions for the primitive variables formulations it is necessary to express the Jacobian J_l in Eq.(4.16) by the staggered velocity components u_i. The operators N_{l1} and N_{l2} should satisfy the following relationship

$$\frac{\partial N_{l2}}{\partial x_1} - \frac{\partial N_{l1}}{\partial x_2} = J_l \qquad\qquad i = 1, 2, 3 \qquad\qquad 4.18$$

where l indicates the kind of Jacobian and the second index is the velocity component. These expressions permit to find the discrete operators for each velocity component. For J_1 and J_2 the task is quite simple i.e. substituting in Eq.(4.16) the discrete kinematic relationships

$$\omega = \overline{(u_2)_{x_1}}^{x_1} - \overline{(u_1)_{x_2}}^{x_2}, \qquad u_2 = -\overline{(\psi)}_{x_1} \qquad u_1 = \overline{(\psi)}_{x_2} \qquad\qquad 4.19$$

The expressions for N_{li} are

$$N_{11}(u_1, u_2) = \overline{(u_1)}^{2x_2}\,\overline{(u_1)_{x_1}}^{x_1} + \overline{\overline{\overline{(u_2)}}_{x_1}(u_1)_{x_2}}^{x_2}$$

$$N_{12}(u_1, u_2) = \overline{(u_2)}^{2x_1}\,\overline{(u_2)_{x_2}}^{x_2} + \overline{\overline{\overline{(u_1)}}_{x_2}(u_2)_{x_1}}^{x_1}$$

$$N_{21}(u_1, u_2) = (\overline{(u_1)}^{x_1}\,\overline{(u_1)}^{x_1})_{x_1} + (\overline{{}_1u}^{x_2}\,\overline{(u_2)}^{x_1})_{x_2}$$

$$N_{22}(u_1, u_2) = (\overline{(u_2)}^{x_2}\,\overline{(u_2)}^{x_2})_{x_2} + (\overline{u_1}^{x_2}\,\overline{(u_2)}^{x_1})_{x_1} \qquad\qquad 4.20$$

These results were obtained by Grammelvedt (1969), though he furnished only the expressions for $N_{12} + N_{22}$ and $N_{11} + N_{21}$.

The expression related to J_3 is more cumbersome. By long and tedious passages based on the algebra of discrete operators, it can be shown that the following identity holds

$$J_3 = -\overline{[\frac{\Delta x_1{}^2}{2}(\psi_{x_1}\omega_{x_1})_{x_1} + \omega\overline{\psi_{x_1}}^{x_1}]}^{x_2}_{x_2} + \overline{[\frac{\Delta x_2{}^2}{2}(\psi_{x_2}\omega_{x_2})_{x_2} + \omega\overline{\psi_{x_2}}^{x_2}]}^{x_1}_{x_1} \qquad\qquad 4.21$$

J_3 is given by the third expression in Eq.(4.16). From the right hand side of Eq.(4.21) using the continuity equation and the discrete kinematic relationship $\omega_{x_1} = \nabla^2 u_1$ and $\omega_{x_2} = -\nabla^2 u_2$ the final expressions is:

$$N_{31}(u_1, u_2) = \frac{\Delta x_2{}^2}{2}\overline{(u_1\nabla^2 u_1)_{x_1}}^{x_1} + \frac{1}{2}\overline{[(u_1)^2]_{x_1}}^{x_1} + (\overline{u_1}^{x_2}\,\overline{u_2}^{x})_{x_2} + u_1\overline{(u_1)_{x_1}}^{x_1}$$

$$N_{32}(u_1, u_2) = \frac{\Delta x_1{}^2}{2}\overline{(u_2\nabla^2 u_2)_{x_2}}^{x_2} + \frac{1}{2}\overline{[(u_2)^2]_{x_2}}^{x_2} + (\overline{u_2}^{x_1}\,\overline{u_1}^{x_2})_{x_1} + u_2\overline{(u_2)_{x_2}}^{x_2} \qquad\qquad 4.22$$

The combination of the three N_{i1} and N_{i2} shows that the expression in primitive variables conserving the same invariants as for ω, ψ is very onerous. It requires several terms with an increase of the number of operations. Furthermore the stencil is wider and it requires approximations near the wall. As it is shown later on, even in 2D the most important requirement is to conserve energy, since the discretization of the nonlinear terms requires only the terms in Eq.(4.20).

4.8. Viscous terms and factorization

The discretization of the viscous terms in Eq.(4.2) does not necessitates any particular treatment except of the accuracy, which for the present applications can be limited to second order. In the case of uniform and non-uniform Cartesian coordinates the expression of the second derivatives, described in chapter 2, applied to the right hand side of Eq.(4.2) gives

$$\frac{d\omega_{i,j}}{dt} = H_{i,j} + \frac{1}{Re}\left(a_i^{p1}\omega_{i+1,j} + a_i^{c1}\omega_{i,j} + a_i^{m1}\omega_{i-1,j} + a_j^{p2}\omega_{i,j+1} + a_j^{c2}\omega_{i,j} + a_j^{m2}\omega_{i,j-1}\right) \qquad 4.23$$

In this equation $H_{i,j}$ represents the nonlinear term discretized by the Arakawa scheme. As mentioned for the Burgers equation, it is convenient to treat the linear terms implicitly and the nonlinear terms explicitly. The time advancement discretization in the general form which includes Runge-Kutta of 3rd order or Adams-Bashfort leads to the following relationship:

$$\left[a_i^{p1}\Delta\omega_{i+1,j} + a_i^{m1}\Delta\omega_{i-1,j} + a_j^{p2}\Delta\omega_{i,j+1} + a_j^{m2}\Delta\omega_{i,j-1} - (a_j^{c2} + a_i^{c1})\Delta\omega_{i,j}\right]\frac{0.5}{Re}\alpha^h\Delta t$$

$$+\Delta\omega_{i,j} = \Delta t(\gamma^h H_i^k + \rho^h H_i^q) +$$

$$-\left[+a_i^{p1}\omega_{i+1,j}^n - (a_i^{c1} + a_j^{c2})\omega_{i,j}^n + a_i^{m1}\omega_{i-1,j}^n + a_j^{p2}\omega_{i,j+1}^n + a_j^{m2}\omega_{i,j-1}^n\right]\frac{1}{Re}\alpha^h\Delta t \qquad 4.24$$

This expression has been obtained following straightforwardly what was done for the Burgers equation. From the quantity $\Delta\omega_{i,j}$ the vorticity $\omega_{i,j}^{n+1}$ at the new time is calculated. On the left of Eq.(4.24) the matrix is pentadiagonal. This matrix can be inverted by the *fishpack* solver. However a factorization procedure leads to a saving of computational time. The factorization consists of substituting the pentadiagonal matrix by matrices more easy to invert; this algorithm has been widely used in solving compressible and incompressible Navier-Stokes equations (Briley Mc Donald 1975, Beam & Warming 1976). In the two-dimensional case the algorithm consists of writing the system of equations

$$[I - A_{i,j}]\,\Delta\omega_{i,j} = RHS_{i,j} \qquad 4.25$$

as

$$[1 - c(L_1 + L_2)]\,\Delta\omega_{i,j} = RHS_{i,j} \qquad 4.26$$

where L_1 and L_2 are simple discrete operators. The operators more often used are those associated to tridiagonal matrices; others are those associated to lower and upper diagonal matrices. For any operator chosen, the property of the factorization is based on the following identity

$$[1 - c(L_1 + L_2)]\,\Delta\omega_{i,j} = [1 - cL_2]\,[1 - cL_1]\,\Delta\omega_{i,j} + O(\Delta t)^2 \qquad 4.27$$

The second order accuracy results from the proportionality between c and the time step Δt. This accuracy is the same as the time discretization for the vorticity equation. Eq.(4.27) is split into two steps

$$[1 - cL_1]\,\Delta\omega_{i,j}^* = RHS_{i,j}$$

$$[1 - cL_2]\,\Delta\omega_{i,j} = \Delta\omega^*_{i,j} \qquad 4.28$$

$\Delta\omega^*_{i,j}$ is an intermediate variable without physical meaning, which is related to the physical quantity $\Delta\omega_{i,j}$ trough Eq.(4.28).

Tridiagonal operators lead, for the first step, to

$$\left[a_i^{p1}\Delta\omega^*_{i+1,j} + a_i^{m1}\Delta\omega^*_{i-1,j} - a_i^{c1}\Delta\omega^*_{i,j}\right]\frac{\alpha^h\Delta t}{2Re} + \Delta\omega^*_{i,j} = RHS_{i,j} \qquad i = 1, n_1, \quad j = 2, n_2 - 1 \qquad 4.29$$

and, for the second step, to

$$\left[a_j^{p2}\Delta\omega_{i,j+1} + a_j^{m2}\Delta\omega_{i,j-1} - a_j^{c2}\Delta\omega_{i,j}\right]\frac{\alpha^h\Delta t}{2Re} + \Delta\omega_{i,j} = \Delta\omega^*_{i,j} \qquad i = 1, n_1, \quad j = 1, n_2 \qquad 4.30$$

The operators, related to the upper and lower matrices for the first sweep give

$$\left[a_i^{m1}\Delta\omega^*_{i-1,j} + a_j^{m2}\Delta\omega^*_{i,j-1} - a_i^{c1}\Delta\omega^*_{i,j}\right]\frac{\alpha^h\Delta t}{2Re} + \Delta\omega^*_{i,j} = RHS_{i,j} \qquad 4.31$$

and for the second sweep

$$\left[a_i^{p1}\Delta\omega_{i+1,j} + a_j^{p2}\Delta\omega_{i,j+1} - a_j^{c2}\Delta\omega_{i,j}\right]\frac{\alpha^h\Delta t}{2Re} + \Delta\omega_{i,j} = \Delta\omega^*_{i,j} \qquad 4.32$$

The first approximation needs more CPU time than the second one; the second approximation, however, cannot be applied to periodic boundaries. The largest part of the codes described in this book use the first approximation because of its wide range of applications.

Particular care must be taken for the boundary conditions in the first step. The flows dictate in which direction to perform the first sweep. For instance, flows with one periodic direction, force the first sweep in this direction. A free-slip condition $\omega = 0$ gives the simple condition $\Delta\omega = \Delta\omega^* = 0$, hence the first sweep is convenient in this direction.

4.9 Boundary conditions

Periodicity in one or two directions do not require any boundary since $J(\omega, \psi)$ and $\nabla^2\omega$ can be calculated at each point of the grid. To avoid IF statements that need a greater CPU time than a floating point operation, it is convenient to evaluate and store at the start of the calculation the $i \pm 1$ and $j \pm 1$ indices. The periodicity enters in the indices 1 and $n_1 - 1$ for the i direction and 1 and $n_2 - 1$ for the j. Tridiagonal solvers with periodic direction conditions must be used.

The simulation in the presence of free-slip or symmetry boundaries requires the conventional tridiagonal solver in which the coefficients at the boundary $a_l^{pk} = a_l^{mk} = 0$ and $a_l^{ck} = 1$ for $k = 1, 2$ and $l = 1, n1$ and $l = n_2$. The value of the right hand side should be set equal to zero.

At no-slip walls the boundary conditions for the vorticity equations are derived from Eq.(4.4), that should be recast in terms of $\Delta\omega$. Eq.(4.4) at point $i, w + \frac{1}{2}$ and at time step n is

$$\frac{\omega^n_{\xi_2^+}}{x_{2\xi_2^+}} = (u_{2_{\xi_1}})_{\xi_2^+} - \frac{\delta^2 u_1}{\delta x_2^2} \qquad 4.33$$

The expression of $\frac{\delta^2 u_1}{\delta x_2^2}$ is given in chapter 2 Eq.(2.31) and the mixed derivative can be expressed without any approximation. Each velocity component can be related to the streamfunction. However, the velocity itself could be introduced, without increasing memory occupancy because near the boundaries the velocity can be stored in a one-dimensional array. In any case Eq.(4.33) becomes

$$\omega_{i,w+1}^n - \omega_{i,w}^n = \Omega_{i,w}^n \qquad 4.34$$

with the right hand side expressed in terms of $\psi_{i,j}$ at the interior and on the boundary. In boundaries without transpiration the streamfunction derivatives at the wall are zero. Eq.(4.34) must be written in terms of $\Delta\omega = \omega^{n+1} - \omega^n$ where Ω^{n+1} is a function of ψ^{n+1} (that is unknown). By Taylor expansion it turns out that $\Delta\omega_{i,w} = \Delta\omega_{i,w}^{n-1} + O(\Delta t^2)$; then from Eq.(4.34) it follows

$$\Delta\omega_{i,w+1} - \Delta\omega_{i,w} = \Omega_{i,w}^n - \Omega_{i,w}^{n-1} + O(\delta t^2) \qquad 4.35$$

Once the quantity $\Omega_{i,w}^n$ is calculated it is stored for the subsequent step. If the direction x_2 is the direction normal to a no-slip boundary the coefficients in Eq.(4.8) e.g at the boundary indicated by the index $j = 1$ are

$$a_1^{p2} = 1, a_1^{m2} = 0, a_1^{c2} = -1 \qquad RHS_{i,1} = \Omega_{i,1}^n - \Omega_{i,1}^{n-1} \qquad 4.36$$

A similar expression holds for the opposite no-slip wall at $j = n_2$, with now

$$a_{n_2}^{p2} = 0, a_{n_2}^{m2} = -1, a_{n_2}^{c2} = +1 \qquad RHS_{i,n_2} = \Omega_{i,n_2}^n - \Omega_{i,n_2}^{n-1} \qquad 4.37$$

When radiative conditions are used Eq.(4.5) at the boundary, for instance indicated by the indices i and $j = 1$, is

$$\omega_{i,1}^n - \omega_{i,1}^{n-1} = -\Delta t C_{i,1}^n \omega_{x_2^+}^n + O(\Delta t)^2 \qquad 4.38$$

The radiation velocity at the new time step is unknown and it should be calculated accurately, hence $C_{i,1}^n$ is replaced by $C_{i,2}^{n-1}$, which at $j = 2$ has second order space accuracy. The time accuracy is first order since the accuracy is reduced by assuming that C at the new time step is extrapolated from that at the previous time step. However this reduced accuracy is not so important because of the slow variation of the radiation velocity in time.

4.10 Test of ω, ψ formulation

To show the qualities of the vorticity streamfunction formulation in solving flow fields in a regular domain, as that depicted in figure 4.1, two different codes are furnished. The first one covers a large spectra of flows with different boundary conditions. In this code there is the possibility to treat geophysical flows with bottom topography and/or the β effect (i.e. the variation of potential vorticity due to the curvature of the earth). The model is quasi-geostrophic and it is restricted to a single fluid layer.

The second code deals with periodic boundary conditions and permits the simulation of 2D turbulence at low Reynolds number. This second case has two goals: the first to get the scholar acquainted with the calculation of turbulent energy spectra and the second to give a tool to whom is interested in 2D turbulence, a flow quite difficult to reproduce in a laboratory. However, cleaver researchers can withdraw interesting studies from these simulations as recently Jimenez *et al.* (1996) did. This code is not explained here but in chapter 6, which is devoted to 2D turbulence.

The codes in this chapter includes the solution of a transport equation for more than one passive scalar. The solution of more than one equation for the passive scalar and only one equation for the velocity field permits to investigate the effect of the diffusivity on the passive scalar evolution by reducing the CPU time. In addition if every scalar is located, at $t = 0$, in different locations some considerations on combustion related problems can be made. The equation for the passive scalar has a form similar to the vorticity equation; instead of Re appears the Peclet number $Pe = ReSc$, with $Sc = \nu/k$ the Schmidt number. In this case the passive scalar has the role of a flow visualizer. Although the realization of a perfect two-dimensional flow in a laboratory is difficult, it is important to remember that $Sc = O(1)$ reproduces visualization by smoke in air. On the other hand, visualizations in water by fluorescine require $Sc = O(10^3)$ and simulations with a very fine resolution are requested. Since the amount of memory and CPU for a full resolution with $Sc = O(10^3)$ could be prohibitive it is important to point out that the largest changes on the passive scalar evolution occur when the Sc number varies between 1 and 10 and at $Sc = 10$ the simulation is affordable.

4.11 Code PSOMGBC

The name of the code *PSOMGBC*, is the abbreviation of ψ ω and general boundary condition. As previously mentioned, it permits one to study the evolution of different kinds of flows. The code is given in the directory CODESCH4/PSO and it is divided in several files that are shortly described. I hope that the comments in the Fortran code, even if these are never enough, help to understand the method.

In the file *psomgbcma.f* there is the principal part of the code given in the subroutine *SOLVE* calling all the other subroutines. In the routine *TSCHEME* the time advancement scheme is coded; it consists of different routines related to the evaluation of the nonlinear terms, of the evaluation of the vorticity and that of the streamfunction. In this file there are also two subroutines which permit to introduce time disturbances on the inlet.

In the file *psomgbcnn.f* there are a bunch of routines necessary, at the beginning of the simulations, to evaluate the grid size, the indices for the derivatives, and the coordinates. Furthermore in these files there are the routines to write the output fields, requested by the graphics programs. These fields are written for a very large increment of time; at shorter intervals, however are written certain global quantities as circulation, energy, enstrophy and maxima of the passive scalars as well as of the vorticity. The routines to evaluate the maxima, the CFL and to invert the tridiagonal matrices are also in this file.

In the file *psomgbtn.f* there are the subroutines where the vorticity and passive scalar equations are discretized. In the routines with the name starting with *INVP* there is the

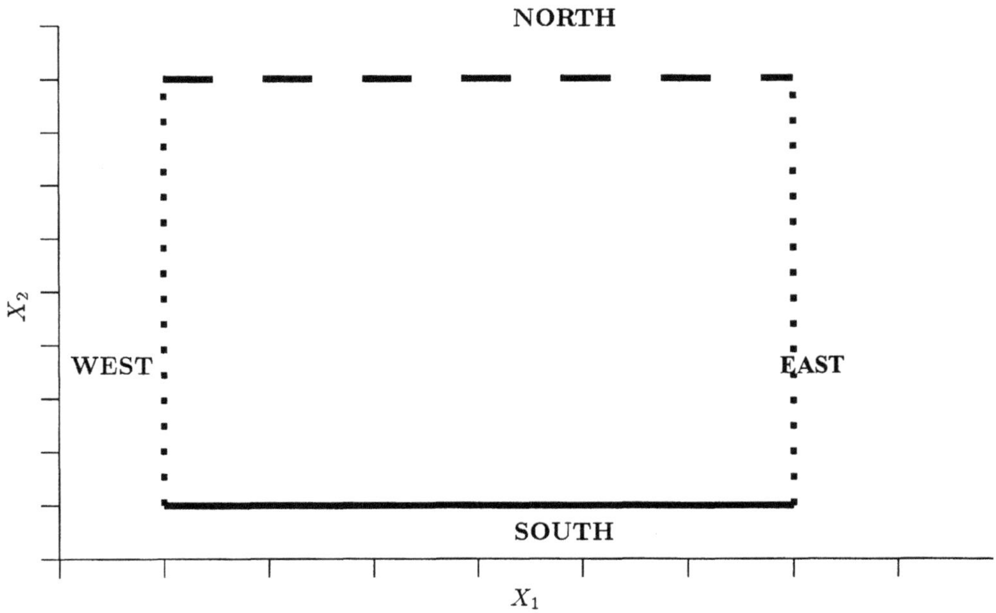

Fig. 4.1 Computational domain

factorization scheme with one direction of periodicity. Those starting with $INVT$ deal with the factorization without any periodicity direction. The passive scalar is considered in the routines with the name ending with PS and the vorticity in those ending by the letter V. The Arakawa scheme for the passive scalar and vorticity is coded in the routine $HDNL$. If geophysical flows are considered the potential vorticity enters in the Jacobian and this is evaluated in the routine $HDNLH$. In this code there is no simplification to deal with inviscid flows hence it is necessary to check this property at a very high value of the Reynolds number. If there is an interest to save computational time for this check a further routine can be derived from $INVTV$ by eliminating the part relative to the tridiagonal inversions.

The file *psomgbps.f* describes the part of the code relative to the streamfunction. There are two different routines to evaluate ψ; $PHCALC$ is called in presence of the general boundaries and $PHCALP$ when there is a periodic direction.

The boundary conditions and the initial conditions in the files *psomgbbc.f* and *psomgbin.f* require a longer description given in the following sections.

4.11a Input data

A uniform mesh in the x_2 direction allows one to make advantage of the FFT routines to get ψ; this coordinate varies between $alx2i$ and $alx2f$. At the $NORTH$ boundary ($x_2 = alx2f$) free-, no-slip and radiative conditions can be imposed. On the $SOUTH$ boundary ($x_2 = alx2i$) in addition an inlet condition is possible. In the direction x_1 a non-uniform grid is possible. The grid clustering depends on the inlet condition on the *SOUTH* boundary. For instance, if at the inlet there is a velocity profile to simulate a space developing mixing layer, the grid is clustered at the center and stretched near the *WEST* and *EST* boundaries. The opposite occurs in the the simulation of flows in a channel with solid boundaries at *WEST* and *EST* boundaries.

On the *EST* boundary radiative, free-slip and no-slip boundaries can be assigned, for instance the latter permits to study the impact of dipoles with solid walls. An enhanced grid clustering near the *EST* boundary allows to investigate the interactions at very high *Re* numbers. On the *WEST* boundary, in addition to the previous conditions, an inflow is possible. Radiative conditions apply to ω, ψ and T.

To chose any of these possibilities it is necessary to change the values in the input data file of the integer variables with names starting with "inbc", and with the next letter indicating the variable (v for vorticity, p for streamfunction and s for the passive scalar). The final letter indicates the boundary (n north, s south, e east and w west). Simulations of flows periodic in the direction x_1 are also possible and in this case the grid is uniform. Flows above a bottom topography (itop=1) have been restricted to topographies with constant levels parallel to the x_2 direction.

The code requires the file *psomgbc.d* as input file. A typical input file is

```
97 97  3     n1  n2  nsst(=3 for Runge-Kutta 3rd order,=1 A-B)
0            nread  (0 generate ic, otherwise read ic)
1            nmolp(multiplier for making names of output files)
.1 401  40 4  0    dt  ntst  nfield nhist nscrn(read ic from file nscrn)
1     1.7       .05          icflm,cflma,dtl
5.0   0.5   40.    0.                       tprin,tpin,tfin,tpstar
-0.0 5.0 1.50 0.2 0.7 2  alx1i  alx1f (xmin,xmax) stretc xc et istr
0.0   5.0                alx2i  alx2f (ymin,ymax)
4.0e+02  .0   .0            re   beta ekmn
.3333 .3333 .33333   chal chbe chga(for Arakawa  each=1/3)
2 1 1 -1 inbcvs,vn,vw,ve (vort bc) -1 no slip  south -1 west inlet
0 1 1 -1 inbcps,pn,pw,pe (psi)    0 free slip  inbcvs=2 omega=0
0 1 1  0 inbcss,sn,sw,se (scalar)  1 rad  inbc*e=-1  om, psi no-slip
0              ib2per
2    .5       npscf   scla
1.        1.        sch(1)    sch(2)
1 2.5 0.  1.  1.  -1.  inm(1dip,0mod,-1al,-2Gau) yc1 yc2 ra vel vsi
0    thet0 (angle for dipole and modon)
1.      1.        ar1   ar2
1    ipert (=1 normal, =2 sin perturb, =3 1+eps*sin(2*pi*om*t))
0  3.0  0.0  0.0  0.0  0.  +0.0 itop,dim1i,dim1f,dim2i,dim2f,hmax,dhmax
1.00   1.0  2.5      u0, blve, cpp (initial velocity, inlet width)
.5  .25   0.5 .5  0.    tau1,ampltp,timepe,dtau,time0
1.        .1           om,eps
1      0.862     1.      nvor akx,alx1d
```

Some of the quantities have self-explaining names which do not require any further explanation. The fourth line is necessary for simulations at constant Δt; on the contrary the data in the next two lines are necessary for simulations at constant CFL, where $icflm = 1$ should be assigned. In these lines there are data necessary for the output: certain global quantities at equal increment of time ($tpin$) and the output file with the fields at $tprin$ increments. The simulations by $icflm = 1$ stop at $t = tfin$. The seventh and eighth string concern the physical dimensions of the computational domain; the type of grid and non-uniformity in the x_1 direction is also assigned. The same values of $istr$ used in the code *trascor.f* (CODESCH2/COR) for the coordinate transformation are used. In the next line the value of the Reynolds number is assigned. If the equations with the geostrophic approximation are considered the appropriate value of β is necessary. In the next line the values of the α_i parameters in the Arakawa scheme allow to investigate the effects of the conservation properties. In studies related to vortex dynamics large differences do not occur, though $\alpha_i = \frac{1}{3}$ gives the global conservation of energy and enstrophy.

The values of the indices starting with $inbc$ state the type of boundary condition. The value -2 imposes inlet conditions, and these can be assumed only at the $WEST$ and $SOUTH$ boundaries. The index -1 imposes a no-slip wall. This condition cannot be applied on the $NORTH$ boundary. The index 0 is necessary for a free-slip wall where $\omega = 0$ holds; if zero vorticity flux is assigned, the value 2 should be assigned. The index equal to 0 for the passive scalar gives zero flux. Radiative conditions are reproduced by the index set equal to 1. When $ib2per = 1$ a flow periodic along x_2 is simulated.

The next line permits to solve more than one equation for passive scalars, with different values of the diffusivity given in the next line. The line starting with $itop$ introduces a bottom topography for geophysical applications. The extension and location of the topography are assigned by the 6 parameters in this line. The successive lines permit the different initial conditions explained in the next section.

4.11b Initial conditions

To run the code it is necessary to copy the file, written in the title of the sections between square parenthesis, and given in the directory CODESCH4/DAT into the file *psomgbc.d*

4.11b.a Lamb dipole

Data file to read [*dipwa.d, dippe.d*]

This initial condition is obtained by $inmod = 1$ in the input data. The Lamb dipole is an exact solution of the Euler equations (cf. Lamb (1932), Batchelor (1967)). The vorticity distribution in polar coordinate r, θ is

$$\begin{aligned} \omega &= -2Uk\frac{J_1(kr)}{J_0(ka}\sin(\theta-\theta_0), \qquad (r<a) \\ &= 0, \qquad (r>a) \end{aligned} \tag{4.39}$$

and the streamfunction is

$$\begin{aligned} \psi &= \frac{2Uk}{k}\frac{J_1(kr)}{J_0(ka}\sin(\theta-\theta_0), \qquad (r<a) \\ &= U(-a^2r+r)\sin(\theta-\theta_0), \qquad (r>a) \end{aligned} \tag{4.40}$$

In the code the streamfunction is not assigned because it is obtained by solving Eq.(4.3). U (*velmo* in *psomgbc.d*) is the propagation speed of the dipole and it is set as reference velocity. It can be ± 1 depending on the propagation direction. a, the reference length, is the radius of the dipole (*ramo* in *psomgbc.d*) and it is set equal to 1. k is a constant such that $ka = 3.81$ is the first non-trivial zero of the Bessel function J_1^1. This is a stationary solution in the comoving reference frame. The dipole could propagate in any direction $\theta = \theta_0$ by changing the value of the angle θ_0 (*thet0*), given in degrees in the file *psomgbc.d*. To let the dipole move along the x_1 direction θ_0 should be 0 or 180. Cavazza *et al.* (1992) studied numerically the stability of the dipole by different perturbations; one of them was given by assuming that the initial vorticity was confined in ellipses of different aspect ratios, rather than in a circle of radius a. The parameters $ar1$ and $ar2$ represent the two semi-axes of the ellipse; to get the circle $ar1 = ar2 = 1$.

The subroutine where the dipole is assigned has the name dipole and it is very simple to understand. The Bessel functions were taken from the book Numerical recipes by Press *et al.* (1993) and are given in the file *matfun.f*.

4.11b.b Stern dipole

Data file to read [*modwa.d*]

In geophysical applications related to flows in a β plane Stern (1975) found a dipolar structure, named modon, that, as the Lamb dipole for $\beta = 0$, is an exact solution of the Euler equations. This initial condition can be assigned by *inmod* = 0 in the input data. The vorticity distribution in polar coordinates is

$$\begin{aligned} \omega &= -\beta a \frac{J_1(\lambda r)}{J_1(\lambda a} \sin(\theta - \theta_0), \qquad (r < a) \\ &= -\beta a \frac{K_1(\rho r)}{K_1(\rho a} \sin(\theta - \theta_0), \qquad (r > a) \end{aligned} \qquad 4.41$$

where J_1 and K_1 are the first order and modified Bessel functions, respectively. The border between the interior and exterior regions of the modon is a circle of radius a. The ratio $\rho = \beta/U$, where U is the propagation velocity of the modon, is the inverse length-scale of the vorticity distribution. The interior inverse length-scale, $\lambda = j_1^1/a$, is such that the relative vorticity vanishes at the first zero of the Bessel function J_1. This expression in the limit of $\beta = 0$ gives the Lamb dipole. For the values of β, of interest, the peak vorticity in the external layer is smaller than the peak vorticity at the interior. Also in this case the sign of U changes the velocity propagation and θ_0 the direction. In the input data the names of the variables are the same as for the Lamb dipole, and β is indicated by *beta*.

4.11b.c Isolated vortex

Data file to read [*triprb.d*]

An "isolated" vortex, has net zero circulation. This initial distribution was used by Carton *et al.* (1989) as well as by Orlandi & van Heijst (1992) to simulate the formation of a stable tripolar structure in 2D. The initial distribution is given by

$$\omega = (1 - \frac{\alpha \rho^\alpha}{2}) exp(-\rho^\alpha) \qquad 4.42$$

The value of α controls the steepness of the vorticity gradients, and $\rho = R/\sqrt(C)$. R is the dimensionless radius $R = r/a$. Gent & McWilliams (1986) found that $m = 2$ is the most unstable mode for monopolar vortices; therefore Carton *et al.* (1986) added this perturbation to the distribution in Eq.(4.42). This flow has been studied experimentally in a rotating tank by Kloosterziel in his Ph.D thesis (1990); later on it was published by Kloosterziel & van Heijst (1991). Then, it is also interesting to assign perturbations as close as possible to those in the experiment. The experiment was initiated by inserting a plexiglas cylinder in the rotating tank and by stirring the fluid inside the cylinder. After a certain time, necessary to damp the perturbations inserted by the stirring mechanism, the cylinder was pulled out from the tank. This movement can, with a good approximation, be represented by a random disturbance mainly located in the region near the walls of the cylinder. Thus Orlandi & van Heijst (1992) assigned the following disturbance.

$$\omega^* = \eta e^{(-\alpha\rho^\alpha - 2)^2} - < \omega^* >, \qquad 4.43$$

where $\eta = \eta(x_1, x_2)$ is a random perturbation amplitude, $< \omega^* >$ is the area averaged value of ω^*. The subtraction of $< \omega^* >$ enforces that the disturbance has a zero mean value. The exponential ensures that the perturbation is the largest at the location where the basic vorticity changes sign. It is easy to show that this initial vorticity distribution has zero total circulation. This random disturbance permits the vortex to choose the most unstable azimuthal mode. Carnevale & Kloosterziel (1994) showed that for $\alpha = 5$ the most unstable mode is $m = 3$.

In the input file the following quantities have to be set: $inmod = -1$, *yc1mo* and *yc2mo* are the coordinates of the center of the isolated vortex, *ramo* is the value of C, *velmo* corresponds to α and *vsi* is the amplitude of the random disturbance. The random disturbance, that varies between 0 and 1, can be assigned by the function *rand*(), *ranf*() or others available in the computer libraries. If these function are not in the libraries the function *gerand* in the Numerical recipes book (Press *et al.* (1983)), copied in the file *matfun.f*, can be used. This routine, even if called only in the initial conditions, is very slow and takes much more time that the *rand*() and *ranf*() calls. A file *fcv.c* has been given in the directory *NCARFFT* that, once compiled in c in computers with *LINUX* system, behaves as *rand*() or *ranf*(). In this case the name of the random number generator is *rnd*().

4.11b.d Monopole in a β plane

Data file to read [*modbe.d*]

This initial vorticity is the simplest and basic structure to study the generation of Rossby waves. Hopfinger & van Heijst (1993) in their review paper on rotating fluid vortices dedicated chapter 2.4 to the topography effects and they reported a large number of references devoted to the study of monopolar vortices on a $\beta-$plane. This flow can be reproduced in a laboratory by a suitable topography in a rotating tank spinning at constant speed. A Gaussian vortex can be generated by a sink and its mathematical expression is

$$\omega = Aexp(-\tilde{r}^2) \qquad 4.44$$

where $\tilde{r} = \sqrt{(((x_1 - x_{10})^2 + (x_2 - x_{20})^2)/2\delta)}$ is the radius of the monopole. The monopole can be located in any position by changing the values of x_{10} and x_{20}. The quantities *ramo* and *velmo* permit to vary the circulation and the size of the vortex. To impose the monopole distribution $inmod = -2$ must be assigned in the data file.

The passive scalar has an initial distribution equal to that of ω.

4.11b.e Flow near a coast with topography slopes

Data file to read [*into.d*]

There is an increasing interest to understand the coast dynamics for environmental as well as strategic reasons. Satellites images, of several parts of the earth, have shown a complex vortex dynamics near the coasts. The variations of the bottom topography near the coast can explain the causes of some of the observations. A very simple case can be simulated by letting a stream enter on the west boundary to flow along the south free-slip boundary. In the simulation at $t = 0$, the flow is at rest, and in a preassigned time ($tim0$ in *psomgbc.d*) it reaches the velocity profile.

$$u_1(x_2) = u_0 exp(-x_2/\delta) \qquad 4.45$$

where u_0 is the maximum velocity at the free-slip boundary, and δ is a sort of thickness where the velocity is concentrated. The simulation requires $inbcvs = 0$ or $inbcvs = 2$, to account for the free-slip south boundary. The time dependency is assigned in the function $extflo$ where the inlet velocity is obtained by multiplying $u_1(x_2)$ by $f(t) = 3\tau^2 - 2\tau^3$ with $\tau = t/t_0$. For $t > t_0 f(t) = 1$. In the function $extflo$ there is the possibility to superimpose a perturbation at $t > t_1$,

As in all the previous flows, a passive scalar with a distribution similar to the velocity profile is assigned at the inlet to get a picture close to that from the satellites observations or from laboratory experiments by Serravall *et al.* (1998). These initial conditions in the data file *psomgbc.d* require $inmod = -3$.

4.11b.f Time developing mixing layer

Data file to read [*mlt.d*]

This flow is relevant for engineering applications related to combustion. The time developing mixing layer has been successfully studied in 2D and 3D by Rogers & Moser (1992) and by Moser & Rogers (1992) by solving the Navier-Stokes equations by means of pseudo-spectral methods. To understand the importance of the initial disturbance they preferred to assign clean disturbances rather than random disturbances, as in Lesieur *et al.* (1988). The Rogers & Moser (1992) procedure has been followed here by taking their initial mean velocity profile and their eigenfunctions, calculated by solving the Orr-Sommerfield equations. These functions, given in the file *inmosit.d* in the directory CODESCH4/DAT, were evaluated at their collocation points that do not coincide with the present grid. Finite differences allow a more general distribution of grid points, and in this example the distribution in section 2.4.b has been assumed. This is closer to that in Rogers & Moser (1992); it has a constant mesh in the center of the domain where the highest vorticity gradients occur. These initial conditions are obtained by setting $inmod = -4$ in the input data file.

4.11b.g Space developing mixing layer

Data file to read [*mls.d*]

The space developing is more realistic than the time developing mixing layer; in fact inlet and outlet conditions are requested as in reality. The instabilities can be generated without introducing any disturbances at the inlet. A forcing can also be imposed at the inlet and since these flows are convectively unstable, (Huerre & Monkevitz (1980)) the flow is sensitive to the forcing. In the directory CODESCH4/DAT there are two files *modarm.data* and *modfon.data* with the harmonic and the fundamental eigenfunctions calculated by solving the Orr-Sommerfield equations. I got these functions from Bottaro (personal communication), who evaluated them in certain positions. To get the values in the points of the coordinate x_2 spline interpolations in the routine $UINPER$ were employed. In the present example $C = U_2/U_1$ was set equal to 0.2; consequently $R = \frac{U_1-U_2}{U_1+U_2} = \frac{2}{3}$ and the velocity distribution is

$$U = U_1 + U_2 + (U_1 - U_2)tgh(x_2)0.5 \qquad 4.46$$

where $U_1 - U_2$ is the reference velocity. This profile is the same as that in the Buell & Huerre (1987) study of space developing mixing layers. The simulations are performed for the same conditions, the only difference is that here the vorticity-streamfunction formulation is used and as a further difference different conditions on the west and east boundaries can be used. Buell & Huerre (1987) had to impose a velocity distribution on these boundaries to allow the appropriate growth of the mixing layer and I used the same free-slip conditions. In this code radiative conditions can be assigned and these should better reproduce the real situation; I suggest to the reader to perform this exercise.

As in all the previous cases a passive scalar with a distribution similar to the inlet velocity profile has been assigned to get a picture similar to the flow visualizations in the experiments. There should be differences between the 2D simulation and the experiments (Brown & Roshko 1974) because in the laboratory the flow at the inlet also has a disturbance in the spanwise direction. By setting *inmod* $= -5$ in the data file the inlet conditions of Eq.(4.46) are reproduced; the C parameter coincides with *blve*.

4.12 Results by the code *PSOMGBC*

4.12.1 General suggestion

In the present section some simulations are presented to show the performances of this code. Flows for which there are experimental results have been selected. However, it is important to point out that in a laboratory it is very difficult to reproduce the conditions for real 2D flows, so the comparison should be qualitative . Moreover, these simulations have been performed with relatively coarse grids; therefore if the Reynolds number plays a large role, it has been kept relatively low. For example in the case of a dipole impinging a wall $Re = 500$ has been assumed to compare the distribution of the passive scalar with the picture of dye in the experiment by van Heijst & Floor (1989). As shown by Orlandi (1990) by increasing the Reynolds number the vortex dynamics changes. I am confident that the fantasy of the user in playing with the code will lead him to explore new situations, which I did not consider. However I would like to suggest that, before investigating new cases, at least once the code has to run with the data in the directory CODESCH4. This is an important task, since the code has been tested when it was compiled with the g77 Fortran Linux compiler on a PC, and with xlf compiler on RISC IBM workstations.

4.12.2 Flow phenomena simulations

4.12.2a Lamb dipoles

Before performing simulations of Lamb dipoles impinging free-slip or no-slip boundaries, the good sense suggests to investigate how the global conservation properties are preserved by the numerical scheme. To this purpose the Lamb dipole is adequate since it is a smooth vortex structure preserving its shape and all the invariants in a frame translating with the dipole. The code as it is does not work in the inviscid limit properly. The interested reader could simplify the code for $Re = \infty$ by neglecting in the routines $INVTV$ the part related to the factorization. However, by assuming a very large Reynolds number a substantial inviscid simulation is achieved. Usually the conservation properties should be maintained for a long computational time; hence these tests should be done by periodic conditions in the $SOUTH$ $NORTH$ direction and by letting the dipole to exit and to re-enter in the domain several times. To investigate the spatial accuracy it is also important to perform the simulations with different resolutions. A non-uniform grid for instance the grid in section 2.4.b gives a good resolution in the region where the dipole moves and locates the $WEST$ and EST boundaries far enough from the dipole. It is also important to verify that free-slip boundaries require the location of the $WEST$ and EST boundaries far from the dipole. In fact if these are too close to the dipole the propagation velocity, due to the image vortices, is reduced. Radiative conditions at the $WEST$ and EST boundaries do not present this inconvenient.

As a further suggestion the user is advised to insert into the subroutine *outp* the possibility to write a file with the vorticity and the streamfunction. These can be postprocessed to evaluate the streamfunction ψ_a in the reference frame traveling with the dipole. To get this streamfunction the velocity of the dipole is necessary and it can be obtained, at each instant, by differentiating the values of the coordinate of the maximum vorticity, written in the files *traj*1.*out* and *traj*2.*out*. The locations x_1 and x_2 of the maxima are calculated by the subroutine *fext*, where by interpolation from the 4 nearest neighbour points around the point of the peak, the real location of the maximum is achieved. In the translating reference frame the functional relationship $\omega = f(\psi_a)$ is a further indication of the performances of the numerical model. The Euler equation tends to a steady solution, since $\omega = f(\psi_a)$ gives $J(\omega, \psi_a) = 0$. In the inviscid limit, the Lamb dipole has a linear functional relationship $\omega = k\psi_a$, with k the constant in Eq.(4.39). The user can investigate how viscous diffusion modifies this relationship .

Van Heijst & coworkers, in several publications (for instance that by van Heijst & Floor (1989)) presented nice color pictures of colliding vortex dipoles in a stratified medium. They investigated normal as well as inclined collisions, which can be numerically reproduced by letting the Lamb dipole to collide with free-slip walls. From experimental studies it turns out that the vortex dipole is a very stable structure; that this is indeed true was confirmed by Cavazza *et al.* (1992). They studied the stability of the Lamb dipole after inclined collisions by looking at the functional relationship $\omega = f(\psi_a)$. Since in the laboratory it is quite difficult to measure the vorticity field; it is instructive to investigate the differences between the distribution of vorticity and a passive scalar. The collision with free-slip walls is even possible at high Re since, during the impact, the vorticity distribution is distorted without any vorticity generation. There is thus the possibility to chose a moderate Re number and a high Schmidt number $O(10^2)$ to emphasize the differences between the passive scalar and the vorticity field.

As mentioned before the study of the interaction of vortical structures with solid walls allows to understand the generation of wall vorticity (Lugt (1996, Chapters 12,15)). This generation is important in several practical applications since the new vorticity plays an important role in the vortex dynamics near the surface that determines the forces on the body. The flow past a circular cylinder is the simplest example to understand the wake past bluff bodies. This wake at substantial high Re ($Re = UR/\nu > 75$), consists of Karman vortices produced by the roll up of the thin vorticity layers generated at the surface of the cylinder. Orlandi (1993) studied the interaction of dipoles with circular cylinders of different radii and he found that the vorticity generated, even by bodies several order smaller than the size of the incoming dipole, introduces large perturbations on the trajectory of the dipole. The simulations reproduced the global picture of the experiment by Homa *et al.* (1988), where it was observed that a small eccentricity on the collision was leading to a different vortex dynamics. By smoke visualization it was difficult to distinguish all the small vortices that were generated by the cylinder; on the other hand, the numerical simulation permitted to discern the shape of primary, secondary and tertiary vortical structures and their interaction. Later on Verzicco *et al.* (1995) by numerical and laboratory experiments investigated the interaction with large cylinders. From these studies it turns out that the basic mechanism of the interaction is the same as that occurring in the interaction with plane walls.

These 2D simulations can be also used as simple models to get an idea of the importance of the effects of the near wall vortical structures in turbulent flows. In recent years flow visualizations (see the monograph *Turbulent flow structures near walls* edited by Walker for the Royal Society, 1991, Cambridge Univ. Press) and numerical simulations (see the review article by Robinson (1991)) have shown that the near wall vortical structures are responsible for turbulence production and for turbulent wall friction. Orlandi & Jimenez (1993) by a code similar to $PSOMGBC$ showed that when vortical structures are located near a solid wall there is an increase of the wall friction. Clearly this is not a realistic model since vortex stretching, the key mechanism of turbulent flows, is neglected in 2D simulations. However, 2D simulations are cheap tools to understand drag control as was done by Choi *et al.* (1994). Before performing a realistic simulation they used different wall boundary conditions to show drastic changes in the near wall vortex dynamics. The code $PSOMGBC$ then can be considered as a simple tool to initiate a systematic study of wall turbulence control that in these years is becoming an important topic for experimental and numerical studies. This route has been recently undertaken by Koumoutsakos (1997b) by means of a numerical method based on point vortices.

Contour plots of the distribution of vorticity and passive scalars at different instants obtained by the data file *dipwa.d* in the directory $CODESCH4/DAT$ are shown in figure 4.2 together with the laboratory flow visualization by Van Heijst & Floor (1989) given in Plate 1. figure 4.2 has been obtained by a 90° clockwise rotation during the plotting, in fact the $EAST$ $WEST$ boundaries are oriented vertically. The simulation was performed only in half the box, since the normal collision is symmetric. The results were possible with a reduced number of grid points (97×97) because the thin and strong vorticity layer

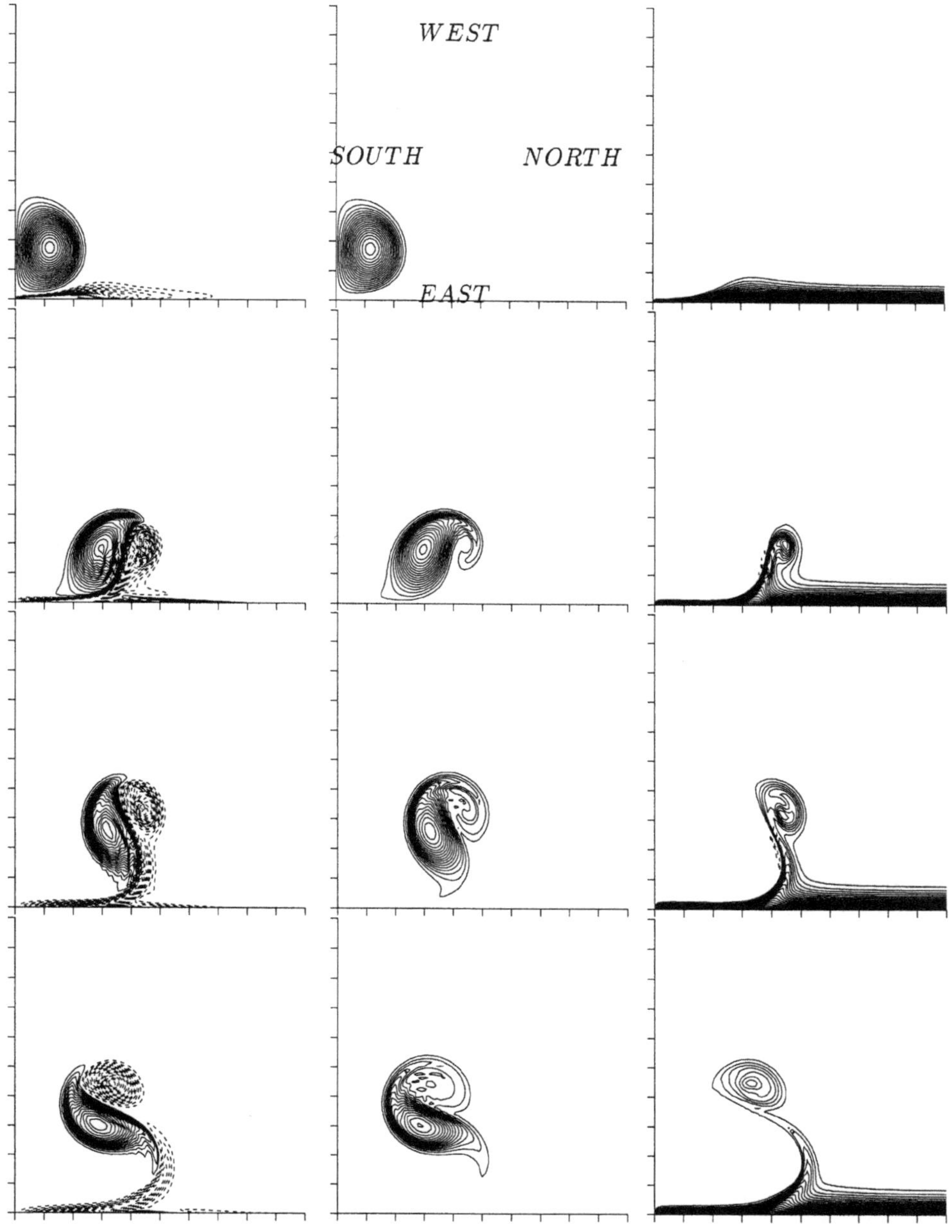

Fig. 4.2 Contour plots of vorticity and passive scalars for the dipole impacting a solid wall at $Re = 400$ At time increments of 2 time units starting from $t = 2$.

near the wall was resolved by clustering the grid points near the $EAST$ boundary. The time evolution of the vorticity shows that while the dipole approaches the wall a boundary layer with an adverse pressure gradient is generated; it detaches from the wall and rolling up forms a secondary vortex patch. This secondary vortex patch together with the initial vortex forms a new dipole with different positive and negative circulations. The unbalance of the circulation causes the circular trajectory. The calculation was performed for a short time to show the first rebound similar to the laboratory experiment. The experiment in the laboratory was setup in a stratified fluid hence the vortex dipole was a pancake rather than a two-dimensional vortex. The difference with a perfect two-dimensional flow produced a further dissipation that did not make it possible to observe multiple rebounds. In the laboratory it was not possible to setup experiments at higher Reynolds numbers to investigate the different flow configurations emerging after pairing of secondary and tertiary vortices. The numerical simulation by increasing the number of grid points permitted to study the dependence of the phenomenon on the Reynolds number (Orlandi (1990)). I would like to point out that in that paper the results were obtained by an uniform grid. Who is interested could reproduce them by a reduced number of grid points because of the non-uniform grid in the new version of the code. Moreover these new simulations have been performed by radiative conditions on the $NORTH$ boundary. These conditions allow to locate the $NORTH$ boundary, if symmetry is imposed on the $SOUTH$ boundary, or both the boundaries, if the full domain is considered, close to the region where the vorticity evolves. Hence the reduced physical domain, with the same number of points, permits a better resolution.

As mentioned before vorticity measurements in the laboratory are rather complicated; however, a qualitative picture of the vortex dynamics can be obtained by inserting dye in the jet that collapsed and formed the dipole. The dye has been inserted also in the numerical simulation. Both approaches show that, after the impact with the wall, part of the marker is transferred by the primary to the secondary vortex. To emphasize the evolution of the secondary vortex, as it was done by Cerra & Smith (1983) for similar experiments of vortex rings impacting on solid walls, a thin layer of dye has been located near the wall. figure 4.2 shows that the thin boundary layer pushes the dye towards the side, and that part of the dye is entrained in the secondary patches and moves with it. The simulation shows that, there is no transfer of marker from the secondary to the primary patch even at this low Schmidt number ($Sc = 1$). This behavior could be of interest in problems related to combustion.

The effect of the Schmidt number has not been investigated. The user has to enjoy the pleasure to play with the Schmidt number. If Sc is increased and the simulations are performed at $Sc = O(10^3)$ close to that of dye in water the number of grid points should be largely augmented. As a rule of thumb the simulation can be defined as fully resolved when the numerical wiggles disappear.

This code has the potentiality to answer the question whether by increasing the Reynolds number the first rebound occurs at a distance from the center that continuously increases. From preliminary relatively coarse simulations it turns out that, even with an infinitesimal amount of viscosity, the distance does not exceed three dipole radii. It should be remembered that for $\nu = 0$ the dipole does not rebound.

4.12.2b Modons

Carnevale *et al.* (1997b) investigated the differences and the analogies between viscous and inviscid rebound. The inviscid rebound occurs when a modon in a β plane approaches a free-slip boundary. This phenomenon is of interest on mesoscale problems in geophysical flows where there is a variation of the Earth's rotation with latitude. In the inviscid case the potential vorticity $q = \omega + \beta x_2$ is conserved. Carnevale *et al.* (1997b) presented laboratory experiments given in Plate.2. to confirm the numerical simulations. figure 4.3 presents the time evolutions of vorticity and passive scalars. As in previous flows the simulations are performed by renaming the data file *modwa.d* in $CODESCH4/DAT$ as *psomgbc.d*. The small amount of opposite sign vorticity is that in the initial distribution of vorticity of the modon discussed in section 4.11b. During the interaction with the free-slip wall this vorticity does not grow, until the modon has moved for a large distance towards the north direction. This is explained by the fact that, from the conservation of the potential vorticity q, the amount of negative vorticity is given by βx_2 and to be comparable to the low vorticity levels of the main patch the modon should be displaced at least two or three modon's radii. Only at this point the modon detaches from the wall and it moves along one trajectory in the north-west direction. The comparison between figure 4.2 and figure 4.3 shows that the rebound is substantially different in the two cases. In the viscous rebound the thin vorticity layer generated on the wall is so strong that it prevents the lateral movement of the dipole. The rebound remains localized near the region of impact and the trajectory has a small curvature radius. In the β rebound the opposite sign circulation grows slowly with the distance and thus the curvature radius is large. Carnevale *et al.* (1997b) investigated the influence of β and found that the duration of the rebound is reduced with the increase of β and that the trajectories, after the displacement from the wall, are, in a satisfactory approximation, parallel to each other.

As in the dipole rebound, the passive scalar contour plots show that the passive scalar within the primary structure is entrapped and, even for a larger value of the diffusivity, does not migrate out. On the other hand the passive scalar near the wall is displaced out from this position and by a spiral motion enters inside the main patch and travels with it far from the coast. This simple simulation can explain why, so often, pollutants discharged near coasts travel long distances on the seas and reach unspoiled regions.

The users interested in comparing the accuracy of the present numerical scheme could perform a set of simulations as those by McWiliams *et al.* (1981) with a pseudo-spectral numerical method. Interactions of modons are also very instructive as shown by Couder & Basdevant (1986); they validated the simulations by qualitative comparisons with soap film visualizations.

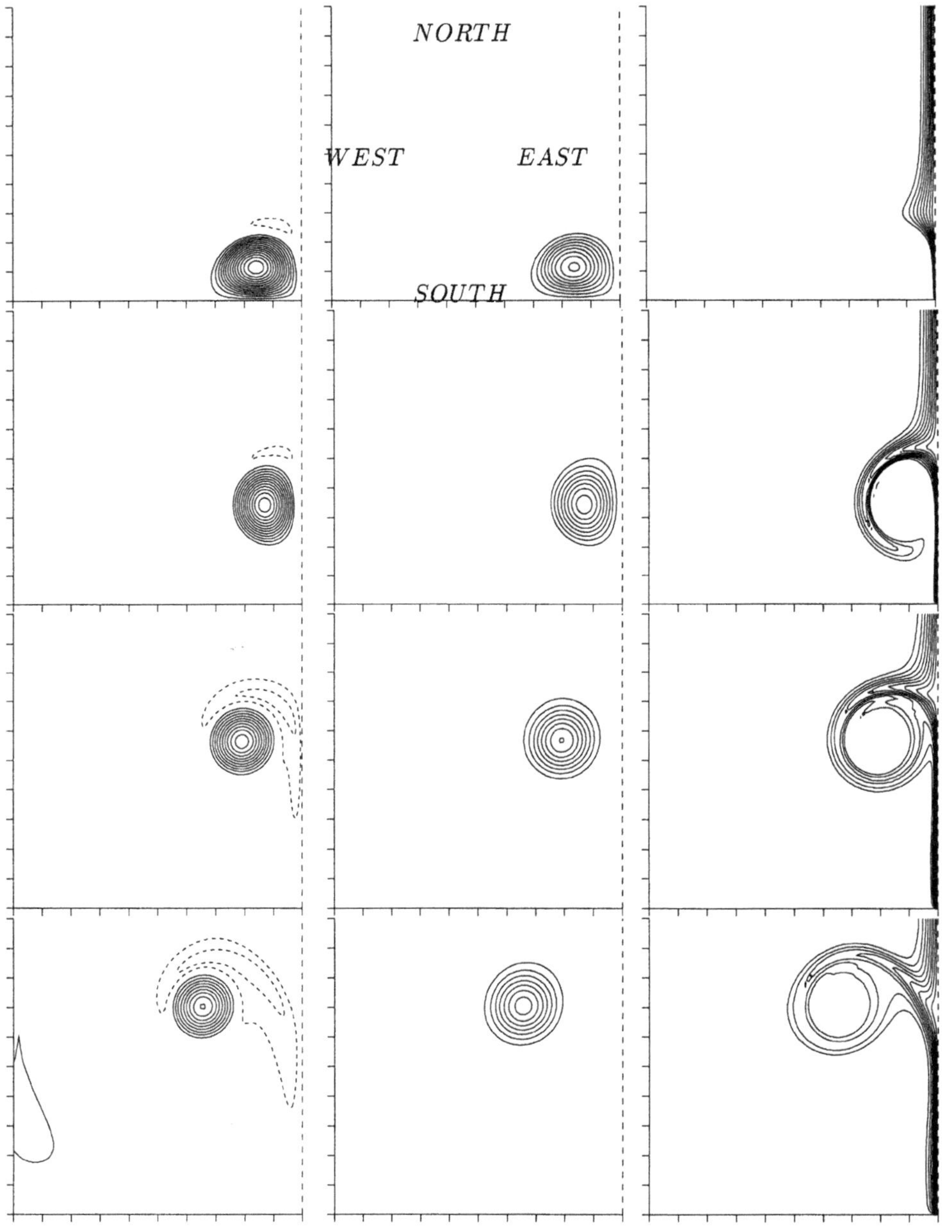

Fig. 4.3 Contour plots of vorticity and passive scalars for the modon impacting a solid wall at $Re = 400$ and $\beta = 1$. At time increments of 2 time units starting from $t = 2$.

4.12.2c Tripoles

The different kinds of long-lived vortical structures in nature can be classified depending on their conservation properties in the inviscid limit. The simplest structure is the monopole: it conserves its position and does not move if $\beta = 0$. A second kind of structures are dipoles or modons, depending whether β is zero or not. These structures, having a skew symmetry, preserve momentum and translate with a constant velocity. A third kind of structures are tripoles, quadrupoles and so on, these structures have null total circulation and preserve angular momentum. Carnevale & Kloosterziel (1994) investigated analytically and numerically the stability of these structures by superimposing on an isolated vortex, with the initial vorticity distribution of Eq.(4.42), clean perturbations. Orlandi & van Heijst (1992) on the other hand preferred to give a more realistic initial disturbance to be as close as possible to the laboratory conditions setup by Kloosterziel & van Heijst (1991,1992). Orlandi & van Heijst (1992) performed their simulations by a code with periodicity in both directions and the time evolution of the rotation angle θ compared well with that measured. The simulation was also intended to evaluate the functional relationship $\omega = f(\psi)$, characteristic of tripolar structures.

The same simulation is repeated here for radiative conditions on the four walls which allow us to reduce the computational domain and to save grid points by keeping the same resolution. Contour plots of vorticity and passive scalar at $Re = 1000$ and $Ek = 1.43 \cdot 10^{-3}$ in figure 4.4 show that the initial random disturbance in time gives rise to a $n = 2$ instability. Here Ek is the Ekman number that accounts for a further vorticity reduction due to the Ekman layers on the bottom boundary of the rotating tank. See Pedlosky (1986) for the derivation of this effect added on th the right hand side of Eq.(4.2). The real effect of this thin viscous layer on the boundary of the tank are shortly described in chapter 11 by three-dimensional simulations. The amplitude of this instability grows and at $t = 115$ a well developed tripole is formed. The passive scalar in time is entrapped in the secondary lobes. The comparison with the laboratory photos in Plate 3 is rather good. In the simulations the spiraling motion of the marker is reduced and the thickness of the layers with accumulation of the passive scalar are larger; the reason is that in the laboratory fluorescine in water,(with a $Sc = O(10^3)$) was used, while in the simulations $Sc = 1$ was imposed.

In this book the analysis of the dependence of the vorticity evolution with the parameter α (Eq.(4.41)) is not presented. Orlandi & Van Heijst (1992) found that at $\alpha = 3$ a quadrupole forms, which structure is unstable; in a short time, the small disturbances produce a pairing of two of the negative vorticity lobes and there is the formation of two dipoles which travel from each other. Carnevale & Kloosterziel (1994), on the other hand, by clean initial disturbances found that the quadrupole survives for a long time. However, by introducing together with $n = 3$ also a $n = 2$ disturbance, they found that the quadrupole is unstable. I am suggesting to the user to investigate the evolution of the isolated vortices by changing the value of α and to enjoy the formation of different vortical structures. By using software available on web sites movies can be made by saving the fields at small increments of time. This more sophisticated way to analyze the data is very fascinating and permits to have a picture of the vorticity dynamics. On the analysis of the α effects, the user should keep in mind that when α increases steep vorticity gradients

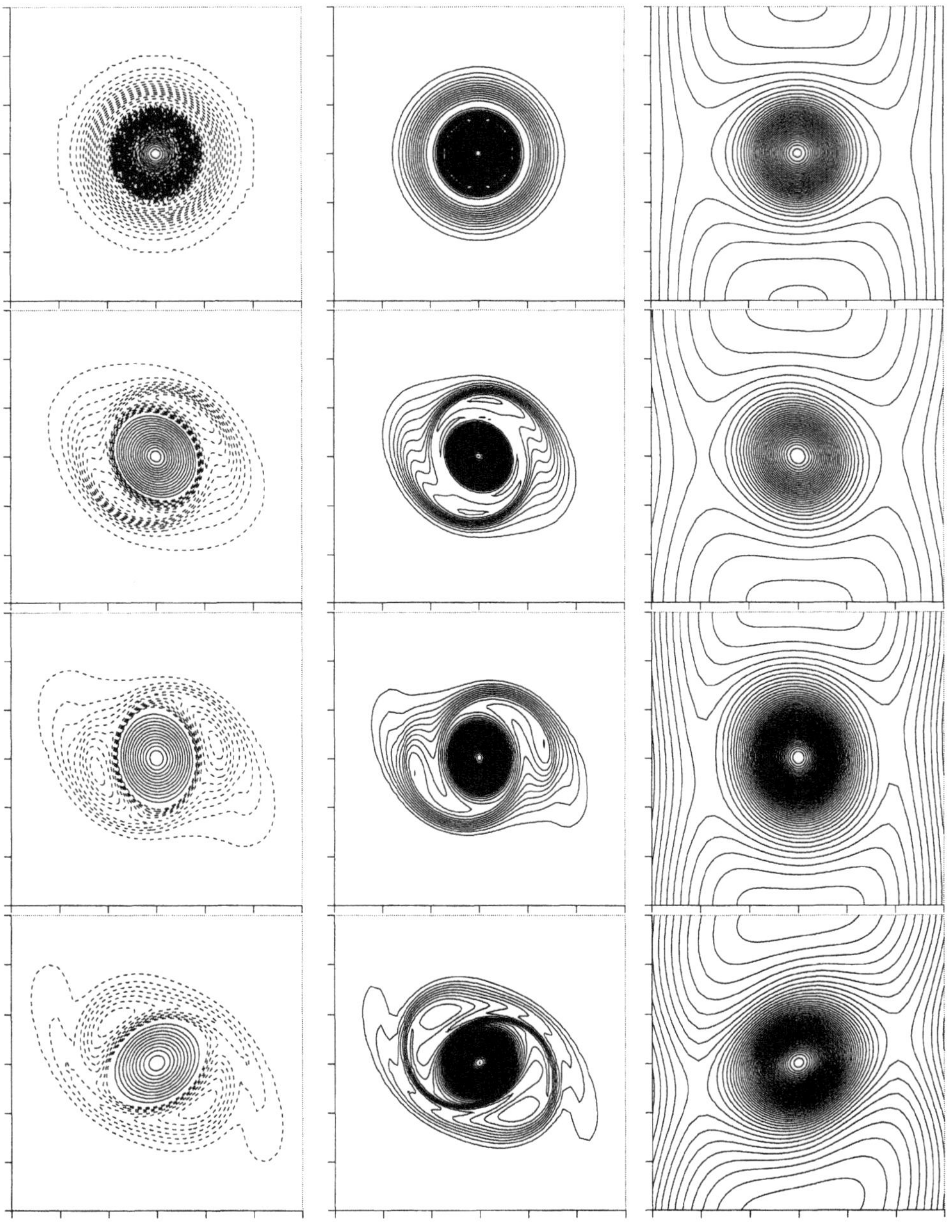

Fig. 4.4 Contour plots of vorticity left, passive scalars center and streamfunction right for the tripole formation at $Re = 1000$ and $Ek = 1.410^{-3}$. The contours increments are the same for ω and the passive scalar, $\Delta = 0.05$ for positive and $\Delta = 0.01$ for negative values. For ψ the full range has been divided into 60 intervals. Each image is distantiated by 45 time units starting from $t = 0$.

occur and hence the number of grid points should be increased. The evolution of the passive scalar distribution and their dependence on the diffusivity was not investigated by Orlandi & Van Heijst (1992) and Carnevale & Kloosterziel (1994). This study is of interest in environmental applications.

4.12.2d Monopols in a β plane

Among the different kind of vortical structures are monopoles, dipoles, tripoles and so on. Numerical simulations of two-dimensional decaying turbulence have shown that tripoles almost never appear. The tripole was observed in the forced simulation by Legras *et al.* (1988) where the forcing was at the wave number corresponding to the size of the tripole. The dipoles, on the other hand, form with a major probability; when they form, they travel inside the domain and hit other vortices. The equal sign vortices pair and thus even dipoles do not survive for the full simulation. Although the ultimate state of decaying two-dimensional turbulence is a dipolar structure with $\omega \approx sinh(\psi)$, as it was obtained by numerical simulations (Montgomery *et al.* (1992)), nevertheless it turns out that monopoles are the more common structures. The simple topology of this structure makes the monopoles very stable structures. In a viscous simulation this vortex shows a continuous growth of the circular vorticity contours. On the other hand when this vortex evolves in a β plane, opposite sign vorticity is generated around the vortex for the conservation of the potential vorticity; it increases since it encounters regions of greater βx_2. This behavior was reproduced by Carnevale *et al.* (1991a) in a rotating tank with a sloping bottom. In the laboratory the β effect is reproduced by putting, on the bottom of the tank, a topography with a ramp of constant slope s. The quantity h, in the potential vorticity expression, is $h(x2) = \frac{f\Delta H}{D}$ with $\Delta H = sx_2$, D the average depth of the fluid and $f = 4\pi/T$, where T is the rotation period of the rotating table. With this setup $\beta = fs/D$.

They investigated two different type of vortices; sink vortices with an initial vorticity distribution as that in Eq.(4.44), and stir vortices with

$$\omega = A(1 - r^2)exp(-\tilde{r}^2) \tag{4.47}$$

They found that both vortices propagate in the northwest direction, but, while the sink vortices propagate in an almost straight trajectory, the stir vortices show curved trajectories with kinks. The explanation was that the sink vortices with a single sign vorticity preserve the round shape, whereas the double-signed stir vortices produce a shedding of anticyclonic vorticity. Carnevale *et al.* (1991a) also made numerical simulations by a pseudo-spectral code with periodic directions with and without viscosity; they reached the conclusion that the northwest drift is essentially nonlinear and inviscid. During this process the generated anticyclonic vorticity radiates as Rossby waves. With a biperiodic code the boundary must be located far from the vortex; on the contrary, the waves reentering within the domain, affect the vortex motion. This unphysical behavior is avoided by assigning radiative boundary conditions, thus the simulations do not necessitate a large number of grid points.

Figure 4.5 shows contour plots of vorticity and passive scalar; the experimental observation for the stir vortex are given in Plate 4. The numerical experiment required a grid of 384×384 to reproduce the roll up of the thin vorticity wake shed by the monopole. In

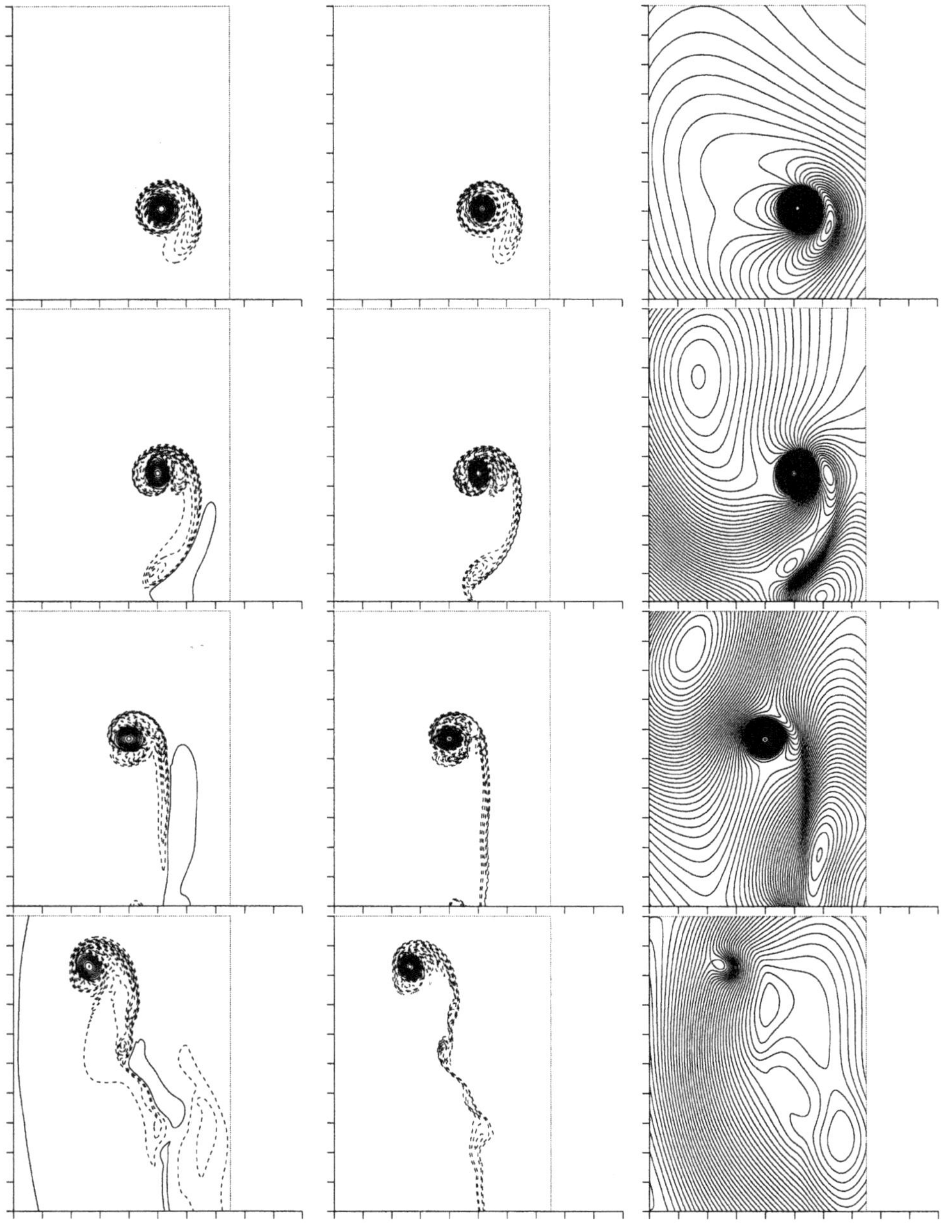

Fig. 4.5 Left: contour plots of vorticity (contours —— positive $\Delta\omega = 0.4$ - - - - negative $\Delta\omega = 0.1$); center: passive scalar (contours —— $\sigma > .25 \Delta\sigma = 0.05$ - - - - $\sigma < .25\Delta\sigma = 0.02$); right: streamfunction (the full range of values has been subdivided in 80 intervals) for a stir monopole in a β plane with $\nu = 0.01$ and $\beta = 0.00914$ the initial monopole has $A = 5$ and $\delta = 3.5$, the dimension of the domain are $L_x = 75$ $l_y = 100$. Each picture is separated by 30 time units starting from $t = 30$.

simulations of sink vortices the negative vorticity is smoother and hence a coarser grid can be used. The deformation of the positive vorticity patches, as described by Carnevale *et al.* (1991a), are reproduced. The opposite sign vorticity layers of small amplitude, behind the monopole, the so called Rossby waves, radiate out of the computational domain, and the radiative conditions describe this phenomenon well. The amalgamation of negative vorticity around the positive patch produces a compact and strong vortex that causes the northwest drift of the dipole. When the amalgamation of negative vorticity becomes very large jumps on the trajectory occur. The contours of the passive scalar resemble very close the wake instability observed in the laboratory experiments. The user could reproduce the different behaviors of monopoles in the presence of conical bottom topographies as studied by Carnevale *et al.* (1991a) in a rotating tank. The present simulation was performed by renaming the file *monbe.d* in $CODESCH4/DAT$ as *psomgbc.d*

4.12.2e Flow over topographic slopes

The effect of topography on flow currents is relevant in meteorology and in oceanography. In reality the stratification plays an important role that can be considered by approximations in the Navier-Stokes equations. In the present section the simulations are performed by the quasi-geostrophic approximation; here the conservation of potential vorticity $\zeta = \omega + h(x_1, x_2)$ holds. As in the previous section, $h(x_1, x_2)$ has the dimension of $1/secs$ and it is defined as $h(x_1, x_2) = fH(x_1, x_2)/D$. In a laboratory $H(x_1, x_2)$ is the topography shape modelled and located on the bottom wall of the tank and D is the averaged depth of the fluid. With this simplified model it is possible to get a first idea of the deviation of the current near the coast due to variation of the bottom topography of the sea. These variations have been observed in the north of California and in the Mediterrane near Sicily (Plate 5).

A stream flowing west to east encountering a topography from shallow to deep and from deep to shallow has been considered. The maximum inlet vorticity, with the inlet profile of Eq.(4.45) with $u_0 = 1$ and $\delta = 0.5$ is about 2.5, the change of topography is ± 1 and this variation occurs in a layer of thickness equal to 1. The amount of potential vorticity produced by the step is, thus, of the same order as that of the incoming vorticity. These conditions are close to that reproduced in a rotating tank experiment by Stern & Austin (1995) and by Serravall *et al.* (1998). Figure 4.6 shows on the right vorticity contour plots for the deep to shallow and in the left those for the shallow to deep case. In the first setup (Fig.4.6 right) there is a weak generation of opposite sign vorticity on the step and for the major part the inlet stream goes through the step and exits from the domain. As a final state the current goes out from the north boundary. These simulations intend to reproduce the transient when the inlet current encounters the step. Carnevale *et al.* (1998) dedicated a detailed and long paper describing this flow where they also made a qualitative comparison between numerical and analytical results.

As shown by Stern & Austin (1995) the second configuration is more interesting since when the current hits the step, there is the generation of a stronger negative vorticity layer that, together with the incoming vorticity, forms a dipolar structure moving backwards at approximately 45^0 with the step. As for the negative h, part of the incoming current goes trough the step but the amount is reduced. These runs were performed with the data file *inltop.d*.

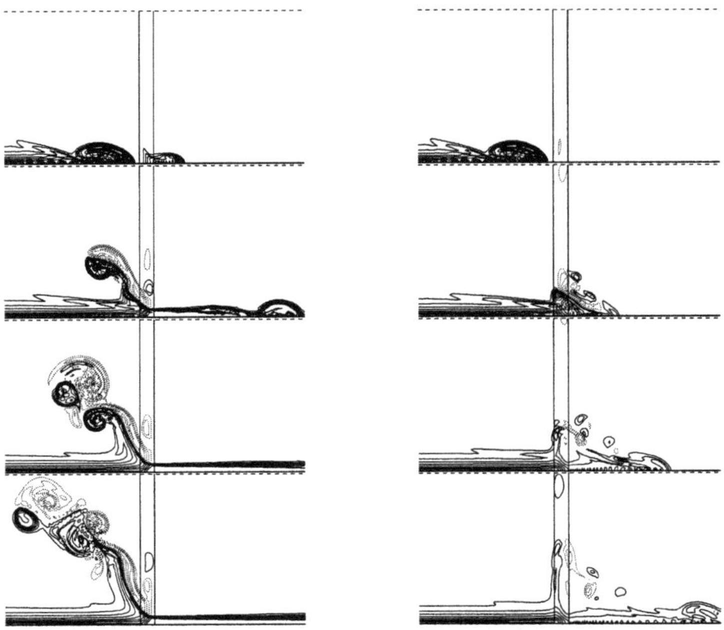

Fig. 4.6 Contour plots of vorticity for a stream near a free-slip coast encountering a topography step:*left* shallow to deep $h = +1$; *right* deep to shallow $h = -1$; $Re = 3000$. The images are separated by 20 time units starting from $t = 20$.

The code in this case provides different output files representing global behaviors. The files *floin.out* and *floou.out* give the time variation of the inlet flow-rate and the outlet flow-rate from the east boundary. In the code there are two passive scalars, the first one has a distribution equal to the inlet velocity profile and the second one has an initial distribution varying linearly in the x_1 direction, the direction normal to the wall. The second scalar allows the evaluation of the amount of fluid carried by the region behind the step to the region ahead of the step. I would like to advice the user that the evaluation of the trajectories of the maximum and minimum vorticity can give funny results depending on the case considered. For example the trajectories in the deep to shallow case show a funny behavior, related mainly to the manner in which the maximum has been evaluated. The user can insert into the subroutine $QUAINL$ a better way to evaluate the maxima and minima.

Serravall in her Ph.D thesis (1999) made a detailed study by parameterizing the flow with the step height. In this study, even without considering the effects of the stratification, it was possible to explain the meandering along the depression in the Adriatic which can bee seen in satellite pictures.

4.12.2f Time developing mixing layers

The flow field generated by two parallel streams which come in contact with different velocities, U_1 and U_2, is a classical problem of fluid dynamics and is the object of many numerical and experimental studies. The thin vorticity sheet at the interface between the two streams rolls up, the Kelvin-Helmholtz instability, forming spanwise vortices which are the driving mechanism for the mixing of the streams. This problem is very important for practical applications, in particular for chemical reacting flows, but the massive attention recently paid in turbulence to coherent structures has generated a new interest in this flow (Corcos & Sherman (1984) and Corcos & Lin (1984)).

The easiest way to generate in a laboratory a mixing layer is by letting two streams with different velocities come in contact at the trailing edge of a splitter plate. The flow thus produced evolves downstream generating vortices of bigger size as the distance from the splitter plate increases. For theoretical and numerical studies, however, a common assumption is to consider the flowfield as an infinitely long sequence of vortices whose structure repeats itself in the streamwise direction. In this case it is sufficient to focus the attention only on a set of fluid patterns and to study its time evolution. The first flow with the streamwise evolution is referred to as 'space developing mixing layer' while the second one is referred to as 'time developing mixing layer'; this problem, therefore, can be numerically studied in its space or time evolution. The former case is more realistic but has the disadvantage of needing the simulation of the whole flowfield (from the splitter-plate up to the outflow) and it is described in the next section. In contrast, in the time-evolving simulation the domain can be reduced to a few, or even one, Kelvin-Helmholz rolls, making possible the use of fine meshes and high Reynolds numbers. In this case the flow can be thought of as the evolution of a set of fluid patterns that are convected downstream and the roles of time and streamwise direction are reversed (Rogers & Moser, 1991).

The main goal of this example is to demonstrate that accurate solutions without unphysical oscillations can be obtained by finite–differences when discretization schemes with the same conservation properties as spectral methods are adopted. The application considered is the two–dimensional temporal evolving mixing layer with periodicity in the streamwise direction as considered by Rogers & Moser (1991). All the simulations have been performed with the same initial conditions as Rogers & Moser (1991) which provided the initial velocity and perturbation profiles. The field consists of a laminar shear flow with a cross-stream monotonic variation whose velocity profile $U = \overline{U}erf(\sqrt{\pi}y/\delta_\omega^0)$ with $\delta_\omega^0 = \frac{2\overline{U}}{(\partial U/\partial y)_{max}}$ the initial vorticity thickness. In this subsection y coincides with x_1 and x, the periodicity direction, with x_2.

To start the roll up a perturbation $\omega'(x, y)$ is superposed to the initial profile: the spatial variation in x is sinusoidal with wavelength ($\lambda_x = 2\pi/0.86 = 7.306$) which, from the linear theory, is the most unstable for the error function profile. The y–variation of the perturbation is given by the Orr-Sommerfeld equation and it is given in the file *inmosit.d* together with the error function profile. The amplitude of the perturbation is assigned requiring that the magnitude of the Fourier mode

$$A_\alpha = \sqrt{\int_{-\infty}^{+\infty} 2\hat{u}_i(\alpha)\hat{u}_i^*(\alpha)dy}$$

(with $\hat{u}_i(\alpha)$ the α–th Fourier mode of the i–th velocity component and $\hat{u}_i^*(\alpha)$ its complex conjugate) at the wavelengths of the perturbation be a given value. In these simulations the same values as Rogers & Moser (1991) have been used and in particular $A_1 = 0.1$ for the single roll up case, and $A_1 = 0.04$, $A_{\frac{1}{2}} = 0.03$ for the pairing case. The initial conditions are computed in the subroutines *UINI* and *VORINI*; in the first the data of *inmosit.d* are read and interpolated by a cubic spline on the computational grid. Once the velocity and the disturbance profiles are known in every grid point the initial vorticity field is computed in *VORINI* and the simulation starts.

The x–dimension of the computational domain is an integer multiple of the chosen wavelength λ_x ($L_x = M\lambda_x$). In this chapter the roll up and pairing events are considered, therefore values $M = 1$ and 2 are used. In the cross-stream direction the dimension of the domain L_y has to be such that the flow is not influenced by the boundary location. It is shown by the results that the value of L_y depends on the flow (roll up or pairing) and on the boundary conditions used. Since in the cross-stream direction the vorticity field is concentrated in a thin central layer, a non-uniform grid with more resolution around $y = 0$ is appropriate. Finite–differences permit a general distribution of grid points and thus the function of section 2.4.b has been used.

4.12.2g Roll-up

When a domain with streamwise length $L_x = \lambda_x$ (M=1) and only a perturbation at this wavelength is imposed, the emerging fluid pattern is that shown in figure 4.7. Initially the disturbance grows and the waviness of the shear layer increases, as shown by figure 4.7b. During the growth of the instability the vorticity is accumulated into the central roller at the expense of the vorticity in the braids and the process continues until the vorticity organizes as a single roller. The decrease of the vorticity in the braids is due to the growth of the vorticity gradients that diffuse rapidly at this low Re.

The growth of the instability is also shown in figure 4.8a by the exponential growth of the fundamental Fourier mode at a rate that Rogers & Moser (1991) found consistent with the linear stability analysis. This exponential growth is stopped by the saturation of the mode and the time at which the maximum A_1 value occurs is referred to as 'roll up time' (τ_r). For the present simulation we have $\tau_r = 10.1$ which is in perfect agreement with the data by Rogers & Moser (1991) which are also reported in figure 4.8a as solid symbols.

A further quantity of interest is the momentum thickness defined as

$$\delta_m = \frac{1}{4}\int_{-\infty}^{+\infty} (1 - \overline{u}_1^2)dy$$

where $\overline{u}_1$ the streamwise velocity component averaged in the streamwise direction †. This quantity confirms the thickening of the layer which has its maximum at $t = \tau_r$, when the roll up is completed. Later on the roller adjusts to an equilibrium elliptical shape and the layer can only thicken because of viscous effects. The data from Rogers & Moser (1991) have been reported also for this quantity and the excellent agreement between the two simulations turns out in figure 4.8b.

---

† Note that in the momentum thickness definition given by Rogers & Moser (1991) the factor 1/4 is missing. However, in the results they present that factor and for this reason it has been reported in the above definition

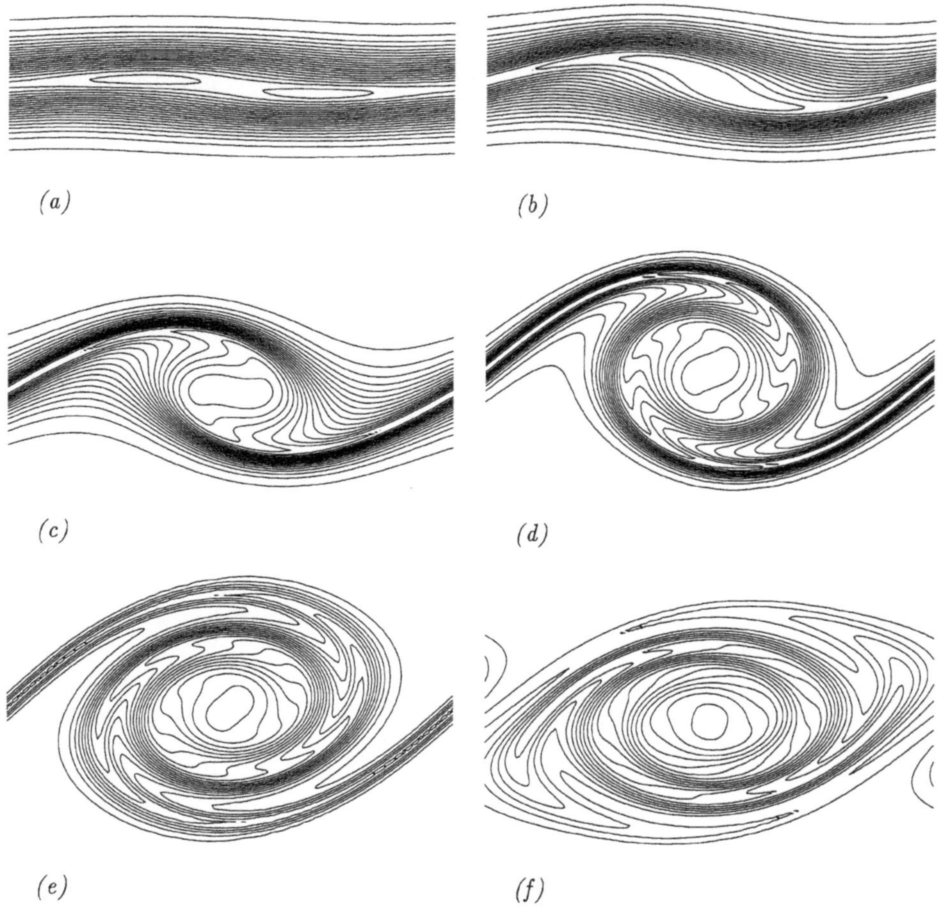

Fig. 4.7. Contour plots of spanwise vorticity at $Re = 500$ *(a)*, $t = 0.$; *(b)*, $t = 3.$; *(c)*, $t = 6.$, *(d)*, $t = 9.$; *(e)*, $t = 12.$; *(f)*, $t = 15.$; vorticity increments $\Delta\omega = 0.1$.

4.12.2h The pairing

In this case a domain of streamwise length $L_x = 2\lambda_x$ ($M = 2$) and a perturbation with two wavelengths (λ_x and $2\lambda_x$) has been introduced. It is important to point out that, since each wavelength has its own solution of the Orr–Sommerfeld problem, the cross–stream variation of the perturbation is different for the fundamental and for the sub-harmonic wavelengths. Even though I am aware of this, I have used the eigenfunction of the fundamental wavelength also for the sub-harmonic. In spite of the difference between the present perturbation and that assigned by Moser & Rogers (1992), the results compare very well.

The time evolution of vorticity contours are shown in figure 4.9 for a single pairing;

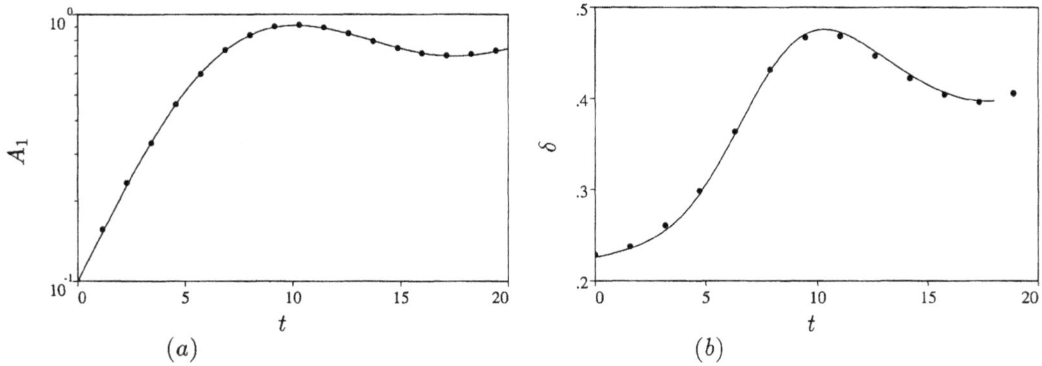

Fig. 4.8. Time evolution of the fundamental mode (*a*) and of the momentum thickness (*b*) for the mixing layer roll up at $Re = 500$; ——— present simulations, • results by Rogers & Moser (1991)

even in this case the shear layer rolls up generating two large rolls and two braids. In this configuration, also the sub–harmonic wavelength develops; therefore, before the rolls can undergo the viscous thickening, the pairing occurs. During the pairing the vortices approach each other, corotate and merge absorbing the braids and a single larger vortex is generated. At this point the evolution is similar to the roll up with the vortex that adjusts to an elliptic shape and with the braids that are depleted of vorticity. As it is shown in figure 4.9, during the pairing spiraling arms of weak vorticity are ejected from the paired structure and these are re–absorbed later on during the viscous thickening of the vortex.

Also in this case the time evolution of the fundamental mode A_1 and of the momentum thickness δ_m are compared with the data from Moser & Rogers (1993). The mode A_1 in figure 4.10a shows two maxima at $t \simeq 12$ and $t \simeq 21$ that are respectively the roll up and the pairing times. The time evolution of the momentum thickness evidences only the second maximum, the first being completely hidden by the growth of the layer. This large growth can be appreciated by comparing figure 4.8b with figure 4.10b where in the pairing a thickness about three times bigger than that in the roll up appears. It could be objected that this thickening is a Reynolds number effect that is lower in the pairing case ($Re = 250$) than in the roll up case ($Re = 500$). However, Reynolds number effects have been carefully studied by Rogers & Moser (1991) and Moser & Rogers (1993), and the chosen values are high enough to ensure that viscous effects are negligible when compared to the shear layer dynamics.

4.12.2i Arakawa scheme effects

Before concluding this example a few words should be spent on the discretization of the non–linear terms. In the previous two cases (roll up and pairing) the initial flow configuration contained energy only at the lowest wavenumbers; these needed few grid points to be adequately resolved. These, however, are particular conditions since in general the initial flowfield has the energy distributed on a broad range of scales and the presence of tiny filaments and structures renders the spatial resolution marginal.

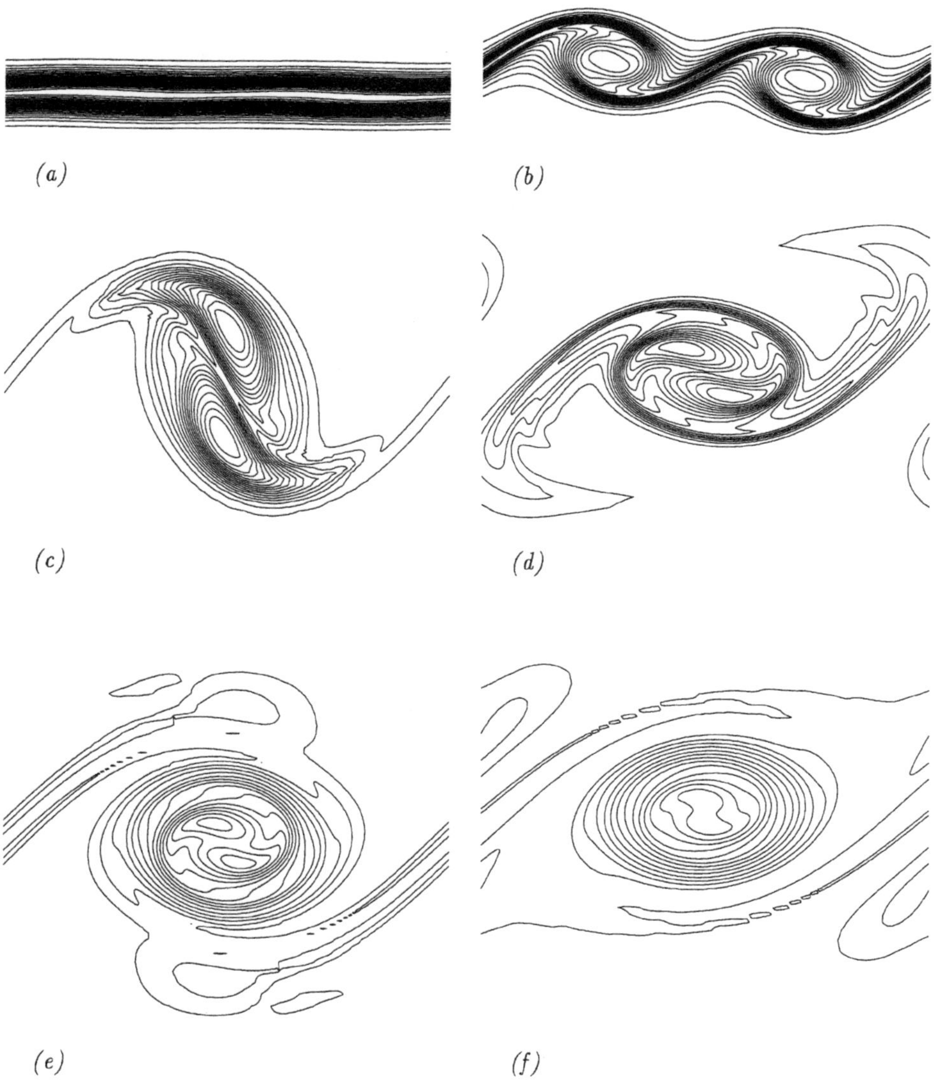

Fig. 4.9. Contour plots of spanwise vorticity at $Re = 250$ *(a)*, $t = 0.$; *(b)*, $t = 10.$; *(c)*, $t = 20.$, *(d)*, $t = 30.$; *(e)*, $t = 40.$; *(f)*, $t = 50.$; vorticity increments $\Delta\omega = 0.1$.

A similar arguments holds for the time step: if one knew *a priori* the range of velocities involved in the flow dynamics given the spatial resolution, one could compute the maximum time step by fixing the CFL number below its stability limit. Unfortunately, this information is not generally available and it might happen to run a simulation with the

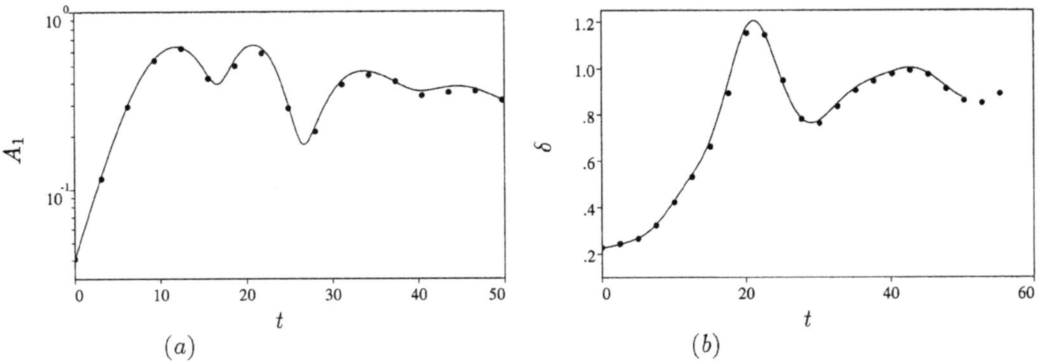

Fig. 4.10. Time evolution of the fundamental mode (a) and of the momentum thickness (b) for the mixing layer pairing at $Re = 250$; —— present simulations, • results by Moser & Rogers (1993)

time step close to or even beyond the stability limit. This can occur also for a short time interval when the flow dynamics becomes more active and the velocities increase. Indeed, this problem can be bypassed by performing simulations at fixed CFL instead of at fixed time step, so that the time step automatically adjusts during the simulation to the CFL limit. However, when fixing the CFL limit one in addition relies on the viscous terms so that the CFL can be slightly increased. In conclusion, the desire of saving computer time often leads to simulations with the CFL close to its limit.

Such a case has been built into this example simulating the mixing layer roll up. The presence of initial small–scales has been mimicked by the superposition of a small amplitude random vorticity over the field described in section 4.12.2f. The simulations were performed at $CFL = 1.9$ which is slightly greater than the theoretical limit $CFL = \sqrt{3}$.

The aim of this example is to show the importance of the discretization of the nonlinear terms together with the effect of the viscous terms. As already mentioned, this discretization is relevant not only for the stability of the calculation but also for the quality of the results. Two values of the Reynolds number have been considered: the first $Re = 500$ is relatively small and gives smooth gradients; in contrast the second $Re = 2000$ produces small scales that can originate the instability. Both cases have been simulated with the different Arakawa discretization schemes of the non–linear terms, J_1, J_2, J_3 and $(J_1 + J_2 + J_3)/3$ that, respectively, preserve the skew symmetry, enstrophy, energy and all the invariants. The last of these schemes is referred to as 'conservative scheme' and should be used when possible.

For $Re = 500$ all the schemes are satisfactory and it turns out that the differences between the peak vorticity compared with that obtained by Rogers & Moser (1991) are never larger than 0.25%. Even if the error of each scheme is larger than that of the conservative scheme the vorticity field is well resolved by all schemes. The simulations have shown that viscous effects smooth the unphysical oscillations and, even if some instability temporarily grows, good results are achieved. In fact, as soon as the unphysical eddies are produced, they are soon diffused and the solution is not affected.

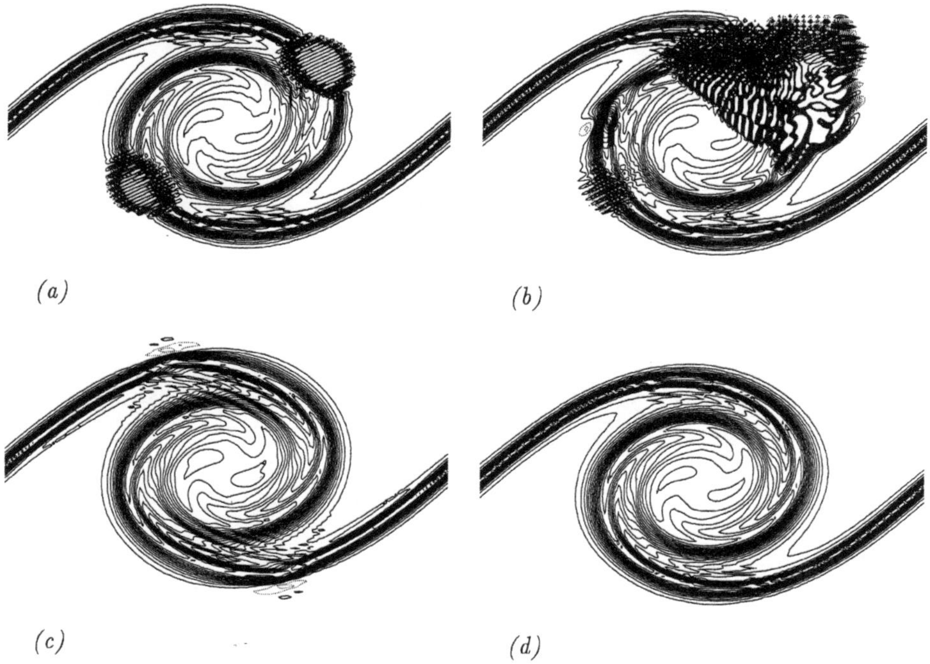

Fig. 4.11. Contour plots of spanwise vorticity at for the mixing layer roll up at $t = 10$ and Re= 2000: *(a)*, J_1 Arakawa scheme; *(b)*, J_2 Arakawa scheme; *(c)*, J_3 Arakawa scheme; *(d)*, 'conservative' Arakawa scheme; vorticity increments $\Delta\omega = 0.1$.

At $Re = 2000$ the viscous contribution is reduced and the discretization of the nonlinear terms becomes relevant. In fact, the unphysical eddies are not rapidly diffused anymore and in few time units the solution is affected by spurious structures.

In figure 4.11 the solutions obtained by the different schemes are shown. It turns out that to get an accurate simulation at $Re = 2000$, the J_2 scheme would require a much finer grid. However, the conservative scheme gives a solution with small wiggles that do not affect the solution. In addition the minimum of the vorticity is predicted with a small error and the large scales are well reproduced; the calculation by the energy conserving scheme (J_3) does not show large oscillations, even if the wiggles appear bigger and small structures of opposite sign vorticity are created. On the contrary, by the J_1 and the J_2 formulation the phenomenon of 'noodling' is large and for the J_2 formulation the skew symmetry is also lost.

Although the J_2 and J_3 schemes both conserve a quadratic invariant, which prevent the solutions to diverge (Arakawa, 1966), the vorticity field is completely different. This occurrence can be explained as follows: the J_2 formulation, conserving the enstrophy, prevents the growth of the nonlinear instability, but since the energy is not conserved it could be an unphysical energy transfer to small scales. On the other hand, by the J_3

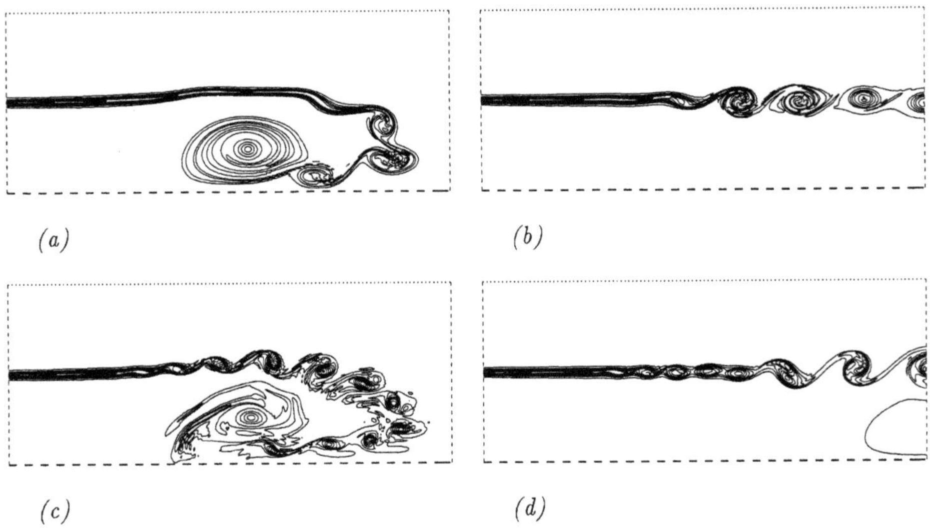

Fig. 4.12. Contour plots of spanwise vorticity for the space developing mixing layer at a), c) $t = 200$ and b), d) $t = 800$, at $Re = 2000$. a) , b) second , c), d) fundamental harmonic perturbation at the inlet. Vorticity increments $\Delta\omega = 0.1$.

scheme false energy transfers are possible, but the conservation of the total energy prevents the small scales to grow indefinitely; then the eddies of limited energy are diffused.

As already mentioned during the description of the Arakawa scheme (section 4.6) the reason for the good performance of the conservative scheme is that the contemporary conservation of enstrophy and energy implies that the nonlinear instability cannot develop and that wrong energy transfers are prevented.

This artificial example has been thought to exaggerate the effect of the non linear term discretization. Usually this effect is smaller but nevertheless it could be dangerous because it cannot always be detected and therefore it can lead to a misinterpretation of the flow physics.

4.12.2j Space developing mixing layer s

The space developing mixing layer has more practical applications and in 2D the numerical simulation is feasible on workstations due to the possibility to use a fine grid reproducing the roll up and pairing events occurring in the region close to the inlet. Simulations of this flow have been performed by Buell & Huerre (1988) that emphasized the role of the outlet boundary conditions. In this short paragraph I do not wish to discuss the physics of the space developing mixing layer since I prefer to discuss in more detail the space developing axisymmetric round coflowing jet that is of more practical application and that received less attention in the past. The aim of this 2D simulations is to show that the *PSOMGBC* code can also be used to solve space developing flows with inlet conditions. I would like to point out that the simulation of plane jets is more interesting, for this case there are also two dimensional experiments and simulations by Cogineni & Shih (1997) on the difference between isolated and wall jets. They analyzed the Reynolds number effects and their results can be reproduced by the present code by putting *inmod* = −6 and by changing the parameters for the inlet profiles. I found that simulations of jets are more difficult than those of mixing layers; in fact for jets the velocity in the far field is zero, while for mixing layers these are different from zero. In the latter case radiative boundary conditions work well but only when at $t = 0$ the velocity at the inlet are given in the full length of the computational domain. On the other hand during the transient a large vortex forms that can exit on the lateral domain and when this happens the flow continues to exit from these boundaries. Thus the two simulations have been done with free slip conditions on the lateral walls.

The simulations of the space developing mixing layer can be performed by copying the file *mls.d* in *psomgbc.d*. This file has been used to perform the simulation with the amplitude of the fundamental harmonic perturbation, with frequency 0.65, equal to 0.005; these data give the results on the right of figure 4.12. The results on the left have been obtained by the same amplitude on the subharmonic at half frequency. The results show that the number of vortices increase with the frequency and that their size is reduced at higher frequency. The pairing mechanism is well described in the higher frequency case. Both simulations show in the transient the formation of a large structure that is convected away and the formation of patches of vorticity resembling very close by the structures observed in the laboratory experiments by Brown & Roshko (1974).

4.13 Conclusions

As a conclusion of this chapter it turns out that the code can be used as a tool to study the vortex dynamics of two-dimensional flows. Some examples were dedicated to geophysical applications which usually occur at high Reynolds numbers. This is the reason why the Arakawa scheme is the key feature in numerical methods devoted to simulations of 2D Navier-Stokes equations in the ω, ψ formulation. The present numerical method consists on putting together the Arakawa scheme, a non-uniform grid, a factorization procedure and an efficient time advancing scheme. Without forgetting the most important requirement to use an efficient and fast method to solve elliptic equations. The results of this combination is a very robust and efficient method that can be used to simulate the evolution of several flows.

I am sure that the user with a deep background in fluid mechanics and with new ideas could insert modifications to get results relevant to practical applications. I am more interested to know whether this code will be used by graduate and undergraduate students and whether they will be enchanted by the beauty of vortex dynamics. This code can be applied to reproduce the flows qualitatively described by Lugt (1983) in his book based on observations rather than on mathematical equations. The reading of that book and the contemporary use of this code will let students understand the mechanisms leading to complex flow phenomena. The more recent book of Lugt (1996) with a complete description of vortical flows, also from the mathematical point of view, could suggest to the students a large number of questions, to which, in part, they could give an answer by the help of this code.

Perhaps someone who has the necessary skill in computer language could transform my primitive programming work and together with animation packages, available on different computers, he could produce computer games generating images that have a real meaning in a 2D world.

5 Two-dimensional flows in general curvilinear coordinates

5.1 Introduction

In the present chapter a numerical method to simulate 2D flows about complex geometries is described. These simulations are necessary in real applications; in fact, often in nature, the bodies have complex shapes and seldom their boundaries lie on a coordinate line of a system of Cartesian or cylindrical coordinates. An example is the flow past an airfoil or the flow above a natural terrain. The number of examples is so large that everyone can think of his own application. Several methods in the past were explored to achieve good results. A large effort, for instance, was devoted to develop finite element methods for their large range of applications. After several years, however, solutions for the major part were restricted to two-dimensional flows. Their extension to 3D flows was limited, for the difficulties related to the inversion of large band matrices. A similar procedure, with a better accuracy, was grounded on spectral elements. In this formulation within each element a pseudospectral method was used. The method is described by Korczak & Patera (1986) for 2D flows and; later on, it was extended to 3D flows by Karniadakis. He got results for the flow past a cylinder (Karniadakis & Triantafyllou (1992)) and for the flow past riblets (Chu & Karniadakis (1993)), to cite only some of the possible applications. Also for this method the resolution was limited by the difficulty to invert the large band matrices occurring in the pressure equation. My personal point of view is that this method requires a good mathematical background and a large programming work.

A simpler method is based on finite difference together with a coordinate transformation, mapping the physical domain onto a Cartesian computational domain. As in the finite element method the boundaries of the body coincide with grid lines. This occurrence allows a good accuracy on the imposition of the boundary conditions. The increase of the memory storage for the metric quantities related to the coordinate transformation is the price paid for this better accuracy. Since in real applications the shape of the bodies can be very complicated, in the eighties, a large effort was made to produce software for the generation of the grid around those bodies. Thompson *et al.* (1985) devoted a book to describe methods for numerical grid generation, and they described also how to solve the Navier-Stokes equations. They cited the difference between conservative and non-conservative formulations for the grid generation and for the flow solution. As in chapter 4 it is important to point out that, even in the case of complex geometries, it is important to deal with numerical methods respecting the global conservation properties.

As shown before the vorticity streamfunction formulation in 2D allows a discretization for the nonlinear terms that, in the inviscid limit, preserves the conservation of the quadratic invariants. On the other hand, it has been shown that with the primitive variables formulation this constraint is difficult or almost impossible to achieve. By reading Choi *et al.* (1993) it turns out how large is the amount of computational time requested to achieve good results; in addition, they never investigated the global conservation properties. How these are important is an open question that no one, to my knowledge, has ever addressed by performing 2D or 3D simulations in the inviscid limit. The primitive variables formulation however is the best choice in 3D flows. Dealing with incompressible flows, in Cartesian or cylindrical coordinates, the best accuracy is achieved by staggering the velocity components. The memory requirement, when a general curvilinear coordinate system is added, increases since the metric quantities are evaluated at different locations of the cell. Memory restrictions thus forced the applications of finite differences, or finite volume, as someone else prefers to name these methods, to flows with complex geometries in two dimensions and with a Cartesian grid in the third direction. This choice was done by Orlandi (1989) and later on by Choi *et al.* (1993) in the simulations of turbulent flows above a grooved surface.

3D flows have been mentioned to introduce a third method that seems to be very efficient. It is based on the introduction in the Navier-Stokes equations of a body force that enforces a no-slip condition on a solid surface and within its internal region. This method was developed by Vieceli (1969) and later on applied by Peskin (1972) in several configurations related to biofluid dynamics. Recently it has received a renewed interest and it has been successively used to solve turbulent flows in presence of riblets (Goldstein *et al.* (1993)) and the 2D flow past a cylinder (Saiki & Biringen (1996)). Goldstein *et al.* (1993) in their introduction documented accurately the history of the immersed boundary method and I am suggesting to whom is interested to read this paper with attention. A method slightly different was developed by Basdevant & Sadourny (1984) and later on applied with success by Briscolini & Santangelo (1989) to simulate flows past square and circular cylinders. In this version, after a first step without the body, a mask was applied to the flow to set equal to zero the velocity field in the region coincident with the body. For the body force and for the mask methods some preferred as flow solver pseudo-spectral methods and others finite-difference pseudo-spectral. Recently Saiki & Biringen (1996) showed that finite difference give more reliable results. Fadlun (1998) introduced the modification suggested by Mohd-Yusof (1997) in this method and in cylindrical coordinates he validated the method by a one to one comparison with an experiment devoted to generate vortex rings. He performed simulations to investigate the space and time accuracy of the method and he showed that by the Mohd-Yusof (1997) improvement simulations were possible at high CFL numbers.

In this introduction I mentioned the large variety of methods with eventual extension in 3D flows. In this book, only 2D flows are considered and solved by the ω, ψ formulation, where the global conservation properties are preserved and the solid boundaries coincide with a coordinate line.

5.2 Equations

In the 2D case the vorticity and the streamfunction can be considered as scalar quantities, and then in a general curvilinear coordinate system the differential operators have simple expressions. Levi Civita (1977, pg.152-155) in his fundamental book on differential calculus, edited for the first time in 1925 long before simulations of the Navier-Stokes equations were possible, furnished several expressions for the divergence operator. At Pg.154 a conservative expression for the divergence is given that the author claimed to be suitable for calculations instead of the equivalent expression in non-conservative form. He defined the covariant derivative of a contravariant vector X^l as $X^l_{|k}$ and the divergence of the vector $\mathbf{X}$ as $X^i_{|i}$. He showed that the divergence becomes

$$X^i_{|i} = \frac{1}{\sqrt{a}} \frac{\partial X^i \sqrt{a}}{\partial x_i} \tag{5.1}$$

where $\frac{1}{\sqrt{a}}$ is the Jacobian of the coordinate transformation. The contravariant derivative is

$$X^l = a^{il} \frac{\partial X}{\partial x_i} \tag{5.2}$$

where the metric coefficients a^{il} are the reciprocals of the coefficients

$$a_{ij} = \frac{\partial y_k}{\partial x_i} \frac{\partial y_k}{\partial x_j} \tag{5.3}$$

The expression of the Laplacian of a scalar Φ is

$$\nabla \Phi = \frac{1}{\sqrt{a}} \left(\frac{\partial}{\partial x_k} \alpha^{kl} \frac{\partial \Phi}{\partial x_l} \right) \tag{5.4}$$

with $\alpha^{kl} = \sqrt{a} a^{kl}$. At Pg.162 of Levi Civita's book the expression of the *curl* of a vector is given. Since the velocity vector can be defined as the *curl* of the stream function vector, the following expression holds

$$u^h = \frac{1}{\sqrt{a}} \left(\frac{\partial \psi_{h+2}}{\partial x_{h+1}} - \frac{\partial \psi_{h+1}}{\partial x_{h+2}} \right) \tag{5.5}$$

where u^h are the contravariant components of the velocity vector. The Cartesian components are linked to the contravariant components by $v^j = u^i c^j_i$, with $c^j_i = \frac{\partial y^j}{\partial x_i}$. If $q^i = u^i \sqrt{a}$, the Navier-Stokes equations in 2D in a general curvilinear coordinate system are

$$\frac{\partial \omega}{\partial t} + \frac{1}{\sqrt{a}} \frac{\partial q^i \omega}{\partial x_i} = \frac{1}{Re} \frac{1}{\sqrt{a}} \left(\frac{\partial}{\partial x_k} \alpha^{kl} \frac{\partial \omega}{\partial x_l} \right) \tag{5.6}$$

The fluxes q^i are related to the stream function by

$$q^1 = \partial \psi / \partial x_2 \quad , \quad q^2 = -\partial \psi / \partial x_1 \tag{5.7}$$

The stream function ψ is related to the vorticity ω by

$$\omega = -\frac{1}{\sqrt{a}} \left(\frac{\partial}{\partial x_k} \alpha^{kl} \frac{\partial \psi}{\partial x_l} \right) \tag{5.8}$$

This system of equations requires a limited number of metric quantities, and it is important to keep in mind that the simple expression in Eq.(5.6) is a consequence of ω and ψ being scalar quantities. Cunsolo & Orlandi (1978) checked that, by evaluating the Laplacian by Eq.(5.8) instead of the non-conservative one reported by Thompson *et al.* (1985) in their

Eq.(103) at pg.126, the errors in a very non-orthogonal grid (for instance near the trailing edge of a cambered airfoil) are reduced to a minimum. A further check showed that the discrete operator $\nabla \times (\nabla\phi)$ is zero. This operation has been performed by evaluating $\mathbf{q} = -grad(\phi)$ in general coordinates as

$$q^k = -\alpha^{kl}\frac{\delta\phi}{\delta x_l} \qquad 5.9$$

and the relation $\omega = curl(\mathbf{q})$ by

$$\omega = \frac{1}{\sqrt{a}}\left(-\frac{\delta q^1 \alpha^{k2}}{\delta x_k} + \frac{\delta q^2 \alpha^{k1}}{\delta x_k}\right) \qquad 5.10$$

In 2D the fluxes q^i are located at the mid points of the sides of the cell, the pressure ϕ at the center and ω and ψ at the corners of the cell. The metric quantities in Eq.(5.9) and Eq.(5.10) are defined at the same locations of the fluxes q^i. When the metric quantities are evaluated at different locations of the cell, the derivatives are evaluated by the values of the coordinates y^i at the corners of the cell. The interested reader can see that by making all these assumptions on the definition of each quantity in the most appropriate location, this finite difference scheme is analogous to the finite volume approach (Vinokur (1989)) that in the past has been widely employed in simulations of incompressible and compressible flows (Rosenfled *et al.* (1991)).

5.3 Laplacian discretization

In Eq.(5.6) and in Eq.(5.8) the discretization of the RHS does not, usually, present any difficulty if the metric quantities are appropriately located. To be clearer, when the terms $\frac{\partial}{\partial x_1}\,\alpha^{1l}$ are discretized, the metrics α^{1l} are defined at the position $i \pm \frac{1}{2}, j$ and consequently the metrics α^{2l} at $i, j \pm \frac{1}{2}$ for the terms $\frac{\partial}{\partial x_2}\,\alpha^{2l}$. This discretization produces a stencil of the Laplacian operator accounting for the 9 points around i, j. If the coordinate system is orthogonal, the metric quantities with $k \neq l$ are zero and the stencil, as for Cartesian coordinates, accounts for 5 points. This choice produces a simplified set of algebraic equations with an easier solution, and with smaller truncation errors. To show the reduction of truncation errors the calculation of $\nabla^2\psi$ with $\psi = y_1^2 + y_2^2$ has been performed in an orthogonal and non-orthogonal coordinate system. The difference between the computed and exact solution, of $\nabla^2\psi = 4$, shows (Fig.5.1) that the errors are greater where the grid is more non-orthogonal. It is interesting to notice that the maximum error in Fig.5.1 is $3 \cdot 10^{-2}$. The error distribution for the orthogonal grid is not given since the maximum was $O(10^{-4})$; for the orthogonal grid the external boundary was a circle.

In a general case it is difficult to find analytical expressions for an orthogonal grid between two boundaries of complicated shapes. Several scholars employed conformal transformations and for certain flows analytical expressions were possible. There are books dedicated to conformal transformations as for instance Walker (1964); however, in a large number of other examples the analytical expression is impossible and the grid is obtained by solving a system of partial differential equations of the same complexity as that of the Navier-Stokes equations. The numerical procedure used to generate the coordinate system

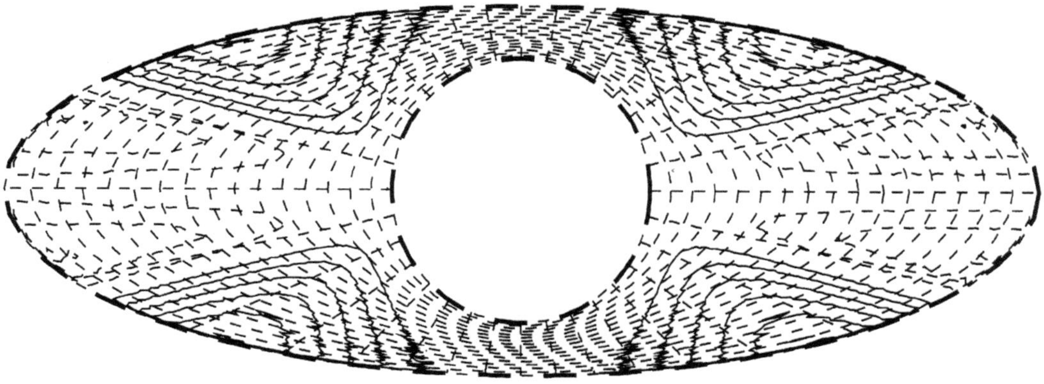

Fig. 5.1 Distribution of errors with respect to the exact solution of $\nabla\psi$. Contour levels $\Delta = 0.01$

are not the subject of this monograph; those who are interested should read chapter 9 in Thompson *et al.* (1985). Moreover, libraries devoted to the generation of a complicated system of coordinates around bodies are now available; see for example the package *GRIDGEN* developed at NASA.

5.4 Nonlinear terms discretization

If the fluxes q^i in Eq.(5.7) are substituted in the second term of Eq.(5.6), the convective terms become $\frac{1}{\sqrt{a}}J(\omega, \psi)$; these take a Jacobian form, as in Cartesian coordinates. In a general curvilinear coordinate system the global conservation property of a general quantity h is fulfilled when the surface integral

$$G = \int\int h dy_1 dy_2 = \int\int \sqrt{a} h dx_1 dx_2 \qquad 5.11$$

remains constant in time. Thus in the inviscid limit, the right hand side of the transport equation for ω in Cartesian and for $\omega\sqrt{a}$ in general curvilinear coordinates have the same expression $J(\omega, \psi)$. Consequently everything mentioned before in chapter 4 for the nonlinear term discretization, can be extended to general curvilinear coordinates. In particular Arakawa's scheme (described in section 4.5) can be straightforwardly extended to general curvilinear coordinates and so once more it is important to point out that the ω, ψ formulation has the property that the discretized equations conserve total energy, total enstrophy and skew symmetry, in the inviscid limit. The extension of Arakawa's scheme to the primitive variables formulation, in the simple case of Cartesian coordinates (section 4.6), requires a very large number of mathematical operations. The task is more onerous and I did not have the sufficient skill and the required patience to find a formulation that in the primitive variables conserved total energy and enstrophy in a 2D general coordinate system. The expression of the nonlinear terms in Arakawa's scheme is given in Eq.(4.15) and Eq.(4.16).

5.5 Boundary conditions

Periodicity in one direction does not require any conditions, since here it holds what was written in section 4.7 for the discretization of the nonlinear and viscous terms of the vorticity equation. For complex geometries the geometry should also be periodic. With radiative conditions Eq.(4.5), for example, holds at each point on a boundary at $x_1 = const$; it implies that at this boundary the contravariant components are $u^1 = C$ and $u^2 = 0$. The convection velocity C could be evaluated by Eq.(4.6) at each point along the line $x_1 = const$. No-slip or free-slip walls require more attention, in particular because of the non-orthogonality of the coordinates at the boundary.

For primitive variable with staggered fluxes, particular attention is necessary in evaluating

$$\frac{1}{\sqrt{a}}\frac{\partial}{\partial x_l}\alpha^{lm}\frac{\partial q^m}{\partial x_l} \qquad l \neq m \tag{5.12}$$

at the first grid point near the boundary. In the ω, ψ formulation the wall vorticity could be evaluated by Eq.(5.10), but, as it was done in section 4.3, it is convenient to differentiate Eq.(5.10) with respect to x_k, for boundaries at $x_k = const$. This operation has the advantage to link the wall vorticity to the vorticity at the closest point to the boundary. For the x_2 direction it follows

$$\frac{\delta\omega}{\delta x_2} = \frac{\delta}{\delta x_2}\frac{1}{\sqrt{a}}\left(-\frac{\delta\alpha^{k2}q^1}{\partial x_k} + \frac{\delta\alpha^{k1}q^2}{\delta x_k}\right) \tag{5.13}$$

This equation is evaluated at the position $i, j_w + 1/2$, with j_w indicating the boundary. The fluxes q^2 are defined at the point $j_w + 1$ and at j_w if the wall is transpirating. The fluxes q^1 on the other hand are evaluated at $j_w + 1/2$ and $j_w + 3/2$. By these fluxes the right hand side of Eq.(5.13) is calculated along the boundary by the same approximations usually done in Cartesian coordinates; the accuracy of Eq.(5.13) is $O(\Delta x^2)$.

In Cartesian coordinates a free-slip wall has zero vorticity at the wall. On the other hand, for a curved boundary the wall vorticity for non-transpiring walls, is

$$\omega = -\frac{q^1\Gamma^1_{12}}{a_{22}} \tag{5.14}$$

where

$$\Gamma^n_{li} = a_{nj}\frac{\partial y_k}{\partial x_j}\frac{\partial y_k}{\partial x_i \partial x_l} \tag{5.15}$$

This expression implies that the vorticity at the wall is the higher as the smaller is the local curvature radius of the boundary.

5.6 Time discretization

The solution of unsteady flows requires a time advancement scheme with, at least, second order accuracy and not too restrictive stability conditions. As in section 4.6 the three-step time advancement for Eq. (4.1) can be written as

$$\begin{aligned}(\omega^k - \omega^{k-1}) - \alpha_k\frac{\Delta t}{\sqrt{a}Re}L(\omega^k - \omega^{k-1}) =& \Delta t(\gamma_k H^k + \rho_k H^{k-1}) \\ &+ \alpha_k\frac{2\Delta t}{\sqrt{a}Re}L(\omega^k)\end{aligned} \tag{5.16}$$

where $k = 1,2,3$ denotes the substep number; ω^0 and ω^3 are the vorticities at the time step n and $n+1$. The L operator is the second order finite difference operator, in the right hand side of Eq.(5.6), which retains only the $i = j$ derivatives. The cross derivatives $(i \neq j)$ have been included in the H operator. The coefficients γ_k and $\rho_k, k = 1,2,3$, are constants selected such that the total time advancement between t^n and t^{n+1} is nearly third-order accurate in time. These coefficients are equal to those used for the Burgers equation (Eqs.3.4.3-5)

$$\gamma_1 = 8/15 \ , \ \gamma_2 = 5/12 \ , \ \gamma_3 = 3/4 \ , \ \rho_1 = 0. \quad , \rho_2 = -17/60 \ , \ \rho_3 = -5/12 \ ,$$

$$\alpha_l = \gamma_l + \rho_l \quad , \qquad \sum \alpha_l = 1.$$

At each substep H is advanced explicitly and part of the viscous term implicitly. To perform this semimplicit calculation the inversion of a pentadiagonal matrix as that in Eq.(4.25) is required. In general coordinates, however, the coefficients of the matrix $A_{i,j}$, that were indicated by a^{pk}, a^{mk} and a^{ck} are different at each grid location i, j; the *fishpack* solver cannot any longer be used and the factorization is mandatory. The same considerations done in section 4.6 about the speed-up of the tridiagonal inversion hold. I hope that the readers are not confused by the choice of the same letters a for the metric quantities and for the coefficients of the tridiagonal matrices, as well for the letters α for the metric quantities and the coefficients in the time advancing scheme.

The coefficients of $A_{i,j}$ at no-slip walls are evaluated by writing Eq.(5.13) in terms of the unknown $\Delta\omega^k = \omega^k - \omega^{k-1}$. At each substep k this equation is

$$\frac{\delta \Delta\omega^k}{\delta x_2} = \frac{\delta \omega^k}{\delta x_2} - \frac{\delta \omega^{(k-1)}}{\delta x_2} + O(\Delta t^2) \qquad 5.17$$

The right hand side is calculated by evaluating Eq.(5.13) at k and by subtracting the value at the previous step $k-1$. These boundary conditions, second order accurate in time, are required for unsteady flows. Eq.(5.17), e.g. at the boundary $j_w = 1$, is $\Delta\omega_{i,2} - \Delta\omega_{i,1} = RHS_{i,3/2}$; the coefficients of the tridiagonal matrix then are $a_1^{p2} = 1, a_1^{c2} = -1$ and $a_1^{m2} = 0$.

5.7 Code PSOMCUR

The code *PSOMCUR* permits to perform simulations with different kinds of boundaries and initial conditions. Concerning the initial conditions an inlet velocity profile at $i = 1$ varying in j allows the simulation of currents along irregular boundaries, with possible applications in geophysical flows. A second class of flows, always connected to geophysical applications, are those where coherent structures as dipoles interact with irregular domains. In these flows the eventual β or bottom topography effects can be inserted in the code; this has not be done in the present version. Whoever is interested can easily introduce these effects in the subroutine related to the nonlinear terms as it was done in the code *PSOMGBC*. A further application, in particular at low Reynolds number, could be linked to biological flows; in fact often the hypothesis of a Newtonian fluid holds for the blood inside arteries. Of particular interest in medicine is to understand how the stenosis, forming inside the arteries, changes the flow regime.

The name of the code $PSOMCUR$ is the abbreviation of ψ ω and 'curvilinear coordinates'. The code is given into the directory CODESCH5/CUR and it has a structure similar to that for the code $PSOMGBC$. Also here the code is divided in several files that are shortly described.

In the file *psomcurma.f* there is the subroutine $SOLVE$ calling all the other subroutines. In the routine $TSCHEME$ the time advancement scheme is coded. It consists of the call of different routines related to the evaluation of the nonlinear terms, of the evaluation of the vorticity and of the streamfunction. In this file there is also the call of the subroutine $EXTFLO$ which permits to assign at the inlet disturbances varying in time.

In the file *psomcurnn.f* there are the routines necessary at the beginning of the simulations to evaluate the grid size and the indices for the derivatives. Furthermore, in this file there are the routines writing the output fields in the formats requested by the graphics packages. These fields are written at large increments of time; at shorter intervals are written certain global quantities as circulation, energy, enstrophy and the maxima of the vorticity. There are also routines evaluating the maxima of flow quantities and the CFL.

In the file *psomcurco.f* there are subroutines related to the curvilinear coordinates and to the metric coefficients.

In the file *psomcurtn.f* there are subroutines for the Navier-Stokes equations discretization. In the routine $INVTRV$ there is the time implicit solution for $\Delta\omega$ solved by a factorization. In $INVTRI$ instead there is the time evolution of $\Delta\omega$ in the inviscid case. Similar subroutines are used to advance the passive scalar and the name is $INVTPS$. $HDNL$ is the subroutine relative to the nonlinear term Arakawa discretization and in $HDNLVI$ the cross terms of the Laplacian are added to the Jacobian.

In the file *psomcurmg.f* there is the part of the code relative to the streamfunction calculation by the multigrid method that was described in chapter 2.

In the file *psomcurit.f* there is the iterative solver for the multigrid method that is based on a POINT SOR method.

In the file *psomcurin.f* there are the initial conditions that can give a dipole or an inlet velocity profile on the $WEST$ boundary. For this code the Bessel functions have been inserted in this file.

Concerning the boundary conditions given in the file *psomcurbc.f*, this code does not allow to reproduce a large variety of boundary conditions as those permitted in $PSOMGBC$. By calling, as before, $NORTH$, $SOUTH$, $WEST$ and $EAST$ boundaries (respectively those defined by the indices $(i, n2)$, $(i, 1)$, $(1, j)$ and $(n1, j)$) the following conditions can be assigned by changing the values of the indices $insln$ and $insls$ in the file with the input data. Radiative conditions at the $NORTH$ boundary require $insln = 2$; $insln = 0$ indicates a free-slip and $insln = 1$ a no-slip wall. At the $SOUTH$ boundary respectively free- and no-slip conditions occur for $insls = 0$ and $insls = 1$. While at this boundary radiative conditions are not possible, at the $EAST$ boundary only radiative conditions are permitted. $iwest = 0$ assigns an inlet condition at the $WEST$ boundary with a prescribed velocity profile. In the present version a $y^{1/7}$ profile is given where y is the distance from the $SOUTH$ boundary. The code starts from rest and at the inlet the velocity reaches the desired profile within a time τ. This transient is described by a function with continuous derivatives at $t = 0$ and $t = \tau$.

In geophysical applications related to environmental problems, for instance, to reproduce the dispersion of a thermal discharge near a sea coast or in a river, a velocity profile can be assigned on the $SOUTH$ boundary together with a profile for a passive scalar. These profiles are given in the subroutine initial.

The other flows concerning the interaction of coherent eddies with the coastal line have as initial condition a Lamb dipole that can be localized in any point of the field. The vorticity distribution of the Lamb dipole is assigned as in chapter 4.

The code requires an input file called *psomcur.d*; a typical input file is

```
193  193     3                        n1,n2,nsst
0     0     11     1     50           nwrit,nread,nfilr,nplot,mnpi
.02   1001  100    5     1   1        dt,ntst    ,nprint,npin,nmolp,ntra
1     0.1         .08                 icflm,cflma,dtl
2.00     0.5        12.                           tprin,tpin,tfin
2001                                  nstop
.33333       .33333       .33333      chal,chbe,chga
0     0     1                         insls    insln inwes
.3e+09    1.                                                re    schmid
1  0  0.   0.   5.   -5.   1.  0.3    imod,istrea,vampu,vampv,xl1,xl2,tau1,cout
-3.5 3.0  1.0  -1.0   -1.0            yc1mo,yc2mo,ramo,velmo,vsi
180.                                                             thet0
-5.0    5.0  -0.     7.5                alx1i,alx1f,alx2i,alx2f
1.0    .35   0.65 0                   str1,xcr1,etr1,istr1
1.5    .4     4.5  0                  str2,xcr2,etr2,istr2
1                                     itypc
2.0     +1.5     0.                                  bum(1),sigb(1),xp(1)
0.            0.5     1.5                   bum(2),sigb(2),xp(2)
5   0   1  50    5  1.7 .5e-03        mlev,iwmg,mlw,maxcmg,nmaxi,omeg,epsm
```

Some of the quantities are very simple to understand or have the same names as in chapter 4; they are not explained. The third line is necessary for simulations at constant Δt and the line with $icflm$ is necessary for simulations at constant CFL; these require $icflm = 1$. In the next line are prescribed the values of the time intervals at which the outputs files are written. In the seventh line there are the coefficients of the Arakawa scheme. The eighth line assigns the values of the indices selecting the different boundary conditions as described before. The values of the Reynolds and Schmidt numbers are given in the ninth line. If $Re > 10^5$ the inviscid simulation is performed and the passive scalar evolution is neglected. In this case free-slip conditions must be assigned on the $NORTH$ and $SOUTH$ boundaries.

On the next three lines data concerning the initial conditions appear. With $imod = 1$ a Lamb dipole is located at the position $yc1mo, yc2mo$, it has an inclination angle $thet0$ with a horizontal line. With $istrea = 1$ the inlet velocity at the $WEST$ boundary reaches the assigned profile in a time $t = tau1$. $vampu$ and $vampv$ are the maximum value of the inlet velocities on the south boundary in a slot located between $xl1$ and $xl2$. The profile has a

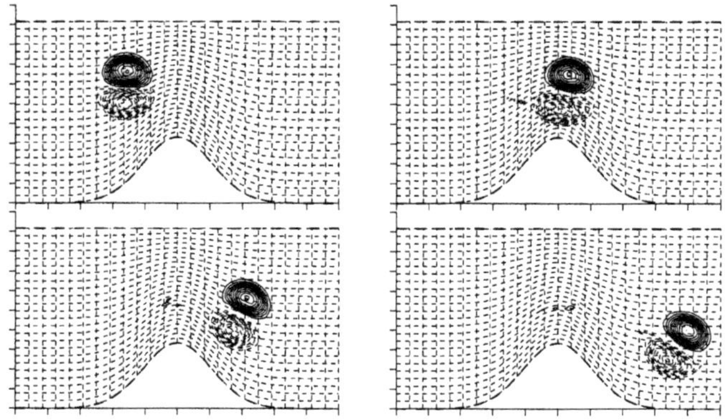

Fig. 5.2 Contour plots of vorticity for the dipole travelling near a bump in the inviscid case. Arakawa's scheme conserves all the invariants.

law given in the subroutine $INITIA$. $cout$ is the value of the outlet velocity, necessary at $t = 0$, to calculate the radiative conditions. In line thirteen the dimensions of the physical domain are assigned. Non-uniform grids can be used in both directions and coordinate transformations as those in chapter 2 are employed by setting the appropriate values of the parameters in the next two lines. $itypc$ allows to change the geometry irregularity on the $NORTH$ and $SOUTH$ boundary. By $itypc = 0$ a semicircular bump of radius $buml$ is located on the $SOUTH$ boundary. It can be concave by setting $sigbl = 1$ or convex by $sigbl = -1$. With $itypc = 0$ the surface on the $NORTH$ boundary is flat. By $itypc = 1$ two Gaussian bumps with deviation $sigb(l)$ and amplitude $bum(l)$ are located at the $SOUTH$ $l = 1$ and $NORTH$ $l = 2$. These perturbations are centered around the positions $xp(l)$. On the last line the parameters necessary for the multigrid solver for ψ are given.

5.8 Results

5.8a Inviscid dipole moving close to a bump

This ideal flow permits to investigate the global conservation properties when a Lamb dipole interacts or moves near a free-slip wall of complex shape. The geometry of the body lying on the south boundary is a Gaussian bump, with a amplitude equal to 2 dipole radii. It is interesting to see whether the dipole, after or during its passage close to the bump, takes its original shape. Orlandi (1992) studied the motion of a dipole encountering a free-slip cylinder showing that, after the collision with the cylinder, the dipole reforms and gets the linear functional relationship $\omega = f(\psi)$. The bump with a free-slip condition modifies the trajectory since this assumption implies that there are image vortical patches on the other side of the boundary to impose a zero normal velocity on the surface of the body. The induced field from those imaginary vortical structures imposes a deviation on the straight trajectory the dipole would have travelled near a flat boundary. Some of the

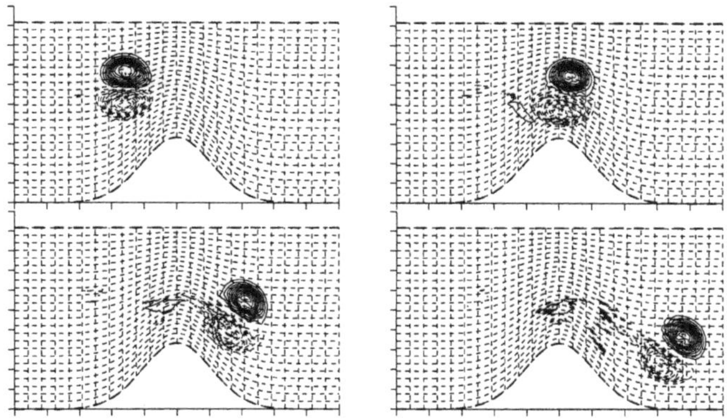

Fig. 5.3 Contour plots of vorticity for a dipole travelling near a bump in the inviscid case. Arakawa's scheme conserving the enstrophy.

simulations showed that when the dipole is too close to the boundary, the image vortex of opposite sign forms with the patch closest to the wall a couple travelling backwards. The inviscid simulations were devoted to investigate the conservation properties of Arakawa's scheme and it was found that the maximum CFL restriction plays an important role.

In a Cartesian grid the dipole moves along a straight trajectory near a flat boundary if the CFL is 1.7. In general coordinates, with the amplitude of the bump of the same order of the radius of the dipole, a large perturbation affects and modifies the shape of the dipole. Large unphysical vorticity gradients, at $CFL = 1.7$, grow and thus the dipole looses its coherence and leaves behind a large amount of vorticity at small scales. By reducing the CFL to 0.5 the dipole, even if perturbed, maintains its coherence. Fig.5.2 shows the vorticity evolution near the free-slip wall when the second order invariants were preserved. In this simulation the upper boundary was located relatively close to the dipole ($L_{x_2} = 5.5$), and radiative conditions were imposed there. A further simulation was performed by imposing a free-slip condition and by locating the boundary at $L_{x_2} = 7.5$ to reduce the effect of the image dipole on the upper boundary. The two simulations predicted equal trajectories.

Simulations performed by conserving only the energy did not differ from those in Fig.5.2. On the other hand, Fig.5.3 shows that when the scheme conserves only the enstrophy small scales are released in the wake of the dipole. To emphasize the differences among the three cases, Fig.5.4 shows the vorticity contours with a low increment $\Delta\omega = 0.25$. The comparison at $t = 6$, when the bump exerts the maximum perturbation on the dipole, shows that if energy is conserved the main structure of the dipole remains. On the other hand, when the enstrophy is conserved the amount of small scale vorticity left behind the dipole increases and even the structure of the dipole changes. By these tests, it can be concluded that in a general curvilinear coordinate system the global conservation properties play an important role. My opinion is that this check should be done in the inviscid case when the viscosity does not dissipate the small scales generated by the perturbation on

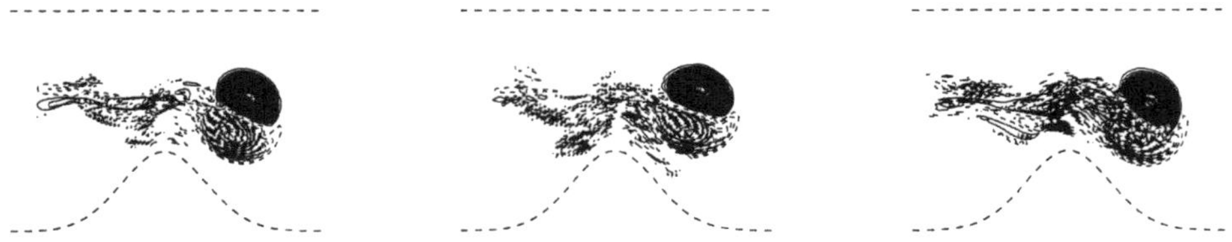

Fig. 5.4 Contour plots of vorticity for a dipole travelling near a bump in the inviscid case at $t = 6$. Left energy and enstrophy conserved, center energy conserved, right enstrophy conserved.

the dipole. To my knowledge, this check is seldom done in developing numerical methods for flows around bodies of complex shape.

5.8b Dipole impinging a semicircular cavity

The validation of the code in the viscous case has been done with the aim to reproduce the experiment of a dipole impinging a semicircular convex boundary. A perfect two-dimensional experiment is difficult to setup since when a vortex couple, with an axial extension greater than the distance between the centers, impacts a wall (even a flat one) it produces long and short axial perturbations. The long waves are the so called Crow instabilities (Crow 1970) and the short waves are the Widnall instabilities (Widnall *et al.* 1974); after a short transient these grow exponentially and affect the main vorticity. Even in a free dipole these instabilities appear as will be shown in chapter 10. An experiment of a vortex ring impinging on a hemispherical surface can be thought to approximately reproduce two dimensional conditions. The flow visualization of this experiment was presented in the Gallery of Fluid Motion by Rambod & Garib (1995).

From these pictures it turns out that the Reynolds number was small enough and the azimuthal instabilities do not grow. Moreover, since this experiment is confined within the hemisphere, the axisymmetric vortex stretching does not play a substantial role; the flow evolution, then, should not be too different from that in two dimensions. The experimental flow visualizations by Rambod & Garib (1995) can be compared with the numerical predictions in Fig.5.5. Despite the different conditions between the experiment and the simulation, both show the formation and the backwards propagation of a secondary structure. The creation of secondary vorticity, as in the impact with a flat wall (described in chapter 4) is due to the adverse pressure gradient upon the solid boundary. The distribution of markers in the experiment and in the simulation (Fig.5.5) permits only to argue that secondary vorticity is generated. In an experiment contours of vorticity can be obtained by PIV which requires sophisticated instruments. On the contrary ω comes out directly from the solution of the equations.

Contours of ω (Fig.5.6), at the same instants of the passive scalar contours (Fig.5.5), show how the secondary vorticity, after its generation in a thin layer, rolls up. Part of ω remains within the hemisphere and part forms the secondary vortex couple that propagates far from the wall. The simulation was performed by a coarse grid with the purpose to show that, even at low Reynolds number, the features of the experiment are captured.

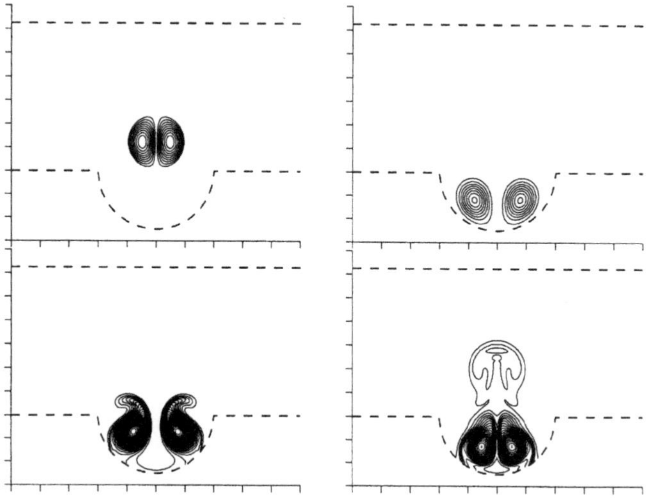

Fig. 5.5 Passive scalar distribution from the simulation at $Re = 300$ and $Sc = 1$; at $t = 0, 4, 6$ and $t = 12$ for a 96×96 grid. At $t = 0$ and 4 $\Delta\sigma = 0.1$, at $t = 6$ and $t = 12$ $\Delta\sigma = 0.025$.

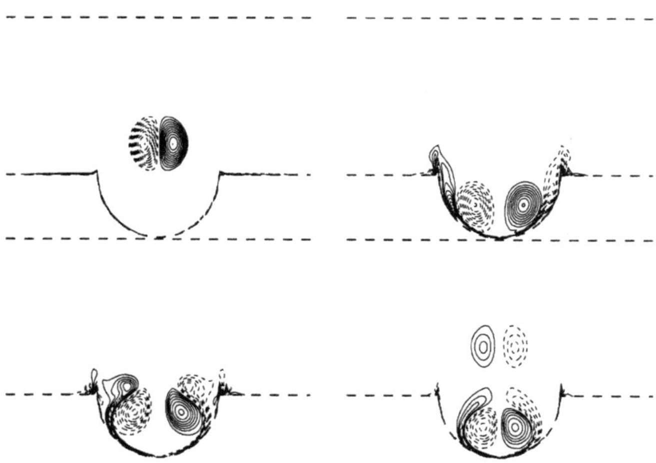

Fig. 5.6 Vorticity distribution from the numerical experiment at $Re = 300$ at $t = 0, 4, 6$ and $t = 12$ for a 96×96 grid. $\Delta\omega = 1.0$.

The calculation required on a portable IBM computer 300 secs of CPU time. Those who are interested to this flow should perform the simulations on more powerful computers by

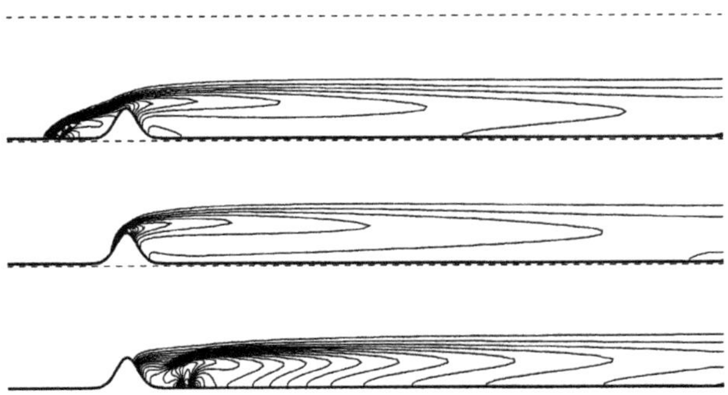

Fig. 5.7 Passive scalar distribution at $t = 80$ for the three different conditions of its discharge a) before, b) on and c) after the bump. The simulations are with a 193×47 grid.

increasing the Reynolds and the Schmidt numbers, and hence the number of grid points. More resolved simulations will produce a better agreement between the passive scalar and the LIF visualizations.

As reported in the comment on the Gallery of Fluid Motion (Rambod & Garib (1995)) this flow has direct implications for the understanding of the flow development inside of the heart's left ventricle during the diastolic filling process. In several other parts of the blood circulation system there are conditions where the blood encounters complex boundaries. Recently Pedrizzetti (1996) and Domenichini & Pedrizzetti(1998) performed simulations with an axisymmetric code similar to the present one with the aim to help the medical doctors during coronary operations. As a result of the simulations they understood that after the operations, devoted to eliminate the blockage of the coronary, a step was forming which gave rise to a very high negative wall stress. After the simulations the medical doctors decided to reduce the cleaning of the coronary so that the separation was moving slightly downstream. The wall stress diminished and the probability of a further operation on the patient was reduced (Pratesi *et al.* (1995)). The present code can be modified to deal with axisymmetric constrictions and this simple exercise is left to whom is interested. More difficult is the transformation of the code to treat complex geometries within 3D tubes. My opinion is that in these more complicated conditions it is better to use methods based in the introduction of body forces on the Navier-Stokes equations. This efficient and easy method was used in the past by Peskin (1972) to understand complex flows of bioengineering interest.

5.8c Stream around a Gaussian hill

This code was used also to simulate flows of geophysical interest where often streams around complicated geometries occur. A case that can be used to predict the distribution of pollutant emitted near a coast was chosen. A laminar flow at $Re = 500$ was simulated and someone could argue that this is a Reynolds number far from the real of the order of 10^7. However, in presence of flow separations around bluff bodies, the main features of

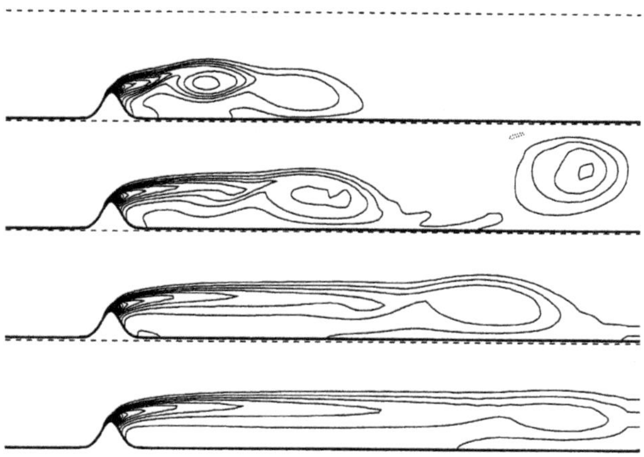

Fig. 5.8 Passive scalar evolution in time each figure is separated by $\Delta t = 16$. The solution was obtained with a 193×47 grid and with the discharge on the bump.

the flow are reproduced by flows at low Reynolds numbers. This has been shown by Van Dyke (1972) in his album of fluid motion; he showed in the pictures 172 and 173 the wake behind a large grounded tank ship, visualized by the oil leaking from the tanker, and that behind a flat plate. In the wake behind the thank the Reynolds number was $Re = 10^7$; instead behind the flat plate it was 4300. The two flows were both turbulent and the differences appear at the small scales. In the present case the difference between reality and the simulation is greater because in reality the flow is turbulent and in the simulation it is laminar. However the gross feature should not differ in a large measure, and for those interested in a simulation closer to reality it will not be difficult to introduce in the code a standard turbulence model based on one-point closures.

Three simulations were performed with an inlet velocity profile $U = y^{1/7}$ with y the distance from the wall. The length of the domain L_{x1} was equal to 12 and the lateral dimension L_{x2} was equal to 3. At the external boundary a free-slip condition has been assigned; at the outlet a radiative condition allows the vortical structures, generated on the bump, to exit from the domain. The three simulations differ on the location of the discharge along the coast line. The size of the discharge was kept constant and equal to 0.3 and the exit velocity was equal to 0.2. In one simulation the discharge was located at $x_1 = 0.85$ before the bump that had height $L_b = 0.5$ and was located at $x_1 = 2$. The other two simulations had the discharge at $x_1 = 2$ and at $x_1 = 3.45$. These three simulations permit one to investigate in which case there is the maximum dispersion of discharged pollutants. The simulations started from rest with the inlet velocity reaching the maximum value in a short time. The simulations lasted for 80 dimensionless time units until the solutions became approximately stationary. The flow fields in the three simulations do not depend largely on the discharge since the exit velocity is low. On the other hand, the passive scalar differs, as is shown in Fig.5.7. The simulations show that the best dispersion is achieved by locating the discharge on the bump; in fact, in this condition the pollutants enter a region with a large velocity produced by the favorable pressure gradient along the bump. The pollutants are, thus, carried out from the coast region. When the discharge is in the recirculating region a large amount of pollutants is trapped in the region after the bump and these are distributed on the coast line. The discharge is also inefficient when it is located before the bump; in this case a large part of the pollutants is accumulated on the bump.

The time evolution of the contour plots of the passive scalar are shown in Fig.5.8 at different instants separated by $\Delta t = 16$. From the figure it turns out that the thin layer of vorticity generated on the bump rolls up and forms a large scale vortex that detaches from the bump and is advected out of the domain. The radiative boundary condition at the exit describes this transient. This fact suggests that the code could reproduce the unsteady flow due to the super-imposition of a disturbance on the mean velocity profile at the inlet; this unsteadiness creates a sequence of vortex shedding. This interesting condition could be pursued by the interested scholar, but, for practical environmental problems, it will be important to add a turbulence model and to increase the Reynolds number.

5.9 Conclusions

This chapter describes a two-dimensional numerical method to solve flows around complex geometries. This is a field that only recently has received large considerations, thus it is still not clear which is the most reliable numerical method. To my knowledge, especially for turbulent DNS or LES, there are few numerical simulations. To cite only few of them, I would like to recall those about the flow past riblets. These were performed by three different methods: the spectral element (Chu & Karniadakis (1993)), the finite differences (Orlandi (1989), Choi *et al.* (1993)) and the body forces (Goldstein *et al.* (1993)). I am suggesting to anyone interested to this topic, before spending a lot of time to develop a code, to consider the qualities and the disadvantages of each method by doing appropriate checks. If the global conservation properties are playing a role, the present code can be used to generate a reference solution. My opinion is that the methods based on the coordinate transformation or those based on finite elements are difficult to code and require a large mathematical skill. On the other hand the methods based on the immersed boundaries seem to have the quality to be simple, versatile and also to have a satisfactory accuracy. Recently improvements have been suggested by Fadlun (1998) to get solutions at high CFL numbers and to have a better representation of the boundary by means of a smoothing.

In the past, I spent a large amount of time to develop a numerical method and to write a code for 3D flows in general curvilinear coordinates in two directions; I was really stressed in those days to get the multigrid solver to converge and sometime it never converged. With this experience, if I should now start again that project I shall put all my efforts to find an efficient way to insert the body forces into the codes that at the base have the solution of the Navier-Stokes equations in Cartesian or cylindrical coordinates. I am currently working on this new methodology with the aim to understand the modification of the vortical structures in a turbulent channel when a bump is inserted. The study with approximations on the bump has been recently described by Lumley (1998). I will be glad to give suggestions to those who are interested to insert the body force in the primitive variables codes furnished in this book.

6 Two-dimensional turbulence

6.1. Introduction

In the past several studies on 2D turbulence were performed to understand this ideal condition that is interesting from the fundamental point of view. Since any basic study should also be linked to real motivations; 2D turbulence is a fundamental topic to understand geophysical flows. The realization in the laboratory of a perfect 2D flow is very difficult. Attempts were done by moving an array of cylinders in a stratified fluid. The 3D wakes behind the cylinders in a short time generated a bunch of 2D vortices that behaved as 2D turbulence (Narimousa *et al.* 1991). In fact, in presence of a strong stratification the fluctuations in the vertical direction are dumped and the flow can approximately be considered two-dimensional. Rotation plays a similar effect; as a consequence in several laboratories rotating tables are used to study the vortex dynamics in 2D (Hopfinger *et al.* (1982)). The difference between the two set-ups is that, in rotating tanks the vortices are spanning the full depth, but, at high rotation rates, the Eckman layers on the bottom wall produce a dissipation that can be greater than the viscous dissipation. However the dissipation it is much less than that acting on the pancake vortices forming in a stratified fluid. Soap films are a further method to achieve experimental conditions close to that of 2D turbulence; in this case the imperfect realization of a 2D flow is related to a slight compressibility (Couder & Basdevant (1986)).

Even if there are differences with the ideal conditions, these experimental set-ups have shown the inverse energy cascade characteristic of 2D turbulence. The inverse energy cascade is the physical phenomenon that makes the difference between 2D and 3D turbulence. In this short chapter a code is described that permits the simulation of the Navier-Stokes equations in the ω, ψ formulation for 2D turbulence. As usual, the simulations are performed in a doubly periodic domain, and it is well known that decaying turbulence depends on the initial conditions. The evaluation of the initial conditions is described because it permits to easily understand the calculation of the initial field in 3D isotropic turbulence. These simulations, in addition, allow to investigate how the evolution of the spectrum depends on the global conservation properties.

The equations of motion are given in Chapt.4, and will not be reported here. Boundary conditions are not necessary, the only requirement is to use periodic tridiagonal solvers in the factorization of the vorticity equation, and to insert one of the routines described in section 3 of Chapt.2 to get the streamfunction. The code *TURB2D* is similar to the code *PSOMGBC* described in section 4.8a. It uses the approximate factorization in the solution of the vorticity equation together with a third order Runge Kutta scheme for the nonlinear terms. In this code the passive scalar has not been considered; its introduction is easy, by following what was done in the code *PSOMGBC*. Biperiodic conditions permit to increase the dimension of the stencil and hence to increase the accuracy. Thus the Arakawa fourth order scheme is discussed.

This chapter can be considered as a first opportunity to get acquainted with the calculation of energy spectra and turbulent statistics.

6.2 Arakawa fourth order scheme

With two periodic directions the accuracy of any finite difference scheme can be increased since the stencils of the differential operators can be as large as desirable. The accuracy plays a greater role in the discretization of the nonlinear terms than in that of the viscous terms. Arakawa (1966), together with the second order accuracy scheme given in Chapt.4, furnished a fourth order scheme. Even this more accurate scheme expresses $J(\omega,\psi)$ by a combination of the three Jacobians given in section 4.5; $J_I = \frac{1}{3}(J_1 + J_2 + J_3)$, with a new expression J_{II} combination of the following three Jacobians

$$\begin{aligned}
J_4(\omega,\psi) = &\frac{1}{8\Delta x_1 \Delta x_2}[(\omega_{i+1,j+1} - \omega_{i-1,j-1})(\psi_{i-1,j+1} - \psi_{i+1,j-1}) \\
&- (\omega_{i-1,j+1} - \omega_{i+1,j-1})(\psi_{i+1,j+1} - \psi_{i-1,j-1})] \\
J_5(\omega,\psi) = &\frac{1}{8\Delta x_1 \Delta x_2}[\omega_{i+1,j+1}(\psi_{i,j+2} - \psi_{i+2,j}) - \omega_{i-1,j-1}(\psi_{i-2,j} - \psi_{i,j-2}) \\
&- (\omega_{i-1,j+1}(\psi_{i,j+2} - \psi_{i-2,j}) + \omega_{i+1,j-1}(\psi_{i+2,j} - \psi_{i,j-2})] \\
J_6(\omega,\psi) = &\frac{1}{8\Delta x_1 \Delta x_2}[\omega_{i+2,j}(\psi_{i+1,j+1} - \psi_{i+1,j-1}) - \omega_{i-2,j}(\psi_{i-1,j+1} - \psi_{i-1,j-1}) \\
&- (\omega_{i,j+2}(\psi_{i+1,j+1} - \psi_{i-1,j+1}) + \omega_{i,j-2}(\psi_{i+1,j-1} - \psi_{i-1,j-1})]
\end{aligned} \qquad 6.1$$

Arakawa proved that $J_{II} = \frac{1}{3}(J_4 + J_5 + J_6)$ also conserves enstrophy and energy and that

$$2J_I(\omega,\psi) - J_{II}(\omega,\psi) = J(\omega,\psi) + 0((\Delta x_1 \Delta x_2)^2) \qquad 6.2$$

has fourth order accuracy.

The scheme can be easily inserted in a code with a biperiodic domain; on the contrary for flows with free- or no-slip walls it can be used in the inner points but near the boundaries the accuracy is reduced. For the applications described in Chapt.4, at relatively low Reynolds numbers, second order accuracy is sufficient. On the other hand, in the simulation of 2D turbulence at sufficient high Reynolds number, it turns out that the accuracy plays an important role. The results are not presented and I suggest the reader to prove by himself that this is indeed true.

6.3 Initial conditions

As mentioned before, the evolution of 2D turbulence depends on the initial conditions; the physical reason is related to the inverse energy cascade. This process consists on the merging of vortices that depends on their size and space distribution. In 3D turbulence, on the other hand, the energy goes from large to small scales; hence for isotropic turbulence, at high Reynolds number, there is an universal behavior independent of the initial conditions. Energy spectra permit to characterize turbulent flows by showing the energy distribution among the different scales. The energy spectra are calculated by the Fourier series which transform the velocity field from the physical to the wave number space. When the simulations are performed by a pseudospectral code the spectra are directly computed by the amplitude of the Fourier modes. Visualizations of the vorticity field permits to understand qualitatively the flow dynamics; these are straightforward by finite difference simulations once a visualization software package is available.

To assign the initial streamfunction which has the desired energy spectrum $E(k,0)$ the amplitude of the real and imaginary part of the Fourier coefficients should be assigned. The discrete Fourier transform is

$$\psi(x_1,x_2) = \sum_{k_1=0}^{N_1-1} \sum_{k_2=-N_2/2}^{N_2/2-1} \hat{\psi}(k_1,k_2) e^{\iota k_1 x_1} e^{\iota k_2 x_2} \qquad 6.3$$

From the relationship between velocity and streamfunction it turns out that the energy, in wave number space, is

$$E(k_1,k_2) = (k_1^2 + k_2^2)\hat{\psi}(k_1,k_2)\hat{\psi}^*(k_1,k_2) \qquad 6.4$$

At each wave number $k = \sqrt{k_1^2 + k_2^2}$ the energy $E(k)$ is given by the contribution of each $E(k_1,k_2)$. The real $\psi_R(k_1,k_2)$ and the imaginary coefficients $\psi_I(k_1,k_2)$ can be expressed as

$$\psi_R(k_1,k_2) = a(k_1,k_2)\cos(2\pi b(k_1,k_2)), \qquad \psi_I(k_1,k_2) = a(k_1,k_2)\sin(2\pi b(k_1,k_2)), \qquad 6.5$$

where $a(k_1,k_2)$ and $b(k_1,k_2)$ are random numbers with amplitude between 0 and 1. The energy spectrum $\tilde{E}_0(k)$ is different from the desired spectrum $E_0(k)$. To obtain at each k the coefficients of the streamfunction $\psi_R(k_1,k_2)$ and $\psi_I(k_1,k_2)$ these should be multiplied by $E_0(k)/\tilde{E}_0(k)$. The inverse Fourier transform gives then the steamfunction in physical space $\psi(x_1,x_2)$; its Laplacian gives the vorticity.

The initial field is assigned by the subroutines in the file *turb2din.f* in the directory *CODESCH6/TUR*. By the function *specin* two kinds of initial spectra can be chosen; the user who wishes to have a different spectrum has to modify this function. The initial spectrum used in the present simulations is $k^4 exp(-k/k_0)^2$, with the energy peak at k_0. By setting $k_0 = 10$ the vortices are mainly at small scales, and a certain number of vortices merge within a dimensionless time $t = 50$. At this time several compact vortices are still left within the computational box. The time necessary to reach the final condition with only a single dipolar structure, as found by Montgomery *et al.* (1992), is much greater. A different initial spectrum with the energy peak at large scales shows that, in 50 dimensionless time units, the state of a single vortex couple is reached.

The code has the possibility to start the simulation with the initial field, generated for instance in a box of dimension L_{x1} and L_{x2}, and to evolve into a larger domain. In each of the simulations the resolution is maintained by increasing of the same ratio the size of the box L_{xi} and the number of grid points N_i. These simulations permit to investigate the time necessary for an initial field to propagate into an external empty space. As expected, the number of mergings decrease and by increasing the box size the rate of inverse energy cascade is reduced.

6.4 Code *TURB2D*

The code *TURB2D* in the directory *CODESCH6/TUR* has been splitted into several parts for an easier handling and to get a faster compilation if only one part is corrected.

Each file is named with the first six characters *turb2d* and with the following two characters distinguishing each part of the code. The initial conditions are assigned in the file *turb2din.f* The main part of the code is in the file *turb2dma.f*. A bunch of subroutines to print the field, to evaluate the *CFL*, to assign the indices for the periodic conditions, only to cite some of them, are in *turb2dnn.f* The name of the subroutines in *turb2dnn.f* are similar to those used in the previous codes; in addition the content of these subroutines is so simple that its description can partly be avoided.

In the file *turb2dtn.f* the time advancement scheme is coded together with Arakawa's scheme. In this code there is the option to choose the second or the fourth order Arakawa scheme. Regarding the solution of the vorticity equation, there is the option to solve the inviscid Euler equation avoiding the inversion of the Laplacian. The subroutine *INVTRI* for the inviscid simulation is called when $Re > 10^7$. In the viscous case the approximate factorization is composed from two periodic tridiagonal routines. These routines have been vectorized to get a good efficiency on the CRAY computers. In the file *turb2dph.f* the elliptic equation for ψ is solved by the *NCARFFTs*. This part, depending on the machine available, as described in Chapt.2, can be changed to increase the efficiency of the code . It is important to remember that the *NCARFFTs* permit to run the code on any computer, by paying the price of a reduced efficiency.

The code requires as input a file with the data. A typical input file named *turb2d.d* is

```
257  257  3  1.        n1  n2  nsst factpi   (dimensions
.01  4001   500 50 250  dt  ntst  nprint  npin  nprspe
0  0  0     0           nstop  nread  nwrit  nkt(time subsT
0   1.7    .05          icflm  cflma   dtl
5.    .5      50.        5.      tprin  tpin  tfin tprspe
3000.  0.                                re timrea
.33333    .33333   .33333   chal  chbe  chga
0            icut
8.00   .004E-04   3.    2  1.    ak0      en0   gam0   ityp   enede
0            iforth
```

Some of the quantities are self-explaining, and do not need any additional explanation; the names of some variable were used in the codes described in Chapt.4 and 5. The variable *factpi*, together with $icut = 1$ fixes the size of the entire box; it is a multiple of the box where the initial turbulence is located. A better explanation is: if the aim is to study how a turbulent field, initially concentrated in a domain of dimension $2\pi \times 2\pi$, evolves when it is encircled by a region without any motion and limited by an external boundary with dimensions $6\pi \times 6\pi$, $N1 = N2$ should be set equal to 513 and $factpi = 3$. Thus the resolution is equal to that of the initial field.

$iforth = 0$ employs Arakawa's second order accurate scheme $J(\omega, \psi) = J_I(\omega, \psi)$, (section 4.5), $iforth = 2$ retains the same accuracy with $J(\omega, \psi) = J_{II}(\omega, \psi)$ (section 6.2.) The fourth order accuracy is obtained by $iforth = 1$. *ityp* permits to change the spectrum;

ityp = 0 or 1 assigns respectively one of the two spectra used by Santangelo *et al.* (1989)

$$E(k,0) = C_0 k exp(-k/k_0)^2 \qquad 6.6$$

$$E(k,0) = C_1 k[1 + (k/k_0)^{\gamma+1}]^{-1} \qquad 6.7$$

The second spectrum gives more enstrophy at small scales. By *ityp* = 2 the spectrum

$$E(k,0) = C_2 k^4 exp(-k/k_0)^2 \qquad 6.8$$

is given; this is the spectrum used in the present simulations. The constants in the spectrum expressions can be adjusted to have the same values of the peak energy. The expression in Eq.(6.8) gives more energy at large scales than the other two expressions. Even in Eq.(6.8) the scales are small enough to let a large number of merging occur between $t = 0$ and $t = 50$ and at the end of the simulation there is still a relatively high number of vortex patches.

6.5 Results

6.5.1 Resolution check

The simulations were performed at a quite low Reynolds number for two-dimensional turbulence, since the aim of this chapter is devoted to describe the physics of 2D turbulence, that, in the past, attracted the interest of several scholars. In fact, in literature there are a large number of papers; see for instance a special issue of the Journal de Mécanique, edited by Moreau (1983), dedicated to this topic with articles by Lesieur, Hopfinger *et al.*, Legras & Ghil, Mcwilliams, Browand & Ho, Huerre, Corcos, Sommeria, Frisch *et al.*, Sulem, Basdevant & Sadurny, Larcheveque. The simulations at $Re = 3000$ however produced decaying energy spectra with a power law in a wide range of wave numbers. The code *TURB2D*, compiled in single precision, with 1025 points in each direction requires 90 megabytes of RAM. This code was not optimized, and in particular memory is waisted in the calculation of the energy spectra, that could be done as a post-process to diminish the memory occupancy. 512^2 simulations can be performed on portable personal computers, but a large computational time is requested.

Simulations with different resolutions allow to investigate the dependence of the evolution of the vorticity field on the resolution. The coarsest grid with 128^2 points is not sufficient; by doubling the number of points the resolution is marginal and by a further doubling the resolution is sufficient. In the simulation of turbulent flows, sensitive to small perturbations, every realization is different; hence comparisons should be done by averaged quantities. The energy spectra require averages and still more significative are total energy and enstrophy. Contour plots of ω at $t = 5$ in Fig.6.1a shows indeed that the 128^2 simulation is not well resolved, in fact elongated and thin structures are contaminated by numerical wiggles. By increasing the resolution Fig.6.1b shows that elongated vortical regions, generated by the direct enstrophy cascade appear. Even by 256^2 points and in particular at this early time the resolution is marginal. The comparison between Fig.6.1b and Fig.6.1c, obtained by a 512^2 grid, indeed shows that this is true. Thinner vortices, with a higher enstrophy level, in fact appear.

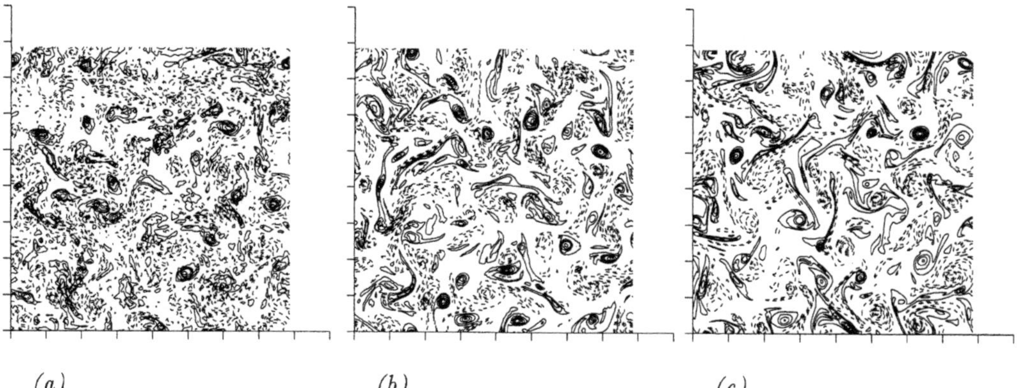

Fig. 6.1 Contour plots of vorticity for 2D turbulence $Re = 3000$ with $\Delta\omega = 2.5$. Initial condition $k^4 exp(-k/k_0)^2$ at $t = 5$; the resolution is: a) 128^2, b) 256^2, c) 512^2.

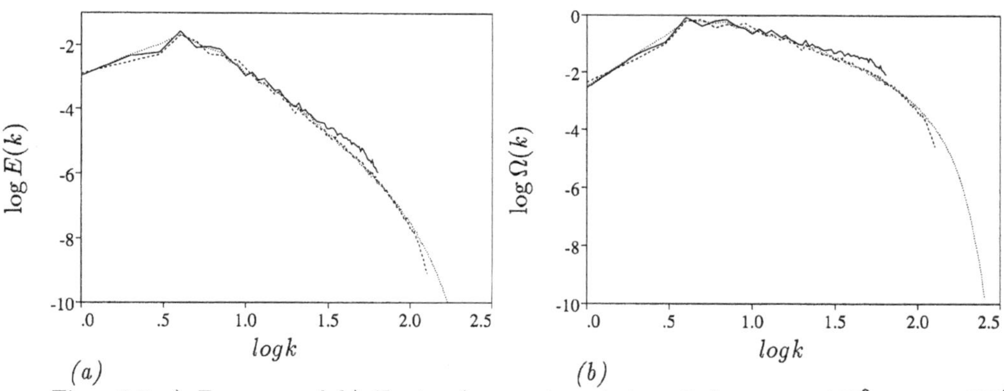

Fig. 6.2 a) Energy and b) Enstrophy spectra at $t = 5$ for —— 128^2, - - - - 256^2, ········ 512^2.

Energy and enstrophy spectra in Fig.6.2 describe these observations. The energy spectrum of the coarsest simulation does not present the exponential decay at small scales; in fact, the spectrum does not deviate appreciably from the inertial range and the small energy bump near the cut-off accounts for the energy associated with the unphysical wiggles due to the insufficient resolution. Since the numerical scheme conserves energy and enstrophy in this simulation there is no piling-up of energy at the highest k, as not dealiased spectral simulations often show. The enstrophy spectrum confirms these considerations. The 256^2 simulations show a marginal decay at high k but still the energy and enstrophy do not exponentially decay the necessary amount to consider the simulation fully resolved. The spectra for the 512^2 grid in Fig.6.2.a-b show that this grid permits to fully resolve the thin and elongate structures generated by the straining field of the large vortices. At this

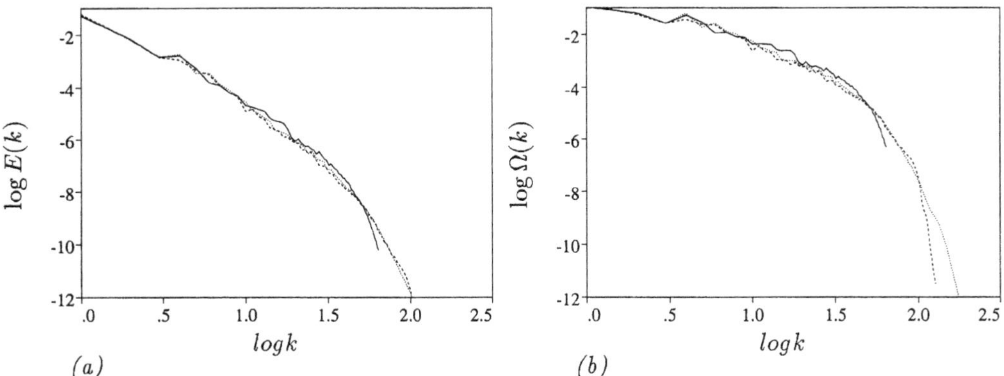

Fig. 6.3 a) Energy and b) Enstrophy spectra at $t = 50$ for —— 128^2, - - - - 256^2, ········ 512^2.

time the palenstrophy, defined as

$$P = \int\int [(\frac{\partial \omega}{\partial x_1})^2 + (\frac{\partial \omega}{\partial x_2})^2] dx_1 dx_2$$

that in the first part of the simulation was increasing, now begins to decrease. This is a clear manifestation that the growth of the spectrum, at high k, ended.

The features of 2D turbulence include viscous effects dissipating energy at small scales and merging between close vortices. The latter phenomenon produces the inverse energy cascade; at $t = 50$ Fig.6.3, see the growth of energy at low k, shows that several mergings occurred. At this time still there is a sufficient number of vortices with a range of scales generating a scaling law. For a better comprehension of this behavior I suggest the reading of the paper by Carnevale *et al.* (1991c). Fig.6.3.a and Fig.6.3.b shows that at $t = 50$ the energy and enstrophy spectra, in the three simulations, differ less; Fig.6.4 confirms that at small scales the vorticity field is better resolved.

From the energy spectrum and the vorticity contours it turns out that the low resolution calculation, in the first 5 units of time, generated unphysical small scales of sizes larger than those viscosity could dissipate. These unphysical scales went through several mergings with the final results: to produce larger scales and in a reduced number with respect to those remaining in the more refined simulations. At $t = 50$, Fig.6.4 shows that dipoles and monopoles are the more common vortical structures; the dipoles propagate inside the box facilitating the interaction with other vortices. Even if the simulations ended at this time it has been proved, by increasing the viscosity, that any simulation reaches the final condition with only one dipole (as shown by Montgomery *et al.* (1992,1993) by means of very long and accurate simulations at much higher Reynolds number than that in the present simulation).

6.5.2 Global conservation properties

Arakawa's scheme gives the opportunity to investigate whether the solution is affected by the global conservation properties. In the first period of the simulation the effects are

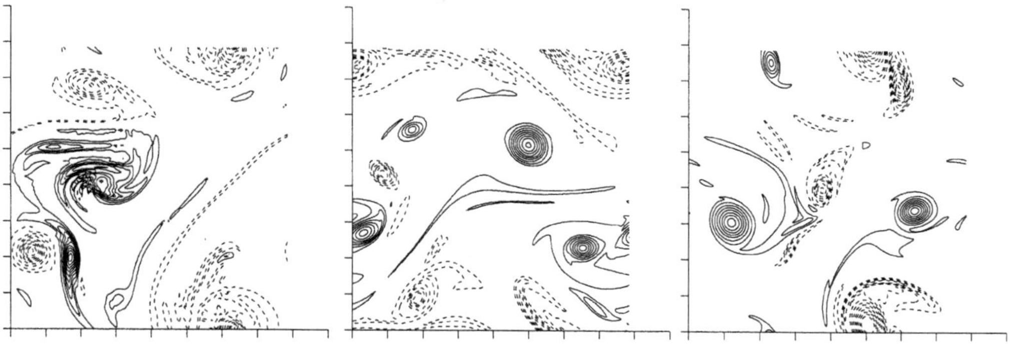

(a) (b) (c)

Fig. 6.4 Contour plots of vorticity for 2D turbulence $Re = 3000$ with $\Delta\omega = 0.5$. Initial condition $k^4 exp(-k/k_0)^2$ at $t = 5$; a) 128^2, b) 256^2, c) 512^2.

marginal, hence to find substantial differences it is necessary to look at the flow after a long time evolution. As shown flow visualizations could give insights, but since each realization is different, a better comprehension is obtained from the spectra. Fig.6.5 shows that when both quantities are conserved there is a major amount of energy at large scales and this is a clear evidence of the greater number of mergings during the evolution. The most important issue, from the spectra in Fig.6.5, is that it is mandatory to conserve the energy to have realistic simulations. In fact, when only the enstrophy is conserved, Fig.6.5 shows a large transfer of energy and enstrophy towards small scales. It can be concluded that some sort of aliasing errors produce the growth of unphysical small scales. Without energy conservation there is a preference for the energy to go from large to small scales. Vorticity contours at different instants have shown that this energy transfer to small scales occurs only at certain instants, as was observed in the time developing mixing layer (Chapt.4). These peaks of energy at small scales occur in regions where the strain field is large and in particular where two vortices of opposite sign get close to form a dipolar structure. This unphysical energy, however, is immediately dissipated and it takes some time before this process appears again.

6.6 Conclusions

This short chapter on 2D turbulence has been inserted to allow those who are interested in this ideal flow to investigate the properties of two-dimensional turbulence. The one-to-one comparison between results obtained by finite differences and pseudospectral methods were not performed because I did not have at my disposal a pseudospectral code. Even in this 2D case the rule of the thumb should be valid that a DNS by finite differences should require a number of computational points twice that of a pseudospectral code. The present scheme however has better CFL restrictions for the time advancement, hence in conclusion the two methods should use the same amount of CPU to get satisfactory results. However, concerning the resolution, a numerical experiment is the only way to draw solid

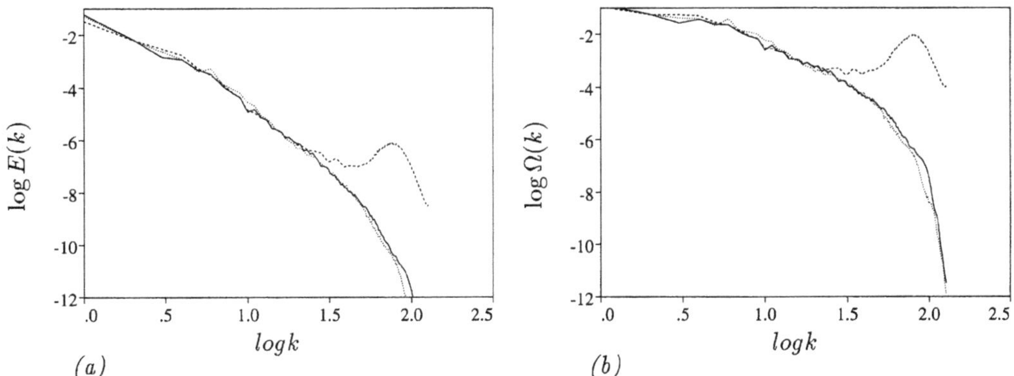

Fig. 6.5 a) Energy and b) Enstrophy spectra at $t = 50$ for 256^2 simulations; —— energy and enstrophy conserved, - - - - enstrophy conserved, ········ energy conserved.

conclusions; I spent the time to perform this check for 3D isotropic turbulence, a flow that, I think, is more important for the understanding of the complexity of turbulence.

I was never interested in the investigation of the properties of 2D turbulence; however several years ago Cavazza presented at the first European Fluid Mechanics conference in Cambridge a numerical simulation to investigate how long a coherent structure, as a dipole, survives when it is inserted in a turbulent background. This study was not completed and published because of the large number of parameters playing a role. The preliminary results can be found in the Ph.D. thesis of Cavazza (1992). The Italian bureaucracy was enforcing the students to write their thesis in my native language that is very pleasant to listen to but that instead does not have any utility in the scientific community. For example, Cavazza showed that the evolution depends on the ratio between the circulation of the dipole and that of the energy containing eddies of the turbulent field. She did not have the time to complete this study, and later on I lost interest in this flow. However, if in this simulations a tracer or even better a large number of particles are added, it could be a basic study to understand the transfer of pollutants in environmental problems. The concentration that, initially, could be supposed to be localized within the dipole, is transferred to the external fluid during the interaction between the dipole and the turbulent eddies. How much of this concentration is transferred and how it depends on the fluid properties diffusivity or viscosity is a fundamental problem that can be addressed and solved by the code described in this chapter.

7 Axisymmetric flows

7.1 Introduction

In this section the equation necessary to solve flows in the axisymmetric case are considered. Axisymmetric simulations are a first step towards flows with a greater complexity than in two dimensions. In fact, even if the field evolves in two dimensions ($r - z$ plane), vortex stretching plays a role. The vortex stretching term is, in fact, the cause of the difference between two- and three-dimensional flows. In the latter case the energy goes from large to small scales, while in the former case the energy goes from small to large scales, as has been shown in chapter 6. For sake of generality flows where the azimuthal velocity preserves the axisymmetry have been considered. This condition occurs in rotating flows as those producing vortex breakdown (Lopez (1990), Brown & Lopez (1990)) or flows with swirl (Moffatt (1988), Turkington (1989)). In these flows mean helicity is generated and this quantity, defined as $\mathbf{v} \bullet \boldsymbol{\omega}$, could play an important role.

In this chapter the equations in vorticity-streamfunction are given. Axisymmetric flows can also be simulated by the equations in primitive variables, hence to reduce the number of codes furnished by this book, the code in ω ψ formulation is not given. Those interested to have this code can contact me and I will be glad to furnish the code; however, this code can be rewritten with a small effort starting from the code described in chapter 4. This chapter is focused on the solutions of flows with inlet outlet conditions or impinging solid boundaries normal to the main propagation velocity. These flows can be simulated by the three-dimensional codes given in chapter 11 which efficiently can run in the axisymmetric conditions with and without solid body rotation or swirl. By setting the number of points in θ $N_1 = 2$ the code reduces substantially the CPU since the non-necessary terms and the v_θ equation, are not calculated when the rotation or swirl are neglected. The only price to pay is that, even if the dimensions of the matrices are set equal to 1 in the azimuthal direction, there is a waist of memory to store quantities related to the non-necessary variable. I would like to point out that the code described in chapter 11 requires minor modifications to deal with outlet radiative boundaries. Orlandi & Verzicco (1993a) used in the study of vortex rings impinging a solid wall, in the axisymmetric the vorticity-streamfunction formulation and checked that the results did not differ from those obtained by primitive variables.

7.2 Equations for the vorticity formulation

The vorticity components in the axisymmetric condition are

$$\Omega_\theta = \frac{1}{r}\frac{\partial Q_r}{\partial z} - \frac{\partial Q_z}{\partial r}, \Omega_r = -\frac{1}{r}\frac{\partial Q_\theta}{\partial z}, \Omega_z = \frac{1}{r}\frac{\partial r Q_\theta}{\partial r} \qquad 7.1$$

In these expressions the vorticities in capital letters coincide with the 3D components averaged in the azimuthal direction. Q_θ, Q_r and Q_z are quantities related to the velocity components in cylindrical coordinates by the relationships

$$Q_\theta = r v_\theta, \quad Q_r = r v_r, \quad Q_z = v_z \qquad 7.2$$

These quantities, as in 3D, are introduced for numerical reasons connected to the solution near the axis. The quantity Q_θ has a physical meaning, it is defined as the swirl, that appears in swirling flows or in flows with background solid body rotation. From Eq.(7.1) it turns out that only the azimuthal vorticity component is of interest; the other two components, related to the swirl, can be evaluated in a postprocess. The transport equation for Ω_θ is

$$\frac{\partial \Omega_\theta}{\partial t} + \frac{1}{r}\frac{\partial \Omega_\theta Q_r}{\partial r} + \frac{\Omega_\theta Q_\theta}{r^2}\frac{\partial \Omega_\theta Q_z}{\partial z} = \Omega_r \frac{\partial Q_\theta / r}{\partial r} + \frac{\Omega_z}{r}\frac{\partial Q_\theta}{\partial z} + \frac{\Omega_\theta Q_r}{r^2} - \frac{1}{\mathrm{Ror}}\frac{\partial Q_\theta}{\partial z} - 2\frac{\Omega_r Q_\theta}{r^2}$$
$$+ \frac{1}{\mathrm{Re}}\left[\left(\frac{\partial}{\partial r}\frac{1}{r}\frac{\partial}{\partial r} r\Omega_\theta\right) + \frac{\partial^2 \Omega_\theta}{\partial z^2}\right], \tag{7.3}$$

This equation can be simplified by rearranging some of the terms. In this manipulation, to eliminate the other vorticity components, part of the vortex stretching terms cancel out and the new equation becomes

$$\frac{\partial \Omega_\theta}{\partial t} + \frac{\partial (\Omega_\theta Q_r / r)}{\partial r} + \frac{\partial \Omega_\theta Q_z}{\partial z} = \frac{2}{r^3}\frac{Q_\theta \partial Q_\theta}{\partial z} - \frac{1}{\mathrm{Ror}}\frac{\partial Q_\theta}{\partial z}$$
$$+ \frac{1}{\mathrm{Re}}\left[\left(\frac{\partial}{\partial r}\frac{1}{r}\frac{\partial}{\partial r} r\Omega_\theta\right) + \frac{\partial^2 \Omega_\theta}{\partial z^2}\right], \tag{7.4}$$

As in two-dimensional flows a streamfunction, named the Stokes streamfunction, is introduced; it is related to the quantities Q_i through

$$Q_r = \frac{\partial \Psi}{\partial z}, Q_z = -\frac{1}{r}\frac{\partial \Psi}{\partial r} \tag{7.5}$$

By this assumption the transport equations for Ω_θ and Q_θ read:

$$\frac{\partial \Omega_\theta}{\partial t} + J(\frac{\Omega_\theta}{r}, \Psi) = \frac{2}{r^3}\frac{Q_\theta \partial Q_\theta}{\partial z} - \frac{1}{\mathrm{Ror}}\frac{\partial Q_\theta}{\partial z} + \frac{1}{\mathrm{Re}}\left[\left(\frac{\partial}{\partial r}\frac{1}{r}\frac{\partial}{\partial r} rQ_\theta\right) + \frac{\partial^2 Q_\theta}{\partial z^2}\right],$$
$$\frac{\partial Q_\theta}{\partial t} + \frac{1}{r}J(Q_\theta, \Psi) = \frac{1}{\mathrm{Ro}}Q_r + \frac{1}{\mathrm{Re}}\left[\frac{1}{r}\frac{\partial}{\partial r} r^3 \frac{\partial}{\partial r} Q_\theta / r^2 + \frac{\partial^2 Q_\theta}{\partial z^2}\right], \tag{7.6}$$

The Jacobian in Ω_θ is expressed by the quantity Ω_θ / r. At the axis, where Ω_θ and r are zero Ω_θ / r cannot be defined. In certain flows, for example for a vortex ring traveling along its axis, $\Omega_\theta \to 0$ more rapidly than $r \to 0$, then $\Omega_\theta / r = 0$ is a good approximation. However, in the most general case, this is not true thus it is more appropriate to express the Jacobian by Ω_θ. Since the Arakawa scheme is very efficient in two dimensions it is worth to maintain this scheme even for axisymmetric flows. To this purpose the following expression helps

$$J(\frac{\Omega_\theta}{r}, \Psi) = \frac{1}{r}J(\Omega_\theta, \Psi) + \Omega_\theta J(\frac{1}{r}, \Psi) = \frac{1}{r}J(\Omega_\theta, \Psi) - \frac{\Omega_\theta}{r^2}\frac{\partial \Psi}{\partial z} \tag{7.7}$$

The transport equation for a passive scalar σ can be added to visualize the evolution of the flow field, as in the experiments. The equation is

$$\frac{\partial \sigma}{\partial t} + \frac{1}{r}J(\sigma, \Psi) = +\frac{1}{\mathrm{ReSc}}\left[\frac{1}{r}\frac{\partial}{\partial r} r \frac{\partial}{\partial r}\sigma + \frac{\partial^2 \sigma}{\partial z^2}\right] \tag{7.8}$$

where Sc is the Schmidt number.

The solution of these equations can be obtained by following a procedure similar to that used in the solution of the vorticity-streamfunction formulation, that is by using the Arakawa scheme for the nonlinear terms and the third order low storage Runge Kutta scheme for the time advancement. In order to reduce the amount of codes included in this book I decided to describe the results, related to axisymmetric flows, that were obtained by a code in primitive variables. The comparison between the results of the two different formulations has been performed and they agree very well. In the following section three cases for which experimental results are available are described.

7.3 Results

7.3a Vortex ring formation

The aim of this simulation deals with the reproduction of one of the most fascinating and studied cases in fluid dynamics, the formation of a vortex ring, which in the past was realized in laboratory experiments (Didden (1979), Glazer (1988), Auerbach (1991), Gharib *et al.* (1998)). Only recently numerical simulations were performed; in one case (Nitsche & Krasny (1994)) the vortex dynamics was reproduced by a vortex-sheet method, where a thin vorticity layer injected from an orifice rolls up and forms a vorticity patch resembling the dye pictures taken in a laboratory (Van Dyke (1982) picture 77). The importance of the Nitsche & Krasny (1994) study resides on the almost inviscid simulation, an important issue on the comprehension of vortex dynamics.

The Navier-Stokes equations on the other hand permit to perform simulations in one-to-one correspondence with the laboratory experiments (Verzicco & Orlandi (1995)). In this case it is very simple to simulate the evolution of a vortex ring exiting from a nozzle surrounded by a solid wall. In this set-up the boundary conditions are well defined, that is, at the inlet there is a velocity profile only in the region of the nozzle and no-slip conditions on the side. The importance of the no-slip condition on the side of the nozzle was investigated numerically by Verzicco & Orlandi (1995). A simulation close to the experimental set-up requires the reproduction of the nozzle; this has been recently carried out by Fadlun (1998). Since the interest is to reproduce the formation for a short time the assumption of a free-slip wall on the lateral boundary is satisfactory if it is located at a distance approximately three radii from the axis. The axis treatment is not explained here; since the discretization of the axis is critical for three-dimensional simulations, it is discussed in the chapters 10 and 11.

The formation and the evolution of a vortex ring can be considered as a 'space developing flow', that is a flow with an inhomogeneous streamwise direction. For such flows, inflow and outflow conditions in the axial direction are requested. At the inlet ($z = 0$) for $0 \leq r \leq R$ an axial velocity profile $U_z(r)$ modulated by a time dependence $f(t)$ is prescribed. The time dependence reproduces the injection of fluid through the nozzle of the experimental apparatus. $U_z(r)$ is specified by

$$U_z(r) = U\frac{\tanh[(R-r)/\delta]}{\tanh[R/\delta]} \qquad 7.9$$

where δ is the initial vorticity thickness of the inflow velocity profile and $\delta = R/20$ was assumed. The exact definition of the vorticity thickness is

$$\delta_{\omega_\theta} \equiv [U_z/(\partial U_z/\partial r)]_{\max} = \delta \tanh[R/\delta] \qquad 7.10$$

for values of $\delta \ll R$; however, $\tanh[\delta^{-1}R] \simeq 1$, so that $\delta_{\omega_\theta} \simeq \delta$. $f(t) = 0$ for $t = 0$ and it reaches the unity by a cubic law in a dimensionless time $t = 0.1$. Later on it remains constant and equal to 1 for a time $t = T_i$ and finally it returns to zero in a time $t = 0.1$. Thus the total injection time is $\mathcal{T} = (T_i + 0.2)$ that corresponds to a dimensional time (t') of $\mathcal{T}r/Usec$, according to the dimensional values of R and U in the experiments. For $z = 0$ and $R < r \leq R_f$ all velocity components are set to zero. Identical inflow conditions have been used for the passive scalar concentration σ to simulate the flow visualizations in which dyed fluid is injected from the nozzle.

Boundary conditions at the outlet are important in space developing flows. In these simulations, it has been assumed that the velocity components and the passive scalar are convected out of the domain according to the relation

$$\frac{\partial Q}{\partial t} + C\frac{\partial Q}{\partial z} = 0, \qquad 7.11$$

Q is the convected quantity and C the convection velocity. In this simulation the value $C = 0.6$ has been fixed, although it has been checked that the results are quite insensitive to the value of C. The same result was obtained, in a different flow, by Pauley *et al.* (1992).

The simulations have been performed in an axisymmetric domain of $R_f = 3R$ and $L_f = 4R$ in the radial (r) and axial (z) directions, respectively. The spatial discretization was uniform in z with $\Delta z = 0.031R$ while a non–uniform grid was used in r to achieve a fine resolution close to the edge of the orifice, where the thin vorticity layer rolls up. A coarser grid was used near the external boundary to minimize the number of grid points in this dynamically passive region. The analytical transformation for the radial coordinate is

$$r = R_f r_1(\eta) r_2(\eta), \quad \text{with}$$

$$r_1(\eta) = R_m/R_f \tanh(\beta\eta)\tanh(\beta\eta_m)$$

$$r_2(\eta) = \frac{1}{r_1(1)} + (1 - \frac{1}{r_1(1)})\tanh(\alpha(\eta-1))\tanh(\alpha(\eta_m - 1)) \qquad 7.12$$

where η is the radial computational variable uniformly spaced between 0 and 1. This is the coordinate transformation given in section 2.1.3f. I would like to recall that by the code in chapter 11 the simulation can be repeated by clustering the grid points in z near the nozzle exit. I did not perform this exercise which I leave to the reader.

By the numerical simulation it is possible to understand the difference between vorticity and a passive scalar. In the laboratory this quantitative comparison is more difficult because, even with the new measurements techniques as PIV, it is not possible to trust completely the vorticity measurements. However, I heard from Koochesfahani (Personal communication) that a new experimental technique permits to get vorticity measurements

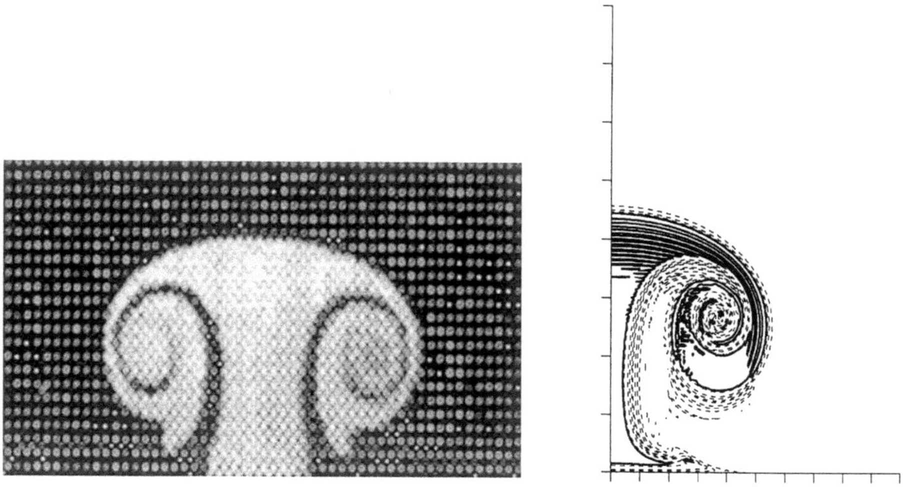

Fig. 7.1 Dye visualization of a vortex ring formation at $Re = 1000$ and contours at Sc = 100 of a passive scalar from a numerical simulation.

in very good agreement with those in the numerical simulations. It is well known that large differences between vorticity and tracers occur in three-dimensional flows where vortex stretching plays an important role. Often it has been noticed that tracers are depleted in regions of intense vorticity (Hussain & Melander, (1991)). In two-dimensional and in axisymmetric flows the differences between tracers and vorticity are related to the different values of diffusivity and kinematic viscosity, that is, to the Schmidt number Sc $= k/\nu$. Axisymmetric simulations permit very fine grids; in addition, since the ring is forming in a short time the simulations permit to investigate the effects of the Schmidt number. The results here presented are taken from the paper by Verzicco & Orlandi (1995) about the mixedness properties at different Sc numbers during the ring formation. In this book I am not interested in describing the mixing process, a very important issue in combustion related problems, but, instead, in investigating the differences between the vorticity field and the laboratory images by flow visualizations. This analysis was done by performing only one simulation with three scalars, each one with a different Peclet number $Pe =$ ReSc. The maximum Sc was 100 close to the value of 1000 of the fluorescine. I would like to point out that, in axisymmetric flows, large variations between vorticity and a passive scalar occur when Sc goes from 1 to 10 and less at higher values.

Fig. 7.1 shows the comparison between a laboratory experiment and the contours of passive scalars at a value of $Pe = 10^5$ close to the orifice where the ring forms. The photo has been taken just after the piston stopped, and it must be remembered that the ring continues its formation later on. The experimental picture has been taken from the paper by Verzicco *et al.* (1996). It turns out that the simulation reproduces very well the dye pattern including also the number of spirals. The dark regions are a plotting artifact related to the value of $\sigma \approx 1$ in the inner region of the ring, that is not diffused at this

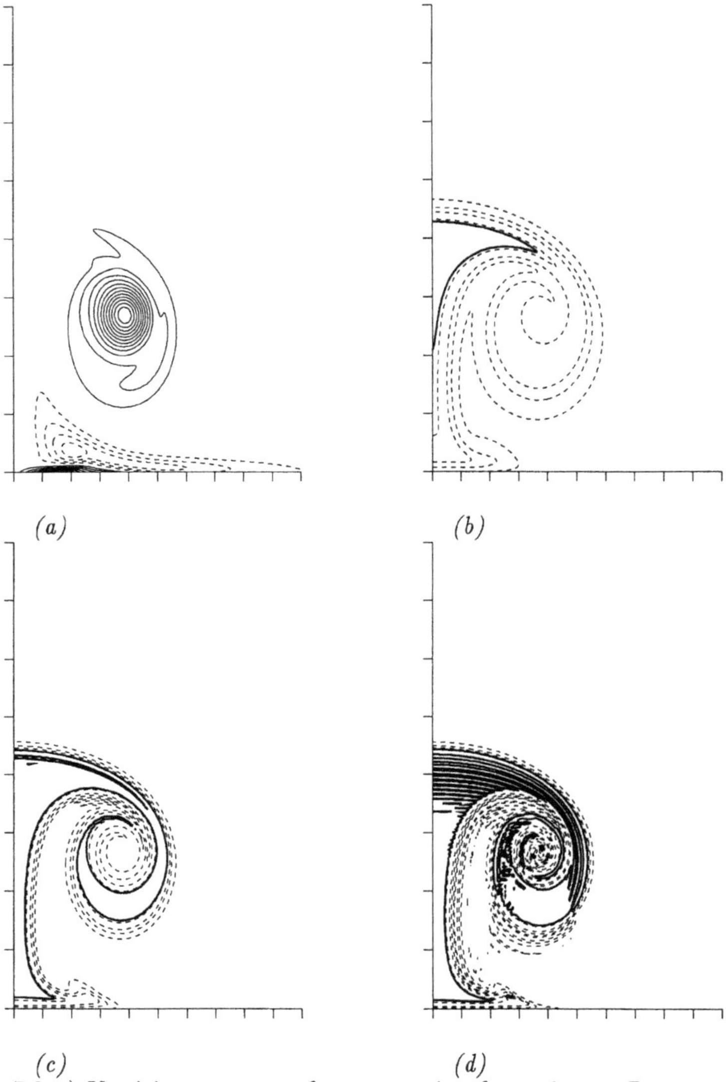

Fig. 7.2 a) Vorticity contours of a vortex ring formation at Re = 1000 compared with passive scalar contours at b) Sc = 1, c) Sc = 10 and d) Sc = 100 from a numerical simulation. a) Vorticity, b-d) Passive scalar.

high value of the Schmidt number.

Fig. 7.2 shows contour plots of vorticity and passive scalar at $Pe = 10^3, 10^4$ and 10^5. It turns out that at this Reynolds number (Re $= 10^3$), the vorticity diffuses more than the passive scalars and that the former does not show any spiraling motion.

The axisymmetric vortex ring, in the inviscid case, reaches its equilibrium solution when the Jacobian $J(\frac{\Omega_\theta}{r}, \Psi)$ is null; at this stage the vorticity field reaches a stationary solution in the reference frame travelling with the vortex ring. To have a null Jacobian there should be a functional relationship between $\frac{\Omega_\theta}{r}$ and the Stokes streamfunction $\tilde{\Psi}$ in the travelling frame. This consideration permits to understand why in the experiments and in the simulations behind the ring a wake forms. To see this phenomenon the reader should perform a simulation without reproducing the piston motion, but by inserting at $t = 0$ at a certain distance from the wall a Gaussian vorticity distribution. This profile is not equilibrated, and in a short while, when the ring begins to propel, a weak wake forms behind. The vorticity in the wake comes from that in the outer layer and mainly from that near the axis. In the inviscid case the tail is very long and the simulation necessitates a long domain, on the other hand in a viscous simulation the low vorticity level in the wake is dissipated and in a short time a single-value functional relationship is achieved. In several simulations and in particular in 2D it has been observed that the viscosity does not have any effect on the shape of the functional relationship. In the axisymmetric case it produces a reduction of the maximum values of $\frac{\Omega_\theta}{r}$ and of $\tilde{\Psi}$. With a good approximation, near the maximum there is a linear relationship that does not hold everywhere as in two-dimensional simulations (Lamb dipole). The study of the functional relationship for vortex rings was performed by Orlandi & Verzicco (1993a); they also investigated the difference between a passive scalar and vorticity. In that study it was shown that the scatter plots between a passive scalar and $\frac{\Omega_\theta}{r}$, at Sc = 1 give a linear relationship. By increasing the Schmidt number the difference between the meridional flow $\frac{\Omega_\theta}{r}$ and the passive scalar increases.

In Fig. 7.3 the laboratory visualization (Fig. 7.3a) and the numerical simulation (Fig. 7.3b) at Sc = 100 are compared at $t = 12$ when the ring has moved approximately 3 radii from the orifice. In this short time after the formation the ring has not had sufficient time to reach an equilibrated solution. The analysis of the functional relationship at $t = 16$ permits to understand why the ring did reach the equilibrated solution. Fig. 7.3c shows by the scatter plot between $\tilde{\Psi}$ and $\frac{\Omega_\theta}{r}$ that even if the spreading diminished with respect to earlier times still the distribution is far from an equilibrated solution. Fig. 7.3d shows that the scatter plot between $\frac{\Omega_\theta}{r}$ and σ at Sc = 1 is rather good; on the contrary $\frac{\Omega_\theta}{r}$ does not correlate at all with the passive scalar distribution at higher Sc (Fig. 7.3e and Fig. 7.3f). From these results it can be concluded, that even in axisymmetric flows quantitative information on the vorticity field is very questionable in water; on the other hand, in air (Sc $\approx$ 1) smoke visualizations give a satisfactory description of the vorticity field.

The vortex ring, after reaching the equilibrium solution, in the inviscid case, propels indefinitely in the axial direction; the viscosity reduces the translation velocity. In reality the ring does not remain axisymmetric; in fact in a laboratory at a certain value of Re azimuthal disturbances grow. The growth of these perturbation observed in experiments (see the pictures in the Van Dyke book at Pg.66-67), was analytically predicted by Widnall *et al.* (1974) for rings with small cores, and was reproduced by direct simulations for rings of thick core by Shariff *et al.* (1994). The simulation was performed by solving the 3D equations with the code described in chapter 10.

Even in the axisymmetric case, the motion of a vortex ring is largely affected by the application of external forces. The forces easily reproducible in a laboratory are those due

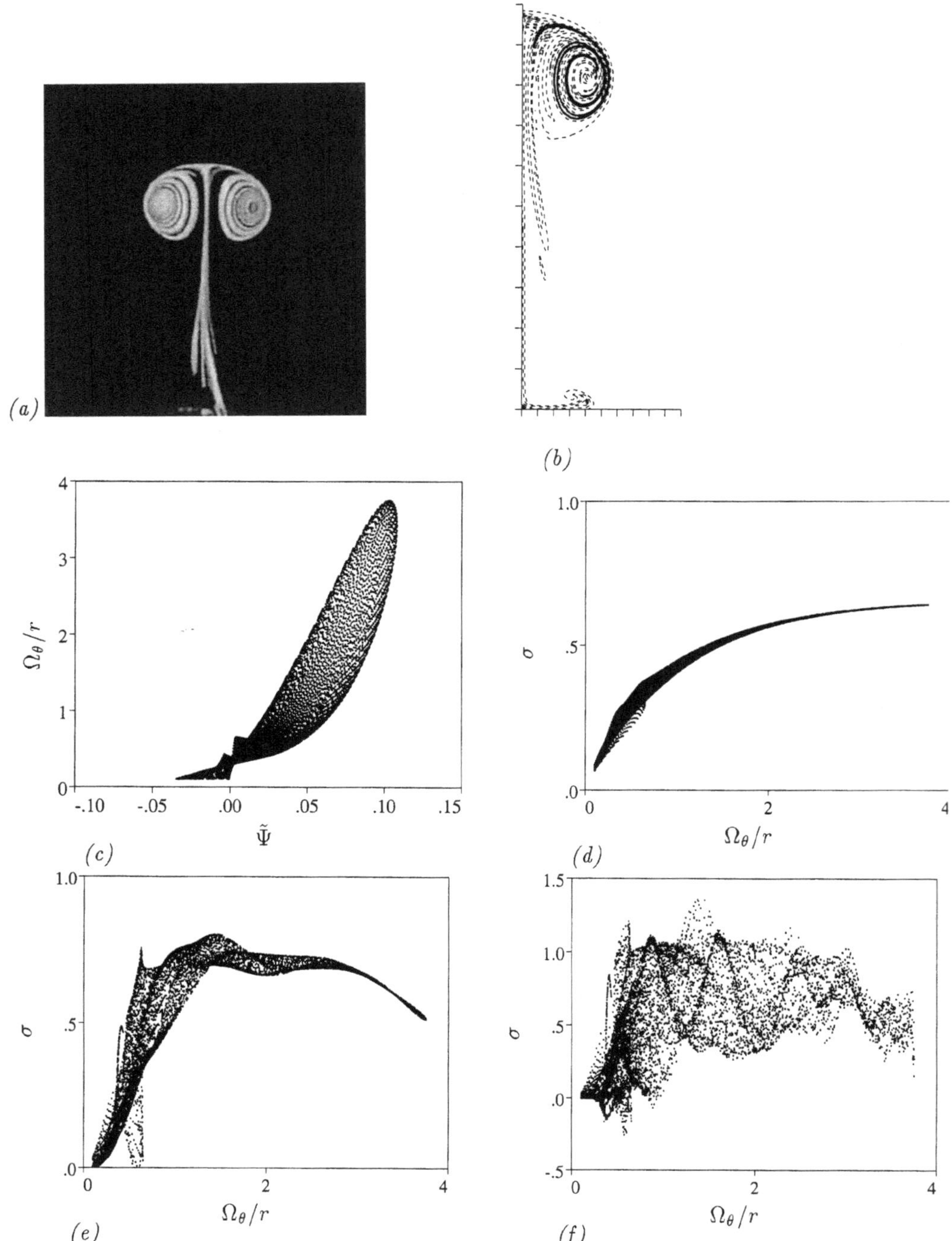

Fig. 7.3 Passive scalar distribution a) experiment, b) simulation at Sc = 100, c) scatter plots between Ω_θ/r and $\tilde{\Psi}$, scatter plots between σ and Ω_θ/r at d) Sc = 1, e) Sc = 10, f) Sc = 100.

to background rotation or to buoyancy. The buoyancy on the vortex ring can increase or decrease the acceleration of the ring. The former condition is achieved by an unstable and the latter by a stable stratification; the latter has a greater practical interest. The stratification gradient can be applied in the direction of the axial motion of the ring or in a direction orthogonal to it. Recently, a study has been devoted to the effects of the stratification on coherent vorticity patches (Sarpkaya (1983), Spalart (1996), Orlandi *et al.* (1998)). These effects can be studied by the set of codes given in this book. These simulations, for the sake of brevity, are not considered here; on the contrary effects of the background rotation are here described. Even for flows with background rotation the external rotation can be directed in a direction parallel to the ring translation or in a direction normal to it. The latter condition was investigated by Eisenga *et al.* (1998) by a combined experimental and numerical analysis. The 3D simulation was performed in Cartesian coordinates with a code similar to that described in chapter 9. The former case was performed by Verzicco *et al.* (1996) and the results for the axisymmetric restriction are reported in the next section.

7.3b Vortex ring in the presence of solid body rotation

From the fundamental point of view the interest of this flow is motivated by the desire to understand vortex dynamics of Beltrami flows, that is flows with mean helicity production. Moffatt (1988) found a class of steady state solutions with a distribution of azimuthal velocity and vorticity. The study of similar kinds of flows was continued by Turkington (1989) but neither investigated whether or not vortex rings with a swirl are stable. The experimental generation of a vortex ring with a swirl is rather difficult because the nozzle should be appropriately designed to introduce the azimuthal velocity distribution. To my knowledge nobody attempted to build such a device, mainly because of the difficulty to eliminate any disturbance. A simpler way to investigate the effects of swirl is by generating the ring on a rotating environment; in this way the initial and boundary conditions are well defined and the flow parameters can be easily assigned. In the laboratory it is necessary to have a rotating table to locate the tank where the ring moves. Vortex rings in a rotating environment have also the further quality to be close to flows of real applications. For example in environmental applications there is the interest to estimate the changes of the penetration depth of a round jet as a result of the Earth rotation (Etling & Fernando (1994)).

In the numerical simulations the equations in a rotating frame are solved. These equations in primitive variables are given in chapter 10; in Eq.10.1-4, the rotation number N appears. In the equation for the azimuthal vorticity instead the Rossby number is used. These numbers are linked each other by $N = 1/\mathrm{Ro}$ with Ro the Rossby number defined as $\mathrm{Ro} = U/2\Omega L$ with Ω the rotation rate of the background motion and U the reference velocity of the flow. In the vortex ring U is the maximum exit velocity at the nozzle. In applications related to geophysics usually the Rossby number is used and it tends to infinity for Ω tending to zero. In engineering applications the rotation number $N = 2\Omega/L$ is preferred.

Verzicco *et al.* (1996) presented experiments and simulations with and without background rotation. The experimental set-up for the ring generation is described in their paper; the velocity of the injected fluid was controlled by the stepmotor-driven traversing

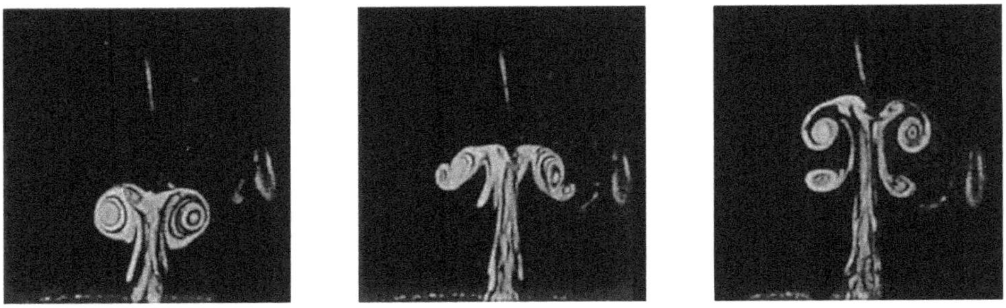

a) b) c)

Fig. 7.4 Dye–visualizations of a vortex ring evolution at Ro = 4.8 and Re $\simeq$ 1000: *(a)* $t = 7$, *(b)* $t = 10$, *(c)* $t = 13$.

system which pushed the piston of the syringes. A cylindrical column of fluid was injected through the orifice of length $L_o = 2.4$ cm and mean velocity $U = 2.4$ cm/s. The value of Re was thus fixed at Re $= UR/\nu = 484$, with $\nu = 0.01\text{cm}^2/\text{s}$. Fig. 7.1a and Fig. 7.3a are taken from Fig. 7.3 in Verzicco *et al.* (1996;, the whole sequence was showing that, in absence of background rotation, the vortex 'atmosphere' is a closed region of fluid steadily propagating. The motion of the vortex ring was laminar in all experiments.

If a background rotation is added many additional effects arise, making the dynamics of the flow more complex. In Fig. 7.4 pictures at successive times for a vortex ring in a rotating fluid (Ro = 4.8) are shown. Initially, the vortex ring is created in the same way as that without rotation. However, after the formation of the vortex ring is almost complete, a small depression is observed ahead of the vortex (Fig. 7.4a). From this region a thin layer of dye rolls up into a secondary vortical structure with circulation of opposite sign. This secondary vortex is advected by the primary vortex, increasing in strength during stretching around the vortical core of the primary ring (Fig. 7.4b). Later on, the secondary structure is shed from the primary vortex while its localized vorticity diffuses due to viscosity (Fig. 7.4c).

The impulse of the primary ring is reduced by the shedding of secondary structures and this is evidenced by the shorter distance travelled by the ring at $t = 12$ (Fig. 7.4c) with respect to the case without rotation. These conjectures on the vorticity field have been inferred by flow visualizations, that show the evolution of a passive scalar; the real behavior may present a different behavior.

Laboratory experiments and numerical simulations have shown that the flow dynamics can be roughly divided into two regimes: 'low rotation' (Ro $\gtrsim$ 3) and 'high rotation'. In the former regime, the rotation constitutes only a perturbation on the basic state, while in the latter the rotation dominates the flow and the vortex ring hardly evolves or even forms. The results of the low rotation regime are summarized here, while the second regime is discussed in Verzicco *et al.* (1996).

In Fig 7.5 the time evolution of the axial coordinate of the center of the core is shown

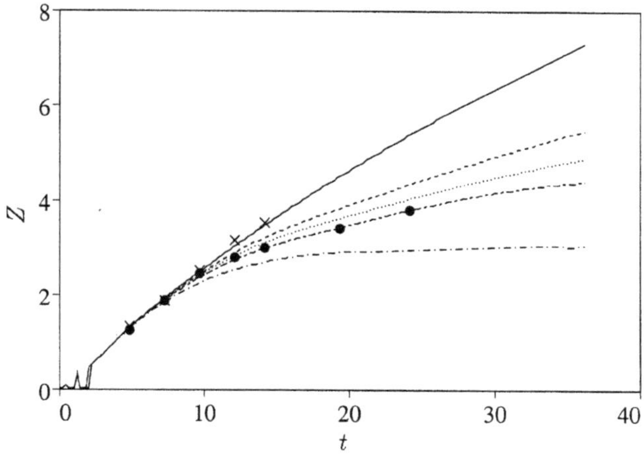

Fig. 7.5 Time evolution of the axial position of the core center of the ring (Z) for Re = 484 and —— Ro = ∞, - - - - Ro = 10, ········ Ro = 7.4, — — — Ro = 5.5, —·— Ro = 3. Experimental results: $\times$ Ro = ∞, $\bullet$ Ro = 5.5.

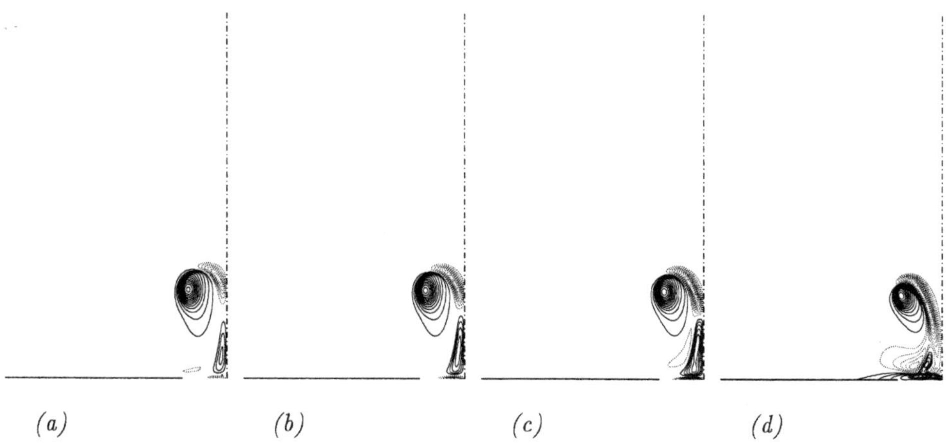

Fig. 7.6 Contour plots of azimuthal vorticity at t = 7.26, Re = 484 and *(a)* Ro = 10.0, *(b)* Ro = 7.4, *(c)* Ro = 5.5, and *(d)* Ro = 3.0, vorticity increments $\Delta\omega = \pm 0.25$, —— for positive, - - - - for negative values.

for several values of Ro; for comparison the trajectory of the ring without rotation is given as well. Although data are presented only for Re = 484 the same behavior has been observed for different Reynolds numbers that, even if changed by a factor of three, produced no changes in the trajectories comparable with those induced by rotation effects.

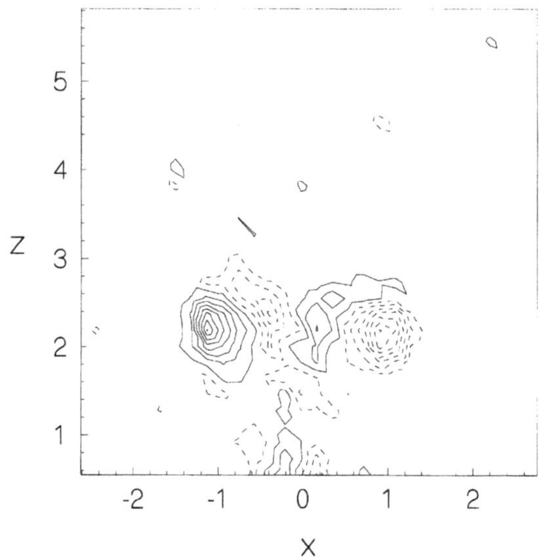

Fig. 7.7 Azimuthal vorticity of a vortex ring at Ro = 3.5 and Re = 484 and $t = 6$. Vorticity increments $\Delta\omega = 0.5$, —— for positive and ········ for negative values.

It can be noticed that only a small part of the velocity decrease is due to the viscosity, while a larger effect is induced by the rotation. The graph also contains data plots of experimental trajectories obtained for two different cases with parameter values as in two of the numerical simulations. The agreement is very good, confirming that the numerical set–up is well suited to mimic the laboratory experiments.

Fig. 7.6 presents some azimuthal vorticity contour plots for the cases shown in Fig. 7.5. It appears that as the Rossby number decreases, the oppositely–signed structure, inferred from flow visualizations, strengthens, and thus causes a deceleration of the primary ring. The same phenomena have been observed by particle tracking experiments, and these were shown by Verzicco *et al.* (1996, Fig. 7.8). Fig. 7.7 shows the contours of the azimuthal vorticity obtained in the laboratory, and by comparing this figure with those in Fig. 7.6c-d it turns out that although some low–level noise is shown, the secondary structure ahead of the primary ring is clearly discernible. Also note in Fig. 7.7 the vorticity in the tail, which was absent in the flow without background rotation.

A simple explanation for the appearance of these additional vorticity structures is obtained by inspection of the azimuthal vorticity transport equation. In Eq.(7.6) the second term on the right-hand side represents the effect of the system rotation. Axial gradients of

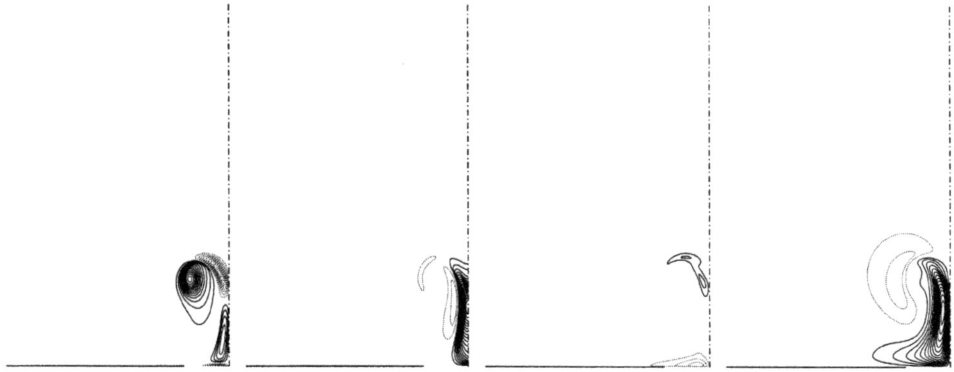

(a) (b) (c) (d)

Fig. 7.8 Contour plots of ω_θ *(a)*, ω_z *(b)*, ω_r *(c)* and u_θ *(d)*, at $t = 7.26$ for Re = 484 and Ro = 7.4. Vorticity increments $\Delta\omega = \pm 0.25$, velocity increments $\Delta v = \pm 0.02$, —— for positive, ---- for negative values.

the azimuthal velocity constitute a source for azimuthal vorticity. This effect is illustrated by Fig. 7.8d which shows a map of azimuthal velocity for a vortex ring at Ro = 7.4. Evidence of the strong positive swirling motion is also given in Fig. 7.9b, which shows an intense axial vortex on the axis of symmetry. The presence of strong positive swirl below the rising vortex structure was clearly observed in the experiments as well. The negative swirl, in contrast, is less evident and Fig. 7.8d shows that it is much weaker.

If one moves along a line parallel to the axis and passes through the maximum of u_θ ($r \simeq 0.25$ in Fig. 7.8d), one first encounters a positive axial gradient of azimuthal velocity ($\partial u_\theta / \partial z > 0$), whereas further upwards it becomes negative. According to Eq.(7.6), this implies a region of positive azimuthal vorticity generation close to the axis and in the rear part of the ring, while in the front there is a region of negative azimuthal vorticity generation, consistent with the flow map of Fig. 7.8a. For completeness in Fig. 7.8c also the radial vorticity component is reported which remains always small in magnitude and does not contribute significantly to the flow dynamics.

An alternative way of explaining the deceleration of the ring is obtained by considering the behavior of the pressure due to the presence of an anticyclone ($\omega < 0$) in front of the ring and a cyclone ($\omega > 0$) at its rear. Since pressure is lower in cyclones than in anticyclones; the ring during the motion has to counteract the effect of an adverse pressure gradient which exerts on the ring a force opposite to the direction of translation. That the pressure effectively behaves as described above has been verified by looking at its distribution for a vortex ring both in absence of background rotation (Ro = ∞) and for the case at Ro = 5.5 (Fig. 7.9a,b). The corresponding passive scalar distributions are shown in Fig. 7.9c,d. In the former case the front of the ring is concave according to the classical picture given in many studies; in contrast, in the case with rotation the head of the ring shows the depression, already shown in flow visualizations, that is due to the

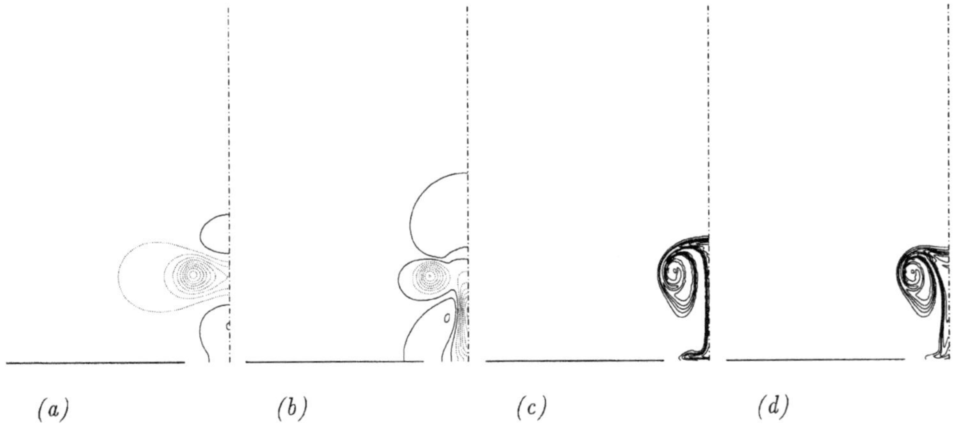

Fig. 7.9 Contour plots of pressure ((*a*),(*b*)) and passive scalar ((*c*),(*d*)) at $t = 7.26$, $\mathrm{Re} = 484$ and $\mathrm{Sc} = 10$: *(left)* $\mathrm{Ro} = \infty$, *(right)* $\mathrm{Ro} = 5.5$. $\Delta T = 0.2$, $\Delta p = 0.015$ ——— for positive, ········ for negative values.

pressure being higher in the front than in the rear part.

All cases presented in this section have been computed using a fixed Reynolds number ($\mathrm{Re} = 484$) since this was the value at which most of the experiments were performed. However, viscous effects play a large role in the flow dynamics and therefore it is important also to examine how the flow changes for increasing Reynolds numbers. Verzicco *et al.* (1996) presented results at $\mathrm{Re} = 1500$ and $\mathrm{Ro} = 10$.

7.3c Vortex rings impacting free-slip walls

A further case that shows the modification of the structures of a vortex ring is the impact with a non-deformable free surface or with a solid wall. The interaction is very different in the two cases and these flows are basic for the comprehension of several practical applications, which goes from ship hydrodynamics (Sarpkaya (1993)) to turbulent drag (Orlandi & Verzicco (1993b)). The easier case from the computational point of view is a ring impinging a free-slip wall. Lugt (1996) in his book reported this example even in the case that the boundary deforms due to the ring (Bernal *et al.* (1989), Song *et al.* (1991), Song *et al.* (1992), Lugt & Ohring (1994))

This case has been considered for a close comparison with the results of spectral calculations (Stanaway *et al.* (1988)) and for a qualitative comparison with flow visualizations by Bernal *et al.* (1989), who let rings to interact with free surfaces with different amounts of surfactants. In pure water, the interaction is similar to the centered collision of two rings of the same strength. While an exact alignment between the two rings is difficult in the experiment, the numerical simulation is simplified by the imposition of symmetry at the collision plane. However Lim *et al.* (1991) produced an interesting movie of the collision of two vortex rings of different colors. The effect of the interaction was the generation of a large number of smaller rings traveling far from the axis. The number of rings is

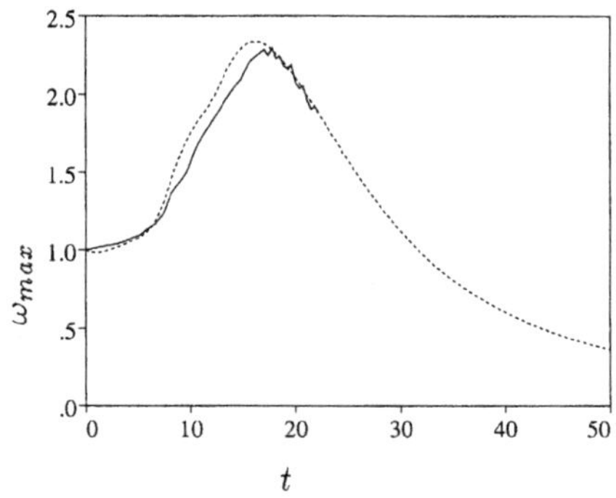

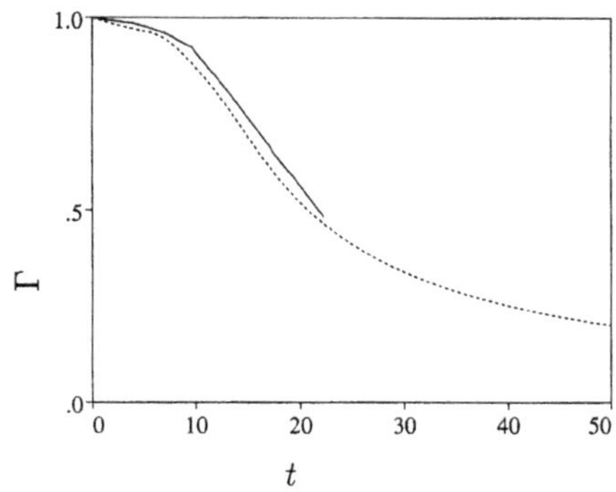

Fig. 7.10 Time evolution of peak vorticity and circulation: —— Stanaway *et al.* (1988); ...present.

related to the azimuthal instabilities of the primary structure, but in the early stage of the collision the interaction by a good approximation is axisymmetric. Stanaway *et al.* (1988) considered two different rings: one is a thin-core ring with Gaussian distribution, and the other is a Hill vortex with a smoothing to overcome the abrupt jump between vortical and irrotational regions. The initial vorticity distribution is

$$\omega = \frac{3}{2r} \sin\theta \begin{cases} 1 & r \leq r_i \\ 1 - f(\xi) & r_i \leq r < r_e \\ 0 & r_e < r \end{cases} \qquad 7.13$$

where $f(\xi)$ is the smoothing function $[f(\xi) = 0$ for $r = r_i$ and $f(\xi) = 1$ for $r = r_e]$, that connects the inner to the outer values of the vorticity. Stanaway *et al.* (1988) considered collisions at low and at high Re. At low Re ($\mathrm{Re}_\Gamma = 300$) viscous dissipation overcomes the stretching effects; this, therefore, is not a satisfactory condition to test a numerical scheme designed to describe the evolution of the thin vorticity layers generated during the impact of a vortex with no-slip walls. At $\mathrm{Re}_\Gamma = 1073$, on the other hand, vortex stretching produces a significant increase in the vorticity level, after the collision. Fig. 7.10a-b show the comparison between the present peak vorticity and circulation time history and the spectral results by Stanaway *et al.* (1988). The simulations, after the collision, show an increase in the peak vorticity. Vortex stretching causes this increase since, when the vortex comes close to its image vortex, its radius increases, and, to conserve $\hat{\omega} = \omega_\theta / r$, ω_θ must increase.

Two-dimensional calculations at high Reynolds numbers, performed with the code in chapter 4, do not show any substantial modification in the shape of the Lamb dipole after the wall collision. On the contrary the axisymmetric simulation shows that the vortex has a head and an elongated tail (Fig. 7.11). The tail elongates and the head becomes smaller and, when the size of the head becomes sufficiently small, the peak vorticity is rapidly dissipated (Fig. 7.10a). The evolution of the circulation (Fig. 7.10b) during the approach of the vortex to the wall reveals that the diffusion of vorticity is negligible. After

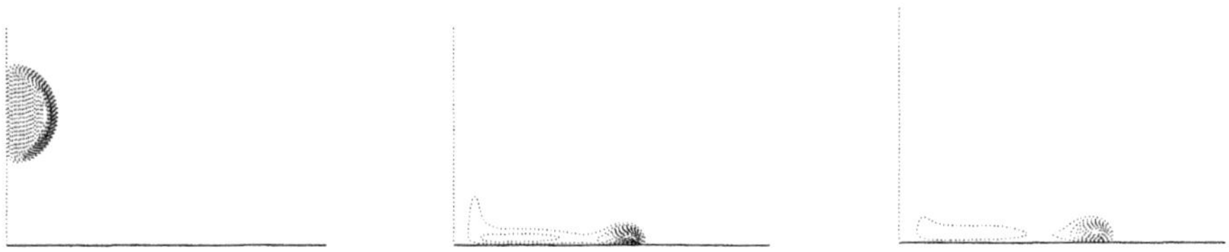

Fig. 7.11 Vorticity contour plots of a Hill vortex colliding with a free-slip wall at $\mathrm{Re}_\Gamma =$ 1073: *(a)* $t = 0$, *(b)* $t = 40$, *(c)* $t = 50$. Contour increment is ± 0.07.

the collision, on the other hand, the gradients at the wall increase (Fig. 7.11) and the vorticity diffuses appreciably. During the formation of the head-tail structure (Fig. 7.11), the diffusion reaches a maximum and maintains this level for a short time. Later on ($t > 20$) the gradients at the wall decrease and the circulation diminishes at a lower rate. The tail is dissipated faster than the head since the vorticity intensity in the tail is smaller than that in the head.

The simulations were carried out for a longer time than that of Stanaway *et al.* (1988), to investigate whether the small oscillations of the spectral calculation were physical or not. Simulations by grids up to 256×256 never showed any oscillations in time, and, therefore, we suppose that the oscillations in the spectral simulation have a numerical origin related to a lack of grid resolution. The authors, in fact, asserted that oscillations occurred in the tail region and that these were related to the small number of modes used (64×64). The authors furthermore pointed out that more accurate simulations require a memory larger than the in-core memory of a CRAY XMP48. In the present simulation, on the other hand, grid refinements are affordable since a 128×128 grid, which has twice the resolution of the spectral calculation, requires only 0.5 megawords of memory.

To confirm the accuracy of the present method, Figs. 7.12b-c show that the shape of the vortex, in the first period after the collision, resembles very close to visualizations by Bernal *et al.* (1989). In their experiment, however, at a later stage a secondary vortex ring was created far from the symmetry line. The formation of this secondary ring is a consequence of the surfactants present even in clean water, which change the surface tension and create interactions similar to those with a no-slip wall described in the following paragraph. The lack of detailed measurements by Bernal *et al.* (1989) does not allow a quantitative comparison with the present results.

A vortex ring impinging on a free-slip surface can be used to study the evolution of the functional relationship $\Omega_\theta / r = f(\Psi)$ in a reference frame translating with the ring. A single-valued relationship is the steady solution of the inviscid vorticity equation and $\Omega_\theta / r = const$, for $r < a$, is the simplest relationship possible, which is useful for theoretical considerations but is not suitable for numerical simulations. The smoothing function $f(\xi)$ does not give a single-valued functional relationship $\Omega_\theta / r = f(\Psi)$, as shown in Fig. 7.12a. The calculation, at a very high Reynolds number ($\mathrm{Re} = 10^4$), reduces the effect of diffusion;

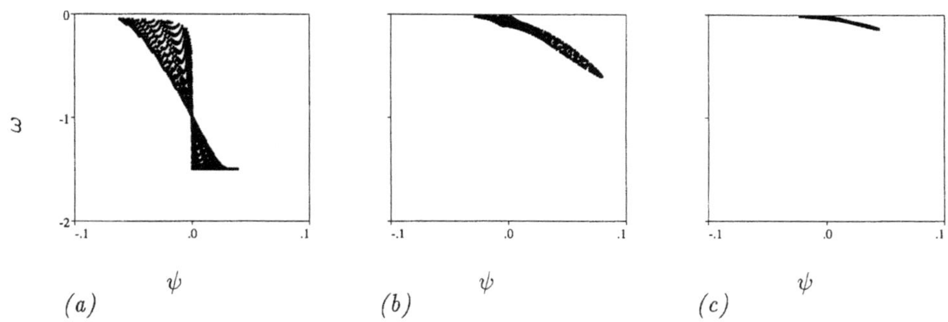

Fig. 7.12 Vorticity *vs* Stokes streamfunction scatter plots at the same times as Fig. 7.2.

since the ring was located only a short distance from the wall, the time was not long enough to reach the equilibrium condition of $J(\Omega_\theta/r, \Psi) \approx 0$ before the collision. Shariff *et al.* (1994) performed this test for a free ring and they found that, at steady state, the functional relationship had an almost linear relationship with a steep slope. After the collision, the functional relationship becomes single-valued and linear, and Fig. 7.12c shows that it tends to the characteristic form of two-dimensional flows with a slope weaker than in the free ring. The tendency to the two-dimensional configuration is not surprising since, far from the axis, vortex stretching is not significant anymore in the ring. In this viscous calculation the linear relationship holds but, as far as the vorticity reduction is concerned, a true steady configuration is not achieved.

7.3d Vortex rings impacting no-slip walls

The satisfactory results for a vortex ring colliding with a rigid free-slip wall is an indication that also the impact with no-slip walls can be simulated. In this case, the process consists of the generation of a very thin vorticity layer at the wall that rolls up and forms a secondary vortex ring. The interaction between the two patches is initially inviscid; later on, the interaction leads to different scenarios depending on Re. Walker *et al.* (1987) have visualized this mechanism; moreover, they attempted the explanation of the experimental observations by numerical methods based on viscous-inviscid interaction; this simplified model could fulfill its aim only partially. The full Navier-Stokes simulation, on the other hand, has the capability to describe the complex features of the flows and to explain causes of events difficult to detect in the experiments.

In the experiment Walker *et al.* (1987) reported the trajectories of the center of each vortical structure together with several visualizations, making the comparison with the numerical simulations possible. The experiment did not furnish the vorticity distribution at $t = 0$; thus the difference between the initial fields could lead to quantitative discrepancies between the numerical and the experimental trajectories. The trajectories of the peak vorticity obtained by Orlandi & Verzicco (1993a) were compared with the experimental trajectories at $\mathrm{Re}_v = 564, \mathrm{Re}_v = 1250, \mathrm{Re}_v = 1680, \mathrm{Re}_v = 2840$ which correspond to $\mathrm{Re}_\Gamma = 1306, \mathrm{Re}_\Gamma = 2895, \mathrm{Re}_\Gamma = 3890, \mathrm{Re}_\Gamma = 6577$ respectively. Here we are showing the

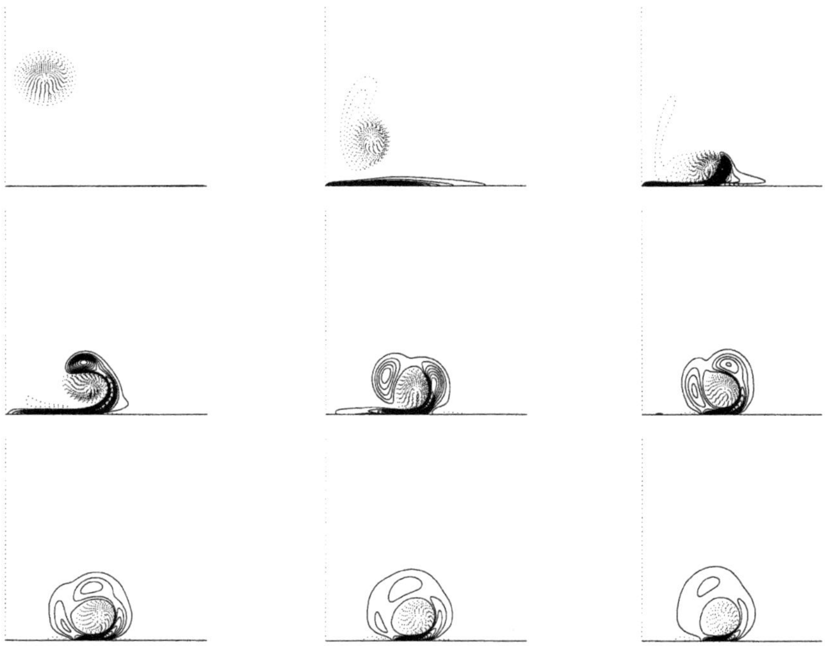

Fig. 7.13 Vorticity contour plots at $\mathrm{Re}_\Gamma = 1306$: *(a)* $t = 0$, *(b)* $t = 10$, *(c)* $t = 20$ *(d)* $t = 30$, *(e)* $t = 40$, *(f)* $t = 50$ *(g)* $t = 60$, *(h)* $t = 70$, *(i)* $t = 80$. The sequence in time is from top-left to bottom-right and contour increment is ± 0.07, $\omega(t = 0)_{min} = -1.9$.

results for the smallest and the highest Reynolds numbers.

The interaction with a no-slip wall differs greatly from that with a free-slip wall; however, Fig. 7.13 shows that until $t = 10$ the thin wall layer does not affect the primary ring; therefore, it does not differ from the shape it had in the free-slip collision or in the absence of walls. The thin wall-layer is the boundary layer generated by the velocity field of the primary ring. Since there is an adverse pressure gradient, the boundary layer separates and, at the same time, the velocity field of the primary ring detaches it from the wall. The lifting of this positive layer stops the radial motion of the primary ring. At $t = 20$ the layer has significantly detached from the wall, which, in turn, induces a further thin layer of opposite-sign vorticity at the wall. Walker *et al.* (1987) observed the generation of the secondary layer, but the dye did not display the creation of the other thin layer. Although this layer does not affect the interaction of large scale vortices, its appearance is a further proof that the numerical simulation can highlight structures not easily detected in the experiments. The trajectory of the center of the primary ring (Fig. 7.14b) agrees with that in the experiment (Fig. 7.14a). Experimental and numerical observations show that the lift of the primary vortex by the secondary one causes a kink in the trajectory. At this Re, the primary vortex moves slowly to the right until the secondary layer stops the lateral translation characteristic in free-slip walls.

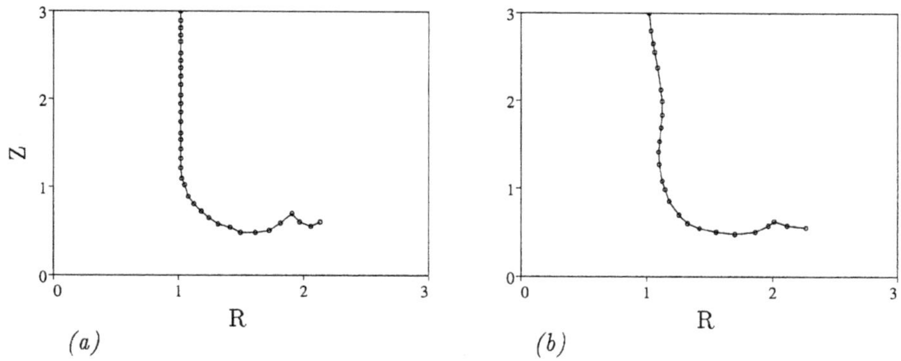

Fig. 7.14 Trajectory of the center of primary ring at $Re_\Gamma = 1306$, *(a)* Experiment of Walker *et al.* (1987) *(b)* present.

Since Fig. 7.13 is self-explanatory, it is not necessary to describe the vortex dynamics in detail. However, the primary ring remains near the wall, its initial strength is reduced, and it is encircled by secondary vorticity with high values at the wall. The two structures have an equal amount of positive and negative circulation. In the two-dimensional collision of a vortex dipole with a solid wall, although the primary vorticity did not increase due to vortex stretching, the same qualitative behavior occurs. The interested reader can perform a close comparison by using the code in chapter 4.

In the paper by Orlandi & Verzicco (1993a) the interaction was described at the intermediate $Re_\Gamma = 2895$ and in that case the grid refinement check was presented proving that a 128×128 grid in a domain with radial extension $r_f = 4r_0$ and vertical extension $z_f = 3.5r_0$ is sufficient to capture the formation of secondary and tertiary vortices, and to describe the interactions among these structures. The grid refinement at this Reynolds number also proves that the direct simulation is adequate and that it is not necessary to use methods based on viscous-inviscid interactions (Peridier *et al.* (1983)). These methods require cumbersome programming efforts and the introduction of approximations makes the solution valid only for the first rebound but not for the successive ones. This numerical technique could be appropriate at very high Reynolds numbers, where the direct simulation cannot be applied due to computer limitations. However at very high Re, as shown by the experiments, the incoming ring becomes turbulent and the viscous-inviscid interaction is even more difficult to compute.

The grid refinement was very important to show that a 128×128 grid in the $r - z$ plane is satisfactory to study the more complex three-dimensional interaction and to investigate whether the ejection velocity of the new vortex ring is different from that in the axisymmetric case. The results of the three-dimensional simulations are described in chapter 11. In this chapter the interaction by increasing the Reynolds number in the axisymmetric calculations is described.

Fig. 7.15 shows the vorticity evolution, at $Re_\Gamma = 6577$, obtained on a 256×512 grid and in a region of different extensions in the vertical direction. In fact at $Re_\Gamma = 2895$, the height was fixed at $z_f = 3.5r_0$; this height, with 128 points in z, gives sufficient resolution to resolve the small scales generated in the full simulation. At $\mathrm{Re}_\Gamma = 6577$ the vertical size must be increased up to $z_f = 5r_0$ to describe the stronger ring ejection and thus 512 points are necessary in z. The results in Fig. 7.15, during the approach of the ring and immediately after the collision, show analogies with Fig. 7.13. The reduced viscosity leads to secondary structures that break and form patches of vorticity with increasing circulation because of the merging process. The vortex merging, initiating approximately at $t = 60$ (Fig. 7.15g), explains the elongated loop in the trajectories of the primary ring observed in the experiments (Fig. 7.16a) at this Reynolds number. In fact, due to the merging, vorticity disappears from the region close to the axis, letting the primary ring to be advected by the secondary one. At $\mathrm{Re}_\Gamma = 6577$, secondary and tertiary structures have higher vorticity levels that produce a very compact vortical structure which continues to entrain tertiary vorticity; its circulation becomes so strong that it produces its fast ejection from the wall.

This event is described by Walker *et al.* (1987) who evaluated the trajectories of the centers of the vortical structures. Figs. 7.16a-b show the good qualitative agreement between the trajectories of the numerical simulation (Fig. 7.16b) and experimental measurements (Fig. 7.16a). Both studies show the double loop of the primary vortex with the second loop, elongated for reasons previously described. The differences between the values in Fig. 7.16a and in Fig. 7.16b, as already mentioned, could be ascribed to differences in vorticity distributions at $t = 0$. The Gaussian distribution, however, captured the relevant features of the phenomenon; differences between the experiment and the numerical simulation could also be ascribed to three-dimensional effects. This is a further reason which pushed us to furnish the code in chapter 11. Axisymmetric simulations can be performed on personal computers with a reasonable memory and CPU time; on the other hand three-dimensional simulations require personal computers with performances similar to those of workstations.

7.3e Space developing coaxial jet

Coaxial jets are basic flows occurring in combustion related problems; in practice these are confined and this configuration has been considered by Aksevol & Moin (1996) by LES. Here I am interested to simulate axisymmetric free jets that with free-slip conditions at the external radial boundary give satisfactory results when the radial extension of the computational domain is large enough. By this assumption the impermeability does not largely affect the entrainment of the external flow in the jet. The importance of the radial boundary is not only the peculiarity of the numerical simulations; in fact even in laboratory experiments the dynamics of the jet is affected by the location of the external confinement. To test the sensitivity of the results on the location of the radial boundary, two simulations were performed for a single jet, with the radial size respectively equal to $5R$ and $3R$ (R is the radius of the jet) at $\mathrm{Re} = 2000$. In the narrow domain the instabilities due to the impulsive start-up die out, while in the wider domain these are sustained. The same qualitative behavior was observed for coaxial jets and these checks can be repeated by the reader.

The solution is more sensitive to the axial length of the domain even if at the outlet radiative conditions are imposed. In chapter 4 the Orlansky boundary conditions were

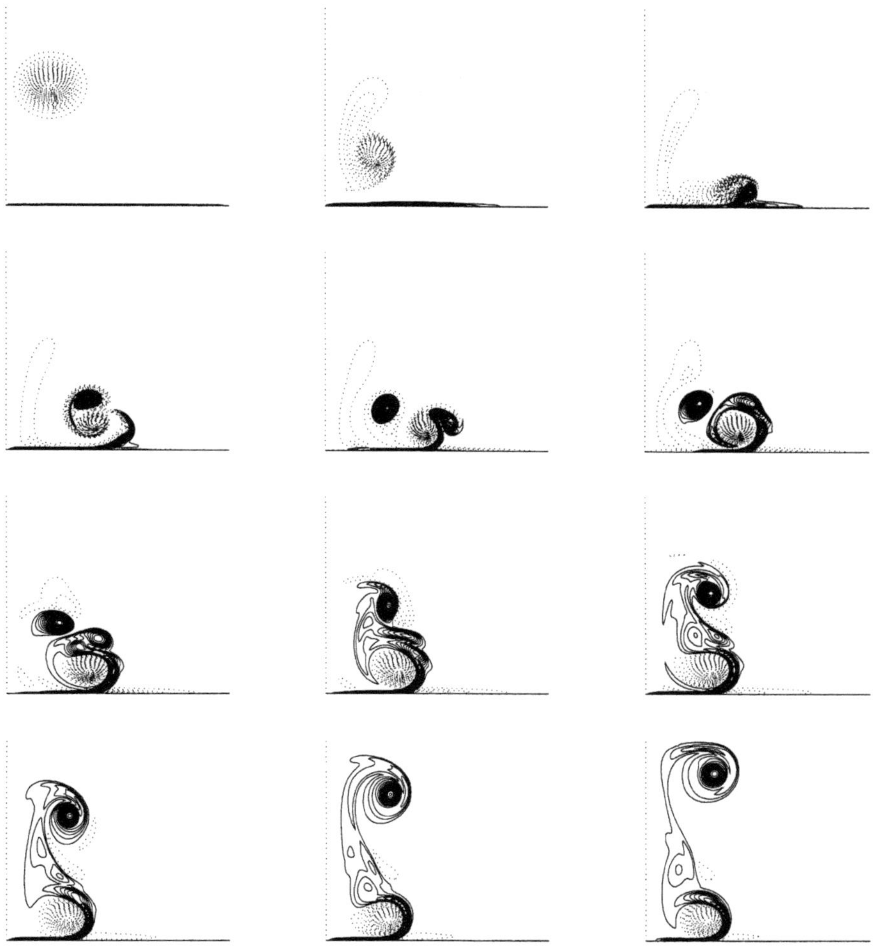

Fig. 7.15 Vorticity contour plots at $\mathrm{Re}_\Gamma = 6577$: *(a)* $t = 0$, *(b)* $t = 10$, *(c)* $t = 20$ *(d)* $t = 30$, *(e)* $t = 40$, *(f)* $t = 50$ *(g)* $t = 60$, *(h)* $t = 70$, *(i)* $t = 80$. *(g)* $t = 90$, *(h)* $t = 100$, *(i)* $t = 110$. The sequence in time is from top-left to bottom-right and the contour increment is ± 0.07, $\omega(t = 0)_{min} = -1.9$.

described and the convection velocity was calculated at each time step and at each grid point. In geophysical applications, for instance, when waves are radiating out from the domain the calculation of the exact outflow velocity is important. In confined flows, where vorticity must radiate out, the calculation of the convection velocity is less important and usually it is assumed constant across the whole boundary. This assumption was done by Buell & Huerre (1988) for the mixing layer, by Akselvoll & Moin (1996) for the 3D confined coaxial jet and by Le *et al.* (1997) for the three-dimensional turbulent backward

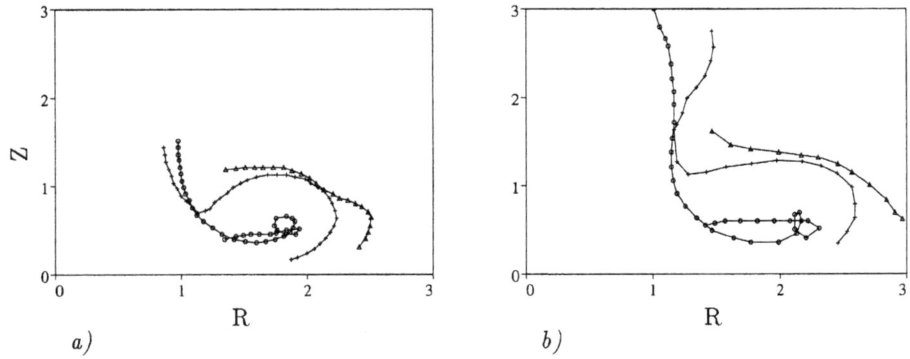

Fig. 7.16 Trajectory of centers of primary (∘), secondary (+) and tertiary (△) vortex rings. *(a)* Experiment of Walker *et al.* (1987) *(b)* Orlandi & Verzicco (1993a) at $Re_\Gamma = 6577$.

facing step. Two simulations have been performed by Salvetti *et al.* (1996) and a feed-back from the outlet to the inlet boundary was observed. Figs. 7.17a-b show that in the longer domain the pressure perturbation at the inlet has a smaller amplitude and frequency than in the shorter domain; it means that the feed-back travels backwards with a finite velocity. The feed-back between the inlet and the outlet is a realistic phenomenon; in fact, also in the laboratory free evolving shear layers depend on the length of the test section.

At the inlet of the computational domain, the axial velocity is set to zero at t=0, and evolves to a prescribed profile $Q_z(r)$ in a time τ:

$$q_z(t,r) = f(t)Q_z(r) \quad t < \tau \quad ; f(t) = 3\left(\frac{t}{\tau}\right)^2 - 2\left(\frac{t}{\tau}\right)^3 \qquad 7.14$$

where $Q_z(r)$ is the basic stationary profile. The law for $f(t)$ in Eq.(7.14), together with a very small value of τ, permits the reproduction of the transient generated by the opening of a valve or the time required for a motor to reach its final speed in an experiment. The impulsive start-up of the jet and its initial development was simulated, by looking at the evolution of the large vortex forming at the start-up. This study is interesting to help understanding the propagation of a flame front and the dynamics of pulsating jets. To compare our results with the visualizations by Dahm *et al.* (1992) the axial velocity profile $Q_z(r)$ at the inlet of the computational domain should reproduce as close as possible the experimental profile of case 2 studied by Dahm *et al.* (1992). Salvetti *et al.* (1996) described accurately how to reproduce the measured profile. They performed two simulation:s one with the assumption that at the end of the transient the velocity profile is constant in time. However this condition does not reproduce the experimental conditions, where the inlet profile varies in time, because of the turbulence of the flow at the exit of the nozzle or of the disturbance induced by experimental devices. It is difficult to identify an inlet condition that could account for both effects. To reproduce the fluctuations of the inlet velocity due to turbulence in the nozzle, as a second case a perturbation with zero mean

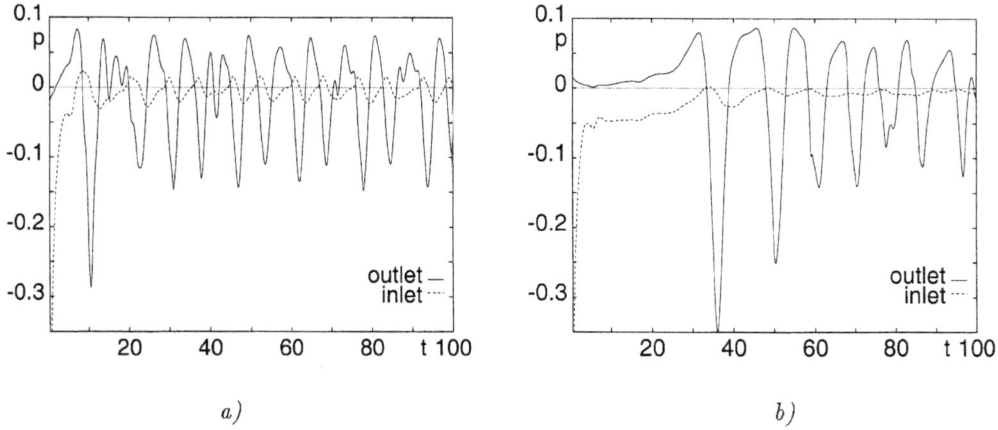

Fig. 7.17 Time evolution of the pressure at the inlet ---- and at the outlet —— :a) $L_z = 4R_i$ and b) $L_z = 16R_i$

has been superposed on the velocity $Q_z(r)$. Since no measurements of the inlet velocity were available in the experiments, a random perturbation $q'_z(t,r) = g(r)G(t,r)$ has been assumed. The inlet profile then is:

$$q_z(t,r) = Q_z(r) + g(r)G(t,r) \qquad 7.15$$

where $G(t,r)$ is Gaussian white noise in t and r, and $g(r)$ is the perturbation intensity accounting for a higher turbulence level near the walls of the nozzle. The analytical expression of $g(r)$ used by Salvetti *et al.* (1996) (that, as they claimed, did not reproduce the real conditions in the experiments by Dahm *et al.* (1992)), is useful to have indications of the effects of inlet perturbations on the evolution of the jets. It should be stressed that this disturbance does not account for the turbulent structures of the real boundary layers. In addition the component q_r is set equal to zero at the inlet.

Numerical simulations were performed at Re = 3000, since at this Re a comparison with the flow visualizations for case 2 in the experiments by Dahm *et al.* (1992) is possible and the validation of the numerical method can be carried out. The axial length of the computational domain has been set equal to 16 R_i. The grid has 193 and 385 points respectively in radial and axial directions.

A first simulation was carried out without any perturbation of the inlet velocity profile. The vorticity field in Fig. 7.18b shows the formation of vortical structures from the roll up of the external shear-layer. These vortices are advected downstream until they pair (this is not visible in Fig. 7.18b). Although the dynamics of the flow is characterized by the repetition of such events, the flow is not perfectly periodic, probably because of the interactions between instabilities of the internal and external shear-layers. The distributions of a passive scalar, having a concentration at the inlet equal to 1 in the external jet and equal to 0 elsewhere, were compared to the sequence of visualizations by Dahm *et al.* (1992).

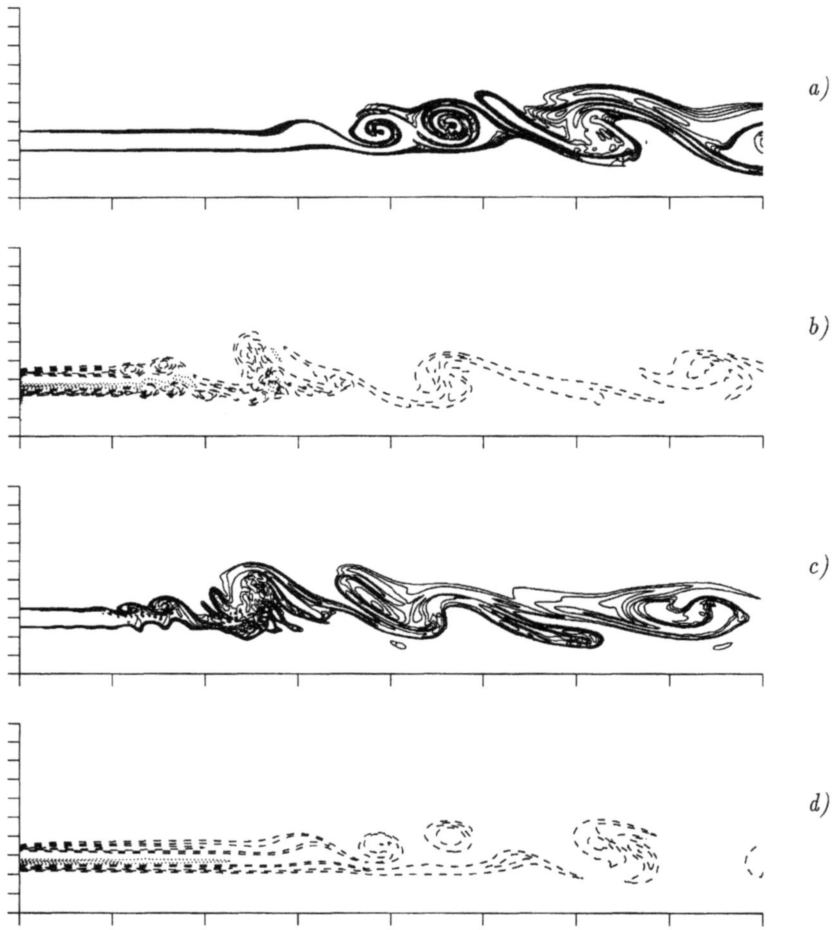

Fig. 7.18 Contour plots at $t = 206$ of: a) passive scalar; b) vorticity in the unperturbed case; c) passive scalar; d) vorticity in the perturbed case.

The distribution of the passive scalar and the vorticity field at $t = 206$ in Figs. 7.18a and 7.18b shows that in the numerical simulation the roll up of the external shear-layer occurs at a distance from the inlet of about $6R_i$, whereas in the experiment it was found at a distance of about $2R_i$.

Different possible explanations for this discrepancy have been examined. The enhanced stability of the external shear-layer in the numerical simulation could be due to the assumption of axial symmetry, even if this assumption is confirmed by the experimental observations in the initial region. With more probability this discrepancy could be related to the absence, in the numerical simulation, of temporal perturbation at the inlet. A reasonable way to introduce the effects of turbulence is by a random perturbation, like that given in Eq.(7.15), superimposed on the inlet velocity profile. Even if a random perturbation does not reproduce the real conditions in the experiments, a better agreement with the experimental observations is obtained. Indeed the roll up of the external shear-layer occurs approximately at a distance of $2R_i$ from the inlet, as shown by the instantaneous picture of the passive scalar concentration in Fig. 7.18c. As in the experimental visualizations in Plate 6, taken from the set of pictures in the paper by Dahm *et al.* (1992), a second roll up of the external shear layer is observed immediately after the first one. Larger structures are observed downstream. The vorticity field in Fig. 7.18d shows that these are the result of the pairing of the vortices from the external shear-layer, and also of their pairing with the structures from the internal shear-layer. Nevertheless the comparison with the experiments is possible only in the initial region, due to the axisymmetric assumption made in the simulations. It was proved that the position of the initial roll up does not depend on the amplitude of the inlet perturbation.

7.4 Conclusions

In this chapter simulations of axisymmetric flows have been presented but the axisymmetric code to produce them has not been given; these results can be obtained by the 3D code described in chapter 11. To perform the axisymmetric simulations it is necessary to set in the data file $n1 = 2$; there will be some waist of memory and computational time, but it is better to have this waste than to deal with a large number of codes. The simulations presented in this chapter were performed with an old version of the code that did not allow non-uniform meshes in the streamwise direction. The new code in chapter 11 does have this possibility and hence the reader, interested in these flows, can repeat the simulations with the new code. I would like to recall that the new code requires minor modifications to account for inlet conditions and radiative conditions at the outlet.

Axisymmetric flows are important because vortex stretching enters and contributes to the vortex dynamics. In a certain sense this condition can be interpreted as a 2.5 dimension because the energy is flowing from the large to the small scales. Axisymmetric simulations permits very fine resolution to describe the very small scales. In this chapter the code has been used to solve a flow which evolves in space and in time and this sort of flow is basic for combustion related problems. As it was shown for the mixednes in a vortex ring, the axisymmetric condition permits to perform the simulations with more that one passive scalar. At this point it is very simple to introduce some sort of chemistry and to allow the scalars to interact. In this way there is the possibility to test ideas about combustion processes, that is to introduce complex chemistry in a realistic flow. With regard to this point it is important to recall that in jet experiments, with a good approximation the first part of the jet can be considered axisymmetric. This is the potential region where the large part of chemical reactions occur.

These incompressible simulations are far from the real applications in combustion where due to the heat the density has large variations, and the simulations should take account of it. Recently Cook & Riley (1996) developed a numerical method accounting for variable density problems. The method is based on a staggered finite-difference method similar to those described in this book. I am sure that several scholars are interested in these conditions. I did not have the time to insert the modifications by Cook & Riley (1996) modifications in this code for cylindrical coordinates, but I realized that this is a very easy task that can be pursued by a graduate student during his Ph.D work. The outcome of this effort is the creation of a tool useful to design practical combustors. However, a real practical tool that could be used by the industry should account for a realistic subgrid model to deal with high Re flows. With regard to the basic principles of LES chapter 12 and 13 have been written but a more specific paper on subgrid models for variable density problems has recently been published by Cook & Riley (1998).

8 Three-dimensional flows with three periodic directions

8.1 Introduction

The numerical simulations of three-dimensional flows until two decades ago was possible only for those working in an institution where there was access to the biggest computers at that time available. It is interesting to recall that a simulation by a 32^3 grid at the end of the seventies required the fastest commercial computer at that time, the CDC 7600. In that period a 64^3 simulation at NASA Ames required the ILIAC IV, a parallel processor machine, with a special computer language that was accessible only to a few scientists. In the eighties, with the spreading of the CRAYs around the world, the simulations of 3D Navier-Stokes equations were accessible to a wide range of institutions. But even in that period, those computers were located in large research institutions and the number of researchers with access to them was still limited. Only in the last years the increase of the speed of the desk workstations is giving to almost everyone the possibility to perform simulations of 3D Navier-Stokes equations. In the review article by Moin & Mahesh (1998) the history of the DNS is reported.

In the first period of DNS, there was a large consensus that in 3D it was necessary to use very accurate numerical methods. The major part of simulations thus were performed by means of pseudospectral methods. In the last years on the other hand there was a large increase of simulations by finite-difference schemes. This choice was motivated by the interest in performing simulations in practical flows where boundary conditions and/or geometry constraints make the extension of pseudospectral methods very difficult. The aim of this book is to show that finite-differences are simple to understand and to use, and that these produce satisfactory results comparable to those achieved by the more elaborate pseudospectral methods. In this chapter a numerical method that permits to solve the 3D Navier-Stokes equations with three periodic directions is described and the relative computer code is furnished. With this chapter I would like to contribute to the prove that it is indeed true, as someone claims, that "codes to solve 3D Navier-Stokes equations are available at the supermarkets".

The most interesting features of 3D flows are linked to vortex stretching. This term causes the energy to go from large to small scales. This physical property motivates the requirement of very fine grids to discretize the small scales with the highest velocity gradients. The kinematic viscosity decreases the velocity gradients, as is shown in the section devoted to the Burgers equation; or, from the energetic point of view, the viscosity dissipates the kinetic energy. Then in 3D the resolution requirements at high Reynolds numbers are important. About the global conservation properties, important to produce an efficient numerical scheme, it is worth to remember that while in 2D energy and enstrophy are conserved, in 3D only energy is conserved. The other basic principle to keep in mind in the development of a numerical method is the choice of the appropriate set of equations to discretize. For 3D flows it is convenient to use the Navier-Stokes in primitive variables because only three parabolic and one elliptic equations are necessary. This is mainly true for finite differences because of the efficient matrix inversion algorithms that require second order as the greatest order of the derivatives in the system of equations discretized . In pseudospectral methods, differentiation does not increase the band size of the matrices, thus Kim *et al.* (1987) by certain variables used in stability theory (Drazin & Read pg.76 (1981)) reduced the number of equations.

The other issue for methods in physical space is the variables location; these must be defined in points of the cell which give the most compact discrete differential operators in the Navier-Stokes equations. In the sixties the group in Los Alamos, lead by Harlow, formulated a scheme based on staggered velocities and with the pressure at the center of the cell. This scheme, named MAC (Welch & Harlow (1965)) was slow due to the pressure solution. Later on Kim & Moin (1985) added to this scheme a pressure solver based on trigonometric expansions and accelerated the implicit scheme by a factorization procedure. The scheme since then became popular and it was extended to equations in general curvilinear coordinates by Orlandi (1990) and by Choi (1993). The scheme has been applied, with small differences, to cylindrical coordinates by Verzicco & Orlandi (1996a) and it is described in chapter 10. This scheme was applied to laminar flows (Kim & Moin (1985)) and in the last few years it was extended to direct simulations of turbulent flows by such a large number of scholars that for sake of brevity it is better not to mention them at all. In this chapter the numerical method has been used to understand whether the Lamb dipole is stable to three-dimensional disturbances and to simulate isotropic turbulence. For these simulations periodicity in the three directions is a good assumption.

8.2 Governing equations

The Navier-Stokes equations, for an incompressible flow, in dimensionless and in conservative form are:

$$\frac{\partial u_j}{\partial x_j} = 0$$

$$\frac{\partial u_i}{\partial t} + \frac{\partial u_i u_j}{\partial x_j} = -\frac{\partial p}{\partial x_i} + \frac{1}{\mathrm{Re}} \frac{\partial^2 u_i}{\partial {x_j}^2} + \epsilon_{ijk} \frac{1}{\mathrm{Ro}} u_j \Omega_k \qquad 8.1$$

where u_i are the velocity components, t is time, p is pressure. The variables were adimensionalized with respect to a referent length L and a velocity scale U; the Reynolds number is $\mathrm{Re} = \frac{UL}{\nu}$. In the momentum equation the effect of background rotation has been added and a further dimensionless number, the Rossby number, appears; it is defined as the inverse of the rotation number $N = 2\Omega L/U$. Ω is the background rotation and Ω_k indicates the versor of its direction. In the physical cases here described with periodicity in the three directions, pseudospectral methods are convenient, because Fourier series are the natural choice. However, pseudospectral simulations require a dealiasing procedure that could produce inaccuracies in the solution. Although the finite differences are appropriate in complex domains with nonuniform grids or with non-simple boundary conditions as for instance radiative conditions it is important to show that, even for three periodic directions, finite differences work satisfactorily well.

8.3 Space and time discretization

Fig.8.1 shows the location of the variables in the cell; the pressure at the center of the cell permits to evaluate the pressure gradient, with second order accuracy, by two

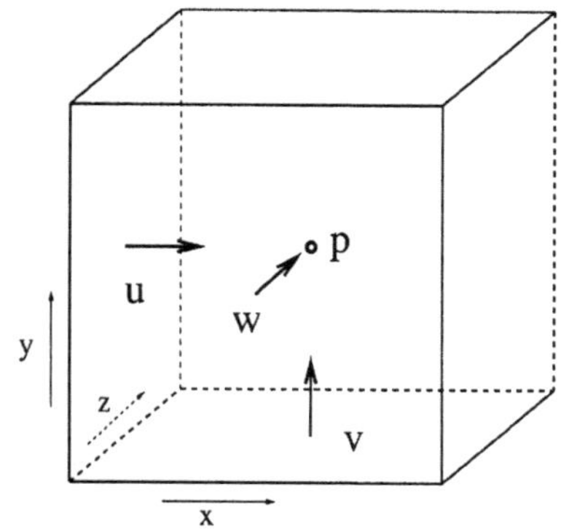

Fig. 8.1. Cell and variables location.

neighboring points for each velocity component. The gradient operator is thus the most compact discrete operator that can be imagined by finite differences. In this chapter the notation used in chapter 2 has been substituted with that where the discrete operators are indicated with $\frac{\delta}{\delta x_i}$. The uniform grid produces the second order accurate discrete operators in Eq.(2.7) and Eq.(2.10).

The space discretization of the system of Navier-Stokes equation produces the following set of differential equations

$$\frac{\delta u_j}{\delta x_j} = 0$$

$$\frac{\partial u_i}{\partial t} + H_i = -\frac{\delta p}{\delta x_i} + \frac{1}{\mathrm{Re}} L_{jj} u_i \qquad 8.2$$

L_{jj} is the discrete Laplacian operator. H_i accounts for the non-linear terms and the Coriolis forces. The former is important at high Reynolds number and its discretization is shortly described in the next section.

8.4 Discretization of the non-linear terms

H_i indicates the non-linear terms together with those due to the external body forces, thus

$$H_i = \frac{\delta u_i u_j}{\delta x_j} + \epsilon_{ijk} \frac{1}{\mathrm{Ro}} u_j \Omega_k \qquad 8.3$$

The calculation of the non-linear terms needs the product $u_i u_j$ in points where the velocities u_i are not defined. The interpolation operators in Eqs.(2.2) and (2.3) permit their calculation. In this operation it is important to verify that the discretized operators of the non-linear terms maintain the global energy conservation property of the Euler equations. This check can be done theoretically by a lengthy and tedious calculation. Even if it is done on paper, in any case it must be verified by the computer code. Then the theoretical check can be avoided; on the other hand the heuristic check permits to investigate the effects of the integration scheme and to find eventual bugs in the coding of the discretization operators.

The convective terms are evaluated in subroutines called $HDNLi$ with i varying between 1 and 3 depending on the velocity components. The terms are stored in arrays that later on are used for other operations. For example the term H_1 is stored in the three-dimensional array rhs, H_2 in the array phi (that later on is used for the ϕ calculation) and H_3 in the array $qcap$ (where later on is stored the divergence of the intermediate field $\hat{u_i}$). The evaluation of the Coriolis force is simpler and it is easy to see that the terms cancel out in the energy equation. From these considerations it turns out that the present method requires 10 arrays of dimensions $n_1 \times n_2 \times n_3$ and this is the cause of the limitations in resolution and hence on the Reynolds number. I would like to mention, to this purpose, that pseudospectral methods are more efficient. The Rogallo code (Rogallo (1981)), in fact, necessitates of only 7 arrays allowing on the same computer a more refined simulation.

8.5 Fractional step

This set of equations is integrated by a fractional step method. The theoretical formulation of this method was developed by Yanenko (1971) and since then there have been several formulations, slightly differing from each other. In their preliminary version, Kim & Moin (1982) did not introduce the pressure gradient in the first step, and hence there was the necessity to make some elaboration to assign, for instance in wall bounded flows, the exact boundary conditions to the intermediate solution $\hat{u_i}$. Later on Mansour (Personal Communication) modified the scheme by introducing the pressure gradient in the first step. With three periodic directions boundary conditions are not necessary, hence the early version of Kim & Moin (1982) could be used; in any case it is worthy to insert the pressure gradient in the first step. In the general case the three momentum equations can be discretized in time by the scheme described in chapter 3 for the Burgers equation. It consists of discretizing the explicit non-linear terms and the terms related to the external body forces by a Runge-Kutta third order or by an Adams-Bashfort scheme and the linear terms by an implicit Crank-Nicolson scheme. Eq.(8.2) in the first step is

$$\frac{\hat{u_i}^k - u_i{}^n}{\Delta t} = \gamma^k H_i^n + \rho^k H_i^{n-1} + \frac{\alpha^k}{2\mathrm{Re}} L_{jj} \left(\hat{u_i}^k + u_i{}^n \right) - \alpha^k G_i p^n \sigma_1 \qquad 8.4$$

$G_i p$ indicates the discrete pressure gradient operator $\frac{\delta p}{\delta x_i}$.

$u_i{}^n$ denotes the solution at the time n; $\hat{u_i}^k$ is the intermediate solution, where k indicates the time substep. Δt defines a time step and the integer i denotes the directions x_i. The constants γ^k, ρ^k and α^k have been previously used and the values are given in chapter 3. These constants indicate either a third-order Runge-Kutta or an Adams-Bashfort scheme.

It can be proved that the intermediate velocity field $\hat{u_i}^k$ is globally divergence-free but not locally. Eq.(8.4) can be written in terms of the increment $\Delta u_i = \hat{u_i}^k - u_i{}^n$, hence the equation in the computer code is

$$\Delta u_i - \beta^k L_{jj} \Delta u_i = \left(\gamma^k H_i^n + \rho^k H_i^{n-1} - \alpha^k G_i p^n \right) \Delta t + 2\beta^k L_{jj} u_i \qquad 8.5$$

where $\beta^k = \frac{\Delta t \alpha^k}{2\text{Re}}$.

The second step is necessary to get a divergence-free velocity field. To this purpose a scalar quantity ϕ permits to evaluate the velocity at the new time step $n+1$ by

$$\frac{u_i{}^{n+1} - \hat{u_i}{}^k}{\Delta t} = -\alpha^k G_i \phi \sigma_2 \qquad 8.6$$

If the discrete *div* operator D_i is applied to Eq.(8.6) and $D_i u_i^{n+1} = 0$ is imposed an elliptic equation permits to get the scalar ϕ. The equation is

$$L_{jj}\phi = \frac{\sigma_2}{\alpha^k \Delta t} D_i \hat{u}_i \qquad 8.7$$

If the velocity, at the intermediate step, from Eq.(8.6) is substituted in Eq.(8.4) one obtains the equation that is solved by the two successive time steps. This operation is necessary to derive the relation between the pressure p and the scalar ϕ and to get the accuracy of the method. For this demonstration the Adams-Bashfort scheme with only one substep for each Δt is used for sake of simplicity. The values of the constants in Eq.(8.4) are $\gamma^k = 1.5$ and $\rho^k = 0.5$ and hence $\alpha^k = 1$. A similar procedure holds for the Runge-Kutta of third order; it requires to add the velocities at each of the three substeps (the demonstration was done by Rai and Moin (1991)). These substitutions give

$$\frac{u_i{}^{n+1} - u_i{}^n}{\Delta t} = H_i^{n+\frac{1}{2}} + \frac{\alpha}{2\text{Re}} L_{jj} u_i{}^{n+\frac{1}{2}} - G_i \left[p^n \sigma_1 + \sigma_2 \left(\phi - \beta L_{jj}\phi\right)\right] \qquad 8.8$$

The constants σ_1 and σ_2 can be chosen to have the pressure gradient at the desired time and location. For example in the first version of the fractional step Kim & Moin (1982) assumed $\sigma_1 = 0$. The pressure p hence was not necessary and its calculation was neglected. If there was the necessity to analyze the pressure field, it was evaluated in a post-process and it was related to ϕ through the relationship $p^{n+1} = \phi - \beta L_{jj}\phi$. In the successive improvement of the method the pressure at the new time is defined as

$$p^{n+1} = p^n + \phi - \beta L_{jj}\phi \qquad 8.9$$

thus the pressure gradient in Eq.(8.8) is

$$-G_i \left[p^n(\sigma_1 - \sigma_2) + \sigma_2 p^{n+1}\right] \qquad 8.10$$

If $\sigma_1 = \sigma_2 = 1$ the pressure gradient is evaluated at $n+1$; on the other hand for $\sigma_1 = 1$ and $\sigma_2 = \frac{1}{2}$ the pressure gradient is evaluated at $n + \frac{1}{2}$. The accuracy is different in the two cases: when the term is evaluated at $n+1$ the accuracy is $O(\Delta t)$ and at $n + \frac{1}{2}$ it is (Δt^2). A check has been done, and though it is strange to believe, it has been found that the reduced accuracy gives a more stable scheme which is more appropriate in time dependent simulations where stability is an important issue.

These successive two steps allow the solution of the Navier-Stokes equations. At this point the important question arises how to get a fast solution with a minimum storage. These requirements are very important dealing with three dimensional flows. Implicit schemes for the linear viscous terms help to reach the goal to speed-up the solution, but the associated matrix on the left hand side of Eq.(8.5), in 3D, has a very large band and hence cannot be directly inverted. A similar matrix is associated with the Poisson equations (Eq.(8.17)) and to get the overall speed-up of the numerical method is even more important to avoid iterative schemes. To overcome these difficulties, for Eq.(8.5) an approximate factorization is used and for the ϕ equation trigonometric series are necessary.

8.6 Factorization in the $\hat{u_i}$ equations

The matrix associated to the left hand side of Eq.(8.5) is sparse and its direct inversion requires by the standard methods, such as Gauss elimination, a large number of operations that makes impossible the solution with a large number of grid points. Fine resolutions are necessary dealing with turbulent flows or with vortex dynamics. In fact, in three-dimensions, vortex stretching generates vorticity components of the same order in each direction. This is the direct energy cascade transferring energy by non-linear interactions to small scales. The smallest scales are linked to the kinematic viscosity and the interest is to get solutions for the highest Reynolds numbers compatible with the available computer. To get the solution, in a reasonable computational time, it helps to approximate the large sparse matrix, associated to L_{ij}, by a combination of tridiagonal matrices, that are easily inverted as is shown in chapter 2. Tridiagonal matrix solvers are not well suited to computers with a vectoral architecture. The speed of these routines is increased by inserting the inner do loops inside the routine instead of calling the routine inside the loop. These procedures were also used in two-dimensional flows. In 2D this computational time saving is important, but in 3D it is mandatory; without it the simulation is impossible.

The left hand side of Eq.(8.5) can be explicited by the centered second derivative operator (Eq.(2.10)). If the right hand side of Eq.(8.5) is indicated for simplicity by RHS_i, it becomes

$$\left[1 - \beta^k(L_{i1} - L_{i2} - L_{i3})\right] \Delta u_i = RHS_i \qquad 8.11$$

The approximate factorization technique consists on the replacement of the left hand side by the product of three matrices. By defining $A_{ij} = \beta^k L_{ij}$ Eq.(8.11) becomes

$$(1 - A_{i1})(1 - A_{i2})(1 - A_{i3})\Delta u_i = RHS_i \qquad 8.12$$

It can be shown that, at order Δt^3, the left hand side of Eq.(8.12) approximates the sparse matrix in Eq.(8.11). This approximation leads to the advantage of inverting tridiagonal matrices.

In the case of three periodic directions it is indifferent in which order each direction is performed. On the other hand, if there is an inhomogeneous direction this should be the last. This is discussed in the chapter devoted to wall flows. Eq.(8.11) is written as

$$(1 - A_{i1})\Delta u_i^{**} = RHS_i$$

$$(1 - A_{i3})\Delta u_i^{*} = \Delta u_i^{**}$$

$$(1 - A_{i2})\Delta u_i = \Delta u_i^{*} \qquad 8.13$$

where Δu_i^{**} and Δu_i^* are intermediate quantities without any physical meaning.

Even in this code, as in the vorticity-streamfunction formulation, the subroutines where the Δu_i are calculated are named $INVTRi$, where the letter i indicates the u_i component. The non-linear terms, that were calculated in the subroutines $HDNLi$, enter in these subroutines. The right hand side is evaluated by applying the operator G_i to the pressure at the old time step, and the operators L_{jj} to u_i^n; the non-linear terms H_i at n, multiplied by γ_k, to those at $n-1$ multiplied by ρ_k are then added. The contribution H_i is stored in the arrays ru_i immediately after the RHS_i are evaluated. In the $INVTR_i$ routines all the arrays necessary for the solution are used and the total number is ten. A first estimate of the memory occupancy can be done; for example, the memory requested by a workstation RISK IBM for a 160^3 calculation is 183.6 Mbites. Some extra memory is necessary to store the two-dimensional arrays for the tridiagonal solvers, hence the size of the code increases up to 190 megabytes. However, when the code is running on the machine it necessitates 167 megabytes.

$\hat{u_i}$ is evaluated from Δu_i. The right hand side of Eq.(8.7) permits the calculation of ϕ which is described in the next section. ϕ produces the solenoidal velocity field u_i^{n+1} by Eq.(8.6).

8.7 Solution of the Poisson equation for ϕ

Eq.(8.7) for the scalar ϕ in the discretized form is

$$(L_{11} + L_{22} + L_{33})\Phi(i,j,l) = F(i,j,l) \qquad 8.14$$

Periodicity permits to use the Fourier series. If $N1$ and $N3$ are the number of intervals in x_1 and x_3 with the series one goes from the physical to the wave number space and vice versa by the relationship

$$\Phi(i,j,l) = \sum_{k_1=-N_1/2}^{-N_1/2+1} \sum_{k_3=0}^{N_3} \hat{\Phi}(k_1,j,k_3) e^{\frac{\iota 2\pi k_1 i}{N_1}} e^{\frac{\iota 2\pi k_3 l}{N_3}} \qquad 8.15$$

by substituting this expression in the left hand side; by a similar expression for $F(i,j,l)$ and by defining the reduced wave numbers

$$k_3'^2 = \frac{2\left(\cos\frac{2\pi k_3}{N_3} - 1\right)}{\Delta x_3^2}$$

$$k_1'^2 = \frac{2\left(\cos\frac{2\pi k_1}{N_1} - 1\right)}{\Delta x_1^2}$$

Eq.(8.14) becomes

$$\sum_{k_1=-N_1/2}^{-N_1/2+1} \sum_{k_3=0}^{N_3} \left[\left(k_3'^2 + L_{22} + k_3'^2\right)\hat{\Phi} = \hat{F}\right] \cdot e^{\frac{\iota 2\pi k_1 i}{N_1}} e^{\frac{\iota 2\pi k_3 l}{N_3}} \qquad 8.16$$

As shown in chapter 2 at each wave number k_1 and k_3 the left hand side of Eq.(8.16) is a tridiagonal matrix in the x_2. The coefficients of the matrix are

$$AP(j) = AM(j) = \frac{1}{\Delta x_2^2} \quad , AC(J) = -\frac{2}{\Delta x_2^2} + k_1'^2 + k_3'^2 \qquad 8.17$$

In the code furnished in this book $\Phi(i,j,l)$ has been evaluated by the Temperton routines. It is not difficult to write the subroutine using the ESSL library. This more efficient way for the RISC IBM machines is used in the code described in the section about wall bounded flows. The calculation of Φ is done in the subroutine $PHCALC$; it requires, at the start of the simulation, the calculation in the routine $PHINI$ of the reduced frequencies and of some arrays for the Temperton routines. The real $fft99$ routine transforms the right hand side of Eq.(8.14) from the physical x_i to the wave number space k_i. The real FFT furnishes $N_1/2$ complex coefficients which, by the complex $cfft99$ are transformed in the l direction. Thus, after the successive application of the real and complex FFT, the $N_1/2+1, N_3$ real coefficients are stored in rhs and the imaginary in $qcap$ and these arrays constitute the right hand side of Eq.(8.16). The $DSOLV$ routine performs the calculation of $\hat{\Phi}$ by inverting the tridiagonal matrix in the direction x_2 indicated by the index j. Before the call of the tridiagonal periodic solver $TRIPVV$ $\hat{F}$ at $K_1 = 0$ and $K_3 = 0$ is set equal to zero.

After application of $DSOLV$ by the $cfft99$ and the $fft99$ routines with $isign = +1$, $\hat{\Phi}$ in the wave number is transformed in Φ in the physical space. Usually the calculation of ϕ necessitates a computational time of the same order as that necessary to evaluate a velocity component.

8.8 Initial conditions

The code has been used to simulate isotropic decaying turbulence and to analyze whether a Lamb dipole, if perturbed along the direction of the main vorticity, is stable or not. The difference between the two flows is related to the initial conditions. The outputs of the code are also different and these can be changed depending on the wishes of the user.

8.8.1 Isotropic turbulence

The initial energy spectrum is assigned by fixing an analytical expression. In this case the energy spectrum has been taken from Mansour & Wray (1994); it is

$$E(k,0) = \frac{q^2}{2A} \frac{1}{k_p^{\sigma+1} k^{\sigma}} exp\left[-\frac{\sigma}{2}(\frac{k}{k_p})^2\right] \qquad 8.18$$

where k_p is the wavenumber at which $E(k,0)$ is maximum, σ is a parameter, and $A = \int_0^\infty k^\sigma exp(-\sigma k^2/2)dk$. The energy spectra is related to the amplitude of the Fourier modes of the velocity components and not to the phases. In Temperton's routines the wave numbers are defined as

$$K_i = j - 1 \quad for \quad 1 \le j \le N_i/2+1$$

$$K_i = N_i - j + 1 \quad for \quad N_i/2+2 \le j \le N_i \qquad 8.19$$

The index i indicates the three directions. In this space the velocity components can be assigned, with some attention, to the zero wave number, that corresponds to $j = 1$. For this value the amplitudes are not calculated because, by definition, the energy of the zero wave number is null. To avoid the origin in the wave number space there are two do loops, as it is shown in the routine $TOURBIS$ in the file *isointu.f*. In each do loop the velocity components are generated by three sets of random numbers that permit to uniformly distribute on the interval $(0, 2\pi)$ the angles θ_1, θ_2 and ϕ. The velocity components ar. (the $\wedge$-symbol now indicates complex quantities and should not be confused with the symbol used to indicate the intermediate velocity in the fractional step):

$$\hat{u_1} = (\alpha|k|k_2 + \beta k_1 k_3)/(|k|k_h)$$

$$\hat{u_2} = (-\alpha|k|k_2 + \beta k_3 k_2)/(|k|k_h) \qquad 8.20$$

$$\hat{u_3} = (-\beta k_h^2)/(|k|k_h)$$

This expression fulfills the requirement to have a solenoidal velocity field in wave number space $k_i \hat{u}_i = 0$; moreover it can be shown that if

$$\alpha = \frac{E(k,0)}{4\pi k^2} exp(\iota\theta_1)\cos(\phi), \qquad \beta = \frac{E(k,0)}{4\pi k^2} exp(\iota\theta_2)\cos(\phi) \qquad 8.21$$

the three dimensional energy spectrum of this field is equal to $E(k, 0)$. These initial conditions were used in the famous Rogallo code (Rogallo (1981)) used worldwide to get insights into homogeneous turbulence. The code permits even to read the components in wave numbers, for instance those obtained by the Rogallo code. This initial condition is important to perform a direct comparison between the results by finite differences and those by a pseudospectral method.

To have the field in the physical space the $fft3d$ routine is applied at each component. The routine is similar to the second part of the subroutine $PHCALC$ that evaluates ϕ from $\hat{\Phi}$. Although the velocity field is divergence-free in wave number space it is not in the staggered grid. Even if not strictly necessary, to start the simulation with a divergence-free velocity field, in the subroutine $INITUR$, the non-soleinodal velocity field can be projected into a solenoidal field by Eq.(8.6). This operation necessitates the call of the routine $PHCALC$ to get ϕ and $UPDVP$ to get the divergence-free velocity.

8.8.2 Lamb dipole stability

The stability of the Lamb dipole was studied, to our knowledge, only in the two-dimensional case by Orlandi *et al.* (1993) and by Nielsen & Rasmussen (1997). Orlandi *et al.* (1993) have shown by two dimensional simulations, using a code similar to that described in chapter 4, that the Lamb dipole is a very stable structure; even if it is very strongly perturbed it sheds part of its vorticity and then it goes back to a stable structure. This process can be explained by the $\omega = f(\psi)$ relationship; in fact the perturbation disrupts the single value relationship $\omega = f(\psi)$. The vorticity that does not respect this relationship is convected out, and in the region bounded by the zero streamline, only the vorticity redistributed to get the linear relationship remains. The slope of the linear relationship depends on the distance between the two centers of the dipole. Leith (1984), by arguments based on minima of energy and enstrophy, argued that the Lamb dipole is a very stable structure. A theoretical proof about the stability of the Lamb dipole has not yet been found.

In the 3D case it is expected that vortex stretching produces vorticity components in the other two directions comparable to the vorticity of the dipole and these break this compact structure. The linear stability calculation of the Lamb dipole is a difficult task and, to our knowledge, it was never analytically attempted to know which is the most unstable wave number in the axial direction. However Leveke & Williamson (1998) confirmed by an experiment that the theoretical analysis by Waleffe (1990) for localized disturbances in an unbounded domain, holds even in bounded elliptical flows. If the analytical treatment is difficult, it is convenient to linearize the set of equations and to find the solution numerically. However, the computational effort requires the same amount of work as that necessary to perform the direct simulation. The present code can thus be used to predict the size of the secondary structures generated by a non-linear random disturbance. The spanwise size of the computational box should give sufficient resolution to reproduce the growth of vortical structures directed in the direction normal to that of the primary vorticity. As usual the grid independence of the solution must be verified.

The initial velocity field in the 1 and 2 directions are calculated, in the file *isoindip.f*, by assigning the vorticity ω_3 in Eq.(4.39) in the subroutine $DIPOL$. In this subroutine the streamfunction is evaluated by the solution of Eq.(4.3); the subroutine $PSCALC$ solves Eq.(4.3). From ψ the velocity components u_1 and u_2 are obtained. To this velocity field is added, only in the region where $\omega_3 > \omega_s$ (with $\omega_s = 2$), a random velocity field with an amplitude uniformly distributed in the interval $(0, v_\epsilon)$ with $v_\epsilon = 0.05$. Even in this case the velocity is not divergence-free, thus the same steps, described above for isotropic turbulence, are done.

8.9 Numerical checks of the code ISO

The code ISO is given on the diskette in the directory $CODESCH8/ISO$ with the make files for the IBM RISC (*iso.makibm*) and DIGITAL (*iso.makdig*) computers. The user, once more, can observe that the name of the different files are constructed as in the previous chapters. In each of these files there are subroutines with names similar to that previously used.

Before we describe the results it is important to evaluate the performances of the code, which on CRAY computers are obtained by calling the $PERFVIEW$ routine and on the workstation RISK IBM by calling the $PROF$ routine in the make file. In here the performances have been evaluated on the RISK IBM machines; these do not give the megaflop speed as the CRAY computers do. In the CRAY C90 at NASA Ames it has been checked that this code, with the Temperton FFTs for a 128^3 simulation has a performance of 250 Mflops; on the contrary with the CRAY's FFTs the performance increases to 500 Mflops. The performances on the RISC IBM have been obtained by a 32^3 grid simulation without the call of the routine with the outputs. The $PROF$ routine furnished the performances of each part of the code. As I have pointed out several times, I am not an expert in computer language and in codes optimization. I am sure that several users have the appropriate skill to improve this code to reach the optimal performances in order to get the results

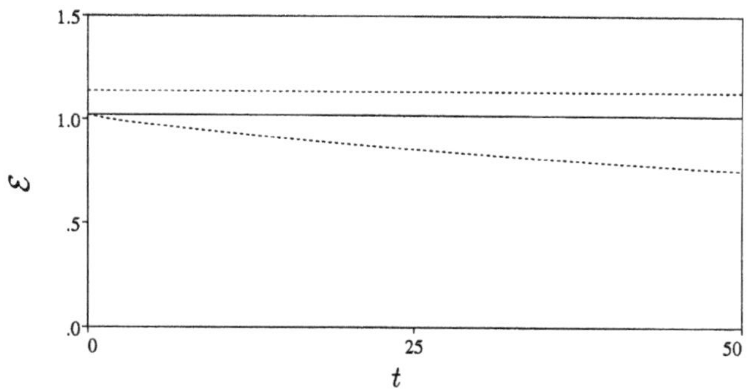

Fig. 8.2 Time evolution of total energy, ---- $CFL > 1.7$, —— $CFL \approx 0.5$ with $k_p = 9$, ········ $CFL \approx 0.5$ with $k_p = 3$.

in the minimum CPU time. The code *ISO*, compiled by the make file *isoper.mak* in the directory *CODESCH8/PERF* to run for 11 time steps including the generation of the initial condition for isotropic turbulence, requires 15.34*secs* on a RISK 3CT and 20.34*secs* on a RISK 390. The compilations were done by using the power2 architecture. Of these 15.34*secs*, approximately 38% was spent in the tridiagonal periodic solvers in the *INVTRi* routines. The rest of the calculations on these routines, that is the evaluation of the RHS_i, took 28, 1% of the CPU time. The evaluation of the non-linear terms, in the subroutines *HDNLi*, took 6.8% and finally 16% of the CPU was required by the Poisson solver. From these results it turns out that, to save time, changes should be made in the tridiagonal routines. Attempts have been done by evaluating some of the coefficients only once, or by increasing the length of inner loops. These changes increase the memory occupancy. A further saving could be reached by the ESSL routines instead of the Temperton ones.

The check of the capabilities of a numerical method to discretize the non-linear term as in 2D even in 3D must be done by inviscid simulations. In the truncated space, the total energy must be conserved and equidistributed in wave number space. By assigning in the data *iso.d_inv* in the directory *CODESCH8/DAT* a value of Re $> 10^7$, instead of the *INVTRi* routines the *INVISC* routine is called in the routine *TSCHEME*. Therefore the simulation is faster because the factorization procedure is neglected. Fig.8.2 shows the time evolution of the total energy $\mathcal{E}$ by two different time steps. These simulations indeed confirm that the energy conservation depends on the time step and that $\mathcal{E}$ is exactly conserved only in the limit of $\Delta t \to 0$. The largest time step was set equal to .1 and the relative CFL was slightly larger than 1.7, the value compatible with the third order Runge-Kutta scheme. A Δt three times as large indeed shows that the percentage of energy dissipated by the time integration scheme is approximately 0.3%. A further simulation with an initial spectra with a smaller extension of the k^2 range shows that the energy is equidistributed on the spherical shell, independently of the initial spectrum. Fig.8.3 indeed shows that with $k_p = 9$ and $k_p = 3$ the flow evolves towards a spectrum with a k^2 law. The achievement of the k^2 law is reached in a short time, which can be quantified by the evolution of the total enstrophy Ω. The total energy $(\mathcal{E})$ is conserved, but the non-linear

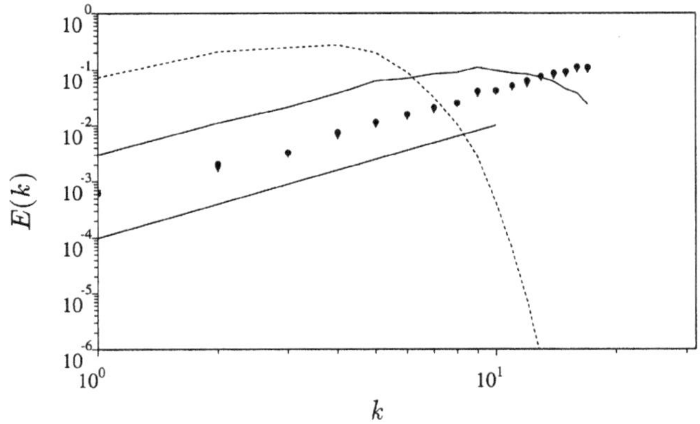

Fig. 8.3 Three-dimensional spectra, $t = 0$ —— $k_p = 3$. ---- $k_p = 9$; $t = 50$ ▵ $k_p = 3$. and • $k_p = 9$.

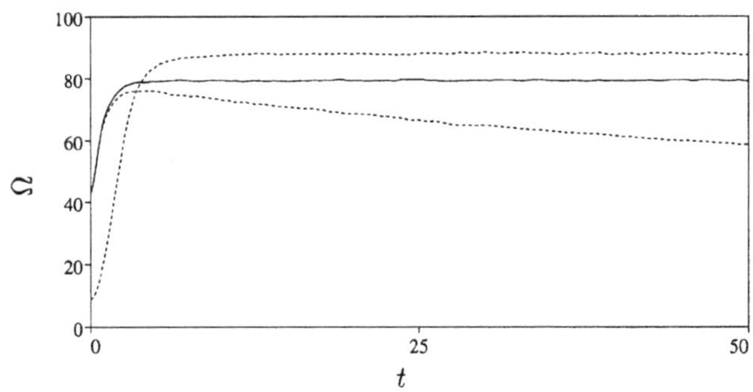

Fig. 8.4 Time evolution of total enstrophy, ---- $CFL > 1.7$ —— $CFL \approx 0.5$ with $k_p = 9$, ········ $CFL \approx 0.5$ with $k_p = 3$.

terms by triadic interactions transfer energy to small scales and thus the enstrophy grows. When an inertial range is reached, in the whole spectrum, the energy transfer stops and the enstrophy remains constant, as is shown in Fig.8.4.

From this inviscid simulation it turns out that the DNS at low wave numbers produces a k^2 spectrum. In real simulations the viscous dissipation at low wave numbers is very small hence the behavior of the energy spectra at low k can be extrapolated. About this subject assumptions were made in the past different that permitted to make assertions on the time law decay of the turbulent energy. For the energy spectrum holds

$$E(k,t) \approx k^{\sigma} exp\left(-2\nu k^2 t\right) \qquad 8.22$$

Batchelor & Townsend (1949) assumed $\sigma = 4$, that gives a law $q^2 \approx t^{-5/2}$; on the other hand Saffman (1967) by assuming $\sigma = 2$ obtained $q^2 \approx t^{-3/2}$. The results from DNS show that the low wave number spectrum tends preferably to the k^2 law at intermediate Reynolds number, and this can be related to what has been observed in an inviscid simulation.

8.10 Definition of turbulent quantities

On analyzing turbulent flows different order statics of the fluctuating quantities are considered. Averaging procedures can be defined in different ways and the concepts of averages are described in several books as Batchelor (1953), Monin & Yaglom (1971), Tennekes and Lumley (1972) and Hinze (1975) to cite the most famous. This chapter deals with isotropic turbulence and since only one simulation is performed for each initial condition, the averages are in the three homogeneous directions. As recently done by Carati *et al.* (1996) for LES, a set of turbulent simulations could be advanced together on parallel computers with slightly different initial conditions. In this case the number of samples increases but also the computer requirements increase and in particular computers with parallel architecture are necessary.

The first question on making the averages is related to the number of samples necessary to get a statistical quantity of real meaning. The present simulations were performed on a 160^3 hence the statistics are calculated by $4 \cdot 10^6$ samples, which are reasonable at least for low order statistics. In isotropic turbulence the mean velocity is null thus the simulation furnishes the fluctuating field. To verify that the code maintains this condition the mean velocities are evaluated in the subroutine $ENERCA$ and they remain null during the evolution. In this routine the normal turbulent stresses $< u_i'^2 >$ are calculated together with the skewness s_i and the flatness f_i. The averaging procedure for a generic quantity q is defined as

$$< q >= \frac{1}{N_1 N_2 N_3} \sum_{l=0}^{N_1} \sum_{m=0}^{N_2} \sum_{n=0}^{N_3} q \qquad 8.23$$

and s_i and f_i are

$$s_i = \frac{< u_i'^3 >}{< u_i'^2 >^{3/2}}, \qquad f_i = \frac{< u_i'^4 >}{< u_i'^2 >^2} \qquad 8.24$$

On the evaluation of each of these quantities, the velocity u_i' are those at the location of the cell. On the other hand, in the evaluation of the stresses $< u_i' u_j' >$ with $i \neq j$, since each u_i' is defined at different locations, a linear interpolation permits to evaluate the correlations at the center of the cell.

Skewness S_i and flatness F_i of the spatial derivatives $\frac{\partial u_i}{\partial xi}$ are a further important quantity in isotropic turbulence. Frisch (1995) claims at pg.156 that the skewness is a measure of vortex stretching and in isotropic turbulence it is related to the energy transfer term through the relationship

$$< (\frac{\partial u_i}{\partial xi})^3 >= -\frac{2}{35} \int_0^\infty k^2 T(k) dk$$

Following Batchelor (1953) since $< (\frac{\partial u_1}{\partial x_1})^2 > = \int_0^\infty k^2 E(k) dk$ the skewness of the spatial derivative $\frac{\partial u_1}{\partial x_1}$ is

$$S_1 = -(\frac{135}{98})^{1/2} \frac{\int_0^\infty k^2 T(k) dk}{(\int_0^\infty k^2 E(k) dk)^{3/2}} \qquad 8.25$$

The EDQNM theory, described by Lesieur (1990), allows to understand the importance of the energy transfer term. This theory shows that the energy transfer at each wave number is a combination of the transfer by triadic interactions. These are interactions between modes at different wave numbers such that $p + q = k$. It is important to recall that EDQNM (Crocco and Orlandi (1986)) and DNS by Domaradsky & Rogallo (1990) have shown that local interactions, that is interactions with $p \approx q \approx k$, are those contributing the most to energy transfer. From the energy transfer equation (Hinze (1975) pg. 215)

$$\frac{\partial E(k,t)}{\partial t} = T(k) - 2\nu k^2 E(k,t) \qquad 8.26$$

It turns out that $\int_0^\infty k^2 T(k)dk$ is the time-rate-of-change of the enstrophy due to non-linear interactions. Since in 3D the enstrophy is flowing from large to small scales, the numerator of Eq.(8.25) is positive hence the skewness in isotropic turbulence is negative. Skewness and flatness values of velocity derivatives have been measured. Tavoularis *et al.* (1978) showed that S_1 decreases with the Reynolds number starting from a value approximately equal to -0.5, at the highest value of Re.

In the analysis of turbulent flows, energy spectra give insights into turbulence structures. Their calculation is straightforward by pseudospectral codes, on the other hand by finite-differences a passage from physical to wave number space is requested. This passage has been described in the solution of the scalar ϕ. The Fourier transform in all three directions permits to have the three-dimensional energy spectrum. After a real FFT by the $fft99$ routine, two complex $cfft99$ transforms give the energy spectrum $E_i(k)$. The following relationship is used inside the routine $SPECTRE$ to this purpose

$$E_i(k) = \sum_{k_1=0}^{N_1} \sum_{k_2=1}^{N_2/2} \sum_{k_3=0}^{N_3} u_i(k_1,k_2,k_3)u_i^*(k_1,k_2,k_3) \qquad 8.27$$

where the direction 2 is that where the real FFT has been used. For the wave number $k_2 = 0$ the energy is

$$E_i(k) = \sum_{k_1=0}^{N_1} \sum_{k_3=0}^{N_3} 2u_i(k_1,0,k_3)u_i^*(k_1,0,k_3) \qquad 8.28$$

To check that the calculation of the spectrum is correct it should be verified that $< u_i'^2 >= \int_0^\infty E_i(k)dk$. These controls are done in the $OUTP$ routine, where the spectra are calculated and printed. In this routine the spectra are also scaled in Kolmogorov variables, where the wave number is multiplied by the Kolmogorov length scale $\eta = (\nu^3/\epsilon)^{1/4}$ and the spectra $E(k,t)$ by $(\nu^5\epsilon)^{-1/4}$. The dissipation is evaluated in wave number space as $\epsilon = 2\nu\int_0^\infty k^2E(k,t)dk$ or in physical space as $\epsilon = 4\nu < S_{ij}Sji >$ with

$$S_{ij} = 0.5(\frac{\partial u_i}{\partial x_j} + \frac{\partial u_j}{\partial x_i}) \qquad 8.29$$

Also for ϵ the checks of the calculations in wave and physical space have been done. In homogeneous turbulence $\epsilon = 2\nu\Omega$ with $\Omega = 0.5 < \omega_i\omega_i >$ the enstrophy. In the subroutine *OUTH* all these quantities have been evaluated together with other quantities such as the Taylor length scale and hence the Reynolds number based on this length, defined as

$$R_\lambda = \sqrt{\frac{20}{3}} \frac{\int_0^\infty E(k,t)dk}{\nu(\int_0^\infty k^2 E(k,t)dk)^{\frac{1}{2}}} \qquad 8.30$$

R_λ is a fundamental quantity in isotropic turbulence which characterizes the energy spectrum. Chapman (1979) presented a plot with the energy spectra measured at different R_λ scaled by Kolmogorov quantities. At high R_λ the inertial range with $E(k) \approx k^{-5/3}$ was shown. This high R_λ presently can not be achieved by DNS and as shown by Mansour & Wray (1994) the spectra have the maximum close to the dissipative range. If a numerical method is accurate enough it should reproduce well the dissipative region of the spectrum. Inaccuracy and numerical dissipation should give a decay of the spectrum, around $k\eta \approx 1$, greater than that measured.

In direct simulations of turbulence, the good sense suggests to save the fields, and by a post-process to evaluate the quantities of interest; this procedure permits to avoid a new simulation if there is an interest to evaluate a new statistical quantity that was not of interest at the beginning. To save the fields it is necessary to have computers with a large amount of storage memory. In the past this was a large problem but recently the cost of the memory of personal computers dropped and thus the fields can be stored at low cost. What is suggested in this paragraph perhaps is well known, but my experience is that only in few research centers there is experience on dealing with turbulent data and that the transfer of knowledge outside these centers is limited. A book oriented to help scholars to analyze turbulent flows could contain trivial suggestions, but these help the new fellows interested in direct simulation of turbulence to save time. In the sections devoted to the discussion of the results the post-process codes are described.

8.11 Isotropic turbulence with and without rotation

Isotropic turbulence has been studied theoretically, experimentally and numerically; this simple flow has shown indeed the complexity of turbulence and permitted to advance its knowledge to a large measure. There are still a lot of not completely understood phenomena; to cite only one: the intermittency that maintains the interest in this flow alive. From the experimental point of view in recent years a set of measurements were performed by Saddoughi & Veeravalli (1994), on the largest subsonic wind tunnel in the world at NASA Ames Research Center, to validate the local isotropy hypothesis and to evaluate the Kolmogorov constants. Numerically the largest simulations were dedicated to isotropic turbulence since pseudospectral methods are well suited to this flow. Recently the parallelization of these codes for computers with a parallel architecture permitted to perform simulations up to 512^3 and even 1024^3. These simulations are onerous for the CPU time, but perhaps also for storage requirements. A simple calculation shows that one field in single precision for a 512^3 simulation needs a memory allocation of approximately 2 gigabites to store the three velocity components and the pressure. As seen before, the

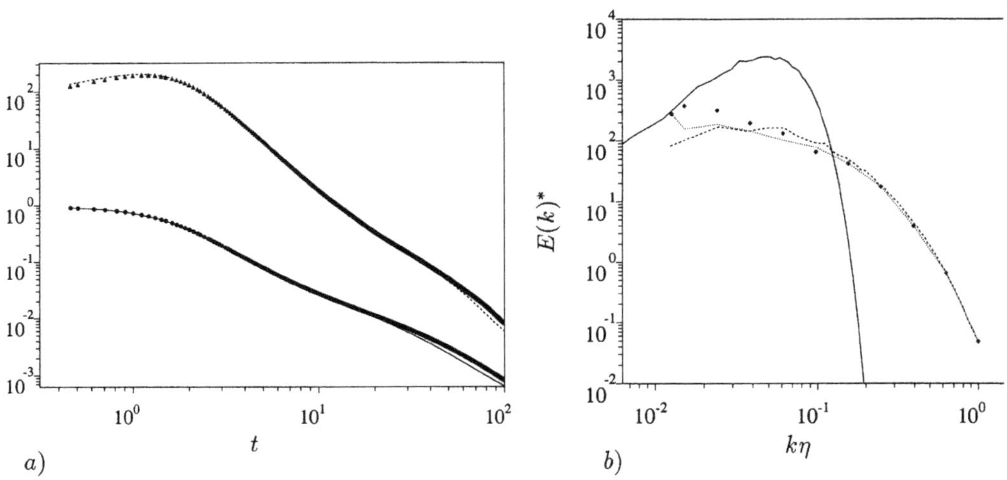

Fig. 8.5 Time history of $\mathcal{E}$ and Ω for finite difference symbols and pseudospectral (Wray) lines; b) Energy spectra (Kolmogorov units) at moderate Reynolds number, —— initial spectrum, ---- present simulation $R_\lambda = 54.3$, ········ Mansour & Wray $R_\lambda = 56.2$, □ Compte Bellot & Corrsin $R_\lambda = 60.7$

storage of the pressure can be neglected because from the velocity field the advancement of a time step produces the pressure. One of the three components can, in addition, be evaluated from continuity by saving only the field in an appropriate 2D plane. These simulations of decaying turbulence produced fields with a marginal extension on the width of the inertial range. To my knowledge the large part of simulations were performed in wave number space (She & Orszag (1991) and Mansour & Wray (1994)).

8.11a Isotropic turbulence

In the summer of 1998 during one of my usual visits at the Center of Turbulence Research I had the occasion to spend a short period of time in performing a one-to-one comparison between finite difference and pseudospectral simulations for isotropic decaying turbulence. To my knowledge this comparison was never done and really I do not know which is the reason. Decaying turbulence is a well defined case because the solution depends only on the initial conditions. The limitation is clearly that the Reynolds number changes in time and at the end the flow has a low energy content at small scales very and hence it cannot be considered as a very valuable example to investigate the qualities or deficiencies of a numerical scheme. Often simulations of isotropic turbulence have been performed by imposing a forcing at low wave number. This is simple to achieve in wave number space and more difficult by finite differences. In addition the forcing could introduce some differences between the two methods and then make the comparison more difficult. Thus at the moment, my opinion is that decaying turbulence could be considered a first and good example to validate the capabilities of finite difference to solve turbulent flows. Several scholars are spending a lot of time in increasing the accuracy of finite differences; by this comparison, it turns out that staggered second order finite differences produce results as good as those by pseudospectral methods.

The comparison has been done by running the Rogallo code on a 128^3 grid, starting from the spectrum in Eq.(8.18) with $\sigma = 2$ and $k_p = 9$. The simulation was stopped at $t = 0.46$, a sufficient time to build the vortical structures characteristic of isotropic turbulence. The velocity field in the wave number space was saved and it was taken as the initial field for the finite difference and the pseudospectral simulations. The simulations were performed for a very long time and finite differences run at constant CFL in one case by fixing it at 1.5 and in a second case at 3.0. The simulations by the pseudospectral code were done by Wray with a modified version of the old Rogallo code whose efficiency was improved by a fourth order Runge-Kutta scheme which runs at $CFL = 4$. Whoever is interested in the pseudospectral code should ask informations to Wray at NASA Ames Research Center. The results by finite difference at the two different CFL values were practically coincident and I am reporting only those obtained by the higher CFL. Fig.8.5a shows that both numerical methods predict the same energy and enstrophy decay. This figure is then a proof that finite differences are reproducing well large and small scales. At this point there is not an interest to understand why at $t = 10$ there is a decrease of the decay exponent of the total energy, but it is interesting to notice that both methods capture this change of slope. The proof that finite differences are as good as pseudospectral methods, in particular, turns out by the analysis of the time evolution of the dissipation rate. In the first part of the simulation this grows; the physical meaning is that in this period energy is transfered to the small scales. The energy content at small scales at this time is not so negligible as it is at the end of the simulations. Thus at $t \approx 1$ the results indeed show that finite difference are as accurate as pseudospectral simulations. I am presenting in this book the results of these preliminary simulations and my intention is to continue this comparison by making a more detailed comparison even at higher Reynolds numbers or in simulations with an external forcing.

8.11b Isotropic turbulence subjected to uniform rotation

Here simulations in physical space on a 160^3 grid at a R_λ close to that of Compte Bellot & Corrsin (1971) experiment are presented. The parameters of the simulation should be assigned to get the highest R_λ compatible with the resolution. Preliminary simulations as in Mansour & Wray (1994) with different spectra at low k permitted to see that with $\sigma = 4$ the spectrum was tending to $\sigma = 2$. Thus the results have been obtained by setting in the file *iso.d_visc* $\sigma = 2$, $k_p = 13$ and $q = 3$. With this spectrum the turbulent energy is equal to 1.138i, the size of the box is 2π, and the simulations were done at Re = 3000. This value was empirically chosen by different simulations; that at which, at $t = 10$, $k_{max}\eta$ was equal to 1 was considered. From these data it turns out that the equations were normalized with a velocity scale 1.138 smaller than the turbulent energy and a length scale 2π smaller than the box size. During the transient the spectrum adjusts and, at $t = 10$, Fig.8.5b shows that it ($R_\lambda = 54.3$), for a large part of the $k\eta$ range, follows that of Compte-Bellot & Corrsin (1971) at $R_\lambda = 60.7$, and that of Mansour & Wray (1994) at $R_\lambda = 56.2$. This field has been taken as an initial condition of different simulations with and without background rotation, to investigate the variations of the turbulence statistics with the rotation rate. In the file *iso.d_visc*, that should be renamed *iso.d*, by $nread = 1$ is imposed to read a

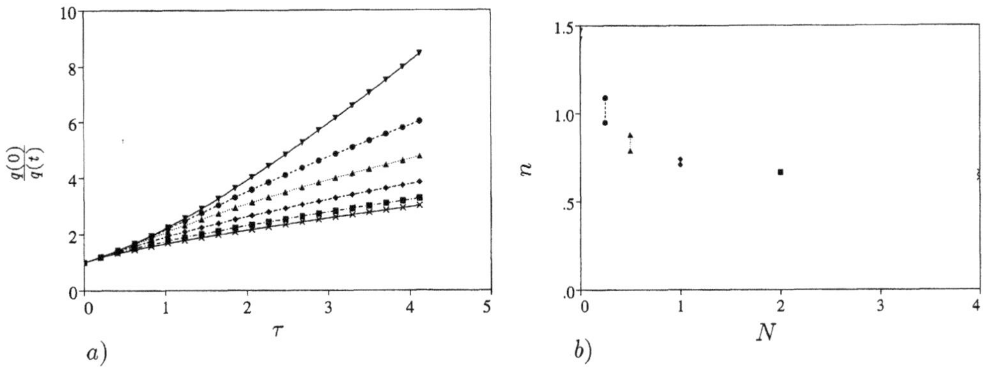

Fig. 8.6 a) Energy decay rate for ▽ $N = 0$, • $N = 0.25$, △ $N = 0.5$, ◇ $N = 1.$, □ $N = 2.$, × $N = 4$ ($\epsilon/q(N = 0, t = 0) = 12.4$; b) decay exponent as a function of N.

restarting file.

The effect of rotation on turbulence is very important in several practical applications as e.g. the flows inside rotating machines. These effects have been analyzed experimentally by Traugot (1958) and Wigeland & Nagib (1978) at low R_λ and numerically by Bardina *et al.* (1985) by DNS and LES. The direct simulation by a pseudospectral numerical method (Rogallo (1981)) permitted to explain the inconsistencies of the experiments that, in some of the cases, did not show the primary effect of the rotation on decreasing the rate of turbulence decay. The DNS demonstrated that the increase in decay rate at low rotational speeds, found by Wigeland and Nagib (1978), was due to the interactions in the initial region of the flow. In this book the aim is to show that the present numerical method reproduces the effect of external rotation on turbulence. Simulations have been performed varying $N = \frac{1}{Ro}$, the parameter in Eq.(8.1). The term proportional to N is discretized in the subroutines $HDNL1$ and $HDNL2$ since the external rotation is directed along direction 3. N was increased by doubling its value starting from 0.25 up to 4. In the experiment of Wigeland and Nagib (1978) the Rossby numbers $Ro_\lambda = q/\Omega\lambda$ varied from 0.7 up to 27; the present simulations were performed in a large range of Ro_λ from 12.5 up to 400. This range is similar to that used by Cambon *et al.* (1997) but the equations were integrated for a shorter time. The simulations were stopped when the turbulence energy decay exponent reached, with a good approximation, a constant value. Comparisons with the previous simulations have not been done since in one case (Bardina *et al.* (1985)) the energy spectrum was not given and in the other case (Cambon *et al.* 1997) the simulations used a subgrid model.

The present simulations indeed reproduce the effect of the background rotation i.e. to decrease the energy decay. Fig.8.6a shows that $q(0)/q(t)$ for $N = 0$ increases as a function of $\tau = \epsilon t/q$ approximately as in Bardina *et al.* (1985). By assuming that the turbulent energy decays as a power law $q(t) \approx t^{-n}$ (the simulations permit to evaluate the exponent n). This exponent depends on σ and on R_λ, as reported by Mansour & Wray (1994). In the

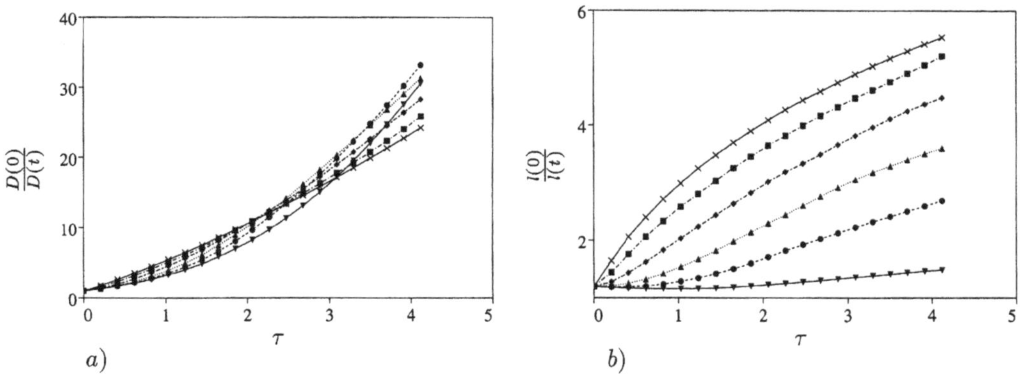

Fig. 8.7 Time history of a) dissipation $D = -2\nu\epsilon$, b) length scale $l = q^{3/2}/D$, for ▿ $N = 0$, • $N = 0.25$, ▵ $N = 0.5$, ◇ $N = 1$, □ $N = 2$, × $N = 4$; at $N = 0$ and $t = 0$ $\epsilon/q = 12.4$.

present simulation the exponent was evaluated in the time interval between $\tau = 2.06$ and $\tau = 4.12$ and their values as a function of N are shown in Fig.8.6.b. The present simulations have been performed with an initial spectrum, (see Fig.8.5b), where the low wave number region has not a well defined slope and thus the decay rate could be different from the values given by Mansour & Wray (1994). The value $n \approx 1.43$ is greater than $n = 1.28$ measured by Compte Bellot and Corrsin (1971) at a slightly greater R_λ. This higher value is supported by the observation that the exponent n decreases with the increase of R_λ. In the equation for ϵ, from dimensional analysis, the rate of dissipation is modeled as follows (Bardina *et al.* (1985)):

$$\frac{\partial \epsilon}{\partial t} = -C_2(R_\lambda)\frac{\epsilon^2}{2q} - C_1(R_\lambda)\epsilon\frac{N}{2} \qquad 8.31$$

The value of C_2 in the present simulations for $N = 0$ is 1.7 in agreement with the value calculated by Mansour and Wray (1994).

Fig.8.6 shows that in the presence of background rotation the turbulent energy decays with a power law. On the other hand Fig.8.7a shows that background rotation affects the dissipation in a more complicated way. The dissipation in a first period increases with the rotation and in a second period, it decreases by increasing N. The simulations predict that ϵ does not tend to zero at a very large rotation rate, in contrast to the major part of one point-closure turbulence models, as was shown by Bardina *et al.* (1985) at low R_λ. The fact that, in high rotating flows, the dissipation is present at a lower level, can be explained by the Taylor-Proudman theorem. It claims that background rotation reduces the turbulent fluctuations along the direction of the rotation axis. In physical space this effect is depicted by elongated turbulent structures. The increase of the length scales was investigated by Squires *et al.* (1993) by performing the simulations in a computational box with a length, in the direction parallel to Ω, four times greater than the other two. In the present simulations the different integral length scales have not been calculated for each N, but the time history of $l = q^{3/2}/\epsilon$ shows, in Fig.8.7b, the growth of the turbulent length

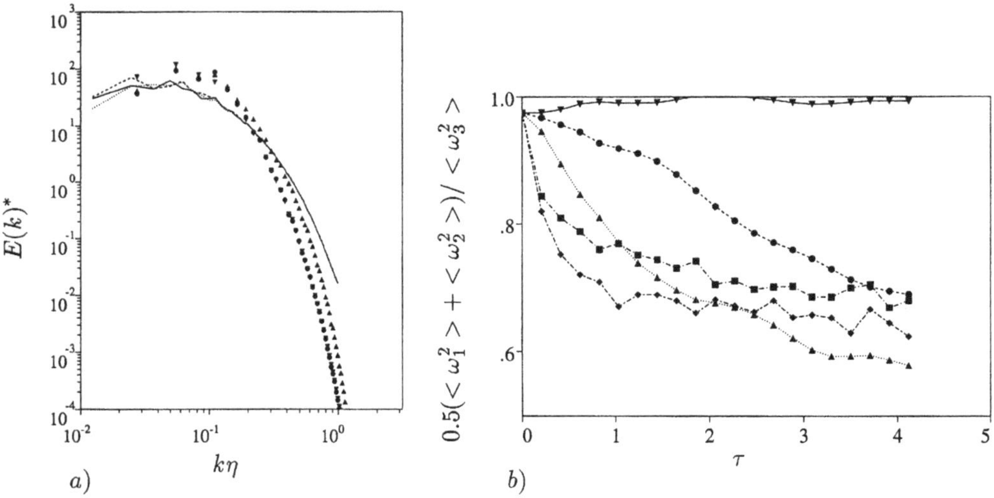

Fig. 8.8 a) Energy spectra in Kolmogorov scales at $N = 4$ at $\tau = 0$ (—— u_1', ⋯⋯ u_2' and - - - - u_3'), and at $\tau = 4.12$ (▿ u_1', • u_2', ▵ u_3'), b) Time history of $0.5(<\omega_1^2> + <\omega_2^2>)/<\omega_3^2>$ ▿ $N = 0$, • $N = 0.25$, ▵ $N = 0.5$, ◇ $N = 1$, □ $N = 2$, × $N = 4$; at $N = 0$ and $t = 0$ $\epsilon/q = 12.4$.

scale with N. In addition it has not been checked whether the results are independent of the length of the box in the direction of the rotation axis. This important check could be performed by those interested to know in more detail the structures of rotating flows.

The changes of the turbulent structures can be observed by the energy spectra. In Fig.8.8a the spectra of the three velocity components at $\tau = 4.12$ at $N = 4$ are compared with those at $\tau = 0$. The higher energy level at low wave number and the steeper descent at high wave number is the manifestation of the fact that background rotation generates an inverse energy cascade and that it reduces the transfer of energy to high wave numbers. This reduction of transfer makes the viscous dissipation at high wave number more effective and it explains why the spectra are steeper than in the non-rotating cases. The spectra show also that rotation increases the turbulence anisotropy. The anisotropy growth is described by the time history of the vorticity ratio $0.5\frac{<\omega_1^2>+<\omega_2^2>}{<\omega_3^2>}$. This quantity accounts for the "cigar-type" shape of the vortical structures along the rotation axis. Fig.8.8b shows that even at low rotation rate there is an anisotropy; it grows with the rotation and at a certain N decreases. Perhaps this decrease, at high N, depends on the marginal length of the computational box in the direction x_3. The measure of the anisotropy for the vorticity, given by $b_{33}^{\omega} = \frac{<\omega_3^2>}{<\omega^2-\frac{1}{3}>}$, is not shown, but it can be calculated by the post-processing code *postspe.f* in the directory $CODESCH8/POST1$. This short code reads the energy spectra, calculated during the simulation, that were saved at certain interval of time. This code in addition requires the data file *post.d*.

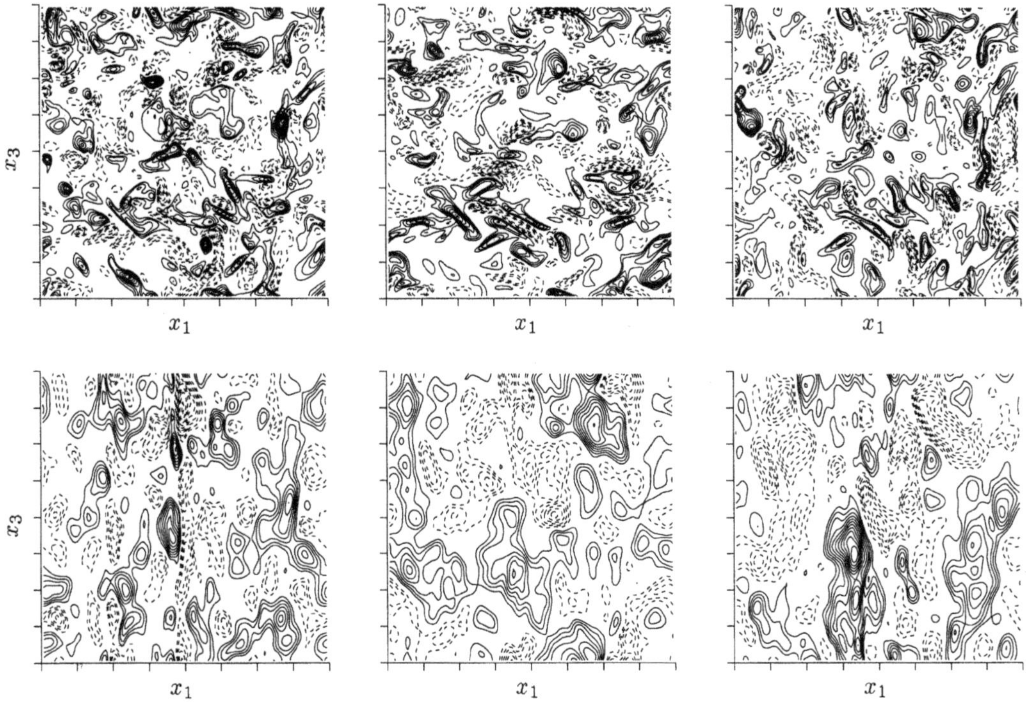

Fig. 8.9 Vorticity contour plots in $x_3 - x_1$ sections at $x_2 = \pi$, three top figures from left to right $\omega_1, \omega_2, \omega_3$ at $N = 0$, below the same quantities at $N = 4$. $\Delta\omega_i = 0.75$.

The vortical structures can be visualized, in physical space, by contour plots of the three vorticity components in sections across the cubic box. Since the interest is in emphasizing the elongation of these structures, the plots in the $x_2 - x_3$ plane passing through the center of the box are given. To generate the vorticity and velocity fields on these planes a post-processing code reads the restarting files written by the routine $CONTWR$ in the file *isoio.f* The code *visiso.f* in the directory $CODESCH8/VISUA$ constitutes of some of the routines in the code ISO and of a routine $OUTPF$, where velocities and vorticities are evaluated at the corner of the cell. The file necessitates the data file *iso.d*, used for the simulation. To visualize the field at $t = tfin$ the value of $tfin$ should be inserted in *iso.d.* Fig.8.9 shows that in the non-rotating case the three vorticity components do not differ from each other; on the other hand for $N = 4$ the anisotropy increases, in fact the ω_3 component is more elongated than the other two components. These flow visualizations show also the inverse energy transfer at large scales; in fact at high N the number of small scales is reduced.

8.12 Physics of turbulence from data

From what has been presented up to now, it has been shown that the generation of a reliable turbulent field is an easy task. It can be fulfilled on the condition that we have a reasonably fast workstation. Once the field has been generated, the scholar has a data set which permits him to answer some of the fundamental questions in turbulence. These data sets, thus, have the merit to facilitate the comprehension of turbulent flows. In the past when powerful computers were not available, experimental data were used, but approximations were done to evaluate quantities difficult to measure directly such as the components of the velocity tensor gradient. The Taylor hypothesis derived from the relationship $\frac{\partial u}{\partial t} + U\frac{\partial u}{\partial x} = 0$ permitted to derive the velocity gradient by a time sequence of the velocity field. The validity of the Taylor hypothesis was confirmed, through direct simulations, by Piomelli *et al.* (1989) and by Kim & Hussain (1993) for channel flows. In numerical simulations any variable is accessible, since at each time the velocity and pressure fields permit to evaluate any derivative. In a simulation of decaying isotropic turbulence by a 160^3 grid the statistical properties of any quantity can be analyzed by taking only one field. The statistics are computed from approximately $4 \cdot 10^6$ samples, a number that makes a large part of the statistics meaningful.

If turbulence would have been a perfectly random process, as in thermodynamics, it was possible to find a theory describing the phenomenon. The complexity of turbulence is related to the fact that it is not a perfectly random process, with the large scales of the flow playing an important role. To understand the difference between the atomic motion in a gas and the motion of a turbulent field, Bradshaw introduced the analogy between the atomic motion in a gas and that of dry peas in a metal can. Turbulence, on the other hand, is analogous to the motion of live worms in a can.

The calculation of the statistics in large part is avoided in the code that solves the Navier-Stokes equations, except for some that are necessary to check the time evolution. Before it was mentioned that the statistics are evaluated by a post-process code, because, if there is the interest to evaluate quantities, that at an earlier time were not in our mind, the Navier-Stokes equations should not be solved again. On the other hand, if a set of velocity and pressure fields, separated by a substantial interval of time to consider each field independent from the other, are saved, these can be handled to evaluate the new statistics. This is the procedure followed at NASA Ames where a large archive of flows is available. The possibility to have access this data base attracted a large number of scholars from other nations to attend the summer schools organized by the Center of Turbulence Research.

The number of fields depends on the type of flows; as mentioned before, for isotropic decaying turbulence only one field can be used. The post-processing is performed by the code $STATISO$ which is composed by different files; these are given in the directory $CODESCH8/POST2$ together with the *make* file and with the file *isostat.d* read in *statisoma.f*. In this code some of the routines are taken from the Navier-Stokes solver (ISO) and are not explained; on the other hand some of the new routines are described, in the next paragraphs, where also the results are discussed.

8.12.a Correlation tensor

As mentioned before the spectra are useful to analyze the properties of the structures of a turbulent field. The correlations are the counterpart of the spectra in the physical space.

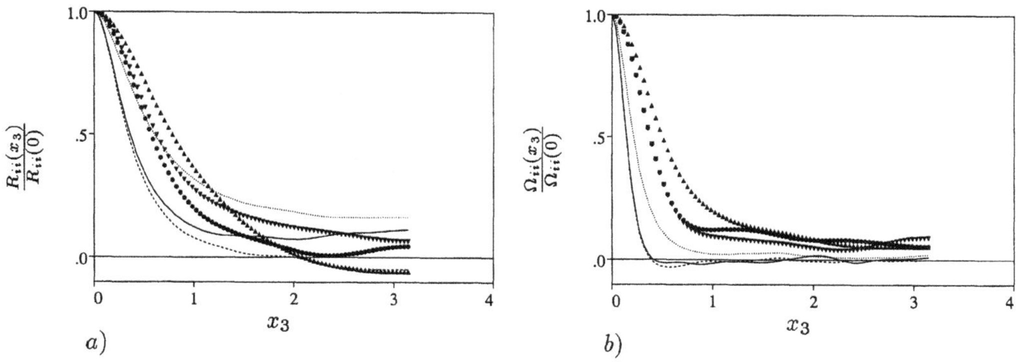

Fig. 8.10 a) Velocity, b) Vorticity, correlation coefficients at $\tau = 4.12$ for $N = 0$ (—— u_1', ········ u_2' and - - - - u_3'), and for $N = 4$ (▿ u_1' , • u_2', ▵ u_3').

The velocity correlation tensor is

$$Rij(\mathbf{r}) = < u_i(\mathbf{x}) u_j(\mathbf{x} + \mathbf{r}) > \qquad 8.32$$

and as reported by Batchelor (1953) it is linked to the energy spectrum tensor $\Phi_{ij}(\mathbf{k})$ through

$$Rij(\mathbf{r}) = \int \Phi_{ij}(\mathbf{k}) e^{\iota \mathbf{x} \cdot \mathbf{k}} d\mathbf{k} \qquad 8.33$$

This relationship suggest that the correlations can be calculated in two ways. Directly in the physical space, but then more computational time is required, since at each point the product $u_i(\mathbf{x})u_j(\mathbf{x} + \mathbf{r})$ necessitates an "IF" statement and an inner "DO" loop. To understand that this is a very long calculation, I am suggesting to see how the PDF of the velocity increments $P(\delta u_j(r)) = P((u_j(x_j + r) - u_j(x_j))$ are evaluated in the routine $STFUNN$. The correlations can be calculated in a similar way, and by looking at the structure of the routine $STFUNN$ it turns out that indeed the computational time is large. To save time it is convenient to evaluate the correlation by FFT routines. This second choice is used in the routine $CORREL$ for the correlations of the normal components. The routine $CORREL$ is called in the routine $OUTPUT$.

From the velocity field in physical space $u_i(x_1, x_2, x_3)$ a real FFT and two consecutive complex FFTs give the complex velocity components in the wave number space $\hat{u}_i(k_1, k_2, k_3)$. In isotropic turbulence the correlations are independent of the direction. Since it is interesting to investigate how the correlations change with the background rotation and since it is expected that the changes are in the direction parallel to the rotation axis, the real FFT is in the direction 3. The dimension of the arrays with the complex numbers $\hat{u}_i$ are then $N1$, $N2$ and $N3/2 + 1$. The summation of all k_1 and k_2 components of $\hat{u}_i(k_1, k_2, k_3)u_i^*(k_1, k_2, k_3)$ gives the real quantity $\Phi_{ii}(k_3)$. From this quantity the direct real FFT furnishes the correlation components along the x_3 direction $R_{ii}(x_3)$. These normalized by $R_{ii}(0)$ give the correlation coefficients. These coefficients show whether the box size is sufficiently large for a correct simulation

The correlation $\Omega_{ii}(x_3)$ for the vorticity components have also been calculated. Interpolations are avoided and each staggered vorticity component enters in the subroutine $CORREL$ to give the correlation coefficients of the vorticity components. In evaluating spectra or correlation coefficients, as a rule, interpolations in the direction of the FFTs should be avoided; in fact, interpolations filter the signal and produce an artificial dissipation at high wave numbers.

In Fig.8.10a-b the velocity and vorticity correlations at $\tau = 4.12$ without rotation are compared with those at $N = 4$ at the same τ. In isotropic turbulence longitudinal and transversal correlations are defined; these are indicated in text books as $f(r)$ and $g(r)$ respectively. The convection used here suggests that R_{33} coincides with $f(r)$ and the other two with $g(r)$. From $f(r)$ the theoretical relationship $g(r) = f(r) + \frac{r}{2}\frac{\partial f}{\partial r}$ permits to evaluate $g(r)$. I would like to point out that small oscillations on $f(r)$ could produce large oscillations on $\frac{\partial f}{\partial r}$; hence this check of the isotropy is more restrictive than the other where $f(r)$ is yielded by an integration from $g(r)$. At $\tau = 4.12$ the correlations $R_{11}(x_3)$ and $R_{22}(x_3)$ do not perfectly coincide with the $g(r)$ evaluated from $R_{33}(x_3) = f(r)$. On the other hand the same calculation at $\tau = 2.06$ shows a better agreement. From this observation it turns out that the growth in time of turbulent scales increase the degree of anisotropy of the flow. The value different from zero of R_{ii} at $r = \pi$, in Fig.8.10a, confirms that the scales become large. The occurrence of energy containing eddies at scales of the same order of the box size is also clear, in Fig.8.8a, from the energy spectra, at low wave number where, even at $\tau = 0$, the k^2 scaling law is barely observed. This fact is related to the choice of $k_p = 13$ in the initial spectrum, that is, to the decision to fix the peak of energy at large scale. The vorticity correlations for $N = 0$ drop to zero at shorter distances, by proving that the degree of isotropy of vorticity is higher, as was also shown by contour plots in Fig.8.9.

By adding the solid body rotation the velocity correlations change and show the increase of the length scales in the direction parallel to the axis of rotation. The occurrence of a negative minimum at $r = \pi$ is an indication that the turbulent structures span the full box. This behavior confirms the speculation, by the Taylor Proudman theorem, that rotation reduces the variations of the fluctuations along the rotation axis. The vorticity correlations quantitatively reproduce the pictures of Fig.8.9 where it was shown that the vorticity scales increase in the x_3 direction, that ω_1 and ω_2 do not largely differ from each other and that both differ from ω_3.

8.12.b Probability Density Functions

In turbulent flows the PDF are a further tool to characterize the flow by showing how the flow deviates from pure randomness. In the code $STATISO$ the PDF of several quantities can be evaluated. The subroutines related to the PDF calculations are grouped in the file *staisopd.f*, where the routine $PDFQUA$ shows how the PDFs are calculated. The probability $P(f)$ that a quantity has a value f is evaluated by fixing the range of variability of the quantity $(-f_M < f < f_M)$; the range $2f_M$ is subdivided into a number of intervals indicated by *lipdf*. The value of f is related to an index l of an array of integer numbers. When f lies between $(l+1)\Delta f$ and $l\Delta f$ the value of an integer quantity $n_f(l)$ corresponding to l increases by one. The number of the total events is $n_T = \Sigma n_f(l)$. The probability density function $P(f)$ is given by $P(f) = n_f(l)/n_T$. In the analysis of

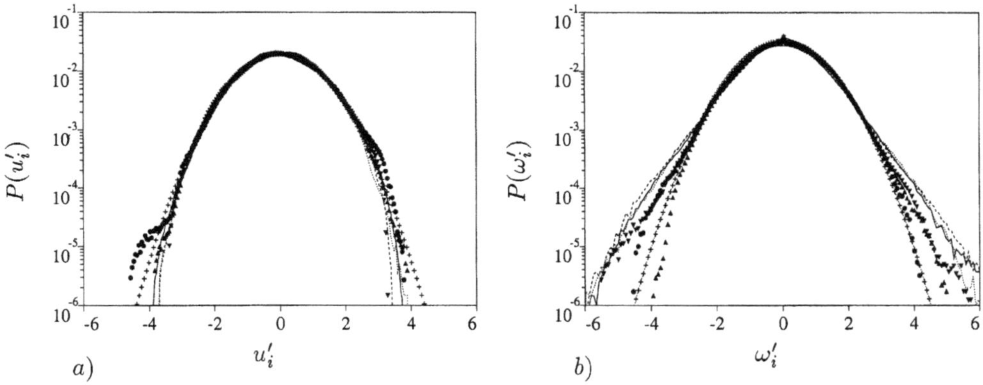

Fig. 8.11 PDF of a) Velocity, b) Vorticity, at $\tau = 4.12$ for $N = 0$ (—— u_1', ⋯⋯ u_2' and ---- u_3'), and for $N = 4$ (▿ u_1' , • u_2', ▵ u_3'), + Gaussian distribution. The same symbols hold for the vorticity components.

turbulent flows usually the fluctuating quantities are defined in such a way that they have zero mean value, and in evaluating the PDF the fluctuations are scaled with its *rms* value. This scaling leads to the relationships

$$\int_{\infty}^{\infty} P(f)df = 1, \quad \int_{\infty}^{\infty} fP(f)df = 0, \quad \int_{\infty}^{\infty} f^2P(f)df = 1 \qquad 8.34$$

It is useful, in addition, to remember that a Gaussian probability distribution is

$$P(x) = \frac{e^{-x^2/2}}{\sqrt{2\pi}}$$

From Dwight (1961) (Pg.330) it turns out that

$$\int_0^{\infty} x^{2a} e^{-x^2 r^2} dx = \frac{1 \cdot 3 \cdot 5 \cdot \cdot (2a-1)\sqrt{\pi}}{2^{a+1} r^{a+1}}$$

hence it is easy to show that the fourth moment is equal to 3, the sixth to 15 and so on. The symmetry of $P(x)$ gives all the odd moments equal to zero.

If a turbulent quantity, for example the rate of energy dissipation ϵ, varies between low $O(10^{-3})$ and high values $O(10^3)$ it is convenient to relate the index l to $log(f)$. This PDF is calculated in the routine *PDFQLO*.

The PDF of the three velocity and vorticity components, of the pressure dissipation and of the cos and sin of the angle between the velocity and vorticity vector have been evaluated. Moreover, to investigate some of the characteristics of turbulence linked to intermittency, the PDF of the velocity and vorticity differences have also been calculated. The routines related to PDFs are called in the subroutine *OUTPUT* in the file *staisoou.f*. In this routine the results are written on separated files with names indicating the quantity.

8.12.b.1 PDF of velocity and vorticity components

In Fig.8.11a the PDF of the fluctuations of the velocity components scaled by its *rms* values, shows that (Batchelor (1953, Pg.170)) the velocities have a Gaussian distribution. With background rotation the figure shows that the velocities to a large measure follow the Gaussian probability with some deviations at the extrema. The Gaussian probability distribution of the velocities resulted from the values of the skewness and the flatness of each component. At $\tau = 4.12$ the values are $s_1 = 0.017, s_2 = 0.018, s_3 = 0.099$ for the skewness and $f_1 = 2.837, f_2 = 2.748, f_3 = 2.738$ for the flatness; these values indeed are close to those expected by a Gaussian distribution. When the background rotation is added (the turbulent energy decay being reduced) the *rms* velocities $< u_1'^2 >= 6.0 \cdot 10^{-4}, < u_2'^2 >= 5.6 \cdot 10^{-4}, < u_3'^2 >= 6.2 \cdot 10^{-4}$ for $N = 0$ become $< u_1'^2 >= 1.72 \cdot 10^{-3}, < u_2'^2 >= 1.54 \cdot 10^{-3}, < u_3'^2 >= 1.75 \cdot 10^{-3}$ for $N = 4$. The values of the skewness and flatness instead do not change, remaining close to zero and to 3.0 respectively. The PDF in Fig.8.11a confirms this behavior and shows minor deviations for the rare and intense events. From the statistics of the fluctuating velocities it could be argued that isotropic turbulence is a random process. Indeed it is well known that this is not true and the confirmation comes from the analysis of the vorticity field.

Fig.8.11b shows that the PDF of the vorticity field deviates from a Gaussian probability distribution, since the probability to have events with a high value increases. These are the events related to the intermittent nature of turbulence. The analysis of the regions where the vorticity intensifies has recently attracted the interest of several scholars. Douady *et al.* (1991) performed the experiments in laboratory. The number of studies based on numerical data were based on DNS of decaying and forced isotropic turbulence. (Ashrust *et al.* (1987), She *et al.* (1990), Vincent & Meneguzzi (1991), Jimenez *et al.* (1993)). The structures associated with a high level of vorticity have been called "worms" in analogy with the picture given by Bradshaw. These structures have a coherence along their axis and their length is comparable to the integral length scale of the turbulence. Verzicco *et al.* (1995) have shown, by axisymmetric simulations of compact vortices subjected to an external rate of strain, that if these structures break or if they are diffused, they immediately reform due to the waves travelling along the axis.

In the presence of solid body rotation the PDFs show that the anisotropy of the vortical structure increases, as depicted in Fig.8.11b, where the number of locations with a high peak of ω_3 is largely reduced with respect to those of the other two vorticity components. In addition, Fig.8.11b shows that a symmetry breaking on ω_3 occurs. The reason of the reduction of the number of events with negative ω_3 depends on the sign of the background rotation. The symmetry breaking suggests an investigation in turbulent flows of the distribution of the angle between velocity and vorticity vectors.

8.12.b.2 PDF of pressure field

The pressure received less attention in the analysis of turbulent flows since it is difficult to measure, in particular far from solid walls. In direct simulations, on the other hand, pressure appears in the equations then and hence it is accessible. By applying the divergence operator to the momentum equation, if the background rotation is directed along

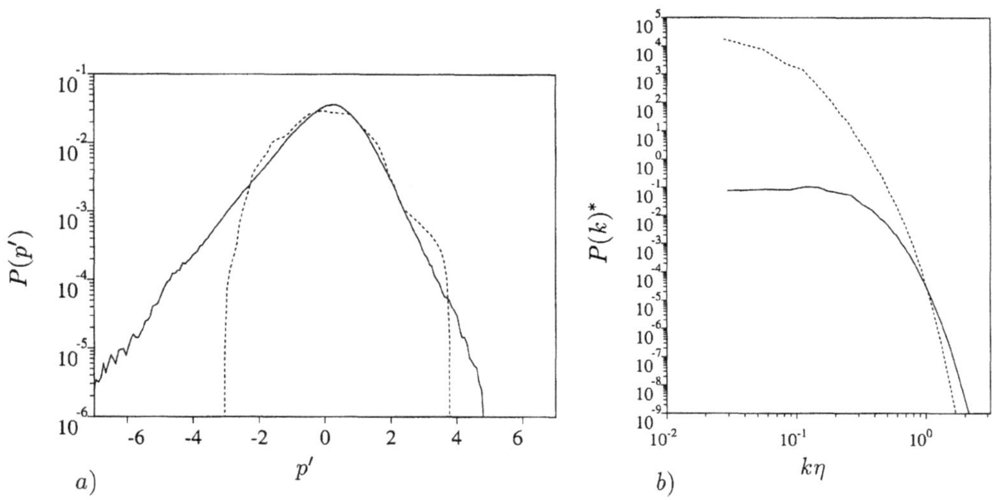

Fig. 8.12 a) PDF of pressure field b) spectrum in Kolmogorov scaling, —— $N = 0$, ---- $N = 4$.

x_3, the following relation holds

$$\nabla p = \frac{\omega_i \omega_i}{4} + \frac{N\omega_3}{4} - S_{ij}S_{ji} \qquad 8.35$$

This relation shows that when the vorticity prevails on the strain rate there is a minimum pressure. Since regions of intense vorticity are more dissipative than regions of high shear, it turns out that in decaying turbulence the PDF of the pressure should show a skew profile with a greater number of negative values. It is important to recall that in the locations where vorticity prevails the structures are rod like and in the regions where the shear prevails the structures are ribbon like. To have a deep knowledge of the importance of this analysis I suggest to read the papers by Ashurst *et al.* (1987) and by Kerr (1985), who evaluated the rate of shear on the principal axes and the PDF of the alignment of the vorticity vector. These two papers can be considered as the basic lectures to understand the topology of the structures in isotropic turbulence. I decided to neglect this type of analysis in isotropic turbulence and to perform it in wall turbulent flows where the variety of structures is larger. Whoever is interested in the topological studies could easily introduce the parts related to the evaluation of eigenvalues and eigenvectors for the channel in the code $STATISO$ for isotropic turbulence.

Fig.8.12a shows that background rotation reduces the number of locations where the local pressure largely overcomes its rms value. The pressure goes towards a Gaussian distribution very different from the skewed form in absence of rotation. Background rotation reduces the rate of dissipation, hence the structures are larger. An equilibration on the right hand side of Eq.8.35 produces a reduction of the skewness and flatness values of the

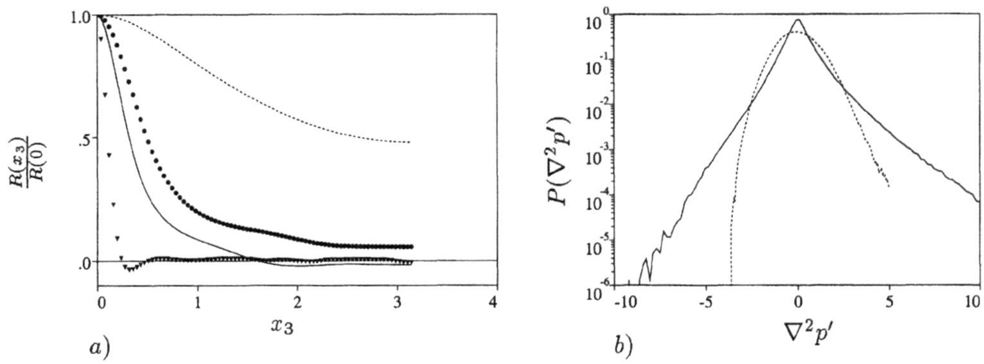

Fig. 8.13 a) Correlation of p (—— $N = 0$, ---- $N = 4$) and $\nabla^2 p$ ($\triangledown$ $N = 0$, $\bullet$ $N = 4$) b) PDF of $\nabla^2 p$ —— $N = 0$, ---- $N = 4$.

pressure, which pass from $S_p = -0.64$ and $F_p = 4.74$ in the non-rotating case to $S_p = 0.013$ and $F_p = 2.92$ at $N = 4$. The rms value of the fluctuating pressure increases with the background rotation, from $< p'^2 >^{\frac{1}{2}} = 2.4 \cdot 10^{-7}$ to $< p'^2 >^{\frac{1}{2}} = 7.47 \cdot 10^{-3}$. Fig.8.12b shows the increase of the rms of the pressure by its spectrum in Kolmogorov units; it turns out that the pressure spectrum increases at large scale. This occurrence is a further indication that background rotation produces large and elongated structures in the direction of rotation.

The correlation coefficients of pressure and ∇p along x_3, the rotation axis, in Fig.8.13a show that the pressure is highly correlated in this direction when the rotation rate is sufficiently high and that, instead, it drops to zero in the absence of rotation. It is important to remember that the correlation of pressure, velocity and vorticity should account for the fact that each quantity is related to the other through a differentiation operator. The differentiation emphasizes small scales, hence it follows that the pressure indicates large scale, velocity intermediate and vorticity the smallest. The correlation of $\nabla^2 p$ in Fig.8.13a, that is the right hand side of Eq.(8.35), shows that indeed this correlation drops to zero faster than the pressure correlation. The PDF of the right hand side of Eq.(8.35) in Fig.8.13b shows, in agreement with the PDF of the pressure that without rotation there are more points with enstrophy prevailing on the shear. Background rotation reduces the number of high peaks, that is, it reduces the intermittent character of isotropic turbulence. Also in the presence of rotation the PDF of $\nabla^2 p$ is positively skewed, since the turbulence is still decaying. As a major conclusion it can be asserted that background rotation decreases the intermittency of turbulence. This reduction can also be emphasized by contours of pressure in a plane passing through the center of the box. The same plane of Fig.8.9 has been assumed and the visualization is obtained by the code *visiso.f* where also the pressure is evaluated at the vertex of the cell. The pressure is an indicator of the amount of vorticity but accounts for large scales rather than small scales. The comparison between Fig.8.14a and Fig.8.14b, once more, shows that elongated structures in the direction of rotation form and that these span the full length of the domain. The values of the contour levels show

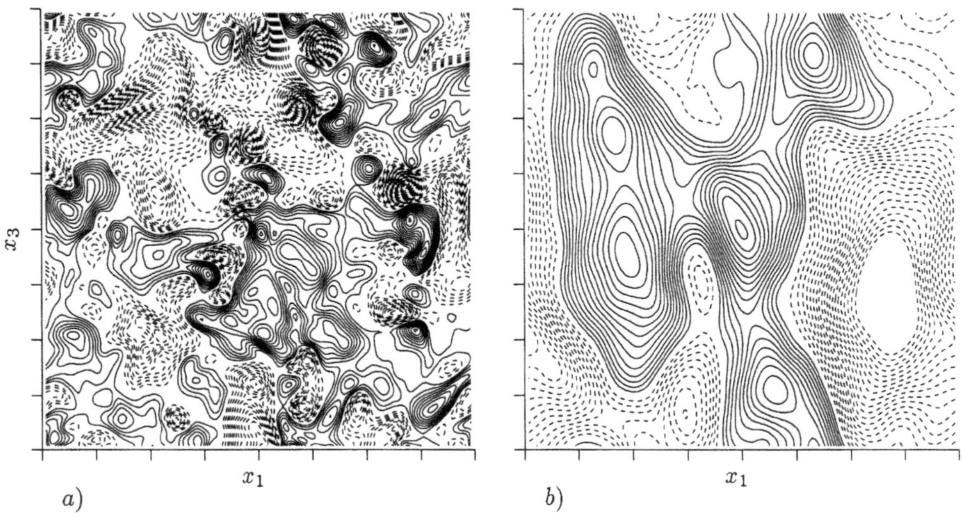

Fig. 8.14 Pressure fluctuation contour plots a) $N = 0$, $\Delta p = .0001$; b) $N = 4$, $\Delta p = .01$

that the pressure increases due to the Coriolis force, and this confirms the increase of the pressure spectrum content at low k in Fig.8.12b.

8.12.b.3 PDF of angle between velocity and vorticity vectors

The Navier-Stokes equations in vectorial form can be written as

$$\frac{\partial \mathbf{u}}{\partial t} + \mathbf{u} \times \boldsymbol{\omega} = -\nabla P + \frac{1}{\mathrm{Re}} \nabla \mathbf{u} \qquad 8.36$$

In these equations the Lamb vector $\boldsymbol{\lambda} = \mathbf{u} \times \boldsymbol{\omega}$ appears. In wave number space it is related to triadic velocity interactions and hence to energy transfer from large to small scales. The orientation between the velocity and vorticity vector then allows to understand the properties of turbulent flows. The orientation can be related to the properties of the tensor $\gamma_{ij} = v_i \omega_j$. It should be pointed out, that this tensor is not Galilean invariant and usually it is preferable to deal with quantities independent of the reference system. However the fluctuating velocity field is Galilean invariant, hence, the tensor γ'_{ij} of the fluctuating variables does not have this inconvenience. The trace of the tensor is the helicity density $\mathbf{u} \cdot \boldsymbol{\omega}$. The components of the Lamb vector $\boldsymbol{\lambda}' = \mathbf{u}' \times \boldsymbol{\omega}'$ are obtained from the other elements of γ'_{ij}. Hussain (1986), as well as Rogers & Moin (1987), observed from the identity $|\mathbf{u}' \times \boldsymbol{\omega}'|^2 + |\mathbf{u}' \cdot \boldsymbol{\omega}'|^2 = |\mathbf{u}'|^2 |\boldsymbol{\omega}'|^2$ that, since the term $|\mathbf{u}' \times \boldsymbol{\omega}'|^2$ indicates a high rate of energy cascade to smaller scales, it might be expected that regions of high helicity density are regions where the energy cascade is inhibited. The PDF of $h' = \cos(\theta') = \frac{\mathbf{u}' \cdot \boldsymbol{\omega}'}{|\mathbf{u}'||\boldsymbol{\omega}'|}$ in Fig.8.15a agrees with that of Rogers & Moin (1987); both simulations show that in isotropic turbulence $P(h' \approx \pm 1)$ is more likely than $P(h' \approx 0)$. However the probability to have an alignment between each velocity and vorticity component is small.

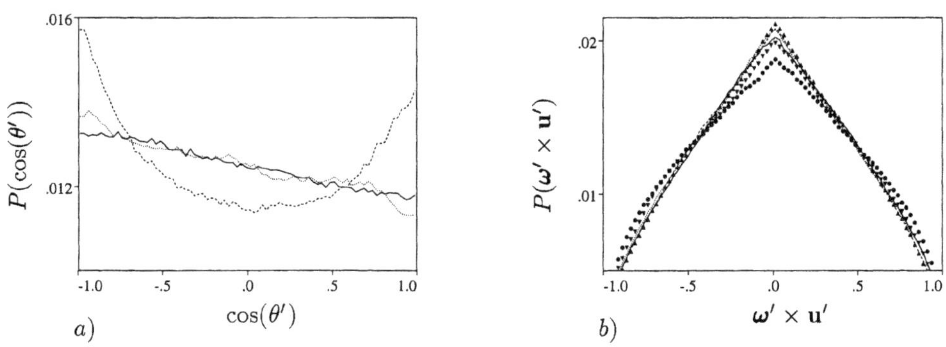

Fig. 8.15 a) PDF of $\cos(\theta')$ —— random field, ---- $N = 0$, ········ $N = 4$; b) PDF of the l components of $\boldsymbol{\omega} \times \mathbf{u}$ (—— $l = 1$, ---- $l = 2$, ········ $l = 3$, $N = 0$) (∇ $l = 1$, • $l = 2$, × $l = 3$, $N = 4$).

The external rotation shows distributions similar to those in strained homogeneous turbulence simulated by Rogers & Moin (1987). The PDF indeed shows that an external field applied to isotropic turbulence reduces the probability of alignment between velocity and vorticity. By comparing the rotating (present simulation), the irrotational strained (Rogers & Moin (1987) Fig.5) and the sheared turbulence (Rogers & Moin (1987) Fig.7), it turns out that an external mean velocity gradient increases the probability to have $h' \approx 0$. As it is shown in chapter 9, devoted to wall turbulent flows, this probability increases even more approaching the wall. The present data contrary to Rogers & Moin (1987) and more in agreement with those by Pelz *et al.* (1985) does not produce a good symmetry of the helicity PDF. As for Pelz *et al.* (1985) the reason is the asymmetry present in the generation of the random field. Fig.8.15a shows that more or less the mean helicity is approximately null in homogeneous turbulence with and without external rotation. It can be then concluded that background rotation does not create any mean helicity when applied to homogeneous flows. On the other hand if there is a wall, as in the laminar flows described in chapter 11 or in the experiment by Hopfinger *et al.* (1982), there is creation of mean helicity by counter-alignment between the velocity and vorticity vector. Fig.8.15.a with the PDF of h' of a random field, shows that with external rotation the PDF becomes similar to that of a random field, and this confirms the previous observation that rotation reduces the intermittency.

The PDF of the three components of $\frac{\mathbf{u}' \times \boldsymbol{\omega}'}{|\mathbf{u}'||\boldsymbol{\omega}'|}$ permit to investigate how the energy transfer is affected by the rotation. Fig.8.15b shows that without rotation the three components have approximately the same distribution; on the other hand the alignment between velocity and vorticity in the direction of rotation (x_3) increases as previously observed. The components of the Lamb vector can be related to turbulence energy production, hence they are more interesting in inhomogeneous turbulence. Profiles and PDF of these components are analyzed in more detail in chapter 9 on DNS of wall bounded flows.

Since helicity is related to the energy cascade it is also related to the rate of dissipation.

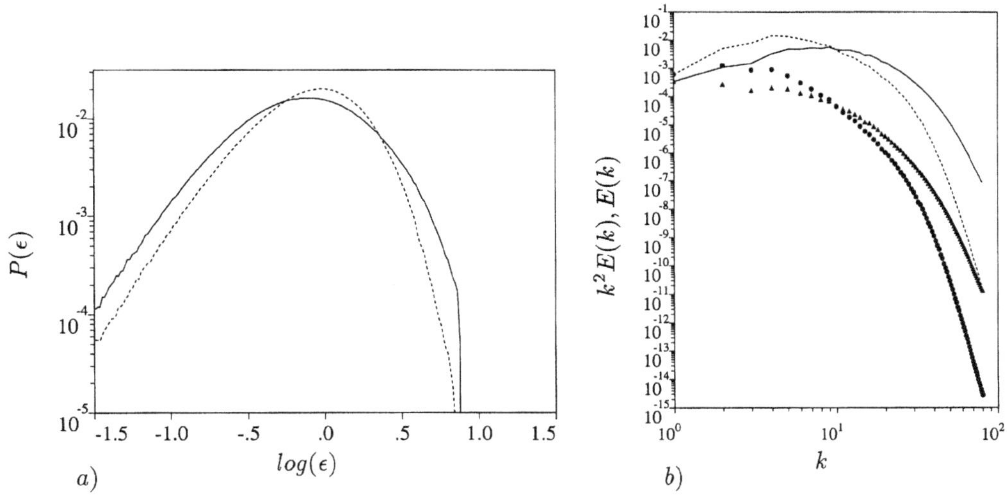

Fig. 8.16 a) PDF of local dissipation —— $N=0$, ---- $N=4$. b) Energy (—— $N=0$, ---- $N=4$) and enstrophy (△ $N=0$, ○ $N=4$) spectra.

Rogers & Moin (1987) showed by the conditional PDF of helicity density and local dissipation that high dissipation occurs in regions of low helicity density, that is where velocity and vorticity are not aligned. In section 9.8d, describing wall turbulent flows, these conditional PDFs are shown at different locations from the wall. In this section it is interesting to see how the external rotation affects the PDF of the rate of dissipation. Fig.8.16a shows that, as observed in the experiments at moderate R_λ (Gledzer *et al.* (1996)), there is low probability to have in the flow high and low values of the rate of dissipation. The rare events of high and low dissipation makes this quantity very interesting to study the intermittent character of turbulence. In agreement with the previous findings the background rotation reduces the intermittency of isotropic turbulence. The intermittency is related to the small scales and hence Fig.16b shows energy and enstrophy spectra where it turns out that with rotation the small scales play a reduced role.

8.12.b.4 Velocity structure functions

The high order statistics of the velocity differences, namely the velocity structure functions, are perhaps the best indicators of the intermittent character of turbulence

$$F_{pj}(r) = <\delta u_j(r)^p> = \int \delta u_j(r)^p P(\delta u_j(r)) \delta u_j(r) \qquad 8.37$$

where $P(\delta u_j(r)) = P((u_j(x_j + r) - u_j(x_j))$ is the probability distribution function of the velocity difference at scale r. As reported in the Frisch (1995) and in several papers as for instance in Briscolini & Santangelo (1994) the theory based on the intermittency is based on the hypothesis that the structure functions $< |\delta u_j(r)|^p >$ scales as $r^{\zeta(p)}$ where the $\zeta(p)$

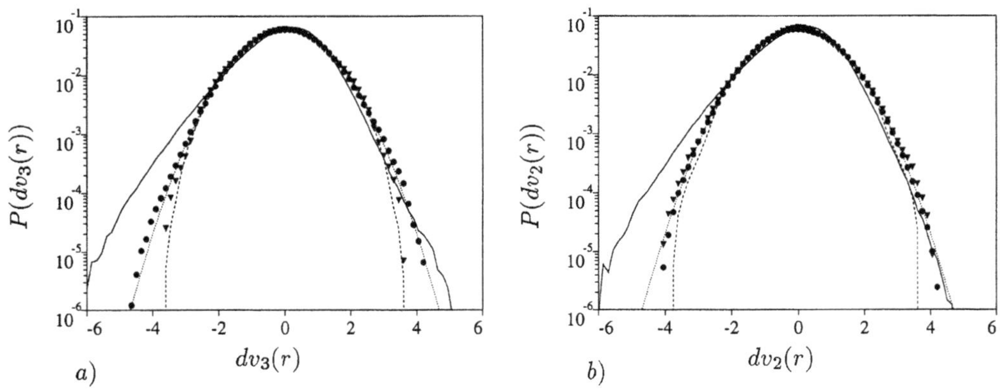

Fig. 8.17 PDF for the normalized velocity a) $dv_3(r)$ and b) $dv_2(r)$ for ——— , • $r = \Delta x$, - - - - , $\triangledown$ $r = \pi$; lines $N = 0$, symbols $N = 4$, ········ Gaussian distribution.

exponent has different laws depending on the theory or on the model. The Kolmogorov-Obukov log-normal model gives

$$\zeta(p) = \frac{p}{3} - \frac{\mu p(p-3)}{18} \qquad 8.38$$

where μ is the intermittency coefficient; it is a measure of the correction to the $\zeta(p) = p/3$ law, given by the K41 theory. The intermittency correction, can also be obtained by the PDFs of the velocity differences. These have a Gaussian distribution at large distances ($r > \eta$) and become increasingly non-Gaussian at small r. This behavior is consistent with the PDFs of velocity and vorticity in a single point shown in Fig.8. The vorticity, in fact is proportional to the velocity differences between points one mesh distant. If the distance is very large the two velocities are independent and thus the PDF of the difference coincides with that in one point.

The PDF of the velocity difference permits to evaluate any moment of high order p through Eq.(8.37). Those who have a large background in statistical mechanics know that high order moments depend on the tails of the PDF. I am recalling some of these concepts because this book is of interest to engineering students that do not receive a deep education in statistic mechanics. Thus, it is important to recall that, the tails of the PDF decay smoothly only if the number of samples are very high. An heuristic way to check whether the number is large enough is to plot the integrand of Eq.(8.37) for each p and to verify that at the extrema it goes to zero. In the present case it can be shown that up to $p = 8$ the integrand behaves satisfactorily well. The calculations of the PDF are performed in the routine $STFUN$ in the file *staisopd.f*. The rms value at each r

$$\sigma_j(r) = <|u_j(x_j + r) - u_j(x_j)|^2>^{\frac{1}{2}} \qquad 8.39$$

of the velocity component difference is necessary to get the dimensionless variable $dv_j(r) = (u_j(x_j + r) - u_j(x_j))/\sigma_j(r)$ and its PDF. The PDF even for the differences are evaluated

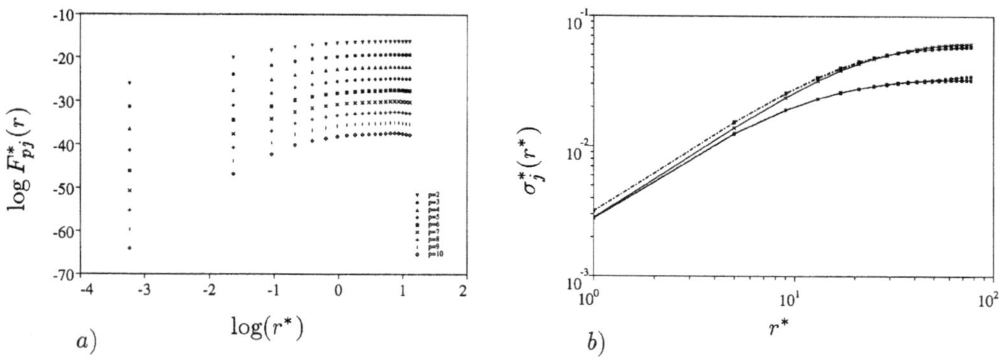

Fig. 8.18 a) Structure functions and b) $\sigma_j(r)$ as a function of r for $N = 0$.

as was explained in section 8.12.b. Figs.8.17a-b show the PDF at $r = \Delta x_j$ and at the maximum distance $r = \pi$ for two velocity components in the rotating and in the non-rotating case. The PDFs for $N = 0$ show a Gaussian distribution at $r = \pi$ and exponential tails at $r = \Delta x_j$ in agreement with other direct simulations (Vincent & Meneguzzi, (1991)). The PDFs of the velocity differences at $r = \Delta x_j$ are negatively skewed, this confirms that the skewness of the velocity derivative in isotropic turbulence is negative. The values of $\frac{F_{pj}(\Delta x_j)}{\sigma_j(\Delta x_j)}$ for $p = 3$ are $-0.48, -0.51, -0.50$ respectively for each velocity component at $R_\lambda = 36$. These values agree with the experimental values given by Tavoularis *et al.* (1978). Vincent & Meneguzzi (1991) reported in their Fig.7 values of $x_i(r)$ higher than here; they found $P(dv) = 10^{-6}$ for $dv = -10$. These differences depend on the flow; in fact they analyzed forced turbulence, thus they were in conditions of a statistical steady state; they could use more fields and hence they had a large number of statistical samples to get values of $P(dv) = 10^{-8}$. In addition, the forcing produces large scale strain rates that generate large velocity differences at small scales. The PDF of $dv_3(r)$ and $dv_2(r)$ show that even at the smallest distance, background rotation decreases the intermittency and the PDF follows the Gaussian distribution, independently of the distance.

A quantitative measure of the intermittency of turbulent flows is given by the exponent of the power-law of the structure function $\zeta(p)$. At this point it is interesting to recall that the Kolmogorov theory of 1941 was derived at high Reynolds numbers (Monin & Yaglom (1971, Pg.395 Vol.II)) hence the $\zeta(p)$ correction holds in the inertial range. Since in this low Reynolds number simulation of isotropic decaying turbulence the inertial range does not exist, this exponent cannot be evaluated as done by Anselmet *et al.* (1984) or by Vincent & Meneguzzi (1991). Their procedure is complicated and was extrapolating the PDF at high values of $dv_i(r)$ to get good values at high p. Fig.8.18a shows for $N = 0$ that at high r^* the velocity structure functions for p going from 2 to 10 do not have a long range with a power-law. For each p, $F_p(r)$, in Kolmogorov units, at small r^* at different R_λ collapse. The flattening of $F^*_p(r^*)$ for $p > 2$ is related to the asymptote reached by the

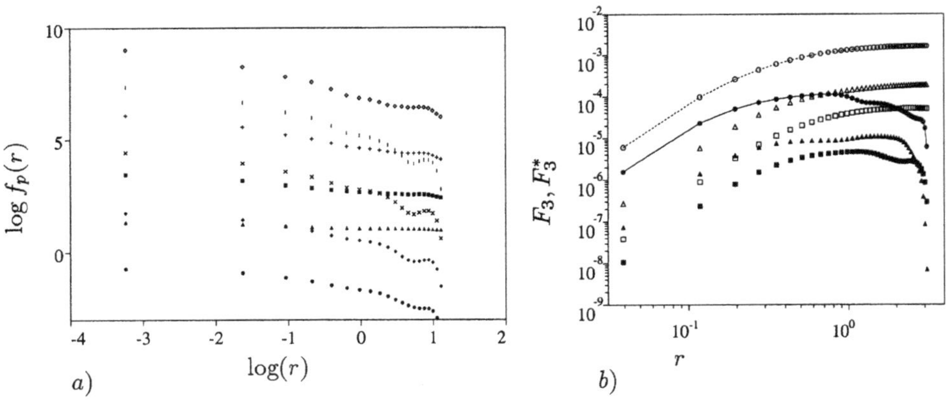

Fig. 8.19 a) Dimensionless structure functions and b) Structure functions; —— $F_p^+(r)$, symbols $F_p(r)$, as a function of log(r) for $N = 0$

$\sigma_j^*(r^*) = \sqrt{F}_2^*(r^*)$ when r increases. $F_{jp}(r)$ can be written as

$$F_{jp}(r) = f_{jp}(r)\sigma_j(r)^p \qquad f_{jp}(r) = \int dv_i(r)^p P(dv_i(r))\delta dv_i(r) \qquad 8.40$$

where $f_{jp}(r)$ is a dimensionless structure function. The subscript + in F_p or f_p indicates that the PDFs are evaluated by substituting $dv_i(r)$ in the integrand of $f_p(r)$ with $|dv_i(r)|$. For $p = 3$ and $p = 4$ $f_p(r)$ corresponds to the skewness and flatness of $\delta u_j(r)$. The comparison between Fig.8.19a and fig.18.8a shows that the even moments of $f_p(r)$ have a good scaling with r.

From this consideration, even without showing clearly this aspect, Vasilenko *et al.* (1975) and Anselmet *et al.* (1984) plotted the even moment versus the second moment and they confirmed the values of the exponent evaluated by the elaborate procedure based on the correction of the PDF. Frisch (1995) in his book mentioned that Benzi *et al.* (1993) re-discovered the Anselmet *et al.* (1984) procedure that since then is called extended self-similarity (ESS). Later on Camussi *et al.* (1996) applied the ESS to low Reynolds number isotropic turbulence and they found for order p up to 8 a good comparison with those at higher R_λ of Anselmet *et al.* (1984). Benzi *et al.* (1993) and Camussi *et al.* (1996) evaluated the exponents by a best-fitting of the slope between the $F_{pj}^+(r)$ evaluated with $|\delta u_j(r)|$ and that with $p = 3$. The rationale (Benzi *et al.* (1993)) in using F_3^+ as the reference in ESS was that, according to Kolmogorov (1941b), $\langle(\delta u)^3\rangle = 4/5\langle\epsilon\rangle r$ in the inertial range. Benzi *et al.* (1993) assumed that F_3^+ is equivalent to $|F_3| \equiv |\langle(\delta u)^3\rangle|$. This equivalence needs to be carefully assessed. Although Boratav & Pelz (1997) found that F_3 is proportional to F_3^+, Pearson and Antonia (1997) noted that $|F_3| \sim {F_3^+}^\alpha$ with the constant of proportionality less than 1 and α, of order 0.9, showing a weak dependence on R_λ. In the present simulation, because of the reduced number of samples, we found that for isotropic turbulence at three different Reynolds numbers, $|F_3|$ and F_3^+ behave as in Fig.8.19b. The proportionality at small r is shown is lost at large r. At small r, the present results are reliable since for $r = 2\pi/160$, $|F_3|/F_2 \approx 0.5$, i.e. the absolute value of the skewness of the velocity derivative found in isotropic turbulence. At high values of r, the disagreement could be ascribed to the reduction in the number of independent samples as r increases, thus causing noisy PDF tails. These oscillations have a large effect on the calculation of the odd F_p but not on F_p^+. The differences between F_p and F_p^+ become larger by increasing p the results are not shown because the interest in the odd moments is relative.

The smooth behavior of F_3^+ with r explains why the adherents of ESS got satisfactory laws for $\zeta(p)$. By the present data it has been checked that good scaling laws are obtained by plotting any moment of order q versus p. To avoid the critics related to the even exponents the scaling exponents have been calculated by plotting moments of order p versus F_2. The following relationship holds

$$\zeta_j(p) = \zeta_j(2)\frac{log(< \delta u_j(r)^p >)}{log(< \delta u_j(r)^2 >)} \qquad 8.41$$

Those who were using F_3^+ imposed that $\zeta_j(3) = 1$. I am aware that Eq.(8.41) replaces one ambiguity by another, since a value for ζ_2 must be imposed; accordingly, we assume the validity of the log-normal model (Eq.(8.38)). $\mu = 0.2$ provides satisfactory agreement with existing experimental data for $p = 2$. It has been noted however that the estimates for ζ_p, obtained in this manner, are equal to those obtained using F_3^+.

The accuracy of ζ_p largely depends on the accuracy with which the slope of F_p vs F_q is estimated. Fig.8.20a shows $\log - \log$ plots of F_p vs F_2 $[ESS(F_p)]$ for isotropic turbulence ($R_\lambda = 40.3$) where the structure functions have been calculated directly and by Eq.(8.40). The figure shows that there is a range of values for which the slope can be accurately evaluated. The differences observed at small r for $p > 4$ relate to the small number of samples causing large oscillations in the tails of $p(\delta u)$. The data for $r < 4\Delta x$ were discarded when estimating ζ_p.

Fig.8.20b shows that for $N = 0$ the exponent ζ_p, evaluated for each velocity component and for all the r are grouped in a narrow region that up to $P = 10$ agree well with the numerical values of Vincent & Meneguzzi (1991) and the experimental values of Anselmet *et al.* (1984) at higher R_λ. The differences could be ascribed to two factors: the first one is related to the insufficient number of samples for high order moments; the second, and more important, to the difference on the kind of flows; here it is decaying isotropic turbulence, in Vincent & Meneguzzi (1991) forced turbulence.

A random field with the same spectrum as that, at a given time, for decaying isotropic turbulence was generated to verify whether ζ_p was equal to $p/3$, that is, to check that the $K41$ theory is satisfied. This is a necessary, though not sufficient, condition for an ESS-type technique to work satisfactorily for isotropic turbulence. It is also a demonstration that the departure from $K41$ reflects the presence of vortical structures, whose presence is a consequence of the dynamics of the Navier-Stokes equations. From the large number of studies devoted to intermittency resumed in the review of Sherivasan & Antonia (1997) there is a large consensus on the attribution of the intermittency to the small scale vortical structures. The simulations described by Jimenez *et al.* (1994) up to 512^3 resolutions

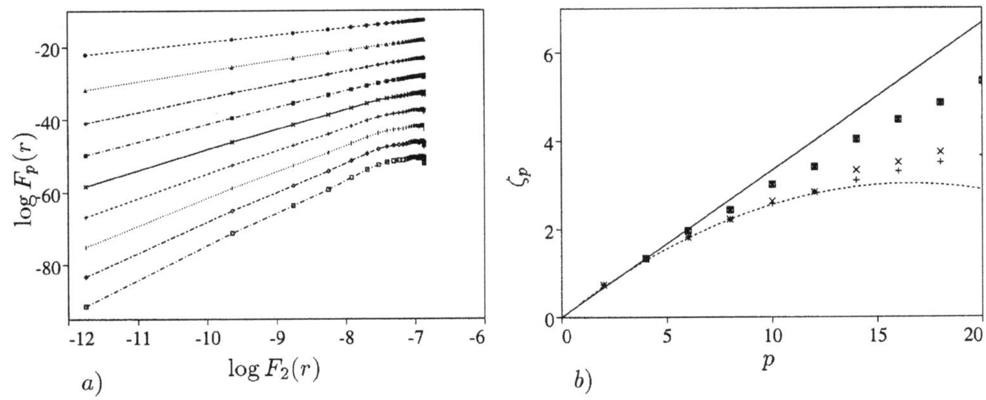

Fig. 8.20 a) Structure functions of even order $p+2$ versus those of order 2 for $N=0$, b) Scaling exponents for $N=0$ at $R_\lambda = 36$, □ present for u_3 compared with experiments of Anselmet *et al.* (+), simulations of Vincent & Meneguzzi (×), —— $p/3$ Kolmogorov law and ---- Log-normal law with $\mu = 0.2$.

investigated the nature of these small scale structures which have a circulation growing proportionally to $R_\lambda^{1/2}$ and a length of the same order of the integral length scale. However they found that even if these structures have a very high local dissipation, 80% of the $< \epsilon >$ is in the large scales and the tubes with high circulation had a volume fraction $< 1\%$. They concluded that these structures ("worms") are not relevant in the overall dynamics of turbulence. On the other hand the "worms" are important to characterize the intermittency, since these are the vortical structures that contribute the most to the high order structure functions and hence to the exponential tails of the small scales (Fig.8.16a). To confirm that the scales and not the energy spectrum produce the intermittency She *et al.* (1990) generated a Gaussian random velocity field with an energy spectrum identical to that obtained by solving the Navier-Stokes equations. They found that the patterns of the vortex lines differ. From flow visualizations they claimed that in the random field the large-scale vortex structures were absent. They did not evaluate ζ_p and this has been done here to quantify the absence of intermittency.

The scaling law has also been evaluated for the flow with background rotation and for the anisotropy of the flow the scaling law has been evaluated for $\delta u_1(r)$ and $\delta u_3(r)$ at two different times of the evolution. Fig.8.21 shows that the intermittency has been largely reduced and that the scaling law in the presence of the rotation coincides with that of the synthetic flow without the structures.

I would like to conclude this section devoted to the simulation of isotropic turbulence by saying that the analysis of turbulent data is very interesting and that it requires more new ideas than the generation of the turbulent fields. The first part of this chapter described the tools to generate the turbulent fields and to perform the post-processing of these fields. I am sure that the users of these codes could add into the post-processing codes the

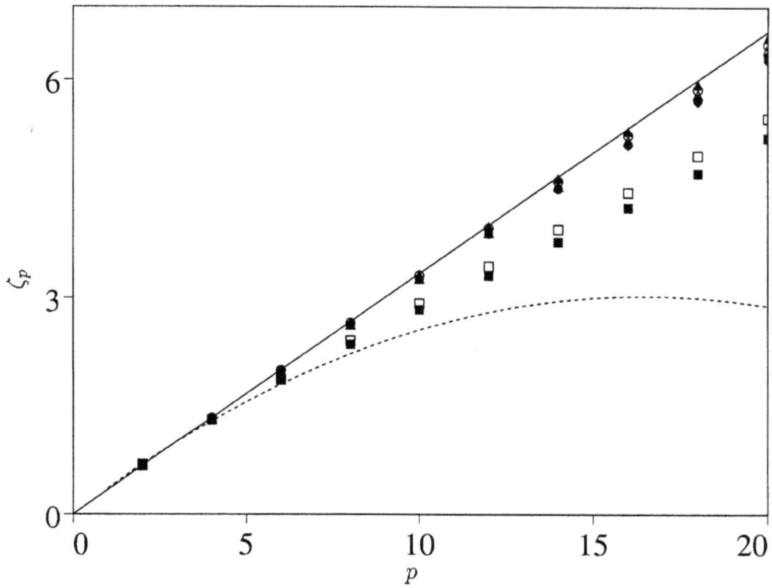

Fig. 8.21 Scaling exponents for ▪$N = 0$, open at $R_\lambda = 36$ and close at $R_\lambda = 40.3$, ▵ $N = 4$ $R_\lambda = 40.3$, open $\delta u_3(r)$, close $\delta u_1(r)$ ∘ $N = 4$ $R_\lambda = 36$, open $\delta u_3(r)$, close $\delta u_1(r)$, —— $p/3$ Kolmogorov law and ---- Log-normal law with $\mu = 0.2$.

calculation of other quantities that could lead to a better comprehension of turbulence. The most interesting feature of the post-process code is that it could help in understanding better the behavior of statistical quantities in isotropic turbulence described theoretically in text books. This consideration is certainly more appropriate to students of engineering faculties that do not have a deep background in statistical mechanics. I am one of them and thus I enjoyed and I learned a lot in writing this section.

8.13 Stability of the Lamb dipole to three dimensional disturbances

Spectral codes with three periodic directions have been used to understand vortex dynamics, and in particular I would like to cite the simulations performed by Melander and his coauthor to study vortex reconnection (Melander & Hussain (1990)) and that by Carnevale *et al.* (1997) to investigate the effects of background rotation on perturbed vortices with a Gaussian vorticity distribution. In this section it is shown that the *ISO* code can be applied to investigate whether the Lamb dipole is stable or not when it is subjected to three-dimensional disturbances. The initial conditions are described in section 8.8.2 and are assigned in the file *isoindip.f*. The routine *INIDIP* is called when $imic = -1$ is assigned in the file *iso.d*. To perform simulations the user should copy the file *iso.d – dipol* in *iso.d*. Preliminary 2D calculations are necessary to fix the minimum number of points in the x_1 and x_2 directions to have a dipole that translates without loosing its shape due to an insufficient resolution. These 2D simulations allow also to investigate at which distance from the center of the dipole, the boundaries parallel to the translation velocity should be located. In fact, periodicity supposes that there are image vortices on the other side of these boundaries which, if the boundaries are too close, reduce the translation velocity of the dipole. Important is also the choice of the Reynolds number to get a 3D fully resolved DNS. Almost inviscid $2D$ simulations show that the dipole maintains its coherence and travels with constant velocity. If the Reynolds number decreases the vorticity diffuses and the dipole slows down. The shape does not change but the centers of the two vorticity patches move apart a very small distance.

All these considerations were done by Orlandi *et al.* (1998b) in their study on the stability of trailing vortices with this code. They focused their study on the short wave instability with the aim to reproduce the experimental observations by Leweke & Williamson (1998). Taking as an initial condition the two semicircular halves of the Lamb dipole separated by some distance, it was found that in the subsequent evolution the two vortices adjusted the presence of each other more smoothly and without the large amount of vorticity shedding observed in the case initialized with two circularly symmetric vortices. In all of the simulations presented below, the unperturbed basic state is taken as the two halves of the Lamb dipole with the vorticity extrema separated by some distance b, and the initial condition is prepared by adding perturbations to this.

In all of the simulations presented here the initial Reynolds number was fixed at $\mathrm{Re}_\Gamma =$ 3400. To initialize the simulations of the laboratory experiments, the perturbation was a randomly generated three-dimensional velocity field. This perturbation was localized to the region were the axial vorticity ω_{z0} was greater in magnitude than a given threshold (set arbitrarily at 20% of the unperturbed vorticity maximum). The basic simulation consisted of the evolution of the counter-rotating vortices for a fixed period of time. Three different values were used for the separation b between the vortices to see how the growth of the instability varied with separation. Grid resolution checks suggested to perform the simulations with $(N_1, N_2, N_3) = (128, 192, 128)$ and $(L_1, L_2, L_3) = (2\pi, 3\pi, \pi)$.

A theoretical estimate of the speed of the dipole can be made based on the speed of a dipole composed of two line vortices. The azimuthal velocity field created by a straight line vortex of circulation Γ at a distance b from the vortex is $v_\theta = \Gamma/(2\pi b)$, where θ is taken as the azimuthal angular coordinate in a cylindrical coordinate system centered on the vortex. Thus, two mutually advecting line vortices of equal strength separated by a distance b will propagate at this speed. For the problem of trailing vortices, it is often convenient to nondimensionalize length by b, the separation between the centers of the vortices, and time by $\tau = 2\pi b^2/\Gamma$, the time it takes the dipole to travel a distance b. In these units, referred as τ units, the speed of the idealized dipole of line vortices is 1. All quantities without the asterisk superscript on the other hand are in the convective units with $a_{\mathcal{L}}$ the length and $U_{\mathcal{L}}$ the velocity units. $a_{\mathcal{L}}$ and $U_{\mathcal{L}}$ coincide with a and U in Eq.(4.39).

Histories of the position of the maximum of the vorticity for three simulations with different values of b are shown in Fig.8.22. The circulation in advective units was the same in each case: $\Gamma = 6.83$. The values of b were measured a short time after the initial adjustment of the dipole. In units of $a_{\mathcal{L}}$ the values of b were 1.0, 1.4 and 1.9. The position

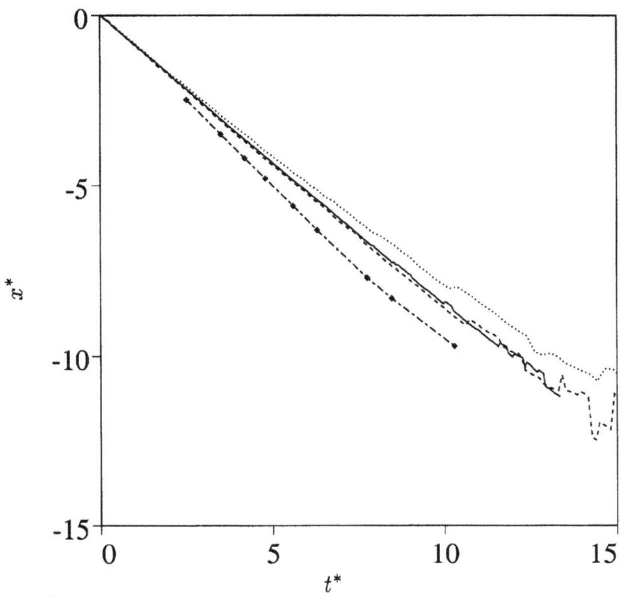

Fig. 8.22 History of the positions of the vorticity maxima. —— $b = 1.9$, ---- $b = 1.4$, ······· $b = 1$, and the line with symbols corresponds to the experimental data of Leweke & Williamson (1998).

and time for the graph are in τ units. Also shown in the diagram are the results from one of the laboratory experiments as given by Leweke & Williamson (1998). The laboratory experiments for the early evolution show a speed almost precisely equal to 1 in τ units. Our simulations, however, show speeds of about 0.85. There are two reasons for the reduced speed. These were discussed by Orlandi *et al.* (1998a). The more relevant is due to the size of the computational domain; in fact, as b increases the effect of the periodic replica of the vortices on the speed of the dipole increases. For example, on a domain with $L_y = 3\pi$ (in units of $a_{\mathcal{L}}$) and with $b = 2$ there would be approximately a 15% decrease in speed due to this effect. A decision about choosing the domain size, that would be large enough to give reasonable values for the speed and yet with a high enough resolution to observe small structures during the breakdown of the vortices has to be made. Some experimentation suggested that $L_y = 3\pi$ was a reasonable compromise.

There are various quantities that can be used to measure the progress of the cooperative instability. For the unperturbed pair of vortices, the only nonzero component of the vorticity is the axial vorticity ω_z. Thus a good indicator of the growth of an instability would be the evolution of the maximum value of the magnitude of one of the other components of vorticity. Figure 8.23a shows the evolution of the maximum value of the spanwise vorticity ω_2 for the three simulations with different values of b. Vorticity and time have been nondimensionalized using τ as defined above. Also plotted is the history of the maximum value of the axial vorticity (chain-dashed line) for one of the simulations. Each curve for ω_2

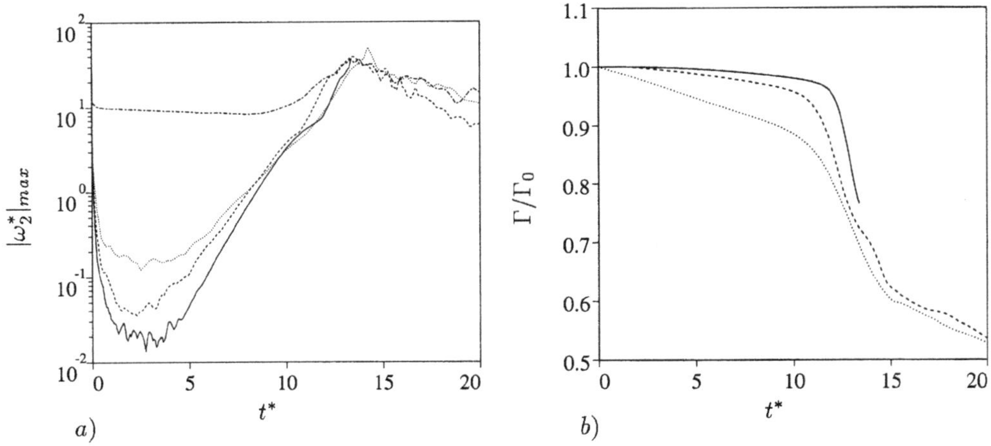

Fig. 8.23 a) History of the maximum value of the spanwise vorticity compared to the maximum value of ω_z b) History of the circulation normalized by its initial value for —— $b = 1.9$, ---- $b = 1.4$ and ········ $b = 1$.

has a section that is roughly linear on this linear-logarithmic plot, indicating exponential growth in time. In the inviscid theory, the growth rate for the short-wave instability is constant when measured in τ units. Thus, the approximate collapse of the data for the three values of b also suggests that this exponential growth is the result of the short-wave cooperative instability.

For the case of the largest b represented (solid line), the growth rate is approximately $\sigma^* = 0.91$ in satisfactory agreement with the theoretical prediction, $\sigma^* = 0.99$ (cf. Leweke and Williamson, (1998)). Note that the value of ω_2 becomes comparable to ω_3 for $t^* \approx 13$. The evolution of ω_1 (not shown) is similar to that of ω_2. The vorticity components in directions 1 and 2 becoming comparable in magnitude to the axial vorticity indicates that the dipolar structure of the vortices may be breaking down. The ω_1 and ω_2 components produce strong deformations associated with small scales as would occur in a transition to turbulent flow. The indication of the destruction of the vortices is given by the history of the circulation Γ in Fig.8.23b. The decay of the circulation can only be due to the cross cancellation of vorticity. By $t^* \approx 13$, in all cases there is a significant drop in the circulation. This drop in circulation occurs at approximately the same time as the values of ω_1 and ω_2 become comparable to that of ω_3.

To visualize the three-dimensional character of the shortwave instability, isosurface plots of vorticity magnitude are calculated. For each of the three simulations with different values of b the vorticity structures observed were qualitatively similar once time was scaled with τ. Visualization by this method shows some structures that are very similar to those observed by dye visualization in the experiments by Leweke & Williamson (see the top panel of Fig.8.24). Figure 8.24 shows the isosurface plots of the magnitude of the vorticity $|\omega|$ for the case $b = 1.9$. The isosurface value is the same at each time shown and is $|\omega/\omega_0| = 0.4$ where $|\omega_0|$ is the maximum magnitude of the unperturbed dipole vorticity

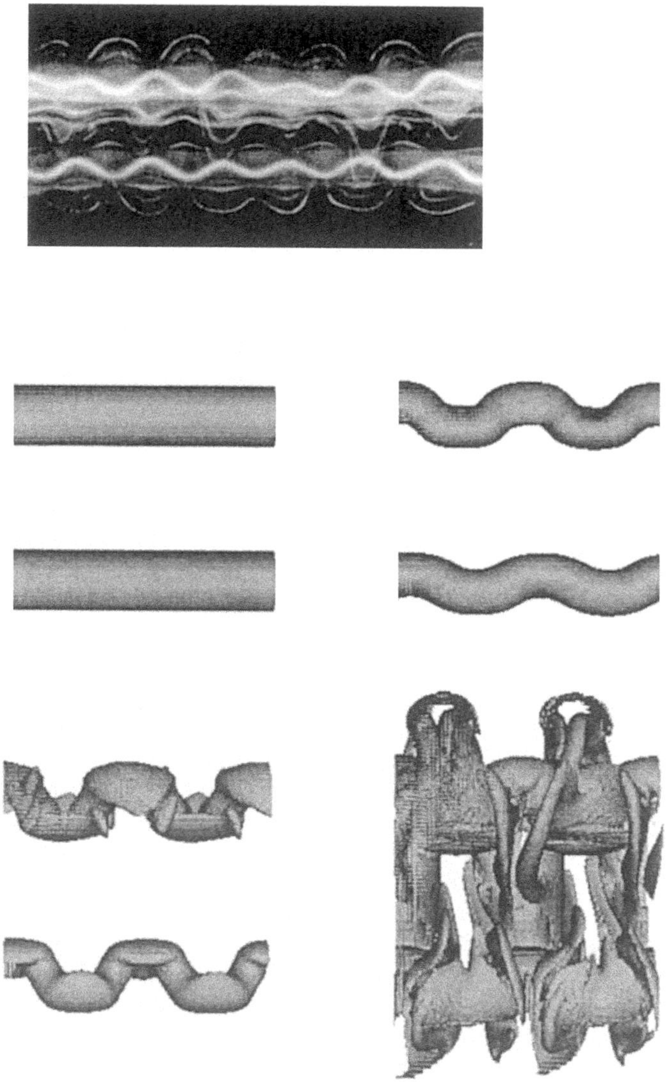

Fig. 8.24. Top flouroscine dye visualization of the short-wave instability in the laboratory by Leweke & Williamson (1998), Isosurface plot with $|\omega/\omega_0| = 0.4$ for the case of the two vortices separated by $b = 1.9$. The times represented from left to right, top to bottom, are $t^* = 1.5$, 9.0, 10.5 and 12.0.

field. Note that in both the laboratory experiments and the simulations the instability is

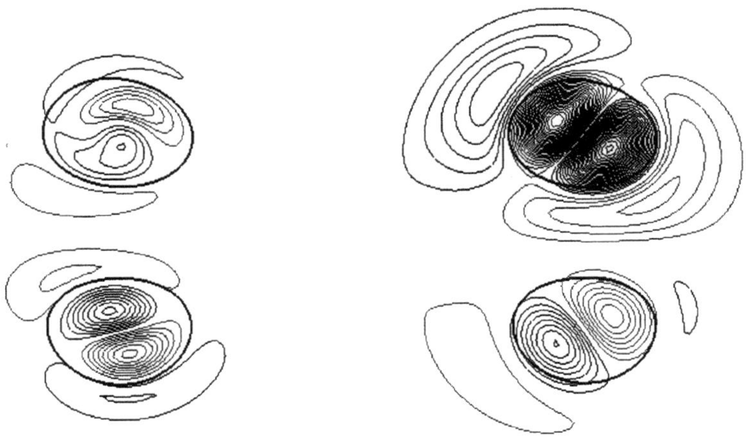

Fig. 8.25. Contours of ω_3' left and of u_3' right in an $x_1 - x_2$ cross section of one vortex in the dipole shown in figure 4 ($b = 1.9$). The time is $t^* = 9$. The thick solid curves indicate the contours of the magnitude of total vorticity $|\omega|$ at a level of e^{-1} times the maximum value.

sinuous, that is the sinusoidal bending of the vortex cores are in phase. This is interesting because in the linear theory there is no mechanism for choosing the phase relationship between the distortions of each vortex. The isosurface plot for $t^* = 9.0$ represents the field at a time in the exponentially growing phase indicated in figure 3a. As it is shown below, the perturbation vorticity and velocity fields at this point match the predictions of linear theory well. By time $t^* = 10.5$, nonlinear effects are evident. The formation of 'caps' on the points where the isosurface is most curved results from vortex stretching in the azimuthal direction. This is followed by the production of the vortices seen at $t^* = 12.0$ which span the two original cores and which begin the cross-diffusion of circulation.

It is interesting to consider the form of the perturbation during the exponential phase of the growth. Theoretical predictions for the fastest growing unstable mode can be found in Leweke and Williamson (1998) and Waleffe (1990). Figure 8.25 shows contour plots of the axial velocity and vorticity perturbation fields in an $x_1 - x_2$ cross section at time $t^* = 9.0$. The cores of the vortices are marked by the thick solid lines, which are vorticity magnitude contours at a level that is a factor of e^{-1} less than the instantaneous maximum. The perturbation fields are qualitatively as predicted by the linear theory. The asymmetry here is probably due to the asymmetry in the original random forcing. According to the theory, there should be a $\pm 45^o$ angular difference in the orientation of the the dipolar perturbation structures on the two vortices. In addition, the orientation of the dipolar velocity perturbation field should be perpendicular to that of the perturbation vorticity field. These relationships hold approximately in this cross section.

From a practical standpoint, it seems from figure 8.23a that random perturbations is an inefficient way of initiating the cooperative instability. The initial perturbation has a maximum vorticity amplitude of about 10% of that of the unperturbed vortices. However, this decays mostly in the initial transient period and, importantly, the larger the value of b the more profound is the initial decay. It took about 5τ periods for the exponential growth to become evident. Apparently the initial perturbation is made of a superposition of normal modes, many of which are decaying.

Orlandi *et al.* (1998a) proposed a different way to initiate the perturbation of the trailing vortices by density perturbation along the axis of the vortical patches. To perform the relative simulations the code was modified by solving also the density equation and by adding the buoyancy force in the momentum equations. I am not going to describe the results in this book.

8.14 Conclusions

In this chapter a finite-difference numerical method has been described to solve the Navier-Stokes equations in a three-periodic box. I think that this is the first example in literature where it was shown that finite-differences can be used to get as good results as those by pseudospectral methods for isotropic turbulence. My opinion is that this is an important results since it permits to use this very simple method to study the physics of isotropic turbulence. I was also astonished to find such a perfect agreement between the finite difference results and those obtained by the best numerical method in literature, the Rogallo (1981) code.

As I just mentioned several times finite-differences can be easier understood by students that do not have a large background in mathematics. In the last years I experienced, in teaching a turbulence class, that the students very rapidly become acquainted to this code. By this tool they could generate realistic turbulent fields and they learned in a simpler way the theory of isotropic turbulence, and, more important, they understood how to evaluate spectra, how to make averages and how to calculate PDF of velocity differences. In addition they were able to generate at home, on personal computers, turbulent fields, to visualize the vortical structures and to understand the relationship between structures, spectra and correlations.

I used this code also to study vortex dynamics. I took one of the most relevant examples that partially affects the life of the researchers during their trips around the world to attend the conferences where they present the results of their work. Everybody, in fact, is annoyed during takeoffs and landing of the planes waiting for the preceding aircraft to finish his manoeuvre. The cause of these delays are related to the trailing vortices shed by the airplanes that must be destroyed or dispersed. There are two strategies to ameliorate this situation. One involves attempting to better quantify the time interval needed for safety given current plane designs. The other considers the possibility of modifying and controlling the vortices to accelerate their dispersal. In either case, an improved knowledge of the possible instabilities of trailing vortices is essential. Thus, in an age of jumbo jets and congested airports, the evolution of vortices shed from the airplanes is a pressing issue and recently has received a great deal of attention. More realistic calculations of this topic at higher Reynolds numbers can be performed by the LES code described in chapter 12.

9 Flows with walls in Cartesian coordinates

9.1 Introduction

The previous chapter was devoted to flows in boxes with three periodic directions; in this case flows of practical interest are limited. However the periodicity permits to study the properties of turbulence in the simplest flow with the isotropy assumption. The introduction of one inhomogeneous direction increases the flow complexity and makes the study of interest for practical applications. These flows in general are classified as shear flows (Townsend (1956)); within this large class there are free-shear flows as jets, wakes, mixing layers and wall bounded flows. Free-shear flows were treated in previous chapters in two-dimension or in axisymmetric conditions with the aim to show the vortex dynamics related to the formation of the large scales, and their interactions. From the point of view of real and practical applications space developing flows are of more interest but, in this case, an enormous amount of grid points are necessary to resolve the transitional and the fully turbulent regions. In addition there are difficulties to assign inlet conditions reproducing those in laboratory experiments. With regard to these difficulties I would like to recall the simulations by Le *et al.* (1997) of the flow over a backward facing step and by Aksevol & Moin (1996) of a coaxial round jet. In the first paper the high cost of the direct simulation is emphasized. The direct simulations of mixing layers are even more expensive; in fact these flows develop for a distance longer than that in a backward facing step (see the pictures in the paper by Brown & Rosko (1974)). Buell (1989) performed one of the earliest direct simulations of a transitional space developing mixing layer but he did not have enough computer power to treat the fully turbulent regime.

The introduction of a subgrid scale model permits more realistic simulations on a reduced number of grid points. These simulations are not in the class of DNS but in that of LES, with other difficulties as it is shown in chapters 12 and 13. Le *et al.* (1997) and Lesieur *et al.* (1995) performed LES for the flow past a backward facing step. Few applications of LES to space developing mixing layers were, to my knowledge, done; one by Vreman *et al.* (1997) recently appeared.

Even more difficult are the simulations of space developing wakes which require also the simulation of the flow past the body. The difficulties increase to find the appropriate coordinate system and a complicated grid is necessary to have the due resolution where the vorticity is concentrated. Karniadakis & Triantafyllou (1992) proposed spectral elements methods, Beaudan & Moin (1994) finite differences and Kravchenko & Moin (1998) B-splines methods. In this last paper a comparison between different methods is also presented.

These difficulties to treat free-shear flows dictated my decision to limit this chapter to wall bounded flows on flat surfaces, which can be simulated by the Navier-Stokes equations in a Cartesian coordinate system. A non-uniform grid in the inhomogeneous directions permits to save computer power. The other two homogeneous direction reduce the number of fields requested to calculate the profiles of turbulent statistics. This is important since the memory occupancy of each field is large and often the allocation memory is expensive in certain computers centers, even if the cost is rapidly decreasing. The simulation of a plane channel is simple; it does not require inlet or outlet boundaries and periodicity in two directions helps to get fast solutions for the elliptic equation. Simulations can be performed at constant R_τ or at constant flow rate. In the first case the flow rate varies in time and the constant pressure gradient balances the wall friction. In the second condition the pressure gradient is evaluated, at each time step by integrating the streamwise momentum equation, to keep constant the flow rate within the machine accuracy.

The plane channel was for the first time simulated by Deardof (1972) by a coarse finite difference LES, and it immediately turned out that such a coarse resolution could not reproduce the complex physics of the wall region. To increase the resolution later on Moin (1979) developed an efficient pseudospectral method based on Chebishev polinomia in the normal direction and Fourier series in the other two directions. In that period I was in Stanford and in the atmosphere there it was the filling of the competition between the group led by Reynolds and the group at MIT led by Orszag. I was really fascinated by this competition, absent in the Italian universities, at that time and still now. These competitions are those which propel research and that drove the American universities in the last century to become the best around the world. Everybody knows that Stanford won the competition for the turbulent plane channel and that the group at MIT used their numerical method to investigate the non-linear transition of Poiseuille flows (Orszag & Patera (1980)).

Moin & Kim (1982) later on adapted this method to the ILIAC IV computer to get a realistic $64\times128\times64$ LES simulation. Even this operation was possible because Rogallo and Wray who wrote the compiler for the ILIAC IV were helping Moin and Kim to get results from that unfriendly computer. In fact it was continuously breaking and it had a very nasty language. However, this large simulation, at that time, permitted to understand that by increasing by a factor of two the number of grid points in each direction the subgrid model was not any more necessary. Thus Kim *et al.* (1987) performed the first direct simulation of wall bounded flows. After this paper, that is worldwide recognized as one of the greatest contribution to comprehension of turbulence, the number of papers concerning the simulation of channel flows increased so largely that it is almost impossible to cite all of them.

The simulations were performed in a domain extending 4π in the streamwise and 2π in the spanwise direction. Later on Jimenez & Moin (1991), to emphasize the regeneration process of wall vorticity, performed a simulation with the pseudospectral code in a smaller domain. They defined their box as the minimal unit where the turbulence is sustained; the reduction of the box size made feasible the direct simulation of turbulent channels even on desk workstations. Rai & Moin (1991) in addition proved that high order finite differences produce results as good as those by pseudospectral methods. Orlandi (1991) showed that second order finite differences with staggered velocities were satisfactory and this outcome allowed the extension of the method to complex geometries. Later on Choi *et al.* (1993) used this method to reproduce the physics of turbulent flows in the presence of riblets.

In the early period of fluid dynamics several studies analyzed the differences between boundary layers, channel and pipe flows (see chapter 7 of Hinze (1975) dedicated to wall flows and chapters 5 and 7 of Townsend (1956)). These flows are similar in the viscous and logarithmic layer (inner layers), but they differ in the outer region where the effects of the pressure gradients are important. The flows inside a pipe and between two parallel walls are described in this book. The boundary layer is not considered here because it changes in the streamwise direction and the periodicity assumption is not valid any longer. From the numerical point of view inlet and outlet boundary conditions should be assigned as for instance was done by Na & Moin (1998) by finite differences. Spalart (1988) developed the fringe method and thus he used the Fourier series even in the streamwise directions. Hilbert polinomia in the vertical direction together with Fourier series expansion in the other two directions produced a full spectral code to perform direct simulations of boundary layers, even with favorable and with mild adverse pressure gradients (Spalart & Watmuff (1993)).

The data base of the plane channel and of the boundary layer permitted to investigate, in great detail, the near wall region. The mechanism of turbulence production and of friction generation emphasized the role of the wall vortical structures, the so called near wall coherent structures. These structures were observed for the first time by Kline *et al.* (1967), and Kim *et al.* (1971) by flow visualizations in a water channel. About this flow it is interesting to read the review article by Cantwell (1982), written before the numerical data were available. From the article it turns out that qualitatively the near wall turbulence was understood but that today's knowledge is much greater. In fact, insights on the vorticity field can be derived by flow visualizations, but a one to one correspondence between marker distribution and vorticity field does not exist. The comprehension of the dynamics of the vortical structures is an easy task by the direct numerical simulation, since the vorticity field can be calculated with the same accuracy of the velocity field. Robinson (1992) using the data base by Kim *et al.* (1987) and by Spalart (1988), described in detail the coherent structures in the wall and in the log region. My opinion is that the most important events near a wall have been studied, but the fantasy of the scholars could produce new ideas to develop simple models to explain the regeneration process of turbulent structures. These models could help in the development of new subgrid models to account for the wall effects in Large Eddy Simulations at high Reynolds numbers. With this aim in mind I am describing in this chapter a code to solve turbulent plane channels.

In engineering simulations it is still necessary to use one-point turbulence models. These models received a lot of attention in the sixties and in the seventies, but, in that early period, little effort was devoted to separate the contribution of each term in the transport equations of the Reynolds stresses. Attempts were done by experimentalists to withdraw turbulent kinetic energy budgets from their measurements as for instance Laufer (1955) did in a turbulent pipe. To my knowledge there is only one paper by Mellor & Herring (1973) that attempted to reproduce the budget in one-point closure for a boundary layer. The reason why the large number of scholars involved in closures did not evaluate the budgets resides on the lack of reference solutions to compare the results. The experimental budgets were not completely reliable due to the difficulty to measure some of the quantities, for instance the rate of turbulent energy dissipation or even for the impossibility to measure others such as those related to the fluctuating pressure. The rate of dissipation was derived by one point measurements by the assumption of the Taylor hypothesis (chapter 8) and of the isotropy assumptions. Near the wall the latter assumption does not hold and hence the measurements could not be taken as a reference solution. However it is fascinating todays to see by a comparison with the numerical data that there was a qualitative agreement in the wall region and in the outer layer and that for some of the terms there was a quantitative agreement. See in this regard figure 5.5 (Pg.145) and 5.6 (Pg.152) in the Townsend (1956) book where the Laufer data are reported and discussed. From figure 5.5 (wall region) it turns out that the pressure gain is completely wrong.

On the other hand, direct numerical simulations permit the calculation of each term in the transport equations and thus new turbulence models can be validated. Mansour *et al.* (1988) calculated each term in the Reynolds stresses and in the rate of dissipation budgets. This study later on was used by Durbin (1991) to introduce the equation of the normal stress to have the right amount of dumping near a solid wall. Previously this dumping was obtained by ad hoc empirical functions, such as that suggested by Van Driest (1956). One of the post-processing codes, here given, permits to calculate these budgets; thus my aim is also to give a tool to those interested in turbulence closures, with which they can check new ideas. My opinion is that today's one-point turbulence models are satisfactory to reproduce wall bounded flows with no-slip walls, but also that these models are not satisfactory to predict flows with some sort of drag control. As an example it is interesting to look at the direct simulations by Perot & Moin (1995a) on shear-free boundary layers. The authors devoted the second part (1995b) to turbulence modeling. In finite differences the changes of boundary conditions are an easy task, thus the present code could be modified to investigate flows with drag control. These studies presently are receiving a lot of interest. From the fundamental point of view these studies are devoted to understanding how to manipulate the vortical structures in order to get the desired output; to this purpose it is interesting to read the paper by Choi *et al.* (1994), by Jimenez (1992), and by Koumoutsakos (1997).

9.2 Adimensionalization

The dimensionless Navier-Stokes equations for an incompressible flow in a turbulent channel are given in Eq.(8.1). Half of the channel height h is the reference length; as velocity scale U can be assumed or the value U_P of the centerline laminar parabolic Poiseuille profile U_P or $\langle U_C \rangle$. U_P is more exactly defined than $\langle U_C \rangle$, in fact $\langle U_C \rangle$ varies with the Reynolds number and with the wall boundary conditions. $Re_P = U_P h/\nu$ is used here. If the variables are normalized in wall units, the friction velocity $U_\tau = \sqrt{(\nu \frac{\partial u}{\partial x_2}|_{x_2=0})}$ appears and the Reynolds number is $Re_\tau = U_\tau h/\nu$.

In the analysis of the turbulent channel the results can be adimensionalised also with respect to Kolmogorov units; these consist of quantities related to the dissipating scales of turbulence. The key quantity for this scaling are the turbulent dissipation rate ϵ and the kinematic viscosity ν. By this scaling the Reynolds number by definition is $Re_K = 1$. The relations among the dimensionless quantities in the simulation to those in wall units (indicated by +) and in Kolmogorov units (indicated by $*$ or by the subscript K) are

$$
\begin{aligned}
L^+ &= lR_\tau & L_K &= l(Re^3\epsilon)^{1/4} \\
k^+ &= k/R_\tau & k_K &= k/(Re^3\epsilon)^{1/4} \\
t^+ &= tR_\tau^2/Re & t_K &= t(Re\epsilon)^{\frac{1}{2}} \\
U^+ &= URe/R_\tau & U_K &= U/(Re/\epsilon)^{1/4}
\end{aligned}
\qquad 9.1
$$

$$\omega^+ = \omega Re/R_\tau^2 \qquad \omega_K = \omega/(Re/\epsilon)^{\frac{1}{2}}$$
$$\epsilon^+ = \epsilon Re^3/R_\tau^4 \qquad \epsilon_K = 1$$

9.3 Space and time discretization

As in the previous chapter the variables are located as in Fig.8.1. Since in the channel a nonuniform grid in the normal direction appears (as explained in chapter 2) a coordinate transformation relates the physical grid $\hat{x}_i$ to a computational grid x_i. The velocities are staggered in the computational grid. In general the coordinate transformation $\hat{x}_i = f_i(x_i)$ can be used in each direction and hence the metric quantity c_i is

$$c_i = \frac{\delta \hat{x}_i}{\delta x_i}$$

As shown in chapter 2 even the second derivative can be expressed by these metric quantities; if $g_i = 1/c_i$ the discretized Navier-Stokes equations are

$$\frac{\partial u_i}{\partial t} + g_j \frac{\delta u_i u_j}{\delta x_j} = -g_i \frac{\delta p}{\delta x_i} + \frac{1}{Re} g_j \frac{\delta g_j \frac{\delta u_i}{\delta x_j}}{\delta x_j} + N\epsilon_{ijk} U_j \Omega_k \qquad 9.2$$

In the channel a coordinate transformation has been used in the direction x_2, hence it turns out that $g_2 \neq 1$. The metric quantities g_2 are evaluated at the j and $j + \frac{1}{2}$ locations by $\hat{x}_2 = f_i(x_2)$ at the grid location j. Eq.(2.20) has been used and coded in the routine $COORDI$ in the file *chadsnn.f* in $CODESCH9/CHA$. The metric quantities g_2 have been calculated in the routine $METRIC$. In direct simulations of turbulent wall flows an enhanced grid clustering near the walls would reduce the resolution in the central part of the channel. In the present simulation the parameter a in Eq.(2.20) was set equal to 2.8 and by 150 intervals in the normal direction the resolution was 0.6 times the Kolmogorov scale η for the nearest interval to the wall and 1.13η at the center of the channel. In wall units the nearest location to the wall of u_3 was at $y^+ = (1 - |x_2|)^+ = 0.8$ ($y^+ = (x_2 - x_{2W})U_\tau/\nu$). The simulations were performed in a domain with streamwise directions $L_3 = 4\pi$ and lateral dimensions $L_1 = 0.5\pi$ with 256 points in the streamwise and with 64 points in the lateral directions. Re_τ was equal to 175.5 hence the grid size in wall units was $\Delta x_3^+ = 8.6$ and $\Delta x_1^+ = 4.3$. The lengths can be scaled either in wall units or in Kolmogorov scales. From the definition of the Kolmogorov scale ($\eta = \nu^3/\epsilon)^{1/4}$ and from that in wall units the relationship between the grid spacing in Kolmogorov units $\Delta^* x_i$ and the grid spacing in wall units Δx_i^+ is

$$\Delta^* x_i = \Delta x_i^+ (\epsilon^+)^{1/4} \qquad 9.3$$

From the present simulation it turns out that ϵ^+ varies from 0.16 near the wall to 0.01 at the center of the channel. The calculation then is resolved up to the Kolmogorov scales in the central region of the channel; near the wall, on the other hand, in the streamwise direction the grid is approximately 5η. Then it should be expected that the spectra in the streamwise direction decay at high k_3 differently from the situation in Fig.8.5.

Each term in Eq.(9.2) is discretized as explained in the previous chapter and the structure of the code is also similar. As for the code with three periodic directions the non-linear terms are evaluated in the routines $HDNLi$ with i varying between 1 and 3 (with 3 the streamwise, 1 the spanwise and 2 the normal direction). In the periodic directions the derivatives are discretized by introducing the arrays $ipv(i), kpv(k)$ and $imv(i), kmv(k)$ indicating $i+1, k+1, i-1, k-1$ accounting for the periodicity. On the other hand in the normal directions in the routines $HDNL1$ and $HDNL3$ the terms near the lower ($j = 1$) and the upper ($j = n2$) walls require the velocity u_1 and u_3 at the walls where these are not defined. These velocities, in flows related to drag control, are assigned in the routine $BOUCDQ$. With shear-free boundary conditions the terms are calculated by setting, in the routine $INDIC$, $jm = j$ at the lower wall and $jp = j$ at the upper wall. These indices in the inner points are respectively $j-1$ and $j+1$.

The viscous terms as well the pressure gradients are calculated in the routines $INVTRi$ where again the index i indicates the different velocity components. The evaluation of the second derivatives in the 1 and 3 directions imply the multiplication of the velocity differences for $\frac{1}{\Delta x_i^2}$; then no care is required to save the number of operations. On the other hand in the discretization along x_2 (Eq.2.7) the metrics appear and the number of operations are reduced by evaluating, only once, in the routine $COEINV$ the coefficients multiplying the velocities at the location $j-1$, j and $j+1$. These integer indices in Fortran coincide with those in the grid for the velocity u_2; instead for the other two velocity components they coincide respectively with $j-\frac{1}{2}$, $j+\frac{1}{2}$ and $j+\frac{3}{2}$. By indicating with $g_{2h}(j)$ the metric $g_2(j+\frac{1}{2})$ and by $g_{2c}(j)$ $g_2(j)$, the coefficients in the second derivative for the components u_1 and u_3 are

$$AP(j) = \frac{g_{2h}(j)g_{2c}(j+1)}{\Delta x_2^2}, \quad AM(j) = \frac{g_{2h}(j)g_{2c}(j)}{\Delta x_2^2}, \quad AC(j) = -(AP(j) + AM(j)) \qquad 9.4$$

and for u_2

$$AP(j) = \frac{g_{2c}(j)g_{2h}(j)}{\Delta x_2^2}, \quad AM(j) = \frac{g_{2c}(j)g_{2h}(j-1)}{\Delta x_2^2}, \quad AC(j) = -(AP(j) + AM(j)) \qquad 9.5$$

In the routine $COEINV$ the coefficients have been evaluated by neglecting Δx_2^2.

Particular care must be devoted to discretize the second derivatives at the points near the walls. For example at the lower wall, if the velocities at $j_W = 1$ and at $j_W = \frac{3}{2}$ are expanded in Taylor series around the point $j_W + \frac{1}{2}$ if $h_m = \hat{x}_2(j_W + \frac{1}{2}) - \hat{x}_2(j_W)$ and $h_p = \hat{x}_2(j_W + \frac{3}{2}) - \hat{x}_2(j_W + \frac{1}{2})$, the second derivative is

$$\frac{\partial^2 u_i}{\partial {x_2}^2}|_{j_w+\frac{1}{2}} = \frac{2}{h_m + h_p}[\frac{u_i(j_W)}{h_m} + \frac{u_i(j_W + \frac{3}{2})}{h_p} - u_i(j_W + \frac{1}{2})(\frac{1}{h_m} + \frac{1}{h_p})] + O(\Delta \hat{x}_2) \qquad 9.6$$

In the routine $COEINV$ the quantities h_m and h_p have been related to the metric coefficients. The second derivative at $j_w = \frac{1}{2}$ has a reduced accuracy, but, since u_3 and u_1 increase linearly with the distance from the wall, the results are unaffected by the reduced accuracy.

In this chapter all the details of the discretization scheme are not repeated, since in large part these can be understood by reading the previous chapter. Some are briefly recalled and only the new features related to the boundaries with Newman or Dirichlet boundary conditions for the velocities are described.

The fractional step method (section 8.5) has been used without any modification, with the non-linear terms explicitly treated by the Runge-Kutta scheme described in chapter 2. The viscous terms have been discretized in time with the Crank-Nicholson scheme and the pressure has been inserted at the non-solenoidal step as shown in Eq.(8.8). The approximate factorization allows the inversion of the large band matrix associated to the implicit viscous terms. In flows with three periodic directions the sequence in the direction to evaluate the tridiagonal matrices is arbitrary because of the absence of explicit boundary conditions. In wall-bounded flows it is convenient to invert the matrices first in the periodic directions and then the direction limited by the solid boundaries. The three successive operations of the fractional steps are given in Eq.(8.13); the intermediate quantities Δu_i^{**} and Δu_i^* do not require any boundary condition. The second derivative along x_2 of the quantities Δu_i for $i = 1$ and $i = 3$ at the inner points are expressed by the coefficients in Eq.(9.4) and the coefficients at $j = 1$ and $j = n2 - 1$ are obtained by Eq.(9.6). If the velocities at the wall vary in time, $\Delta u_i(j = 1)$ and $\Delta u_i(j = n2)$ are calculated in the routine *BOUCDQ* in the file *chadstn.f*. These values are inserted in Eq.(9.6) and modify the right hand side of Δu_i^* at the points near the walls.

The inversion of the tridiagonal matrix for each u_i was performed by the routine *TRVBKJ* in the file *chadstr.f*, that has been written with the inner do loop in x_3 (index K) to increase the efficiency on vectoral computers. For all the u_i the spanwise x_1 (index I) has been solved as the first direction, followed by the streamwise x_3 (index K) direction. The channel simulations to reach a statistical steady state necessitate a mean pressure gradient acting as forcing term. As previously mentioned two choices are possible: the first fixes the wall friction, the second the flow rate. In the second the pressure gradient, at each time substep, is evaluated to balance the friction losses. Theoretically the pressure gradient could be evaluated by $\frac{\partial \langle u_3 \rangle}{\partial x_2}$ at the wall, since the integration of the viscous terms in x_1 and x_3 gives a null contribution for periodicity. Even the non-linear terms give a null contribution for periodicity in presence of impermeable walls. However, for a check of the conservation properties the pressure gradient is calculated in the *INVTR3* routine by the summation of each term. The relationship between the pressure gradient and the wall friction gives u_τ, necessary to scale the profiles in wall units.

Even for the ϕ calculation the major part of the arguments described in section 8.7 hold. The Fourier series expansion of Eq.(8.15) is used and it leads to Eq.(8.16). The coefficients in the x_2 direction for the ϕ calculation, in the inner part, coincide with those for the velocity components 1 and 3.

In wall-bounded flows the treatment of the "pressure" equation at the boundary is delicate from the conceptual point of view. Whoever has experience on this method knows that the solution of ϕ does not require any boundary condition; the others, instead, have difficulties to understand why an elliptic equation does not require boundary conditions. To convince these fellows that this is indeed true, it is necessary to describe in detail, for example, the treatment of one point near the lower boundary. The other points near the boundaries are treated in a similar way. The discretized continuity equation at the new time step $n+1$, by neglecting for simplicity the superscript $n+1$, is

$$\frac{u_{1(i+1,j+\frac{1}{2},k+\frac{1}{2})} - u_{1(i,j+\frac{1}{2},k+\frac{1}{2})}}{\Delta x_1} + \frac{u_{3(i+\frac{1}{2},j+\frac{1}{2},k+1)} - u_{3(i+\frac{1}{2},j+\frac{1}{2},k)}}{\Delta x_3} +$$

$$g_{2h}(j)\frac{u_{2(i+\frac{1}{2},j+1,k+\frac{1}{2})} - u_{2(i+\frac{1}{2},j,k+\frac{1}{2})}}{\Delta x_2} = 0 \qquad 9.7$$

By substituting the relationship

$$\frac{u_i - \hat{u_i}^k}{\Delta t} = -\alpha^k G_i \phi \qquad 9.8$$

into all the quantities in Eq.(9.7) (except into that for $u_2(i+\frac{1}{2}, j, k+\frac{1}{2})$, which is known since it is the velocity at the boundary) it follows

$$\frac{\hat{u}_{1(i+1,j+\frac{1}{2},k+\frac{1}{2})} - \hat{u}_{1(i,j+\frac{1}{2},k+\frac{1}{2})}}{\Delta x_1} + \frac{\hat{u}_{3(i+\frac{1}{2},j+\frac{1}{2},k+1)} - \hat{u}_{3(i+\frac{1}{2},j+\frac{1}{2},k)}}{\Delta x_3} +$$

$$g_{2h}(j)\frac{\hat{u}_{2(i+\frac{1}{2},j+1,k+\frac{1}{2})} - u_{2(i,j+\frac{1}{2},k+\frac{1}{2})}}{\Delta x_2} = F_{(i+\frac{1}{2},j+\frac{1}{2},k+\frac{1}{2})} =$$

$$\frac{\phi_{(i+\frac{3}{2},j+\frac{1}{2},k+\frac{1}{2})} - 2\phi_{(i+\frac{1}{2},j+\frac{1}{2},k+\frac{1}{2})} + \phi_{(i-\frac{1}{2},j+\frac{1}{2},k+\frac{1}{2})}}{\Delta x_1^2} +$$

$$\frac{\phi_{(i+\frac{1}{2},j+\frac{1}{2},k+\frac{3}{2})} - 2\phi_{(i+\frac{1}{2},j+\frac{1}{2},k+\frac{1}{2})} + \phi_{(i+\frac{1}{2},j+\frac{1}{2},k-\frac{1}{2})}}{\Delta x_3^2} +$$

$$g_{2h}(j)\frac{g_{2c}(j+1)(\phi_{(i+\frac{1}{2},j+\frac{3}{2},k+\frac{1}{2})} - \phi_{(i+\frac{1}{2},j+\frac{1}{2},k+\frac{1}{2})})}{\Delta x_2^2} \qquad 9.9$$

If the Fourier series in Eq.(8.15) is applied to $F_{(i+\frac{1}{2},j+\frac{1}{2},k+\frac{1}{2})}$ and to every ϕ Eq.(8.16) is obtained with the coefficients of the tridiagonal matrix at $j=1$

$$AP(1) = \frac{g_{2h}(1)g_{2c}(2)}{\Delta x_2^2},\ AM(1) = 0,\ AC(1) = -AP(1) + k_1'^2 + k_3'^2 \qquad 9.10$$

The same procedure at the opposite wall gives the coefficients at $j = n2-1$

$$AM(n2-1) = \frac{g_{2h}(n2-1)g_{2c}(n2-1)}{\Delta x_2^2},\quad AP(n2-1) = 0,$$

$$AC(n2-1) = -AM(n2-1) + k_1'^2 + k_3'^2 \qquad 9.11$$

With these coefficients together with those in Eq.(9.4) at the inner points $\hat{\Phi}(k_1, j+\frac{1}{2}, k_3)$ is yielded, and hence by calling the *cfft99* and *fft99* routines, with $isign = +1$, ϕ in the physical space is calculated. The calculation of ϕ is performed in the file *chadsphnc.f* where the NCAR FFT routines are used. If the reader has on his computer the ESSL IBM libraries, he could get the solution more efficiently on the RISC IBM machines by the file *chadsphes.f*.

The solenoidal field at $n+1$ is evaluated in the subroutine *UPDV* by Eq.(8.6).

9.4 Initial conditions

The code was used to simulate the flow in a plane channel. Since the interest is to reach a statistical steady state, the initial conditions are unimportant. However, a velocity profile in the streamwise direction and a disturbance are necessary to initiate the turbulent flow. A Poiseuille parabolic profile $\langle u_3 \rangle (x_2) = 4(1 - |x_2|)^2$ has been given, while the other two components $\langle u_2 \rangle$ and $\langle u_1 \rangle$ have been set equal to zero. A random disturbance, with zero mean, has been added to the mean velocity components. To avoid large fluctuations near the wall that could require small time steps for CFL restriction, the random disturbance in the region near the boundary $(1 - |x_2| 0.025)$ has been multiplied by 0.2. This velocity field is not solenoidal, but this occurrence is not relevant, since the fractional step method has the property that at the end of the first substep the velocity field is solenoidal. The initial velocity field is given in the routine $INQPR$.

When the simulation starts from scratch it is convenient to initiate the calculation with a coarse mesh in x_3 and x_1 and to maintain the final grid in x_2. The simulations require several time units to reach a condition where some sort of turbulent structures form. In inhomogeneous flows the shape of the vortical structures depends on the location. From previous results (Kim & Antonia (1993)) structures as those in isotropic turbulence are expected at the center of the channel and tubular structures elongated in x_3 near the wall. A coarse simulation creates structures of incorrect size but still these are linked to the flow inhomogeneity. From a velocity field in the coarse grid the simulation restarts in a finer grid and reaches a statistical steady state. This process should be repeated until the statistics are independent on the grid resolution.

From these considerations, in the routine $INIREA$ where the continuation field, (that was written in the routine $CONTWR$) is read, there are several possibilities to interpolate the old field to restart the simulation in the new grid. In the present version only interpolations between grids with a ratio equal two have been considered. These routines are not explained since they are very simple and are based on second order accurate interpolations. In each routine, with names beginning with *in* to indicate interpolation (coarse to fine) and with *ex* to indicate extrapolation (fine to coarse), u_1 and u_3 are calculated; u_2 is evaluated by the continuity equation.

9.5 Wall boundaries for turbulence control

In the code concentrated regions of transpiration can be located on the lower wall; the routine *slotin* defines the regions with transpiration. This is a very simple task to achieve, but, since simulations with concentrated transpirations are not considered, the routine $SLOTIN$, in the file *chadsin.f* is not described. It is important, however, to point out that to maintain periodicity in the streamwise direction the total transpiration across the boundary should be zero. Thus if in a slot there is a positive flow rate Q_2, in a second slot the same amount of Q_2 is negative. If the interest is to simulate the effects of blowing or suction from a single slot the computational domain in the streamwise direction should be long enough to reduce the interference between the slots. In any case periodicity implies that the simulation concerns the flow in a channel with an infinite number of couples of slots with mass flowing in opposite directions. This flow has been recently computed by a similar code developed at the CTR by Koumoutsakos (1997). A similar procedure permits to simulate a channel with a section of the wall oscillating. This flow was simulated by Jung *et al.* (1992).

I would like to remind that wall turbulence control is a research topic that started in the last years and that several of the possible realizations still have not been investigated. Some of the possibilities as those considered by Choi *et al.* (1994) are easy for numerical simulations and drove the experimentalist to find a very efficient way to realize them in a laboratory (see review article by Ho & Tai (1998)). Others, as those concerning concentrated transpiration, were studied in the laboratory by Antonia *et al.* (1995) and Tardu (1997).

With these considerations it turns out that the topic of wall turbulence control, for its complexity, should be investigated by numerical and laboratory experiments in close connection; each approach should take advantage of the results of the other. In the present code the eventual modifications related to wall control should be included in the file *chadsin.f* and in the routine *BOUCDQ* in the file *chadstn.f*.

9.6 Definition of turbulent quantities

In inhomogeneous flows the averages are defined in the periodic directions and in time. The mean value of a generic variable q at each time is

$$\langle \hat{q}(t, x_2) \rangle = \frac{1}{N_1 N_3} \sum_{l=0}^{N_1} \sum_{n=0}^{N_3} q(t, x_1, x_2, x_3) \qquad 9.12$$

The number of points N_1 and N_3 are not sufficient to infer that this average produces a quantity meaningfull. In addition the channel is driven by a pressure gradient that varies in time. To get averages representative of a statistical steady turbulence a further averaging in time is necessary. In the plane channel some of the mean quantities are evaluated during the simulation, to verifyœ that the calculation is performed for a time sufficient to reach the statistical steady state. The mean velocity profiles together with profiles of turbulent stresses, flatness and skewness are evaluated in the routine *ENERCA*. The routine is called when the time t is a multiple of *tpin* but the averages are calculated in the routine *TAVER* only when $iav = 1$. This index initially is set to 0 in order to avoid the calculation of the averages during the transient. If *tpin* is chosen to be quite small the fields are not statistically independent of each other. This choice does not produce erroneous statistics but increases the CPU time. Since there could be an interest to evaluate the statistics of other turbulent correlations, which at a certain stage of the study were not of interest (for example those related to the vorticity or to the velocity gradient tensor), it is convenient to save several fields and to make the averages by a post-processing code. To save memory space the fields must be statistical independent; these fields are written in the routine *CONTWR* when t is a multiple of *tprint*. The last of these fields can be used as restarting files, if the simulation was not long enough. The average of a general quantity thus is

$$\langle q(x_2) \rangle = \frac{1}{N_t} \sum_{l=0}^{N_t} \langle \hat{q}(t, x_2) \rangle \qquad 9.13$$

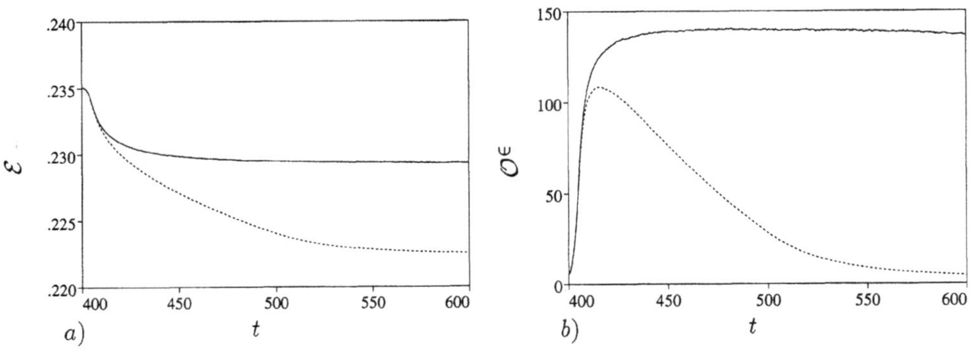

Fig. 9.1 Time evolution of a) total energy, b) total enstrophy; ---- $CFL = 1.7$, —— $CFL = 0.3$.

where N_t is the number of saved fields.

The turbulent statistics converge rapidly or slowly depending on the quantity considered. If the interest is on the statistics of zero order, that is the mean quantities, few fields are sufficient. The normal stresses require a greater number of fields but this is still reasonable because the fluctuations have the same phases. On the other hand the turbulent stresses, e.g. $\langle u_3' u_2' \rangle$, require a large number of fields to reach a statistical steady state. Even the wall stress could converge slowly. Thus a rule to see whether or not the statistical steady state has been reached, is to evaluate the profile of the total stress; scaled with the wall stress, it should have a linear profile varying from 1 at the wall to 0 at the channel center. The dimensionless total stress has been evaluated during the simulation.

One-dimensional spectra are obtained by the NCAR FFT routines. The FFT reduces the number of operations to get the correlations in the streamwise and in the spanwise directions. The correlations can be evaluated during the simulations by the *RUUCAL* routine in the file *chadsru.f* by setting *iruuca* equal to 1. The description of the post-processing code is shortly described later on; I would like to point out that the post-processing is the most interesting part in the direct simulation of turbulence. In the code *CHATRA* only some of the statistical quantities have been calculated to show how to proceed in the calculation of other quantities of interest. In this code at certain selected vertical positions and at the lower and upper walls velocities, vorticities and the components of the strain tensor are written in files. These files are post-processed by the code *statpdfl.f* that evaluates PDFs and joint PDFs of several quantities.

9.7 Check of the code *CHADS*

As mentioned before, the code for the direct simulation of a plane turbulent channel has the same structure of the code *ISO* developed for isotropic turbulence; hence the performances do not vary and what was reported in the first paragraph of section 8.7 about the computational time of each part of the code holds for the channel. Even for the channel it is important to check the conservation properties of the method. The non-uniform grid in x_2 requires this check by means of a simulation without viscous terms and by free-slip conditions for u_1 and u_3 and no transpiration for u_2. The simulations could also be done at a very high Reynolds number, e.g. $Re \approx 10^8$, but then the tridiagonal matrices that employ the major part of CPU are called even if these are not necessary. To save CPU time the routines *INVTIi* in the file *chadstinv.f* solve the Euler equations with the non-linear and the pressure gradient terms on the right hand side of the equations for $\hat{u}_i$. The simulations, performed at two constant CFL numbers, show in Fig.9.1a that even at $CFL = 1.7$ the total energy after a transient is preserved. The time evolution of the total enstrophy emphasizes better the properties of the numerical scheme and the inaccuracies of the time discretization; in fact, with a large Δt the enstrophy increases, reaches a maximum and then reduces for the effect of a numerical viscosity linked to the time discretization errors. Fig.9.1b shows that the numerical viscosity is related only to the time advancement scheme; in fact by a Δt five times smaller, the enstrophy increases and then remains constant in time. Since the enstrophy is related to the small scales these two unrealistic simulations suggest that if the interest is in the statistics of the small scales (for instance in the decay of the high wave number spectra) the simulations should be performed at low CFL. Once more it is important to remind that this method conserves energy only in the limit of $\Delta t \to 0$.

In the channel the energy is not equidistributed due to the inhomogeneity in the normal direction. Even for a free-slip impermeable wall it turns out that $\langle u_2'^2 \rangle = 0$ at the wall. The other two normal stresses, on the other hand, are different from zero. It is also clear that in this ideal case the zero wall friction produces a zero mean pressure gradient. Since at this infinite Re the flow is not resolved, the profiles of the stresses do not have a physical meaning; however, it is interesting to see how they change by starting from a realistic profile. The code performances in the inviscid and viscous conditions were performed by a coarse grid $33 \times 129 \times 65$ in a box of dimension $0.5\pi, 2, 2\pi$ respectively in the $1,2,3$ directions. The statistics obtained by the coarse grid simulations were evaluated during the simulations because the results are obtained in a short time. The simulations started from a turbulent field at $t = 400$ and evolved up to $t = 600$. The averages were initiated at $t = 450$.

Before we describe physical simulations for realistic applications it is interesting to discuss the ideal and not resolved simulation in the inviscid case. The mean velocity $\langle u_3 \rangle$ profile becomes approximately constant across the channel, with a small maximum near the center that is related to the too short lateral dimension of the channel. This profile is not reported; instead the profile of the turbulent kinetic energy is analyzed. The results in the purely inviscid case largely depend on the CFL number, as it was shown in Fig.9.1a; Fig.9.2a indeed shows that this holds also for the time evolution of the total turbulent kinetic energy. Fig.9.2a shows that at $CFL = 0.3$ the turbulent energy remains constant after a rapid growth in 10 time units, that is mainly related to the large increase of the spanwise components depicted in Fig.9.2b. The impermeable wall produces a small increase of the u_2 component with respect to u_1. Even the increase of u_3 is smaller than that of u_1. In summary, Fig.9.2b shows that the initial realistic turbulent profiles evolve towards a more isotropic distribution.

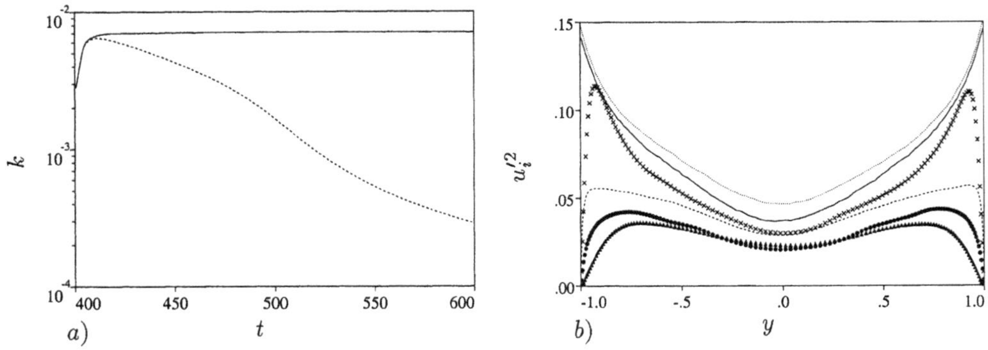

Fig. 9.2 a) Time history of the turbulent kinetic energy, ---- $CFL = 1.7$, —— $CFL = 0.3$, b) velocity rms profiles in wall units, symbols: no-slip viscous case, line: inviscid case. ($\times$ ········ u_3; $\triangle$ ---- u_2; $\circ$ —— u_1).

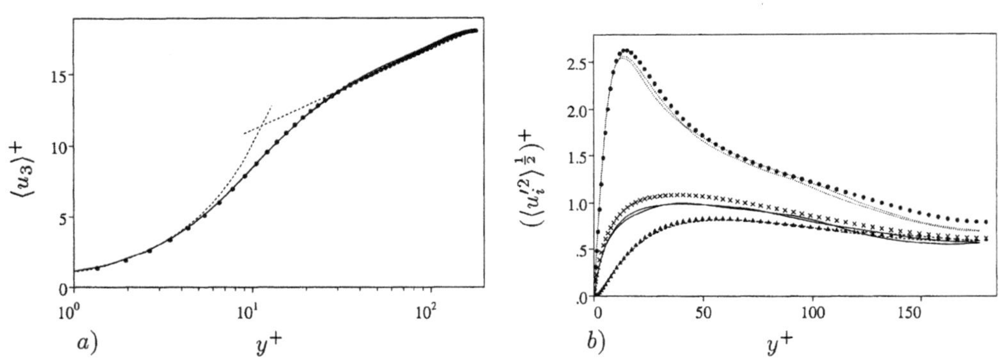

Fig. 9.3 a) Velocity profile in wall units, —— present , $\circ$ Kim *et al.* (1987), ---- $U^+ = y^+$ and $U^+ = \frac{1}{0.41} log y^+ + 5.5$; b) velocity rms profiles in wall units; lines: present, symbols: Kim *et al.*.

In the center of the channel the anisotropy is caused by the different lengths of the computational box in the streamwise and spanwise directions. The integral length scales grow in a different way; the lateral dimension does not permit the required growth hence $\langle u_1'^2 \rangle \leq \langle u_3'^2 \rangle$ at the center. The differences between $\langle u_1'^2 \rangle$ and $\langle u_3'^2 \rangle$ decreases everywhere since the kinetic energy production is zero for the two stresses. From this unrealistic simulation, however, the important physical consideration turns out that the no-slip boundary condition on u_1 plays an important role in wall bounded flows.

The coarse grid simulation in the viscous case at $Re = 4200$ gives $R_\tau = 181$. Fig.9.3a shows that the profile of the streamwise velocity in wall units compares well with that by

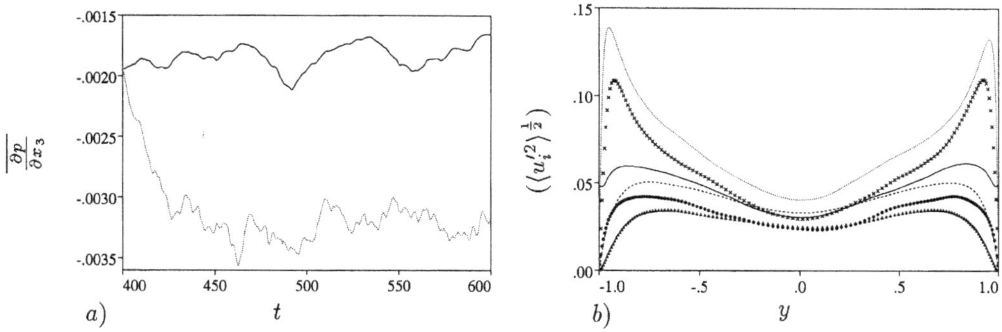

Fig. 9.4 a) Time history of the wall friction, ---- no-slip, —— free-slip for u_1 b) velocity rms profiles; symbols: no-slip viscous case, lines: free-slip.

Kim *et al.* (1987), by reproducing the logarithmic law with the appropriate constants. Even the profiles of the normal stress compare well in Fig.9.2b, despite the marginal resolution. The coarse grid does not capture the very strong gradients in the spanwise direction, in fact the errors occur mainly for the streamwise and spanwise components. On the other hand $\langle u_2'^2 \rangle^{1/2}$ (in which strong fluctuations account less) is well predicted. The results are compared with those by a direct simulation and not with the experimental results for several reason. The first is that Kim *et al.* (1987) showed that their results compare well with those by Eckelman (1974) at the same Reynolds number. The second one is that the Kim *et al.* data base is considered worldwide a validated data base and it was often used to improve and validate new measurement techniques. This approach was recently followed by Durst *et al.* (1995) to get accurate measurements of high order statistics in the near wall region of a turbulent pipe. From these results by a coarse grid it can be asserted that finite differences are a useful and cheap tool to study some of the properties of turbulent flows near a solid wall. The changes of boundary conditions on a wall are quite difficult to reproduce in a laboratory as was shown by Bechert *et al.* (1997); for instance it is almost impossible to build a wall where a free-slip condition is achieved. On the other hand numerical simulations permit a large variety of boundary conditions and their reproduction by finite differences is easier than by pseudospectral methods.

9.8 Wall-boundary conditions effects

From these considerations it turns out that it could be useful to investigate what happens when realistic variations on the walls are imposed. As a first example I am considering the imposition of a free-slip condition for the spanwise velocity. This example was cited by Jimenez (1992) in his study devoted to understand wall turbulence control. Preliminary considerations suggest that a free-slip condition could give a friction reduction. On the other hand the simulation shows in Fig.9.4a that going from a no-slip to a free-slip wall a substantial increase of the wall friction occurs. The changes occur in a relatively short period of time after the condition was changed and, as in the no-slip case, the friction, after the transient, oscillates around a new steady state. Fig.9.4a in addition shows that

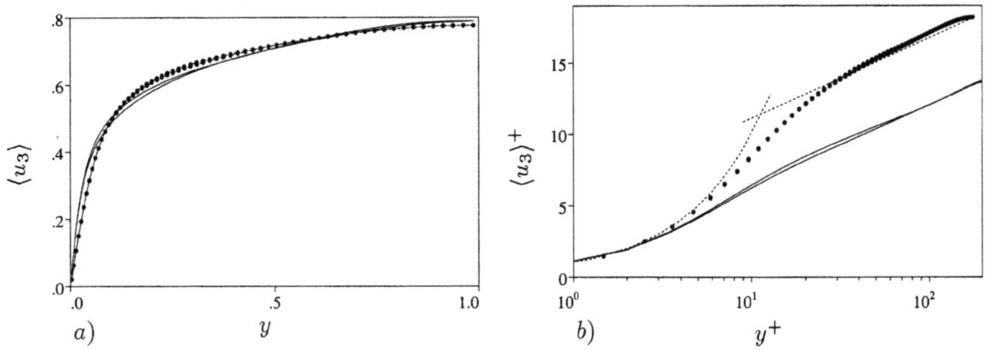

Fig. 9.5 U_3 velocity profile a) dimensionless units, b) wall units; lines: free-slip u_1, symbols: no-slip u_1

the frequencies associated to the wall friction are higher than those in the no-slip case. To understand why these unexpected results occur it is interesting to compare the profiles of the normal stresses in the two conditions. Fig.9.4b indeed shows that each normal stress increases and this occurrence is linked to an increase of the turbulent kinetic energy production.

Wall friction is related to turbulence production, in fact the high turbulence level indicates the presence of the streamwise vortices that at their turn produce high and low speed streaks that cause the wall friction. The free-slip condition permits the vortical structures dominated by ω_3 to translate near the wall; the production of opposite sign vorticity is reduced with a consequent reduction of dissipation. The size of the streamwise vortices are smaller and this occurrence explains why the frequency of the wall friction in the free-slip case is high. A reduction of the viscous effects near the wall produces a thinner viscous region and indeed it occurs as the streamwise velocity profiles show in Fig.9.5a. If the viscous region shrinks the mean velocity gradient increases and thus the turbulence production increases. The velocity profile in wall units (Fig.9.5b) shows a much longer logarithmic profile, and an almost negligible viscous layer. The logarithmic profile shifts parallel downwards, as it should do, by extrapolating from the behavior in drag reducing flows. In fact numerical (Orlandi, (1995)) and laboratory (Lucik and Tiederman (1988)) experiments in dilute polymers solutions show that drag reduction has the log law shifted upwards. The same occurs in flow over riblets as in numerical (Choi *et al.* , (1993)) and in laboratory (Park & Wallace, (1994)) experiments .

By this simulation it turns out that modifications of wall boundary conditions lead to large skin friction variations. In this regard, it is important to cite that Choi *et al.* (1994) performed a complete study about turbulence control by direct numerical simulations. They performed numerical experiments with an optimal control by changing the normal velocity component and the spanwise component. They related the pointwise velocity at the wall to the velocity at a position 15 wall units distant from the wall. This is a difficult control to realize in real applications; however this numerical study pushed the

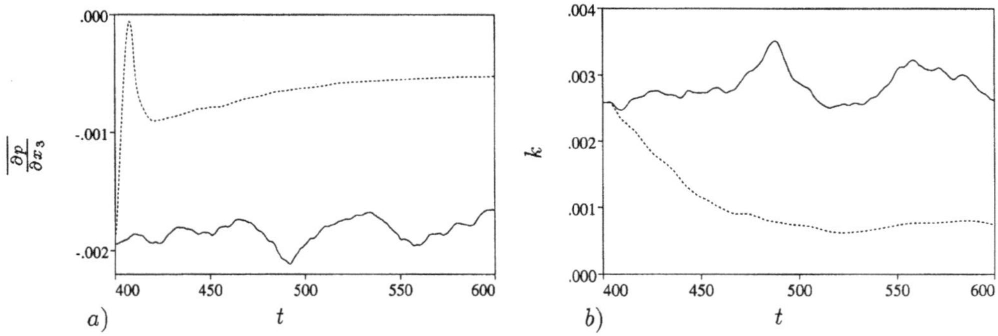

Fig. 9.6 a) Time history of the wall friction, b) total turbulent kinetic energy; ---- no-slip , —— moving wall with $U_w = 0.3$

experimentalists to investigate the possibilities to realize local control by micro actuators (Ho & Tai (1998)). These devices should change the wall velocity according to sensors located on the wall. The pressure gradient can be measured on the wall, it is related to the wall vorticity gradients (Andreopoulos & Agui (1996)) and hence to the wall velocity. How these quantities are linked to each other and how these are linked to the wall friction is still an open question; the direct numerical simulation is a tool to answer these questions. Simulations with changes on u_2 are not performed in this book; however, it is clear that modifications on u_2 are more reproducible in reality than those connected to u_1.

A further way to reduce the drag is by letting the wall to translate in the streamwise direction x_3. This goal in a real experiment can be achieved in two different ways: in the first a motor drives the wall, that is part of a belt fixed to two rotating cylinders. This experiment was set-up by Uzkan & Reynolds (1967) with the aim to investigate how grid turbulence is affected by the wall and not with the intent to understand wall friction. An impermeable wall inhibits turbulence by suppressing fluctuations in their vicinity. If this wall does not move the mean velocity gradient produces u_3' fluctuations; if the wall translates at a certain speed with respect to the external field the mean gradients of u_3 reduce everywhere and mainly near the wall, hence the turbulent kinetic energy production decreases. Bechert and his group in Berlin (private communication) are currently trying to perform a similar study with the aim to see whether it is possible to get a net drag reduction. They do not use a motor to drive the belt at a desired speed, the wall moves at a velocity imposed by the equilibrium between the wall drag and the friction loses of the mechanical system designed to let the wall move. This experiment is difficult to set-up in a laboratory and as a consequence it will be even more difficult to build the device into a commercial vehicle to get a real energy saving. However, this experiment as that by Uzkan & Reynolds (1967) are interesting for a better comprehension of wall turbulence.

A simulation with the velocity of the walls going from zero to $U_w = 0.3$ in 10 dimensionless units allowed to observe a large drag reduction associated also to a reduction of its time oscillations. Fig.9.6a shows that when the belt moves the friction coefficient drops

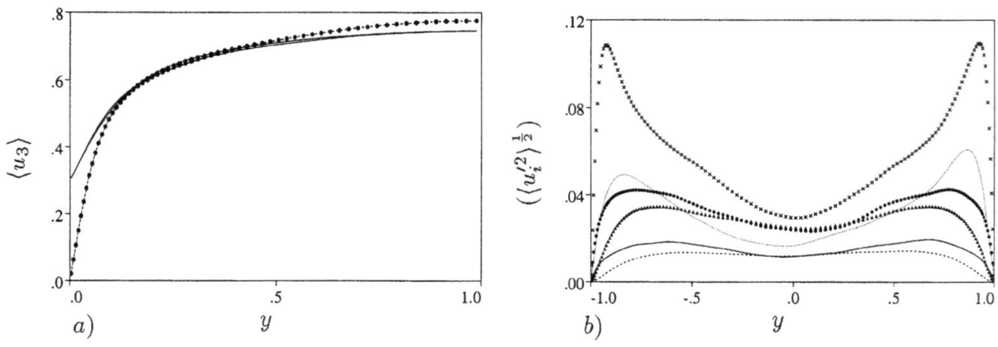

Fig. 9.7 a) U_3 velocity profile in dimensionless units, b) normal stress profiles; lines: wall moving with $U_w = 0.3$, symbols: $U_w = 0$.

almost to zero and after an overshoot it adjusts to a lower value. Flow visualizations were not done during the transient; who is interested could investigate whether the rapid decrease is caused by the complete destruction of the streamwise vortices. If this is indeed true the subsequent growth of the friction could be explained by the formation of new vortical structures of different sizes. I made these considerations to point out that some interesting questions on wall turbulence still do not have an answer. For instance, to my knowledge, nobody investigated how long it takes to the streamwise vortices to reform, once that these have been in any way destroyed. The turbulent energy in Fig.9.6b does not show any longer the time fluctuations and its level has been reduced. Flow visualizations in the $x_3 - x_2$ and $x_1 - x_2$ planes showed that the amount of streamwise and spanwise vorticity has been reduced near the wall and since the wall friction is related to spanwise vorticity fluctuations it explains the friction reduction. Fig.9.7a-b show the comparison between mean velocities and normal stress profiles for the moving and the non-moving wall; it turns out that the reduction of the mean gradient near the wall (Fig.9.7a) decreased the turbulent production and hence the peaks of normal stresses (Fig.9.7b) near the walls. The location of the maximum of $\langle u_3'^2 \rangle^{\frac{1}{2}}$ moved towards the center and this is consistent with the instantaneous visualizations that showed, near the wall, larger vortical structures than those when the wall did not move.

The aim of these two examples was to show that the user with this code could manipulate the wall as he likes to understand which is the better way to achieve drag reduction. A large amount of preliminary tests can be performed by a coarse grid since each simulation requires only eight hours of CPU time on a workstation RISC 3CT. Only when the most efficient way has been found the simulations should be repeated with a finer grid that fully resolves all the scales. These coarse simulations in addition are satisfactory since the vortex dynamics, in drag reducing flows, is smoother than those in the classical turbulent plane channel.

9.9 Analysis of the data

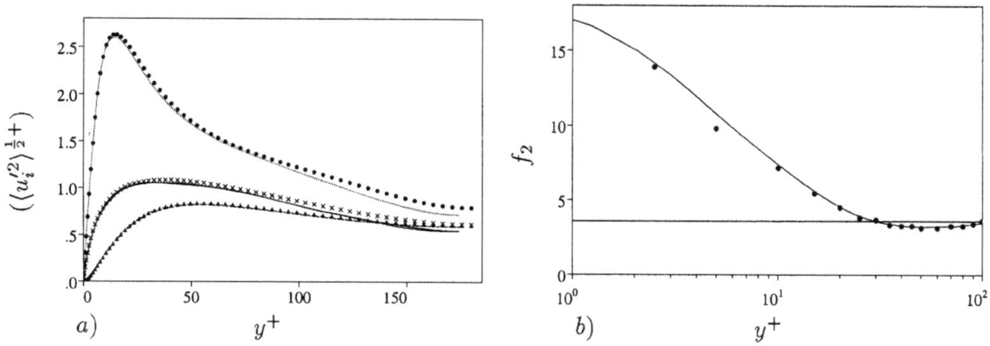

Fig. 9.8 a) Normal stresses in wall units , b)u_2 flatness profiles in dimensionless units; lines present, symbols Kim *et al.* (1987).

9.9a Velocity statistics

From the fields saved any statistics of interest can be calculated and the profiles are meaningful if the number of samples are large. As mentioned before some of the statistics converge very fast and few fields are sufficient, others converge slowly and then the number of field should be increased. In the following analysis 40 fields separated from each other by 10 dimensionless units were used. Each field is independent from the other, since in this interval of time a particle at the center of the channel has travelled half the channel length. 10 dimensionless time units correspond to 73 time wall units ($t^+ = tR_\tau^2/Re$), and since in experiments on boundary layers it was measured (Rao *et al* (1971)) that the bursting frequency is approximately equal to 5 units, thus in 73 units several bursting events occur. Therefore the fields are independent of each other in the wall and in the outer region.

The turbulent statistics vary with the distance from the wall and since the velocities are staggered it is necessary to decide where to calculate the statistics. The optimal choice is to evaluate each variable at the cell center by means of interpolations. The interpolation, on the other hand, must be avoided in the calculation of the one-dimensional streamwise and spanwise energy spectra. In fact, the interpolation is equivalent to a filtering in the wave number space. Averages in the x_2 direction do not introduce any error in the evaluation of the spectra.

The post-processing code evaluates several statistics based on velocity and vorticity fluctuations. The code $CHATRA$ in the directory $CODESCH9/POST2$ has a structure similar to the code $STATISO$ used for isotropic turbulence. In the turbulent channel, particular care is needed to evaluate the quantities at the wall and in particular the vorticity and some of the strains. The vorticity components are staggered, and the averaged velocity and vorticity components are evaluated at the center of the cell and on the location where each component is defined; the former are necessary to evaluate the fluctuations. For any variable q holds $q(x_1, x_2, x_3) = \langle q(x2) \rangle + q'(x_1, x_2, x_3)$. In the routine $VELC$ and $VORC$, in the file *chatratu.f*, the mean velocity and vorticity components are evaluated. In the routine $VELTUR$ the fluctuations, *rms*, skewness and flatness are evaluated together with

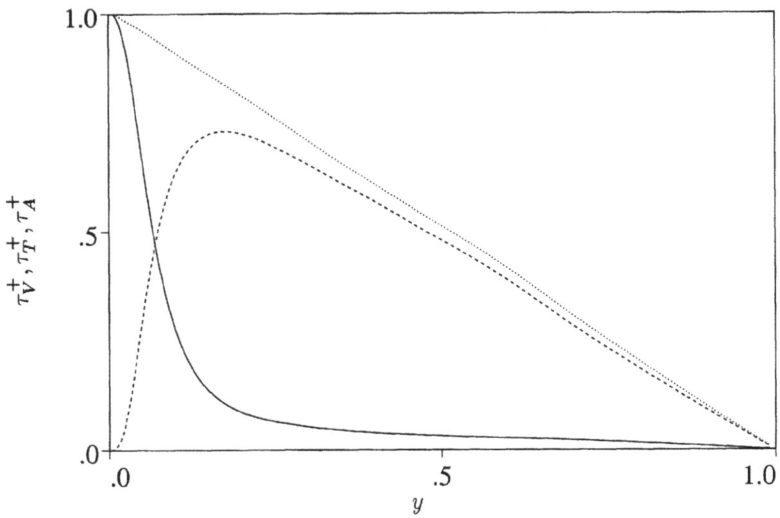

Fig. 9.9 Stresses in wall units , —— viscous stress, ---- turbulent stress, ······· total stress.

the Reynolds stresses and the correlation between pressure and velocities. The profiles of all these correlations are not presented; Fisg.9.8a-b show that indeed the normal Reynolds stresses agree with those calculated by Kim *et al.* (1987). The comparison between Fig.9.3b and Fig.9.8a proves that the unsatisfactory results in the wall region of Fig.9.3b depend on the poor resolution. On the other hand, the disagreement in the central part of the channel depends on the narrow channel assumption. Fig.9.8b shows that the present simulations reproduce the high flatness levels of the normal velocity in the wall region. The values are different from those indicated by the solid line, that should be obtained if u_2' had a Gaussian distribution. These large values indicate that the wall region is highly intermittent compared with the central part of the channel where turbulence is almost isotropic. In Fig.9.9 the profile of the viscous stress $\tau_V = \frac{1}{Re}\frac{\partial\langle u_3\rangle}{\partial x_2}$, the turbulent stress $\tau_T = \langle u_2'u_3'\rangle$ and the total stress $\tau_A = \tau_V + \tau_T$ (each scaled by the wall friction) show the linear profile for the total stress. The fulfillment of this theoretical law proves that the number of fields to evaluate the statistics were sufficient.

9.9b Vorticity statistics

The interested reader repeating the simulation can see that the profiles of the other velocity statistics agree with those in a full channel by pseudospectral simulations (Kim *et al.* (1987)) and by high order finite differences (Rai & Moin (1992)). I would like to draw the attention of the user to the statistics related to the vorticity field that received less attention in the past. The profiles of the vorticity *rms* in wall units show a perfect agreement, in the wall region, with those evaluated by the pseudospectral simulations. Choi & Moin (1994) claimed that the vorticity *rms*, as other statistics, depend on the

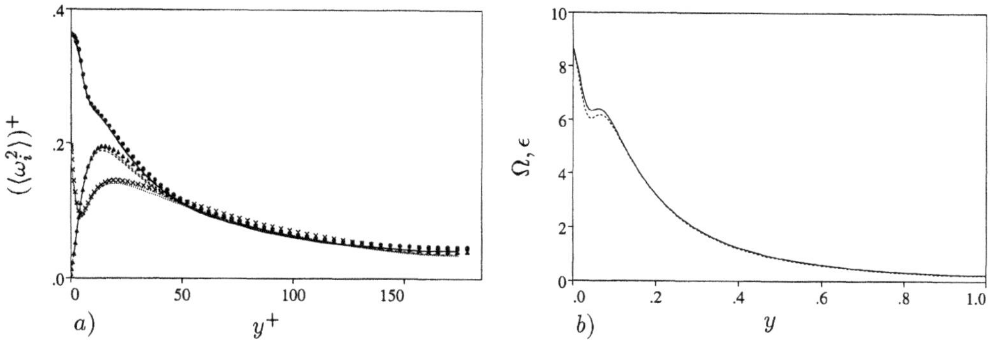

Fig. 9.10 a) Vorticity *rms* profiles in wall units, lines: present, symbols: Kim *et al.* (1987); b) profiles of $\langle s_{ij} s_{ji} \rangle$ ——— and $\langle \omega_i^2 \rangle$ ---- in wall units.

time steps used in their fully implicit scheme. In accord with their findings, $\Delta t = 0.05$ here used (corresponding to $CFL = 1.7$) produces very good results (Fig.9.10a). I have not checked whether the present semi implicit scheme could work at higher CFL; this exercise is left to the user. My opinion however is that in turbulent flows the Δt should be smaller than that necessary to describe the physics of the smallest scales, the Kolmogorov scales. The time in Kolmogorov units is $t^* = t(Re\epsilon)^{\frac{1}{2}}$ and expressed in terms of wall units is $t^* = t^+(\epsilon^+)^{\frac{1}{2}}$. From Mansour *et al.* (1988) it turns out that the maximum ϵ^+ is $\epsilon^+ = .15$, thus $\Delta t = 0.05$ corresponds to $\Delta t^* = 0.17$, which satisfies the physical requirements. In inhomogeneous flows the enstrophy $\Omega = \langle \omega_i^2 \rangle$ does not coincide with $2\langle s_{ij} s_{ji} \rangle$; in fact, if x_2 is the direction of inhomogeneity, it turns out that

$$2\langle s_{ij} s_{ji} \rangle = \langle \omega_i^2 \rangle + \frac{\partial^2 \langle u_2'^2 \rangle}{\partial x_2^2} \qquad 9.14$$

Fig.9.10b shows that the profiles of $2\langle s_{ij} s_{ji} \rangle$ and $\langle \omega_i^2 \rangle$ differ slightly in the region of maximum $\langle u_2'^2 \rangle$ and that in the major part of the channel the two quantities coincide. Thus the enstrophy, in the plane channel, at this low Reynolds number indicates well the amount of dissipation.

9.9c Production and dissipation rate

In inhomogeneous turbulent flows the turbulent kinetic energy budget permits to improve the near-wall turbulence closure based on one-point correlations. The balance was calculated by Mansour *et al.* (1988) and the results for the channel are discussed later on. To show once more that the present simulation produces good results in Fig.9.11 the profiles in wall units and in a log plot of the dissipation rate $\epsilon = 2\nu\langle s_{ij} s_{ji} \rangle$ and of the production $P_K = \langle u_2' u_3' \rangle \frac{\partial U}{\partial x_2}$ are compared with those from the spectral simulation. The log log plot emphasizes that turbulence production in the near-wall region grows with an approximate cubic law and decreases in the log region with $x_2^{-3/2}$. The growth in the wall region is consistent with the observation that near the wall $u_3' \approx x_2$ and $u_2' \approx x_2^2$ (Hinze

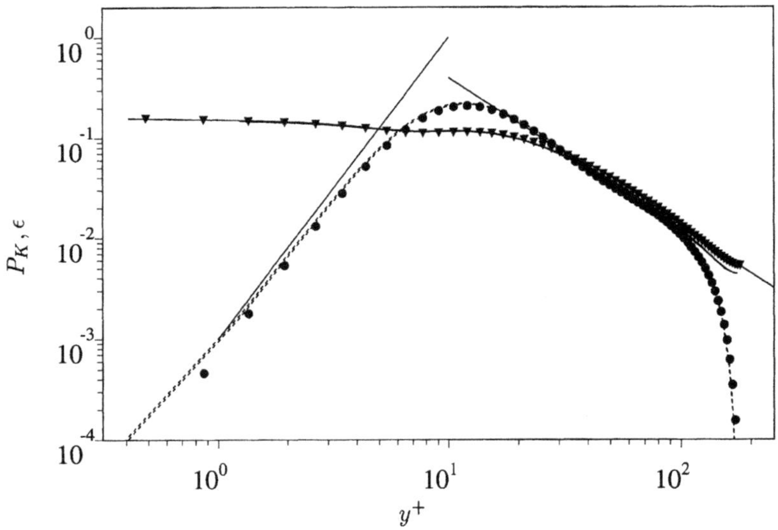

Fig. 9.11 Profiles of turbulent kinetic energy production and dissipation rate; lines: present, symbols: Kim *et al.*. (1987)

(1975), Pg.620), and that the mean velocity profile is approximately linear. The reason for a power law with power smaller than 3 is related to the deviation from a perfect linear profile of the mean velocity near the wall. The log log plot, furthermore, shows that the wall region is dissipating and that the dissipation rate in this region is equilibrated by the other terms. Mansour *et al.* (1988) showed in their figure 5 that the viscous and turbulent diffusion are greater than the production term. On the other hand in the log region there is a balance between dissipation rate and production. This balance between dissipation and production is analogous to what occurs in the inertial range in isotropic turbulence where the energy is transfered from large to small scales.

9.9d Velocity vorticity tensor

In turbulent flows before the advent of DNS the analysis was mainly devoted to velocity correlations because of the facility to measure velocities. On the other hand, a minor effort was devoted to the vorticity correlations despite the difficulties to measure the vorticity components. Recently a large improvement was achieved to measure the vorticity field by new techniques such as PIV (Adrian (1991)) and multiple probes (Wallace (1995)), but still the measured vorticity is not as reliable as that obtained by the numerical simulations. The analysis of vorticity statistics begun with the numerical simulations and recently there was also an interest in velocity vorticity correlations. The importance of these correlations comes from the momentum equations in rotational form,

$$\frac{\partial \mathbf{u}}{\partial t} - \mathbf{u} \times \boldsymbol{\omega} = -\nabla\phi + \frac{1}{Re}\nabla^2 \mathbf{u} \qquad 9.15$$

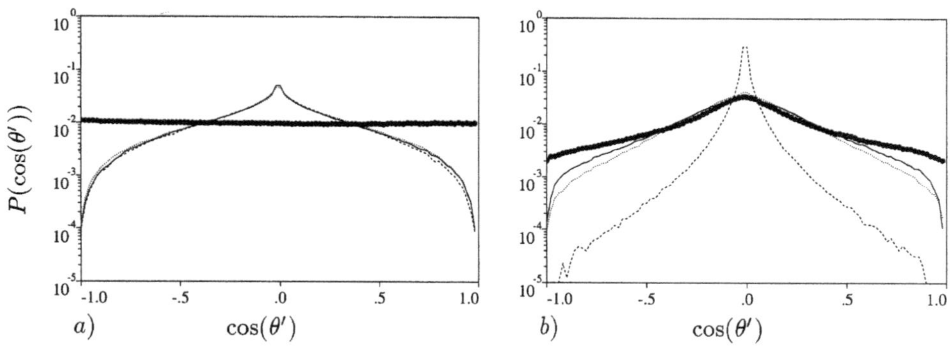

Fig. 9.12 PDF of h' • and of each component (—— 1, ---- 2, ······· 3) for a) $x_2 = 0$, b) $y^+ = 5.8$.

where $\phi = p + \frac{\mathbf{u} \cdot \mathbf{u}}{2}$. In this expression the nonlinear terms, linked to the interaction between turbulent scales and hence to the energy transfer, are expressed by $\mathbf{u} \times \boldsymbol{\omega}$. From Eq.(9.15), by splitting the velocity vector in a mean and in a fluctuating component, and by operating as usual to get the equation for the turbulent kinetic energy $\langle u_i^2 \rangle$, it turns out that the production term is related to $\langle \mathbf{u}' \times \boldsymbol{\omega}' \rangle$; its expression is given later on. This consideration suggests that there could be an interest to analyze the tensor $\gamma'_{ij} = u'_i \omega'_j$. The properties of this tensor were previously described for isotropic turbulence in Sect.8.10.b3.

In the analysis of the turbulent channel the tensor γ'_{ij} indicates whether in certain regions of the flow there is an increase or decrease of turbulent production. In the plane channel the profiles of mean helicity density are irrelevant since this quantity is zero if there is no external condition to produce it. The zero value indicates that there is an equal number of left- and right handed vortical structures. The PDFs of the fluctuating helicity near the wall and in the center of the channel permit to see that this important characteristic of turbulent flows does not depend on the amount of inhomogeneity of the flow. It is important to recall that the helicity density normalized by the module of the velocity and vorticity, i.e. $h' = \mathbf{u}' \cdot \boldsymbol{\omega}' / |\mathbf{u}'||\boldsymbol{\omega}'|$, indicates the angle between the velocity and vorticity vector.

The PDFs of several quantities are calculated in the code *statpdfl.f* in the directory *CODESCH9/POST1* as was explained in section 8.10b. Rogers & Moin (1987) in their analysis of the turbulent channel affirmed that it is questionable whether helicity fluctuations play an important role in three-dimensional incompressible turbulence, in complete disagreement with the conclusions by Peltz *et al.* (1985). The present data base generated by second order finite differences in a narrow channel confirms the conclusions by Rogers & Moin (1987). Recently Orlandi (1997) focused the study on the fluctuating helicity in a pipe rotating around its axis, where the rotation breaks the symmetry between left- and right-handed helical structures. The production of helicity was then related to the dissipation rate reduction and hence to the reduction of turbulence production.

The PDFs of each component of h' and the PDF of h' at the center of the channel are

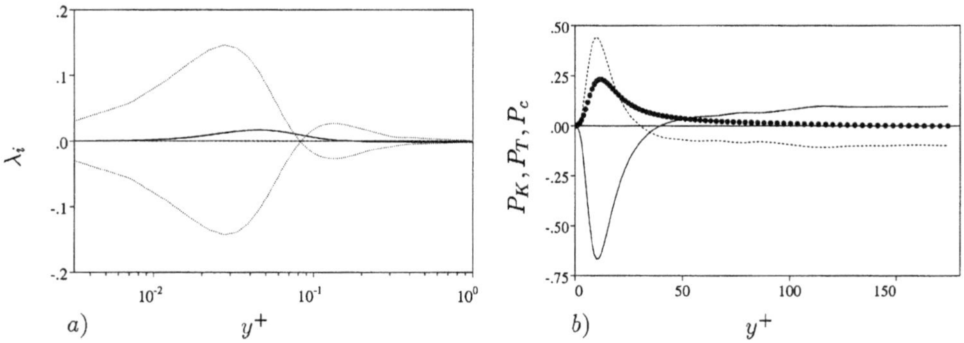

Fig. 9.13 Profiles of a) Lamb vortex components, —— axial, ---- normal, ······· spanwise components, b) turbulent kinetic energy production terms in Eq.(9.19) —— $\frac{\partial \langle U_3 \rangle \langle u_3 u_2 \rangle}{\partial x_2}$, ---- $\langle U_3 \rangle (\langle u_1' \omega_2' \rangle - \langle u_2' \omega_1' \rangle)$,• production.

shown in Fig.12a. In this region, where turbulence is approximately isotropic, the helicity PDF is relatively flat and this indicates that there is an equal probability of alignment of the velocity and vorticity vector. In the field there is the same amount of points where the vectors are aligned and counteraligned as those where the vectors are orthogonal to each other. Fig.9.12a also shows that the probability to have vorticity and velocity aligned in either one of the three directions is very low and this is clear from flow visualizations. In addition the observation that the three components behave in the same way is a further indication of isotropy. Fig.9.12b shows by the same PDFs, at the end of the viscous region at $y^+ = 5.8$, that near the wall the probability that the two vectors are not aligned increases. This confirms, as shown in Fig.9.8a and in Fig.9.10a, that approaching the wall the streamwise velocity rms overcomes the other two and that the ω_1' is the greatest vorticity component. Fig.9.12b in addition shows the flow anisotropy by the different behaviors of each component of h'. This figure shows that the PDF of $\omega_2' u_2'$ differs to a large measure from the others. The low probability to have vorticity and velocity directed in the normal direction is due to the large reduction of the normal velocity components closer to the wall. Fig.9.12b, once more, emphasizes that the variations in x_2 of the normal components characterize wall bounded flows, hence a turbulence model of wall bounded flows should take into account these variations. This explains why an improvement in near-wall turbulence models was achieved by Durbin (1992) by inserting into K, ϵ models the transport equation for $u_2'^2$ to give the right wall dumping without any ad hoc corrections. The symmetric PDFs in Figs.9.12a-b are a proof that the mean fluctuating helicity is null across the channel.

In the previous analysis of Rogers & Moin (1987) the properties of the Lamb vector $\lambda' = \mathbf{u}' \times \boldsymbol{\omega}'$ were not studied but were indirectly derived from those of the helicity density. Orlandi (1997) instead analyzed the modifications of the Lamb vector when the turbulent pipe rotates around its axis. In this paragraph the considerations about the turbulent energy production in the pipe are repeated for the plane channel. From Eq.(9.14) the transport equation for the turbulent kinetic energy $k = \langle u_i'^2/2 \rangle$ becomes

$$\frac{\partial k}{\partial t} + \epsilon_{ijk}\langle U_k\rangle\langle \omega_j' u_i'\rangle + \frac{\partial \langle U_k\rangle\langle u_i u_k\rangle}{\partial x_i} = -\frac{\partial\langle \phi' u_i'\rangle}{\partial x_i} - \epsilon_{ijk}\langle u_k'\omega_j' u_i'\rangle + \frac{1}{Re}\frac{\partial^2 k}{\partial x_j{}^2} + \epsilon \qquad 9.16$$

In a plane channel symmetry gives $\langle U_1\rangle = \langle U_2\rangle = 0$ and $\langle u_1'u_3'\rangle = \langle u_1'u_2'\rangle = 0$. All the derivatives in the homogeneous direction are null, hence the second and third terms in the left hand side of Eq.(9.16) are

$$\langle U_3\rangle(\langle u_1'\omega_2'\rangle - \langle u_2'\omega_1'\rangle) + \frac{\partial\langle U_3\rangle\langle u_3 u_2\rangle}{\partial x_2} \qquad 9.17$$

The following identity

$$\lambda_k = \epsilon_{ijk}\langle \omega_j u_i\rangle = \delta_{ji}\frac{\partial\langle u_i u_k\rangle}{\partial x_j} \qquad 9.18$$

leads to the usual expression of the production term

$$-\langle u_3'u_2'\rangle\frac{\partial\langle U_3\rangle}{\partial x_2} = \langle U_3\rangle(\langle u_1'\omega_2'\rangle - \langle u_2'\omega_1'\rangle) - \frac{\partial\langle U_3\rangle\langle u_3 u_2\rangle}{\partial x_2} \qquad 9.19$$

which can be written as $P_K = P_T + P_c$

This relationship shows that the Navier-Stokes equations in rotational form permit to split the production term P_k in the term $P_C = \frac{\partial\langle U_3\rangle\langle u_3 u_2\rangle}{\partial x_2}$ related to the action of the large eddies advecting the turbulence across the channel and in the term $P_T = \langle U_3\rangle(\langle u_1'\omega_2'\rangle - \langle u_2'\omega_1'\rangle)$ linked to the energy transfer from large to small eddies. This term incorporates the Lamb vector and thus it is interesting to see how the components of the Lamb vector vary across the channel. Fig.9.13a shows that λ_2 is null due to symmetry, and that the λ_1 component is the greatest. The asymmetric distributions of the fluctuation derivatives cause the change of sign on the two sides of the channel. These components, even if the largest are multiplied by $\langle U_1\rangle$, are identically zero across the channel, hence their contribution is null. The terms on the right hand side of Eq.(9.19), plotted versus the distance from the wall in Fig.9.13b, show that the terms convecting turbulence by the large scales near the wall overcome those transferring energy from large to small scales. The convective term gives a total null contribution, in fact the integral of $\frac{\partial\langle U_3\rangle\langle u_3 u_2\rangle}{\partial x_2}$ across half channel is zero, its positive contribution near the wall shows that energy is transfered from the mean to the turbulent motion. Part of this motion is transfered to the very small scales by the term $\langle U_3\rangle(\langle u_1'\omega_2'\rangle - \langle u_2'\omega_1'\rangle)$. The difference between the energy coming from the mean motion and that going to the small scales is the contribution to the kinetic turbulent energy production. On the other hand in the central region of the channel the energy convected is transfered to the small scales and thus the turbulent kinetic energy production decreases and becomes negligible as shown in Fig.9.11.

From these observations it turns out that if the goal is to reduce the turbulent energy production that also has a practical utility in drag reduction, the action should increase the alignment between the velocity and vorticity vector. The PDFs of the three components of the Lamb vector, evaluated near the wall and at the center of the channel show that the

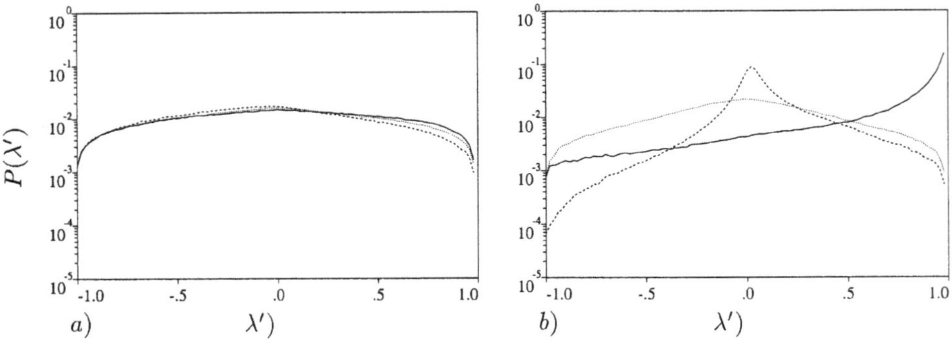

Fig. 9.14 PDF of the components of $\boldsymbol{\lambda}'$, ········ λ'_1, ——— λ'_2, ---- λ'_3, for a) $x_2 = 0$, b) $y^+ = 5.8$.

vortical structures are very different. Fig.9.14a shows, by the PDF of the helicity density, that in the central region, as in isotropic turbulence, there is an equal probability that the angle between velocity and vorticity vectors is pointing in either of the three directions. Fig.9.14b on the other hand shows that near the wall the vector $\lambda'_2 = u'_3\omega'_1 - u'_1\omega'_3$ is pointing out from the wall. The profiles of $\langle u'_3\omega'_1\rangle - \langle u'_1\omega'_3\rangle$ confirm this behavior by a negative value of $\langle u'_3\omega'_1\rangle$ near the wall and by a null value of $\langle u'_1\omega'_3\rangle$ due to symmetry. This counter-alignment in the normal direction is linked to the burst events near the wall.

A burst can be an ejection or an sweep. In an ejection $u'_3 > 0$ is correlated with $\omega'_1 < 0$, and on a sweep the correlation is between $u'_3\langle 0$ and $\omega'_1\rangle 0$. The number of events with negative correlation prevails to those with the same sign. To understand better these considerations the joint PDF between the velocity and vorticity field, evaluated in the routine *statpdfl.f*, can help. Even if these events are very strong they do not contribute to the energy production since these correlations are multiplied by $\langle U_2\rangle$ that is identically null. The PDF of the component of the Lamb vector in the streamwise direction is linked as shown in Fig.9.13 to the energy production; the identity $\langle\lambda'_3\rangle = -\frac{\partial\langle u_3u_2\rangle}{\partial x_2}$ confirms that, because of the increase of $-\langle u_3u_2\rangle$ with the distance from the wall, in the wall region there should be more points were $\lambda'_3 > 0$. The profiles of $\langle u'_1\omega'_2\rangle$ and $\langle u'_2\omega'_1\rangle$, not reported here, show that near the wall the first quantity is positive and prevails to the second one. From this it turns out that the correlations of u'_1 and $\frac{\partial u'_3}{\partial x_1}$, both of the same sign, are high. To achieve an efficient drag control it is important to apply an action that reduces this correlation. Even for this correlation the joint PDFs are very instructive.

Rogers & Moin (1987) investigated the joint PDF of helicity and the rate of dissipation to see whether the alignment between velocity and vorticity is related to the local dissipation; as mentioned before, high dissipation should be connected to low values of the helicity density. In the code *statpdfl.f* the joint PDF between helicity density and dissipation rate is calculated. The dissipation rate (shown in Fig.9.11) at the center of the channel is smaller than near the wall; thus the value at any point has been scaled by the mean value at a distance y from the wall. It was found that $e(x_1, x_2, x_3) = s_{ij}s_{ji}/\langle s_{ij}s_{ji}\rangle$ oscillates

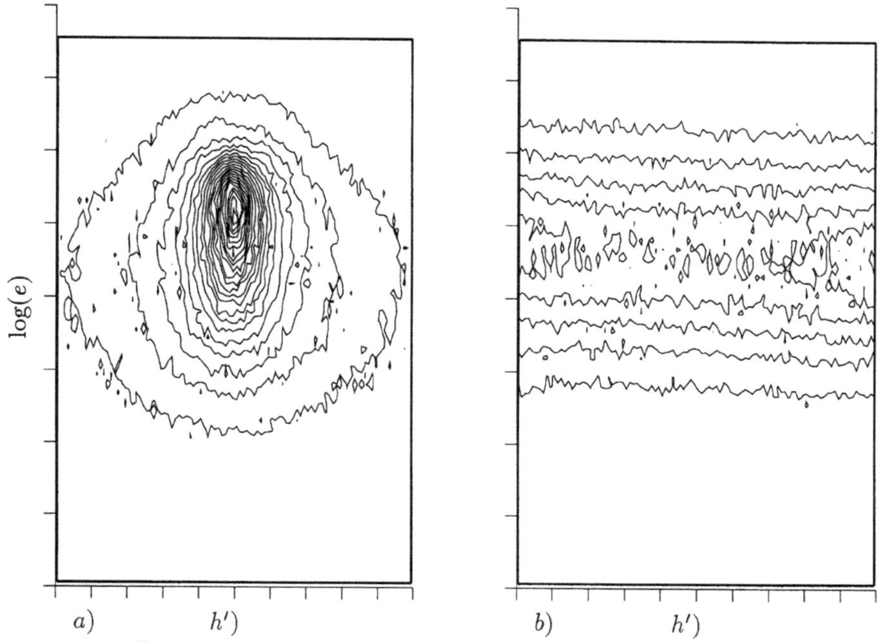

Fig. 9.15 Joint PDF of helicity density and $s_{ij}s_{ji}$ for a) $x_2 = 0$, b) $y^+ = 5.8$.

between a minimum and a maximum value that does not change across the channel. Even by this scaling $e(x_1, x_2, x_3)$ varies between a minimum $O(10^{-3})$ and a maximum $O(10^2)$ hence the ordinate of the joint PDF has a log scale. The joint PDFs in Fig.9.15 confirm the consideration from Rogers & Moin (1987), that in the wall region (Fig.9.15a) the high dissipation is correlated with regions of zero helicity density (velocity and vorticity not aligned). On the other hand in the central region (Fig.9.15b) the dissipation is equally distributed independently of the angle between the velocity and vorticity vector. These figures indeed show that the local dissipation fluctuates between the same extrema when it is scaled with respect to the mean dissipation rate.

9.9e Velocity vorticity spectra

In studying the physics of turbulence the spectra give an idea on the turbulent scales. In the simulation of isotropic turbulence three-dimensional spectra were considered. In the channel one-dimensional spectra in the streamwise and in the spanwise directions are evaluated. The spectra and cospectra are calculated in the code $CHATRA$ in the routines $COSPEX1$ and $COSPEX3$. A real FFT for each velocity and vorticity component in the routines $RUUX1$ and $RUUX3$ furnishes the complex coefficients, functions of the wave numbers k_1 and k_3 respectively. It is important to point out that in the evaluation of the one-dimensional spectra in k_1 $n_3 - 1$ data are used for each field at a given distance from the wall. This is the reason why the spectra have a very smooth profile, taking also in

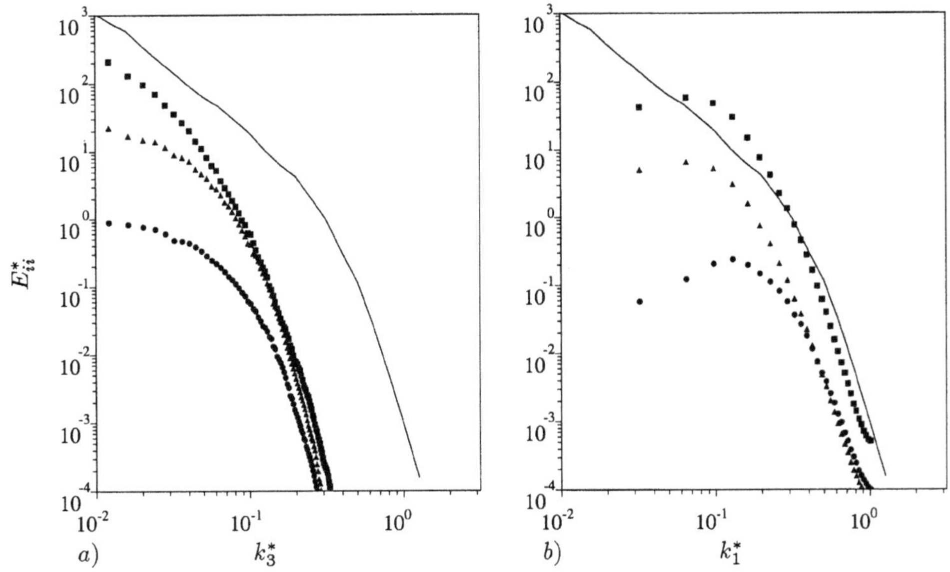

Fig. 9.16 One dimensional energy spectra at $y^+ = 5.8$: □ E_{33}, • E_{22}, △ E_{11}, —— experiments on isotropic turbulence ; a) streamwise, b) spanwise.

account that 50 fields are used. Two different spectra can be defined depending on the normalization scheme. For instance for the energy spectra the following relationship holds

$$\langle u'_i u'_j \rangle = \Sigma E_{ij} = \int \Phi_{ij} dk \qquad 9.20$$

Thus, depending on the question whether E_{ij} or Φ_{ij} are of interest, the passage from the dimensionless spectra as these come out from the routines to those in wall (+) or in Kolmogorov (*) units is:

$$E^+_{ij} = E_{ij}(ReR_\tau)^2 \qquad E^*_{ij} = E_{ij}(Re\epsilon)^{1/2}$$

$$\Phi^+_{ij} = \Phi_{ij}(ReR_\tau)^2 \qquad \Phi^*_{ij} = \Omega_{ij}(Re^5\epsilon)^{1/4}$$

For the vorticity spectra it follows

$$\Omega^+_{ij} = \Omega_{ij}(ReR_\tau)^2 \qquad \Omega^*_{ij} = \Omega_{ij}(Re\epsilon)^{1/2} \qquad 9.21$$

$$\Xi^+_{ij} = \Xi^*_{ij}(ReR_\tau)^2 \qquad \Xi^*_{ij} = \Xi^*_{ij}(Re^5\epsilon)^{1/4}$$

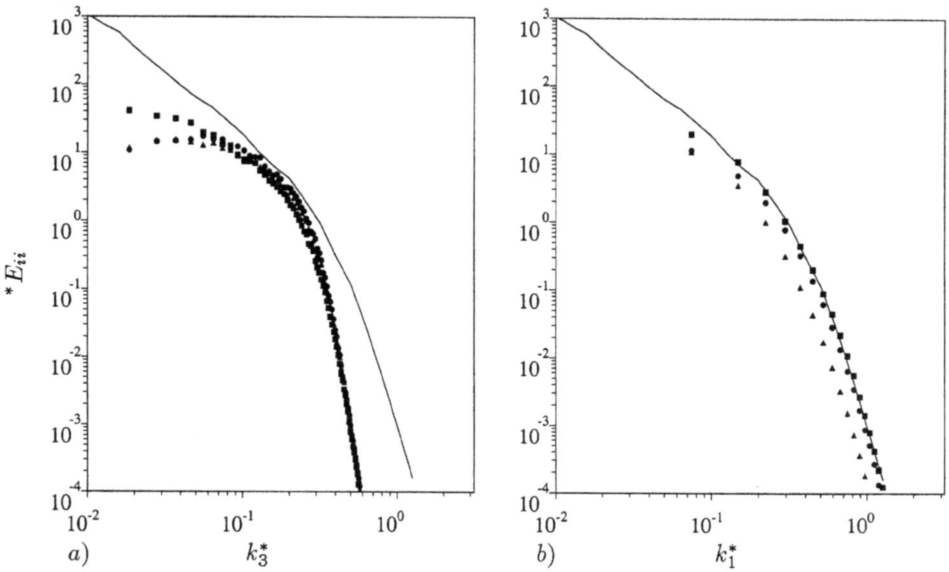

Fig. 9.17 One dimensional energy spectra at $x_2 = 0$: □ E_{33}, • E_{22}, △ E_{11}, —— experiments isotropic turbulence ; a) streamwise, b) spanwise.

The post-processing code evaluates the spectra with $\langle u_i' u_j' \rangle = \Sigma E_{ij}$; thus the passage to the spectra as those reported by Frisch (1995) at Pg.153 should be done by dividing E_{ij} by k_i and then Φ_{ij} should be opportunely scaled. The one-dimensional spectra in the streamwise and in the spanwise directions in Kolmogorov units and at the same positions as those considered by Kim *et al.* (1987) are reported here. These are the same locations where the PDFs in the previous section were calculated. In these figures, as a solid line, a best fitting of the experimental data at high wave number of Fig.8.14 in the Frisch book is given. Fig.9.16a and Fig.9.17a show that the streamwise spectra decay too fast near the Kolmogorov wave number. This behavior could be linked to the poor resolution in the streamwise direction and indeed, this is true; in fact a simulation with more resolution in x_3 shows that the spectrum agrees better with the experimental data. Even near the wall the spectra of the three components should differ less with the experimental data of isotropic turbulence. A further reason for the faster decay of the spectrum as a function of k_3^* than as a function of k_1^* is related to the CFL used in the simulation. The CFL calculated with u_3 was 1.7 and is greater than the CFL evaluated with u_1; thus the Δt is too large to simulate the convective motion in the streamwise direction. On the other hand Fig.9.16b and Fig.9.17b show that the smallest eddies in x_1 are well resolved, in fact the spectra decay as in the experiments. The spectrum E_{11}^* versus k_1^* does not coincide with E_{22}^* and E_{33}^* at the center of the channel (Fig.9.17b) as it should; the reason could be due to the narrow channel assumption and hence to the non-perfect isotropy of turbulence. The comparison of the present spectra with those in Fig.3 in Kim *et al.* (1987) is a proof of the

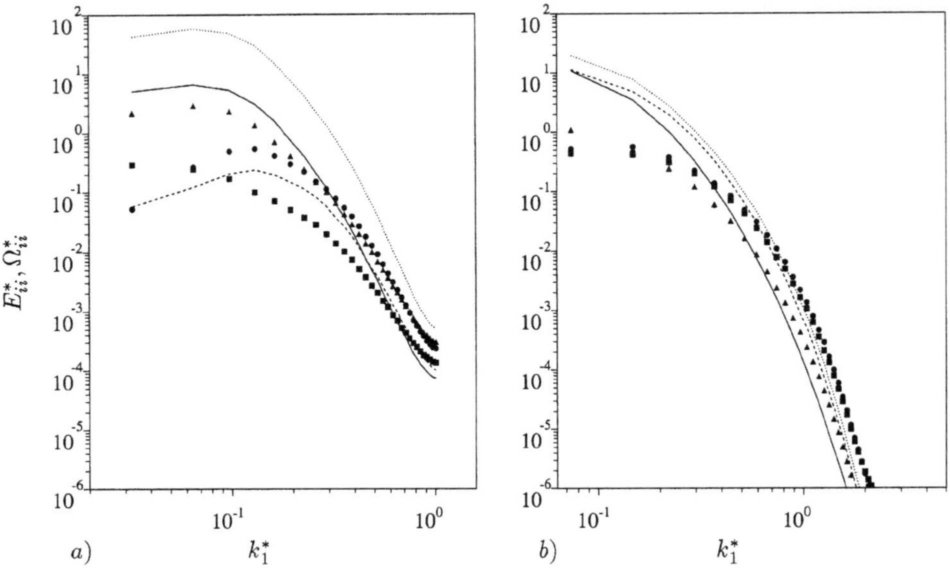

Fig. 9.18 One dimensional spanwise energy: —— E_{33}, ---- E_{22}, ········ E_{11} and vorticity spectra: □ Ω_{33}, ○ Ω_{22}, △ Ω_{11}, at a) $y^+ = 5.8$ and b) $x_2 = 0$.

previous assertion. Kim *et al.* (1987) did not give the scaling units; however a qualitative comparison shows that at the center of the channel the spanwise spectra compare well. Even in the full channel the spectrum E^*_{11} does not coincide with the other two and this occurrence indicates that the small scales at the center are not locally isotropic. The same level of anisotropy in the wall region is reproduced by the two different simulations.

The one-dimensional spectra are a tool to get qualitative information about the structures of a turbulent field. The energy spectra give an idea of the large scales and the vorticity spectra on the small scales; therefore the analysis of both leads to a better comprehension. The vorticity spectra, opportunely scaled in Kolmogorov units, are compared to the velocity spectra in Fig.9.18. For the lack of temporal and streamwise resolution to resolve all the streamwise scales, only the spectra in the spanwise directions are given. At this purpose it is important to point out that the variations along x_1 are those with a great physical content; in fact the variations of u'_3 and ω'_2 in this direction indicate the low and high speed streaks. Near the Kolmogorov scales Fig.9.18b shows at the center of the channel an exponential decay. The explanation is that at the center there is no energy production and the energy is transferred from the large scales to the small scales, where it is dissipated. Fig.9.18a shows, on the contrary, that energy production causes anisotropy at the large scales. Since the wall produces vorticity at small scales, of sizes close to the Kolmogorov scale, the vorticity spectra decay still exponentially, but with a smaller coefficient in the exponent. The spectra show the large scales anisotropy; in fact it, turns out that the most effective eddies are those with ω_1 and with ω_3. The lateral size

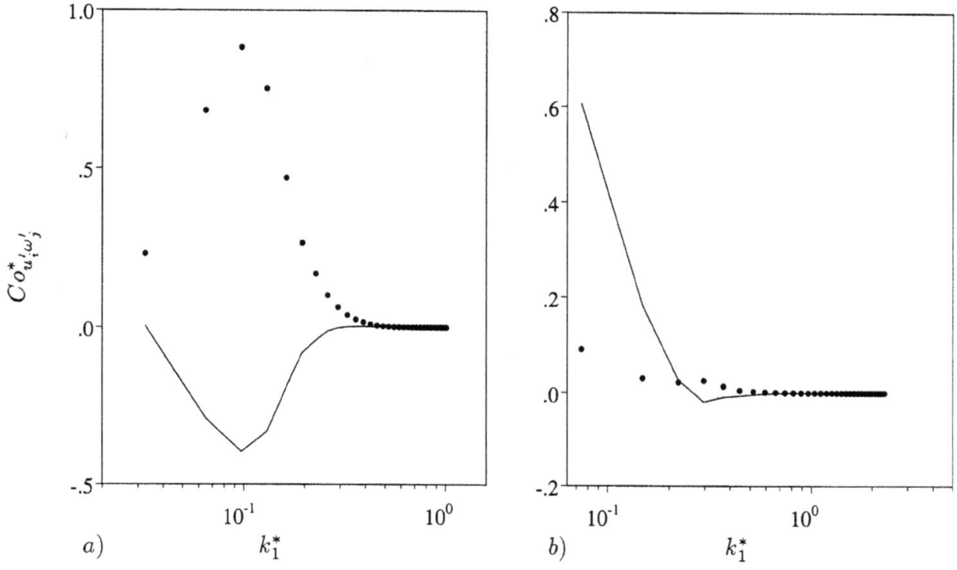

Fig. 9.19 One dimensional spanwise cospectra of • $u'_1\omega'_2$ and —— $u'_2\omega'_1$ at a) $y^+ = 5.8$ and b) $x_2 = 0$.

of these eddies are greater than those with ω_2 vorticity. The maximum content of Ω^*_{22} is in eddies of lateral dimension related to the size of the low and high speed streaks. At very small distance from the wall ($y^+ = 5.8$) the amount of ω'_2 is low; it increases at $y^+ \approx 20$, close to the location of maximum turbulence production.

9.9f Velocity and vorticity cospectra

In the past a minor effort was devoted to evaluate velocity and vorticity cospectra and even less to cospectra of the product between velocity and vorticity. The reason was dictated by the difficulty of accurate measurements of vorticity and thus even frequency spectra were not carried out. Only recently very sophisticated vorticity probes (Zhou *et al.* (1998)) permitted to undertake such measurements and to compare the measured vorticity frequency spectra with those of the present simulation at the center of the channel; the agreement appeared rather good. They reported also the cospectrum of $u'_1\omega'_2$ and $u'_2\omega'_1$ and the agreement here was poor. From these results interesting conclusions on the vorticity probe were drawn. The cospectra of these two quantities in Kolmogorov units are presented in Fig.9.19.a-b, since in the previous section it was observed that these are the terms contributing to turbulence energy production. Fig.9.19.a shows that near the wall the maximum is at $k_1\eta = 0.1$. To this wave number corresponds a wave number in wall units equal to ≈ 14.3 that, at its turn, corresponds to the size of the streaks. Fig.9.19.a moreover indicates that $\langle u'_1\omega'_2\rangle$ is greater than $\langle u'_2\omega'_1\rangle$ and of opposite sign, in agreement with the profiles in Fig.9.13. On the other hand at the center of the channel $\langle u'_2\omega'_1\rangle$ is of the

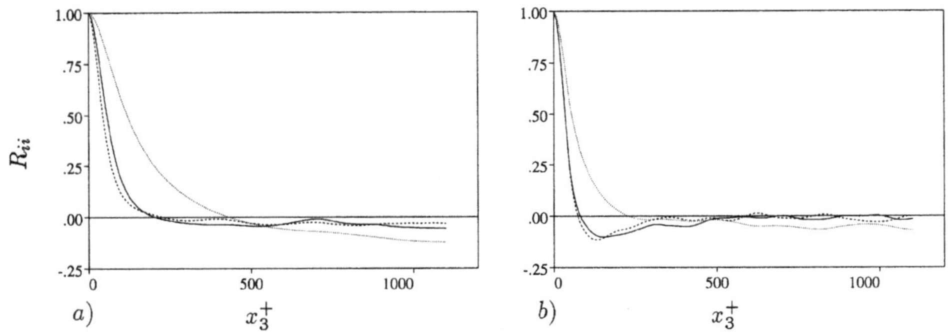

Fig. 9.20 Two-point correlations in the streamwise directions: ——— spanwise, - - - - normal, ········ streamwise velocity components; at a) $y^+ = 5.8$ and b) $x_2 = 0$.

same sign and slightly greater than $\langle u_1'\omega_2'\rangle$. In the production term, that accounts for the transfer to small scales, $\langle u_2'\omega_1'\rangle$ is subtracted from $\langle u_1'\omega_2'\rangle$; hence its contribution is negative and it is balanced by the other term in Eq.(9.17). Fig.9.19b in addition emphasizes that these vorticity velocity correlations are related to large scales.

9.9g Velocity and vorticity two-point correlations

Two-point correlations are the equivalent in physical space of the spectra, and the two are connected by Eq.(8.33). In non-homogeneous flows the correlations, depending on the direction, permit to know the size of the physical structures. As in section 8.10a, to save computational time the correlations were calculated by the FFT routines. In this case the arrays evaluated in the routine $RUUX1$ and $RUUX3$, that give the spectra and cospectra, enter in the routines $COSPX1$ and $COSPX3$ where they are summed respectively $n3-1$ and $n1-1$ times. By a direct FFT ($isign = 1$) the two-point correlations in physical space are derived. The correlations, normalized with respect to the one-point correlation, permit to verify whether the lengths of the computational box are sufficient to impose periodicity. Moin & Kim (1982) in their large eddy simulation looked at the two-point correlations by Compte-Bellot (1965) in a turbulent channel at a higher Reynolds number to fix the length in the streamwise direction. From those measurements it turns out that the correlations drop to zero in a distance of 3 half channel heights; therefore they fixed the length of the computational box equal to 2π. The length in the spanwise direction was large enough to capture several streaky structures. It is useful to remember that the two-point correlations are obtained by averaging several fields and thus are representative of the size of the structures in a statistical sense. Considerations on the physical structures are drawn from flow visualizations of a single realization; these are instructive from a qualitative point of view, but do not give any quantitative insight.

The correlations near the wall and at the center show, in Fig.9.20, that the length $L_3 = 4\pi$ is more than sufficient for the periodicity assumption. From each velocity correlation it results that the structures associated to u_3' are more elongated than those of the other two components. The streaky structures, investigated for the first time experimentally by

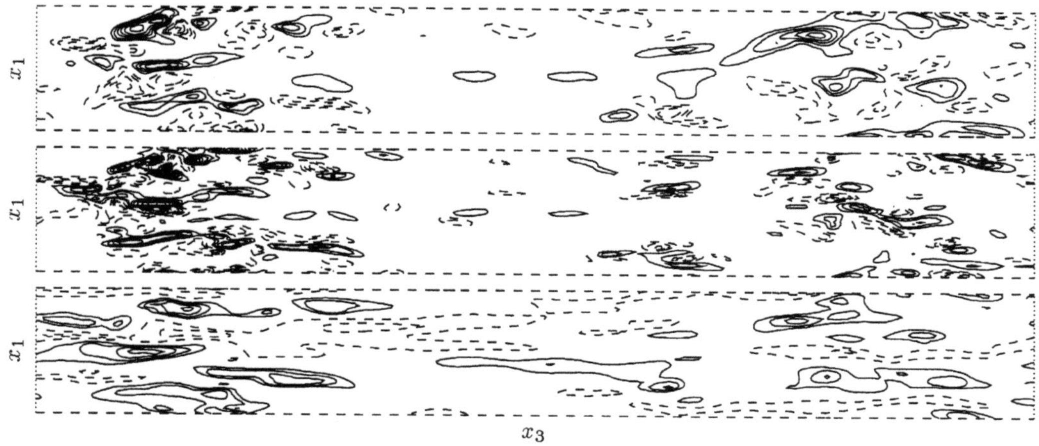

Fig. 9.20c Contours of fluctuating velocity components scaled with their *rms* value at $y^+ = 5.8$; top u_1', center u_2', bottom u_3'

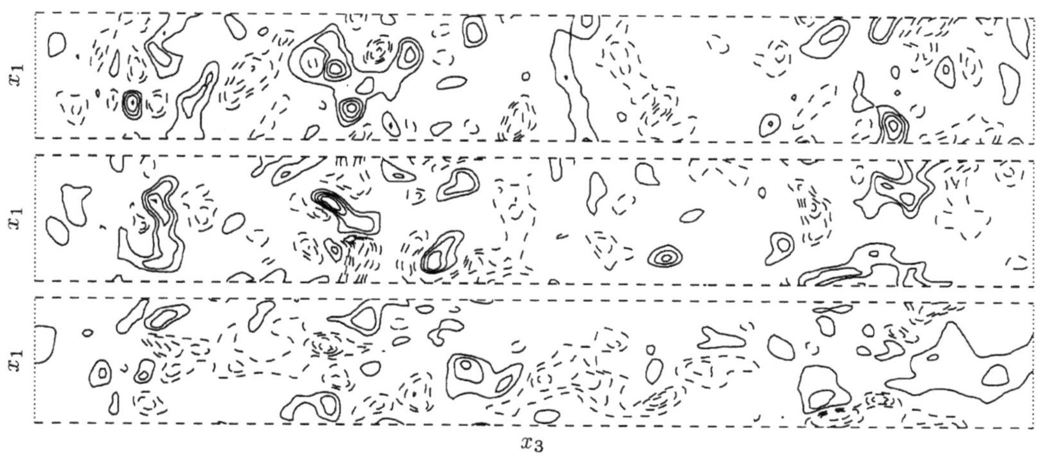

Fig. 9.20d As Fig.9.20c but at the centerline of the channel.

Kline *et al.* (1967), are much longer near the wall than at the center. From the two-point correlations it results that the streaks have a length of approximately 800 wall units, but it is impossible to see that the low-speed streaks are longer than the high-speed streaks and that the latter have higher fluctuations. These considerations can instead be drawn from flow visualizations, made by the code *visua.f* in the directory *CODESCH9/VIS*.

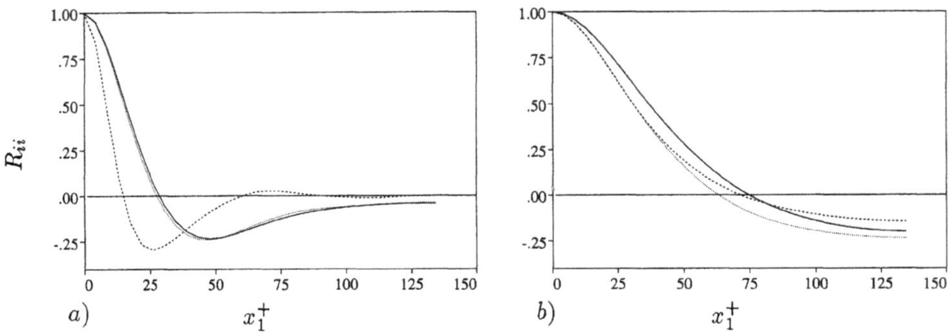

Fig. 9.21 Two-point correlations in the spanwise directions: ——— spanwise, ---- normal, ········ streamwise velocity components: at a) $y^+ = 5.8$ and b) $x_2 = 0$.

This code reads one of the saved fields and evaluates velocity and vorticity components in cross-planes $x_1 - x_2$, in horizontal planes $x_1 - x_3$, and in side planes $x_2 - x_3$. These two-dimensional fields are written in files with names initiating respectively with the words *vis*, *hor*, and *sid*. The fluctuating velocities have been scaled with their *rms* at the given distance from the wall. The contours on horizontal planes are shown at $y^+ = 5.8$ in Fig.9.20c and at $x_2 = 0$ in Fig.9.20d. It turns out, from this picture, that velocity structures at the center of the channel are not completely isotropic as is shown by the longer structures for u_3'. This slightly anisotropy at the center of the channel was previously described at the analysis of the spectra. At the end of the viscous regions the elongated structures cannot so well be characterized as at $y^+ = 15$ where the maximum energy production occurs; in any case also in this region far from the wall the contours of u_3' are more elongated. The intermittent character of the wall region is emphasized by the regions with high positive and negative values of the fluctuating velocity followed by large inactive regions. I have presented these contours to show how much can be learned from flow visualizations. Cleverer studies were performed, in the past, during the CTR summer schools which are reported in the proceedings.

The two-points correlations in the spanwise directions give a quantitative measure of the spacing of the streaks. Fig.9.21a shows that the streamwise component has a minimum at $x_3^+ \approx 50$ which corresponds to a spacing of 100 wall units. Kim *et al.* (1987) observed that the distance between the streaks weakly increase with the distance from the wall; this explains why at $y^+ = 5.8$ it was not found a value equal to 100 that usually occurs at $y^+ = 10$. There is a double change of the normal velocity within a single streak, as it emerges in the left hand side of the channel in Fig.9.20c, thus it is explained why the minimum for R_{22} is at $x_3^+ \approx 25$. Fig.9.21b shows clearly that the simulations were performed in a narrow channel and Fig.9.20d confirms that this is true.

The two-points correlation of the vorticity give the same outcomes as the velocity field on the streaks spacing and on the insufficient lateral dimension of the channel; thus these will not be discussed. The two-points cross-correlations between velocity and vorticity

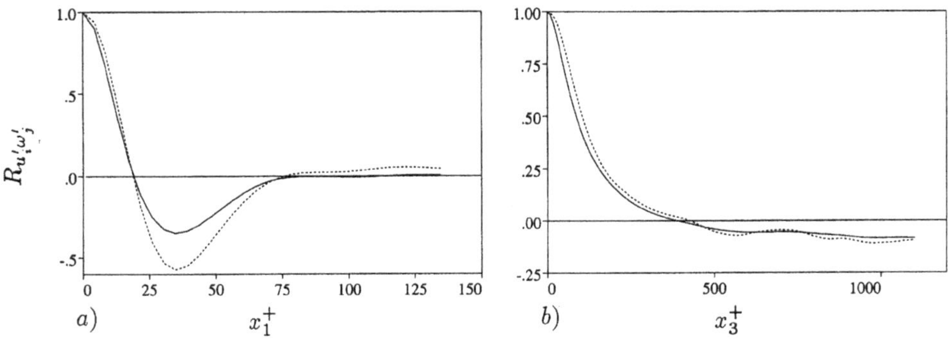

Fig. 9.22 Two-point cross-correlations at $y^+ = 5.8$ in a) spanwise directions b) streamwise direction for —— $\langle u_1'(\mathbf{x}+x_i)\omega_2'(\mathbf{x})\rangle / \langle u_1'(\mathbf{x})\omega_2'(\mathbf{x})\rangle$ and ---- $\langle u_2'(\mathbf{x}+x_i)\omega_1'(\mathbf{x})\rangle / \langle u_2'(\mathbf{x})\omega_1'(\mathbf{x})\rangle$.

have, to my knowledge, not been analyzed in the past, thus I am going to describe them shortly. The link between the Lamb vector and turbulent energy production has been described and it was observed that among the three components the only one playing an important role is that directed in the streamwise direction; this motivates the evaluation of the cross-correlations between vorticity and velocity appearing in λ_3. In addition since the production is large near the wall only those at $y^+ = 5.8$ are presented. The cross-correlations at the center of the channel have a reduced interest since there the production is almost zero. Fig.9.22a shows that in the spanwise directions the cross-correlations have a minimum at the same distance as for the cospectra in Fig.9.19a. Fig.9.22b shows that the cross-correlation in the streamwise directions of $\langle u_1'\omega_2'\rangle$ and $\langle u_2'\omega_1'\rangle$ drop to zero at the same distance. This distance is dictated by the vorticity components ω_2' and ω_1'.

9.9h Joint PDF of velocity and vorticity correlations in one point

To understand better the role of the correlation between velocity and vorticity the joint PDF of all the components of the tensor γ_{ij}' have been calculated in the post-processing code *statpdfl.f*. At the center of the channel the joint PDF between velocity and vorticity has a large degree of symmetry and it explains the low values of the correlations. The fluctuating quantities used for the joint PDF are scaled with respect to the *rms* value; the same scaling is used to get Fig.9.20c-d, and an asterix indicates this ratio.

The symmetry of the PDF of the helicity density in Fig.9.12 explained why the mean helicity density was zero across the channel. The joint PDFs in Fig.9.23 of the three components of the helicity density show different symmetries; $\langle u_1'\omega_1'\rangle$ has a symmetry around $u_1'^*$, while the other two components have a symmetry with respect to the vorticity. This last observation suggests that there is an equal number of left- and right-handed vortical structures. The central figure shows that, in the wall bursts, there are more sweeps ($u_2'^* < 0$) than ejections ($u_2'^* < 0$). The third figure shows that the amount of space with low- is greater than that with high-speed streaks, and that the intensity of the fluctuations in the high- is greater than that in the low-speed streaks. The high-speed

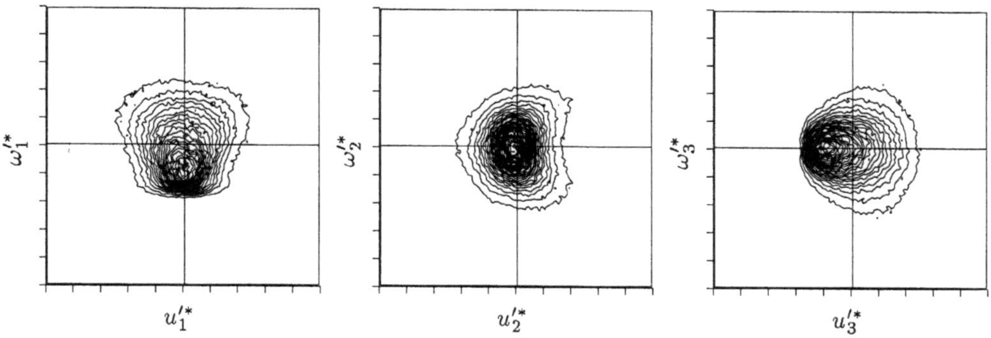

Fig. 9.23 Joint PDFs of the components of the helicity density at $y^+ = 5.8$

streaks contribute to the turbulent friction; together with the low-speed streaks contribute to the bursting events responsible for the turbulence production.

From the kinematic relation $\omega_k = \epsilon_{ijk}\frac{\partial u_j}{\partial x_i}$ it turns out that near the wall $\frac{\partial u_3}{\partial x_2}$ contributes the most to ω_1, and Fig.9.24a shows that in the bursting process (consisting of ejections ($u_2' > 0, u_3' < 0$) and of sweeps ($u_2' < 0, u_3' > 0$)) the high-speed streaks are associated with positive $\omega_1'^*$ and viceversa. Fig.9.24c indeed shows that $\omega_1'^* > 0$ is more correlated with $u_2'^* < 0$ (sweeps) than with $u_2'^* > 0$; this indicates that near the wall the regions of spanwise vorticity are inclined with respect to the wall. Figs.9.24a and c confirms that the number of locations with low-speed streaks overcomes that with high-speed streaks. This consideration is drawn from Fig.9.24a where the contribution of the third quadrant overcomes that of the first, and from Fig.9.24c where the fourth contributes more than the second. These joint PDFs are normalized to give the integral equal to unity, hence from these plots it is impossible to see which of the terms contributes more to the Lamb vector. Fig.9.13 shows that $\langle u_1'\omega_2'\rangle - \langle u_2'\omega_1'\rangle$ is positive; this is confirmed by the cospectra in Figs.9.19a. The vertical profile of the two terms in addition shows that $\langle u_1'\omega_2'\rangle$ is more than ten times larger than $\langle u_2'\omega_1'\rangle$ and that its maximum occurs at $y^+ \approx 5.0$. These considerations suggest that the control of wall turbulent flows is efficient if the action modifies $\langle u_1'\omega_2'\rangle$. It turns out then, that the normal vorticity component is very important in wall turbulent flows as was shown by Blackwelder & Swearingen (1987) in an analogy between Gortler vortices and streamwise vortices. Choi *et al.* (1994) in their study on optimal control found that a large drag reduction was achieved by acting on the spanwise velocity. They did not look at the Lamb vector but their results confirm that the study of the γ_{ij}' tensor enhances the comprehension of wall bounded flows in order to find the best control strategy.

9.9i Vortex stretching

Several times it has been pointed out that 3D turbulence is linked to vortex stretching; the term in the vorticity equation that makes the difference between 2D and 3D flows. In shear flows the generation of the large scales depends on the shape of the boundaries; in the plane channel for instance these scales have a size close to the distance between the

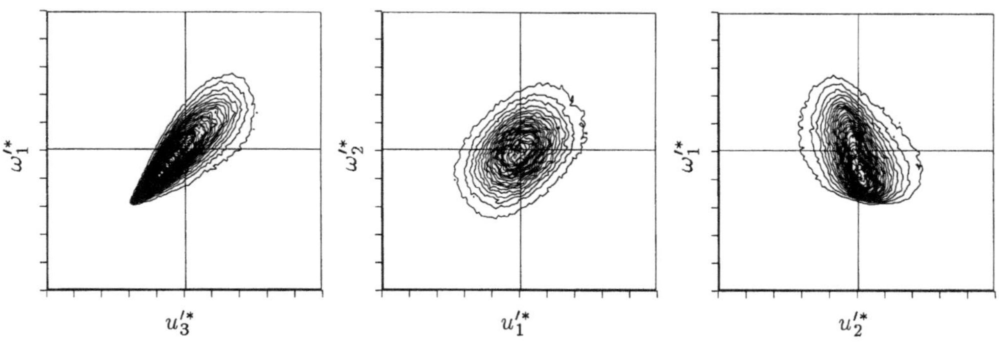

Fig. 9.24 Joint PDFs of some of the components of the Lamb vector at $y^+ = 5.8$

walls. The intermediate scales are generated by vortex stretching. Some of these scales are part of the energy cascade from large to small eddies, but others (for instance, those near the wall) have a dimension comparable to the distance from the wall and thus are smaller than the channel height. These arguments suggest the utility of an investigation on how the vorticity is generated, by means of an analysis of the joint PDF of the quantities contributing to vortex stretching and tilting terms in the vorticity transport equation; it is

$$\frac{\partial \omega_i}{\partial t} + u_j \frac{\partial \omega_i}{\partial x_j} = \omega_j \frac{\partial u_i}{\partial x_j} + \frac{1}{Re} \frac{\partial^2 \omega_i}{\partial {x_j}^2} \qquad 9.22$$

In the plane channel the variables with a mean different from zero are $\langle \omega_1(x_2) \rangle$ and $\langle u_3(x_2) \rangle$. The joint PDFs for the total quantity $q(x_1, x_2, x_3) = \langle q(x_2) \rangle + q'(x_1, x_2, x_3)$ behave as those of the fluctuating quantities but the former are shifted. This consideration suggests to investigate the joint PDFs of the fluctuating quantities as done for the joint PDFs between velocity and vorticity. I would like to remind the reader that the joint PDFs are very useful to describe the physics of turbulent flows by recalling that, from the joint PDFs of u_2' and u_3' it was understood that the second quadrant (ejections) and fourth quadrant (sweeps) contribute to the Reynolds shear stress $\langle u_2' u_3' \rangle$ and hence to the turbulent energy production. In analogy it should be expected that the joint PDF of the stretching and tilting terms in Eq.9.22 could help to understand which are the events contributing for the major part to the mean and rms vorticity generation. Our attention is limited to ω_3' and ω_1' and the joint PDFs have been calculated at different distances from the wall. The PDFs at the center of the channel are not presented; these have the common characteristic of being axisymmetric. On the other hand, near the wall, these reveal the importance of the near wall vortical structures.

In Fig. 9.25, as in the other figures related to the joint PDFs, the asterisk denotes normalization by the local rms value. The joint PDFs of the terms in the ω_3 equation indicate that, for each term, there is symmetry about $\omega_i^* = 0 (i = 2, 3)$ or about the $\frac{\partial u_3}{\partial x_1}'^*$ axis. These symmetries produce a null contribution and thus $\langle \omega_3 \rangle$ remains null. In the wall

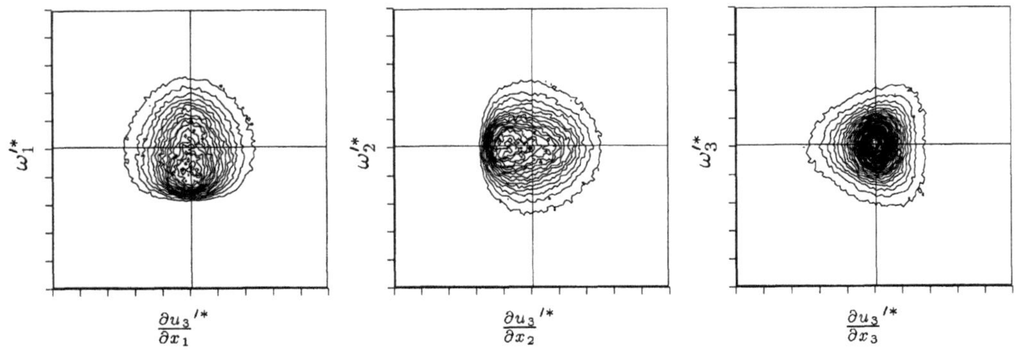

Fig. 9.25 Joint PDFs of the components of the vortex generation terms for the ω_3 component, at $y^+ = 5.8$

region the greatest number of events (the darkest zones in Fig.9.25a-b) are those connected to low-speed streaks where negative values of $\frac{\partial u_3}{\partial x_2}'^*$ (Fig.9.25b) occur together with weak positive or negative values of $\frac{\partial u_3}{\partial x_1}'^*$ (Fig.9.25a). These contribute to the ω_3 generation terms, which have opposite signs (Fig. 9.25); for example, the positive contribution from the third quadrant of Fig. 9.25a is similar to the negative contribution of the first quadrant of Fig. 9.25b. Figs. 9.25ab thus suggest that the same amount of positive vorticity, which is generated by the interaction between ω_1 and $\frac{\partial u_3}{\partial x_1}$ is lost by the interaction between ω_2 and $\frac{\partial u_3}{\partial x_2}$, to summarize the normal and side tilting act in an opposite way. The symmetry of the vortex stretching term $\omega_3'^* \frac{\partial u_3}{\partial x_3}'^*$ (Fig. 9.25c) about the $\omega_3'^* = 0$ axis implies that the amount of positive vorticity stretched is equal to the amount of compressed negative vorticity; in addition this figure shows that there are more stretching than compressing events. The asymmetry around the axis $\frac{\partial u_3}{\partial x_3}'^* = 0$ implies that in the wall region there is a regeneration mechanism linked to the turbulence production. The joint PDFs at $y^+ = 20$ emphasize an increase of the asymmetry; this occurrence was expected because the maximum production occurs around $y^+ = 15$.

The joint PDFs involving ω_1' are more interesting; Fig. 9.26 shows that near the wall first and third quadrant events dominate. The overall contribution thus is to increase the amount of positive ω_1'. The compression events in the third quadrants reduce the amount of negative ω_1' causing in the high speed streaks higher absolute values of ω_1' than the negative in the low-speed streaks (Fig.9.24a). In a study devoted to analyze the intermittency in the wall region Orlandi *et al.* (1997) showed that the side-tilting of ω_2 contributes positively very close to the wall ($y^+ = 2.5$) and negatively in the buffer region ($y^+ = 20$). Fig.9.26b shows on the other hand that at $y^+ = 5.8$ a circular symmetry on the joint PDF produces a value close to zero. It has been observed that across the channel large variations of the side tilting of ω_2 occur. On the overall Fig.9.26c shows that the side tilting of ω_3 gives a contribution of the same sign as the vortex stretching to the spanwise vorticity generation.

The joint PDFs of vorticity and velocity derivatives here presented (to my knowledge

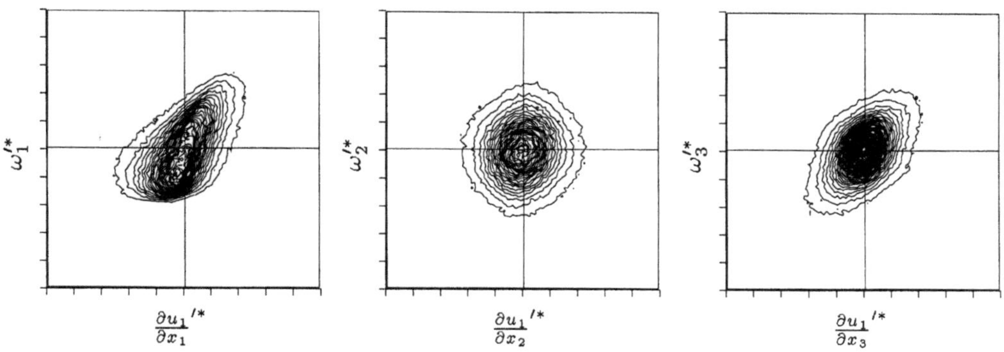

Fig. 9.26 Joint PDFs of the components of the vortex generation terms for the ω_1 component, at $y^+ = 5.8$

never done before) contribute to the comprehension of near-wall turbulence; this study was initiated by Moin & Kim (1985) and was continued by Kim *et al.* (1987), Kim (1989) and Antonia & Kim (1994) and in a number of other papers, increasing so fast that it is impossible to mention them.

The data base can be used to investigate several other quantities in the near wall region, and from those the most important are those linked to the enstrophy production (see Tsinober *et al.* (1995) for pipes). This is a quantity strictly connected to vortex dynamics and it plays an important role in the one-point closure turbulence models. The new trends in turbulence modeling are based on closure that as much as possible account for the vortical structures.

9.9j Budget in the one-point closure equations

The DNS of turbulent flows helps to study turbulence physics, but it is also useful to develop one-point closure models which can be inserted into simulations of flows of practical applications. The necessity of closure (turbulence models) is dictated by the present and even future limitations of DNS to low Reynolds number. However, once the flow is fully turbulent, as it was observed experimentally (Wei & Willmarth (1989)), the Reynolds number dependence is not so large. This is in particular true for engineering applications. Thus the DNS data could be used to validate turbulence models, to use at high Reynolds numbers. LES, discussed later on, requires the solution of the 3D set of Navier-Stokes equations in time. This consideration explains why, even with today's computers, this approach, also for the lack of subgrid models, is not widely used in industries. Industries instead are employing still the averaged Navier-Stokes equations with one-point closures to design their products.

The development of one-point turbulence models was a very important topic of basic research in the sixties and in the seventies. In that period several research groups (for instance at Los Alamos, Imperial college and Stanford) were testing and developing models with some sort of success. The success was mainly dictated by the good results were obtained at the first AFSOR turbulence conference (Kline *et al.* (1969)) on modeling wall bounded flows. In the seventies the effort was then directed to extend the models to complex flows by inserting new constants or flow-dependent constants. In 1980 a new AFOSR conference was organized by Kline (Kline *et al.* (1981)) with the purpose to check the validity of the turbulence models on complex flows. A large number of computors attended the conference and concluded that the models were satisfactory in postdicting the flows and unsatisfactory in predicting the flows; moreover it was clear that the worst performances of the models resulted for homogeneous turbulence. The reason for the good postdictions is related to the evaluation of the constants in the models by experimental data; thus experimental inputs were introduced in the simulations through the constants. Someone concluded that with that conference it was proved that turbulence closure are good interpolation method based on physical rather than mathematical grounds. This explains why some of the results were encouraging in predicting flows where the characteristics of the flow were slightly changed. This for instance append for boundary layers with different pressure gradients. Since the mean shears in the flows depend on the boundary conditions and largely contribute to the turbulence production, it is clear that if a numerical method was satisfactory in describing the geometrical dependence the results were not completely wrong.

On the other hand the inability of the model to simulate homogeneous flows are dictated by the minor role played in these flows by the mean shear, in fact the large contribution to the turbulence dynamics comes from interactions of the turbulent structures. In one-point closures the scale information on turbulence is eliminated; homogeneity in addition eliminates the turbulence gradients and thus the values of the constants are a key factor. The few constants in the models did not permit to take into account the anisotropy of each flow. In the proceedings of the conference it is instructive to look at the results of the effect of rotation on isotropic turbulence which proves the inability of any kind of model. At that conference, for the first time, it was clear to the engineering oriented scholars, that new insights into turbulence could come from LES and DNS. At that conference Rogallo (1981) presented his DNS results of homogeneous turbulence; his paper has been one of the building stones of DNS turbulence. I remember that Rogallo began his talk by claiming that he did not do anything difficult, he was only clever to mix a computer and the Navier-Stokes equations using in this process an efficient numerical algorithm. I think that this is indeed true. I am claiming in this book that the generation of turbulent fields is a very easy task; the good scholars are those who from these data produce ideas to improve our knowledge and our standard of life. Rogallo was one of them; he never was on the stage, he was behind many ideas that were growing in the last fifteen years at the CTR in Stanford.

From these personal considerations (that I presume are not shared worldwide) I would like to remind that at the Stanford conference, named the turbulence's Olympics, was born the idea that DNS is a tool to improve closure. This consideration is dictated by the difficulty to measure velocity gradients and by the impossibility to measure pressure fluctuations in the laboratory. In addition in a laboratory triple correlations require sophisticated electronics equipments and it is difficult to measure some of the turbulent correlations near solid walls due to the physical dimension of the cross wires. These considerations motivated later on Mansour *et al.* (1988) to use the Kim *et al.* (1987) data base to evaluate all the terms in the Reynolds stress budget. Durbin (1992) used these budgets to validate his low-Reynolds number closure, where the $\langle u_2'u_2'\rangle$ transport equation gives the right wall dumping. Before low-Reynolds number closures were based on the introduction of an ad hoc function in the turbulent kinetic equation as in Norris & Reynolds (1975) for one-equation turbulence models. For a review of low-Reynolds number closure see the review article by Patel *et al.* (1985)

The Reynolds stress transport equations can be derived from the Navier-Stokes equations by an ensemble average of the equations for the fluctuating stresses. The transport equations are given in any turbulence book (in Hinze (1975) they are derived at Pg.68 and at Pg.323)

$$\frac{\partial\langle u_i'u_j'\rangle}{\partial t}+\langle U_k\rangle\frac{\partial\langle u_i'u_j'\rangle}{\partial x_k}=P_{ij}+T_{ij}+D_{ij}+\Pi_{ij}-\epsilon_{ij} \qquad 9.23$$

On the right-hand side there are respectively production, turbulent transport rate, viscous diffusion rate, velocity pressure-gradient and dissipation rate; their expressions are

$$P_{ij}=-[\langle u_i'u_k'\rangle\frac{\partial\langle U_j\rangle}{\partial x_k}+\langle u_j'u_k'\rangle\frac{\partial\langle U_i\rangle}{\partial x_k}]$$

$$T_{ij}=-\frac{\partial\langle u_i'u_j'u_k'\rangle}{\partial x_k}$$

$$D_{ij}=\nu\frac{\partial^2\langle u_i'u_j'\rangle}{\partial {x_k}^2}$$

$$\Pi_{ij}=-[\frac{\partial\langle pu_i'\rangle}{\partial x_j}-\langle p\frac{\partial u_i'}{\partial x_j}\rangle+\frac{\partial\langle pu_j'\rangle}{\partial x_i}-\langle p\frac{\partial u_j'}{\partial x_i}\rangle]$$

$$\epsilon_{ij}=2\nu\langle\frac{\partial u_i'}{\partial x_k}\frac{\partial u_j'}{\partial x_k}\rangle$$

Contrary to Mansour *et al.* (1988) the velocity pressure-gradient term has been splitΦlit into two parts, because the calculation is easier and more accurate on a staggered grid. Repeated indices imply summation over 1,2,3. Mansour *et al.* (1988) scaled each term of the budget with respect to u_τ^4/ν. In the post-processing code, this implies to divide each term by the quantity Reu_τ^4. A different code for the budget is furnished to facilitate its comprehension; in fact in the previous code so many quantities were evaluated, that the budget would had lead to an increase difficulty on the comprehension of the code. This new code $CHABUDG$ in the directory $CODESCH9/BUDG$ has been derived from the code $CHATRA$ by inserting the calculation of some new quantity in the routine $VELTUR$. The code is split into different files in the routine $OUTH$; in $chabudgio.f$ the calculation of each term of the budget together with the eventual differentiations of some of the statistics are performed. This code can be used as a base to evaluate the budgets related to the dissipation equations that were computed by Mansour *et al.* (1988) and not here. Perhaps it could be interesting to evaluate the terms in the k,ω model described by Wilcox (1995) in his monograph on turbulence modelling.

Each term of the $\langle u_i'u_j'\rangle$ budget equation has been compared with those evaluated by Mansour *et al.* (1988) and once more this comparison shows that the present data base

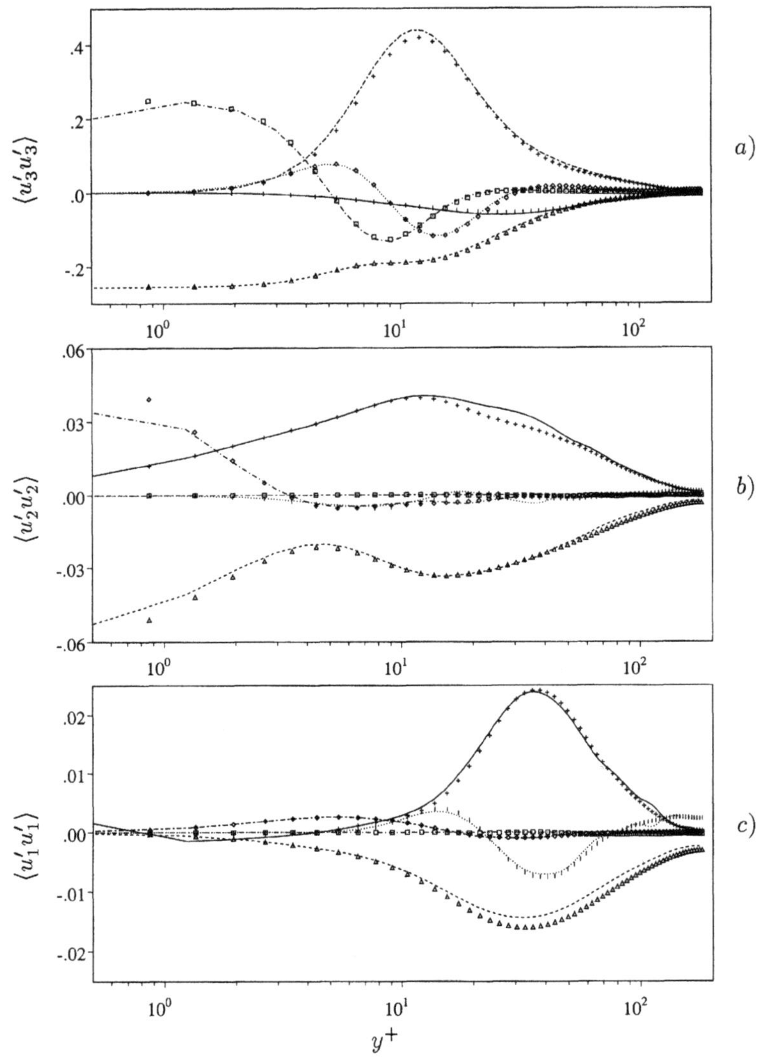

Fig. 9.27 Budget profiles for the three normal stresses.

has characteristics as good as those of pseudospectral simulations. Fig.9.27a-c shows this comparison for the normal and Fig.9.28 for the $\langle u_3' u_2' \rangle$. Direct simulation permits to evaluate profiles of the terms that are implying correlations with the pressure. The Π_{ij} terms for $i = j = 2$ and for $i = 2, j = 3$, can be split into a pressure transport term and a pressure strain term. In this way in the energy budget the pressure strain term gives

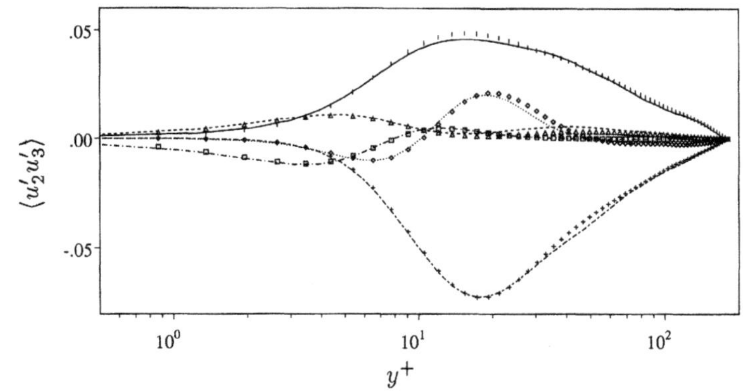

Fig. 9.28 Budget profiles for the stress $\langle u_2' u_3' \rangle$.

a total zero contribution emphasizing its redistributive effect among the three stresses. The splitting (Mansour *et al.* (1988)) shows that the two terms near the wall are equal and hence the velocity pressure gradient term is low. This behavior is confirmed by the present results in Fig.9.29a-b where it appears that the profiles of the pressure gradient terms are in perfect agreement with the Mansour *et al.* (1988) profiles. They analyzed in more details different ways of splitting that could lead to improvements in the modelling of the pressure that could bring to improvements on the modeling of the pressure gradient term. The closure of this term attracted a large amount of effort, starting from one of the most important papers in one-point closure by Launder *et al.* (1975).

9.10 Conclusions

In this chapter a tool has been described and furnished that can be used by those interested in the study of wall bounded flows. The analysis of the turbulent channel flow was performed by the group of scholars at Stanford in several details and thus they left few questions open. For example Kim and later on Mansour (private communications) performed simulations at higher Re numbers to investigate the Reynolds number effects. More simulations at higher Reynolds number should be performed to investigate better the role and the shape of vortical structures in the log layer observed and measured by means of PIV (Needershulte *et al.* (1990)). After more than 15 years of research in this field it seems that the best part on the comprehension of wall turbulence has been done. Perhaps the stage has been reached to direct the simulations towards applications as for instance the transport of tracers. Kim & Moin (1989) and Kasagi *et al.* (1992) looked at the passive scalar but there are still many question on the statistics of the passive scalar left open. The transport equation for the passive scalar can be easily introduced in the code. More interesting could be the introduction of the motion of particles; this effort requires more work, and recently has been pursued by Lombardi *et al.* (1996). The ordinary differential equations for each particle require an accurate time advancing scheme and interpolation procedures that should be tested and validated for these 3D flows in the presence of solid

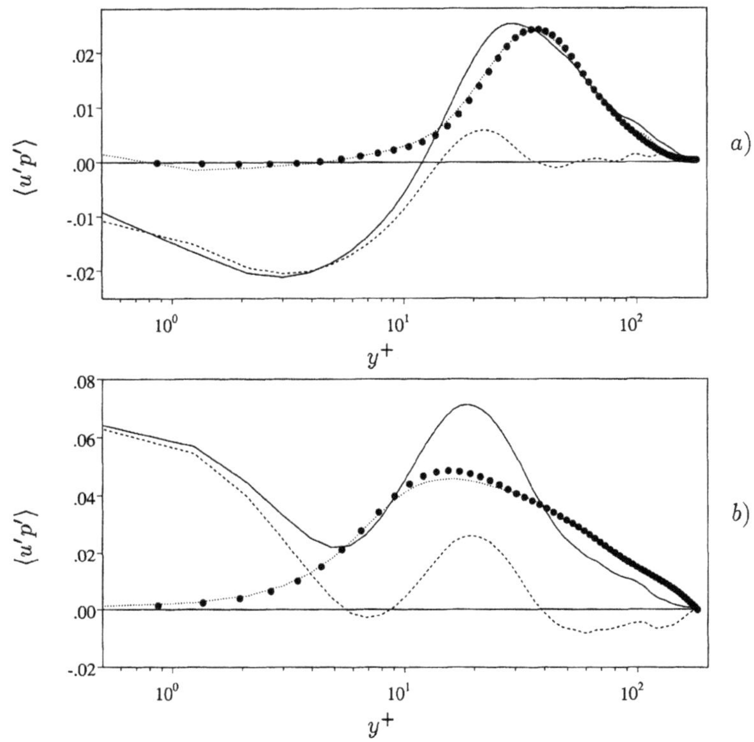

Fig. 9.29 Splitting of the velocity-pressure gradient for a) $\langle u_2' u_2' \rangle$ and for b) $\langle u_2' u_3' \rangle$ equations, —— pressure strain, - - - - pressure transport, ········ • Kim *et al.* (1987) velocity pressure-gradient.

walls. The coupling between particle and fluid flow is of particular interest in combustion related problems.

The largest possibilities, however, are related to turbulence control and mainly to achieve the optimal control (Lee *et al.* (1998)). The common features of flows with drag reduction have not been discovered yet and to this purpose the tool here described could help. This effort leads to a deeper knowledge of turbulence, but, more importantly will lead to more efficient products that enter into everyday life of everyone. This goal will be reached if more scholars will spend their time on the comprehension of the turbulence physics rather than on numerical methodologies. I hope that the codes described here will be used with this purpose.

10 Flows in cylindrical coordinates with one wall

10.1 Introduction

Several flows require cylindrical coordinates. Some of them can be solved in Cartesian coordinates with a detriment of accuracy. For example a three-dimensional simulation of a round jet in Cartesian coordinates gives reasonable results but periodicity or free-slip conditions in two directions introduce an unphysical perturbation with an azimuthal wavenumber $n = 4$. On the other hand in other flows, such as those in pipes, no-slip conditions at the wall drive the flow and enforce the use of cylindrical coordinates.

An efficient spectral method in cylindrical coordinates requires a large effort to approximate the radial derivatives and to treat the singularity at $r = 0$. Even for axisymmetric flows spectral methods with special functions for the radial derivatives are very cumbersome. The complexity of the scheme is illustrated by the statement in the paper by Staneway *et al.* (1988) that 128×128 modes required a memory larger than the in-core memory of a CRAY-XMP48. An accurate method for three-dimensional pipe flows with Jacobi polynomials for the radial direction was developed by Leonard and Wray (1975); they introduced these polynomials to minimize the coupling of the momentum equations. Nevertheless the discretization produced large non-diagonal sparse matrices to be inverted. Buell *et al.* (1983) developed a Galerkin method to simulate the natural convection in a circular enclosure. In the radial direction Bessel functions were used and the condition of finiteness of the solution was applied at $r = 0$. However, the Galerkin method was used to reduce the linearized perturbation equations to an eigenvalue problem and no information was provided about the extension of the method to the full Navier-Stokes equations.

More recently full spectral methods were developed by Zhang *et al.* (1994) and by Loulou (1996). The first authors used Fourier modes in the periodic directions and Jacobi polynomia in the radial directions. Loulou (1996), instead, used B-splines in the radial directions. Perhaps these methods are more accurate than finite difference second order methods but, without any doubt, from the appendices of Loulou's thesis it turns out that these methods are cumbersome and that they require large programming work.

A finite volume method was employed by Eggels *et al.* (1994a) for DNS and later on by Eggels *et al.* (1994b) for LES in turbulent pipes, rotating and non-rotating. The radial momentum equation required an extended grid volume at the axis to avoid the evaluation of quantities at $r = 0$. For the same reason, extrapolations and first-order-accurate differences were used at the axis with a reduction of the overall accuracy. Nevertheless in the pipe the decreased accuracy was assumed to be not significant due to the smooth behavior of the pipe flow near the centerline. Aksevol & Moin (1996) used two different grids: one for the region around the center and the other outside. Even this procedure requires an increased programming work. Verzicco & Orlandi (1996a) followed the idea to build a method that in cylindrical coordinates is as close as possible to that in Cartesian coordinates and this method is described here. Thus even in this chapter the important principle of global conservation has been kept in mind to have a reliable numerical method.

This method is based on a fractional step scheme that uses an approximate factorization. In cylindrical coordinates the treatment of the singularity at the axis requires particular care, that has been simplified here by staggering the velocity components and by introducing the variable $q_r = rv_r$. The stability and accuracy of the method has been validated by Orlandi & Verzicco (1993a) comparing the evolution of an axisymmetric vortex ring with the results from a vorticity-streamfunction scheme. Azimuthal perturbations on a vortex ring, are amplified, at sufficiently high Reynolds number, by the self-induced mean strain as was theoretically found by Widnall *et al.* (1974). Shariff *et al.* (1994) verified that the present numerical method correctly predicted the most unstable wavenumber n^* obtained by the linear stability theory, as well as the growth of a band of modes around n^*. These results were reported by Shariff *et al.* (1994) and are not given here. These can be repeated by inserting a subroutine with the initial conditions for a vortex ring and a subroutine that evaluates the time evolution of the mean kinetic energy at each azimuthal wave number.

The vortex ring has a very small amount of vorticity near the axis hence it is not a severe test of the treatment of the region around $r = 0$. The case of a dipole translating across the center of the axis was chosen by Verzicco & Orlandi (1996a) to check the accuracy at $r = 0$ this is a very delicate test since the dipole preserves its shape during its motion and v_r at $r = 0$ is large.

This numerical method has been also applied to a turbulent flow in a rotating and non-rotating pipe and here the results described by Orlandi & Fatica (1997) and by Orlandi (1997) are reported. With few modification the time-developing round jet can be studied as in Verzicco & Orlandi (1992); in this case the external boundary must be located far from the axis. A coordinate transformation for the radial direction different from that for the pipe and the value of the index that permits to impose free- instead of no-slip conditions at the external boundary are the only variations to insert into the code.

10.2 Governing equations

In the three-dimensional case the Navier-Stokes equations in primitive variables reduce the number of operations necessary to integrate them numerically. In cylindrical coordinates the variables $q_\theta = rv_\theta$, $q_r = rv_r$ and $q_z = v_z$ give the continuity equations

$$\frac{\partial q_r}{\partial r} + \frac{1}{r}\frac{\partial q_\theta}{\partial \theta} + r\frac{\partial q_z}{\partial z} = 0 \qquad 10.1$$

The q_i quantities do not have the same physical dimensions; this could be a matter of confusion, hence it is important to point out that these variables have been introduced only for numerical reasons. In different versions of the code, developed at the university of Roma, in some case the quantity $q_\theta = v_\theta$ and in others the quantity $q_\theta = rv_\theta$ were used. From the practical point of view there is no any difference; however, it is really important to impose $q_r = rv_r$ which helps to treat the singularity at the axis.

The Navier-Stokes equations in terms of the new variables, in a reference frame rotating with the wall and with $q_\theta = v_\theta$, are

$$\begin{aligned} &\frac{\partial q_\theta}{\partial t} + \frac{\partial q_\theta / r q_r}{\partial r} + \frac{1}{r^2}\frac{\partial q_\theta^2}{\partial \theta} + \frac{\partial q_\theta q_z}{\partial z} + \frac{q_\theta}{r}\frac{q_r}{r} + N q_r \\ &= -\frac{\partial p}{\partial \theta} + \frac{1}{Re}\left[\frac{\partial}{\partial r} r \frac{\partial q_\theta / r}{\partial r} - \frac{q_\theta}{r^2} + \frac{1}{r^2}\frac{\partial^2 q_\theta}{\partial \theta^2} + \frac{\partial^2 q_\theta}{\partial z^2} + \frac{2}{r^2}\frac{\partial q_r}{\partial \theta}\right], \end{aligned} \qquad 10.2$$

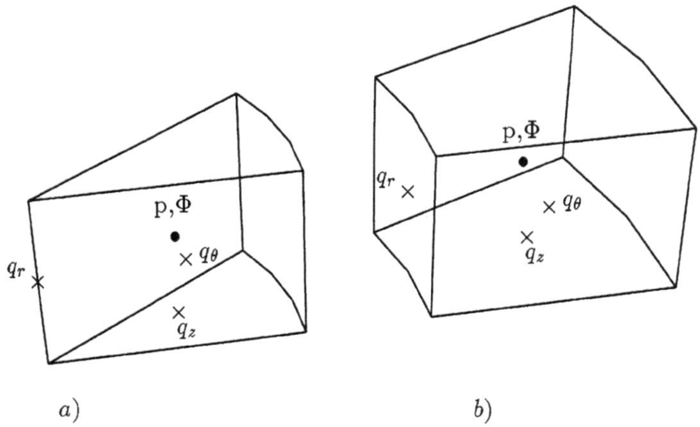

Fig. 10.1 Sketch of the computational cells: *a*) cell at the axis *b*) cell in the domain.

$$\frac{\partial q_r}{\partial t}+\frac{\partial}{\partial r}\frac{q_r^2}{r}+\frac{\partial}{\partial \theta}\frac{q_\theta q_r}{r^2}+\frac{\partial q_r q_z}{\partial z}-\frac{q_\theta^2}{r^2}-Nq_\theta$$
$$=-r\frac{\partial p}{\partial r}+\frac{1}{Re}\left[\frac{\partial}{\partial r}r\frac{\partial q_r/r}{\partial r}-\frac{q_r}{r^2}+\frac{1}{r^2}\frac{\partial^2 q_r}{\partial \theta^2}+\frac{\partial^2 q_r}{\partial z^2}-\frac{2}{r^2}\frac{\partial q_\theta}{\partial \theta}\right], \quad 10.3$$

$$\frac{\partial q_z}{\partial t}+\frac{1}{r}\frac{\partial q_r q_z}{\partial r}+\frac{1}{r^2}\frac{\partial q_\theta q_z}{\partial \theta}+\frac{\partial q_z^2}{\partial z}$$
$$=-\frac{\partial p}{\partial z}-\overline{\frac{\partial P}{\partial z}}+\frac{1}{Re}\left[\frac{1}{r}\frac{\partial}{\partial r}r\frac{\partial q_z}{\partial r}+\frac{1}{r^2}\frac{\partial^2 q_z}{\partial \theta^2}+\frac{\partial^2 q_z}{\partial z^2}\right]. \quad 10.4$$

These equations are suited to perform simulations of rotating pipes with a steady velocity. On the other hand, if the wall of the pipe oscillates in time extra terms should be added to the equations. The equations nondimensionalized by the centerline stream-wise velocity U_P of the laminar Poiseuille flow and by the pipe radius R give the dimensionless numbers $Re = U_P R/\nu$ and $N = 2\Omega R/U_P$. For different flows, the Reynolds number is conveniently defined; for instance, in studies devoted to analyze the growth of azimuthal perturbations of a vortex ring a characteristic length r_0 and a circulation Γ give the velocity (Γ/r_0) and the pressure $(\rho\Gamma^2/r_0^2)$ scales. The Reynolds number is thus $\mathrm{Re} = \Gamma/\nu$, with ν the kinematic viscosity. In this paper rather than the Rossby number $Ro = U_P/2\Omega R$, usually used in geophysical problems, the rotation number $N = 1/Ro$ has been used. This choice has been dictated by the fact that N was previously used in studies of turbulent flows in rotating pipes. In addition $N = 0$ is more understandable than $Ro = \infty$ for non-rotating pipes. Simulations of turbulent channels or pipes can be done at constant mass or at constant friction velocity. As for the channel the first choice is assumed, a mean pressure gradient in the q_z equation maintains a constant bulk velocity $U_b = \int q_z dV$. $\overline{\frac{\partial P}{\partial z}}$ is related to the mean wall friction velocity u_τ by $u_\tau = \sqrt{|\overline{\frac{\partial P}{\partial z}}|/2}$.

Even in cylindrical coordinates the velocity components are located at the center of the faces and the pressure at the center of the cell. The layout of the variables is given in Fig.1 for the general cell within the domain and for that at the axis. From the sketch of the cell in Fig.1a it is clear that q_r has been introduced because it gets a zero value at the axis. In this way the treatment of the axis is simpler than that used by Aksevol & Moin (1996) and by Eggels *et al.* (1994a).

In other chapters of this book we emphasized the need to consider the system of equations in conservative form in solving 1D, 2D and 3D flows. As in 3D Cartesian coordinates dealing with stability or turbulence it is important to conserve the total energy because any production or dissipation of unphysical energy, related to the numerical scheme, could disrupt the quality of the results. Horiuti (1987) showed that the conservative form minimizes, especially near the wall, the truncation errors associated with central second-order finite differences. This is a very important issue in wall-bounded flows. As for the channel also for the pipe it is shown that the energy is conserved even with a coarse grid; this is a further proof that the axis is accurately treated. In inviscid simulations the pipe has a free-slip boundary.

In the z direction periodicity has been assumed and with minor changes it is not difficult to impose in z free- or no-slip boundaries maintaining a uniform grid in z. These changes permitted to simulate the interaction of a perturbed vortex ring with a solid wall (Orlandi & Verzicco (1993a)) or barotropic and baroclinic instabilities (Verzicco *et al.* (1997)). The more important transformation consisted of the change of the pressure solver, where instead of sin and cos trigonometric expansion only the cos transformation is used. The other minor changes consisted of the imposition of the boundary conditions in the evaluation of the viscous terms and in the calculation of the coefficients of the matrices in the factorization scheme. This code is not furnished because more recently a new version of the code has been written to simulate the formation of 3D tripole with particular emphasis on the effects of the Eckman layer; this code permits to have in the radial and in the longitudinal directions non-uniform meshes. This code is described in chapter 11.

10.3. Numerical method

The discretization of the system of equations in a staggered grid in cylindrical coordinates is similar to that in Cartesian coordinates. In this chapter dealing with uniform grids in the azimuthal and axial directions the discretization is very similar to that used in chapter. 9. In the radial direction the same considerations made in the non-uniform direction x_2 for the flow between two walls hold. The θ coordinate coincides with x_1, the axial coordinate z with x_3 and the radius r is a function of the computational coordinate x_2. The first and second derivatives in θ and z are expressed by the usual centered operators given in chapter 2 and indicated by the algebraic operator $\frac{\delta}{\delta x_i}$ and $\frac{\delta^2}{\delta x_i{}^2}$ with $i = 1, 3$. The derivatives in the radial direction are approximated by a coordinate transformation $r = r(x_2)$ which introduces the metric g_2; analogously to chapter 9 $g_2 = \frac{\delta x_2}{\delta r}$. The numeration of the variables q_i is consistent with the coordinate labelling. Since the unknowns q_i are staggered, the radius r and the metric g_2 in the discretized equations appear at the location j and $j + \frac{1}{2}$. The subscript c indicates the nodes and m the middle points on the face of the cell. With these definition the system of Eq.(10.2-4) for $q_1 = q_\theta$, $q_2 = q_r$ and $q_3 = q_z$ becomes:

$$\frac{\partial q_1}{\partial \partial t} + g_{2m}\frac{\delta q_1/r_m q_2}{\delta x_2} + \frac{1}{r_m^2}\frac{\delta q_1 q_1}{\delta x_1} + \frac{\delta q_1 q_3}{\delta x_3} + \frac{q_1 q_2}{r_m^2} + N q_2$$
$$= -\frac{\delta p}{\delta x_1} + \frac{1}{Re}\left[g_{2m}\frac{\delta}{\delta x_2} r_c g_{2c}\frac{\delta q_1/r_m}{\delta x_2} - \frac{q_1}{r_m^2} + \frac{1}{r_m^2}\frac{\delta^2 q_1}{\delta x_1{}^2} + \frac{\delta^2 q_1}{\delta x_3{}^2} + \frac{2}{r_m^2}\frac{\delta q_2}{\delta x_1}\right], \qquad 10.5$$

$$\frac{\partial q_2}{\partial \partial t} + g_{2m}\frac{\delta q_2 q_2}{\delta x_2} + g_{2m}\frac{\partial}{\partial x_2}\frac{q_2^2}{r_c} + \frac{\delta}{\delta x_1}\frac{q_1 q_2}{r_m r_c} + \frac{\delta q_2 q_3}{\delta x_3} - \frac{q_1^2}{r_c^2} - N q_1$$
$$= -r_c g_{2c}\frac{\delta p}{\delta x_2} + \frac{1}{Re}\left[g_{2c}\frac{\delta}{\delta x_2} g_{2c} r_c\frac{\delta q_2/r_c}{\delta x_2} - \frac{q_2}{r_c^2} + \frac{1}{r_c^2}\frac{\delta^2 q_2}{\delta x_1{}^2} + \frac{\delta^2 q_2}{\delta x_3{}^2} - \frac{2}{r_c^2}\frac{\delta q_1}{\delta x_1}\right], \qquad 10.6$$

$$\frac{\partial q_3}{\partial t} + \frac{g_{2m}}{r_m}\frac{\delta q_2 q_3}{\delta x_2} + \frac{1}{r_m^2}\frac{\delta q_1 q_3}{\delta x_1} + \frac{\delta q_3 q_3}{\delta x_3}$$
$$= -\frac{\delta p}{\delta x_3} - \frac{\overline{\partial P}}{\partial z} + \frac{1}{Re}\left[\frac{g_{2m}}{r_m}\frac{\delta}{\delta x_2} g c r_c\frac{\delta q_3}{\delta x_2} + \frac{1}{r_m^2}\frac{\delta^2 q_3}{\delta x_1{}^2} + \frac{\delta^2 q_3}{\delta x_3{}^2}\right]. \qquad 10.7$$

Using this new symbolism it is very simple to figure out how each term is expressed by FORTRAN in the code. Even in cylindrical coordinates linear interpolations allow to evaluate the variables in grid locations where these are not defined.

The above system of equations is a system of ordinary differential equations that is solved by means of a fractional-step method with the introduction of the pressure at the old time step to simplify the boundary conditions for the intermediate velocity field $\hat{\mathbf{q}}$. The third-order low-storage Runge-Kutta scheme in combination with the Crank-Nicolson scheme, described in previous chapter, is used to evaluate the non-solenoidal velocity field $\hat{\mathbf{q}}$. As in chapter 9 the system of Eq.(10.5-7) is schematized as

$$\frac{\hat{q}_i - q_i^l}{\Delta t} = \left[\gamma_l H_i^l + \rho_l H_i^{l-1} - \alpha_l \mathcal{G}_i p^l + \alpha_l (A_{i\theta} + A_{ir} + A_{iz})(\hat{q}_i + q_i^l)\right]. \qquad 10.8$$

H_i contains the convective terms and the viscous terms with velocity derivatives of q_j (with $j \neq i$). $\mathcal{G}_i$, $A_{i\theta}$, A_{ir} and A_{iz} denote discrete differential operators of the pressure gradient and of second derivatives. α_l, γ_l and ρ_l are the coefficients of the time advancement scheme and the same values as in chapter 3 hold. In Eq.(10.8) the convective terms are treated explicitly. The implicit treatment of the viscous terms avoids the viscous stability restrictions particularly heavy at low Reynolds numbers or with enhanced non-uniform grids. It is interesting to note that flows in cylindrical coordinates have a very onerous stability limit if azimuthal velocities are large near the axis. In this region $r\Delta\theta$ is very small and thus the CFL could be very large, consequently the Δt must be reduced and the calculation takes a large CPU. This occurrence was encountered by Verzicco & Camussi (1997) in the simulation of natural convection on cylindrical containers. They thus treated the $\theta = x_1$ fully implicit as was suggested by Aksevol & Moin (1996); they found a very efficient way to program this revised version of the code. This improvement is not described but whoever is interested can contact the author of the book or the authors of that version of the code; they are glad to furnish the improved version of the code.

The solution of Eq.(10.8) requires the inversion of large sparse matrices which can be reduced to three tridiagonal matrices by the factorization procedure. By the increment $\Delta\hat{q}_i = \hat{q}_i - \hat{q}_i^l$ Eq.(10.8) , written in delta form, is

$$(1-\alpha_l A_{i\theta})(1-\alpha_l A_{iz})(1-\alpha_l A_{ir})\Delta\hat{q}_i \\ = \Delta t\left[\gamma_l H^l + \rho_l H^{l-1} - \alpha_l \mathcal{G}_i p^l + 2\alpha_l(A_{i\theta}+A_{ir}+A_{iz})q_i^l\right] = RHS_i, . \qquad 10.9$$

In the general case the solution requires at each time sub-step boundary conditions for $\hat{\mathbf{q}}$ and for the intermediate quantities $(1-\alpha_l A_{iz})(1-\alpha_l A_{ir})\Delta\hat{q}_i$ and $(1-\alpha_l A_{ir})\Delta\hat{q}_i$. The choice to perform the factorization first in the two periodic directions eliminates the boundary conditions for the intermediate variables. For the flow in a pipe the no-slip or free-slip boundary conditions on the wall are imposed during the sweep along x_2. Kim & Moin (1985) and even Verzicco & Orlandi (1996a) in an appendix proved that at the end of the three substeps the intermediate field is linked to the solenoidal field q_i^{l+1} by $\hat{q}_i = q_i^{l+1} + O(\Delta t^2)$. Thus the physical boundary conditions for q_i^{l+1} can be assigned to $\hat{q}_i$. I would like to remind that these arguments have been largely described in the previous chapters; these are shortly repeated here for those who are interested to read this chapter without reading the previous.

The intermediate velocity field $\hat{\mathbf{q}}$ is globally divergence-free but not locally. The global divergence-freeness is linked to periodicity and to wall impermeability. The continuity equation is enforced on each cell by

$$q_i^{l+1} - \hat{q}_i = -\alpha_l \Delta t \mathcal{G}_i \Phi^{l+1}, \qquad 10.10$$

where Φ is a scalar calculated from the equation obtained by substituting Eq.(10.10) in Eq.(10.1) This equation is

$$\mathcal{L}\Phi^{l+1} = +\frac{1}{\alpha_l \Delta t}\mathcal{D}\hat{\mathbf{q}}, \qquad 10.11$$

In Eq.(10.11) $\mathcal{D}$ and $\mathcal{L} = \mathcal{D}\mathcal{G}$ indicate, with the same notation as in Eq.(10.5-7), the following discrete operators:

$$\mathcal{D}\cdot = \frac{1}{r_m^2}\frac{\delta\cdot}{\delta x_1} + \frac{1}{r_m}\frac{\delta\cdot}{\delta x_2} + \frac{\delta\cdot}{\delta x_3}, \qquad \mathcal{G}\cdot = \frac{\delta\cdot}{\delta x_1} + r_c g_{2c}\frac{\delta\cdot}{\delta x_2} + \frac{\delta\cdot}{\delta x_3},$$

$$\mathcal{L}\cdot = \frac{1}{r_m^2}\frac{\delta^2\cdot}{\delta x_1^2} + \frac{g_{2m}}{r_m}\frac{\delta}{\delta x_2} g_{2c} r_c \frac{\delta\cdot}{\delta x_2} + \frac{\delta^2\cdot}{\delta x_3^2}. \qquad 10.12$$

The location of Φ at the center of the cell gives a number of equations equal to the number of unknowns; Eq.(10.11), obtained by dividing Eq.(10.1) by r, does not need explicit boundary conditions. This operation is dictated by the tridiagonal inversion, in fact on doing this division the coefficients of the diagonal elements, after the trigonometric expansion in x_3 depend only on the wave number k_3. The tridiagonal algorithm can be vectorized and thus it is very efficient. This is a very important issue since the efficiency of the method is linked to the solution of the scalar Φ.

Once more it is important to point out that the pressure p^l in Eq.(10.2-4) simplifies the velocity boundary conditions but, on the other hand, it requires the explicit calculation of the pressure through $p^{l+1} = p^l + \Phi^{l+1} - \frac{\alpha_l \Delta t}{2\text{Re}} \mathcal{L}\Phi^{l+1}$.

The present method requires three substeps to advance one time step and at each substep the previous set of equations is solved. Third-order accuracy holds only at Re $\rightarrow \infty$; the definition of $\alpha_l = \gamma_l + \rho_l$ reduces the accuracy to second order. However, the most attractive feature of the third order Runge-Kutta scheme is a stability limit higher than the second-order Adams-Bashforth scheme. The theoretical limit is $CFL < \sqrt{3}$ at Re $\rightarrow \infty$, but the viscous terms permit a peak in CFL up to 2. Moreover, $\rho_1 = 0$ implies that each time step is self-starting; a variable Δt can be used without introducing interpolating procedures. Even for the cylindrical coordinates the codes can run at constant Δt or at constant CFL by setting in the data file $ICFL = 0$ or 1 respectively.

10.4. Treatment of the axis

The discretization of the region around $r = 0$ is an important feature of the present scheme that permits to overcome the difficulties that in the past were encountered in its accurate representation and that caused the lack of simulations in literature. Aksevol & Moin (1996) treated in a different way the region around the axis from that outside and they got good results. In a simple code, accessible to the largest possible number of scholars, it is advisable to treat the entire domain in the same way and as a first consideration it turns out that a staggered grid helps. This layout evaluates only the component q_r at the point $j = 1$, the axis, and there $q_r \equiv q_2 = 0$ by definition (Fig.1). The q_2 equation then is discretized for all $n2 - 1 \geq j \geq 2$ straightforwardly. The equations for q_3 and q_1 at $j = \frac{3}{2}$ require the evaluation of radial derivatives in the region around $r = 0$. Since $q_2 = 0$ at $j = 1$ in the convective terms for q_1 and q_3 the radial derivatives at $j = \frac{3}{2}$ are easily computed.

The viscous radial derivative for q_3, with only the radial indices indicated, is

$$\frac{g_{2m}}{r_m} \frac{\delta}{\delta x_2} r_c g_{2c} \frac{\delta q_3}{\delta x_2} \Big|_{j=\frac{3}{2}} = \frac{(g_2)_{\frac{3}{2}}}{r_{\frac{3}{2}}} \left[r_2 (g_2)_2 \frac{(q_3)_{\frac{5}{2}} - (q_3)_{\frac{3}{2}}}{\Delta r^2} \right], \qquad 10.13$$

where $r_1 \equiv 0$ eliminates the approximation of $\delta q_3 / \delta x_2 \,|_{j=1}$. For flows around an annulus of radius $r_1 \neq 0$ $\delta q_3 / \delta x_2 \,|_{j=1}$ is discretized by first-order approximations as for the first point near a solid wall in Cartesian coordinates (Rai & Moin (1991)). The viscous derivative for q_1, expressed as in Eq.(10.2), is well suited for finite differences; it is

$$\frac{\delta}{\delta r} r \frac{\delta q_\theta}{\delta r} - \frac{q_\theta}{r^2} \Big|_{j=\frac{3}{2}} = (g_2)_{\frac{3}{2}} \left[(g_2)_2 r_2 \frac{(q_1)_{\frac{5}{2}} - (q_1)_{\frac{3}{2}}}{\Delta x_2^2} \right] - \frac{(q_1)_{\frac{3}{2}}}{(r_{\frac{3}{2}})^2}. \qquad 10.14$$

Even for this term at $r_1 \equiv 0$ no approximation is requested for the first derivative and the second derivative maintains second-order accuracy. Verzicco & Orlandi (1996a) discussed several expressions for radial derivatives of q_θ; here the most efficient one is used.

10.5 Code informations

The structure of this code for turbulent flows in pipes is similar to that used for turbulent flows in a plane channel. The code is given in the directory $CODESCH10/PIP$. The non-uniform distribution of grid points in the radial direction, by a coordinate transformation, is given in the subroutine $CORDIN$ where the metric quantities are also calculated. In the turbulent pipe, as in the turbulent channel, good results are achieved only when the first grid point is at a distance less than $y^+ = 1$. y is the distance from the wall defined as $y = r_w - r$, where r_w is the dimensionless radius of the pipe, set equal to 1. The subscript $+$ indicates the wall friction scaling. The pipe in certain circumstances is better than the channel; for instance, the cylindrical coordinates make the grid spacing in the central region smaller and thus a better resolution is reached in the region where turbulence is close to isotropic.

To evaluate the first and second derivatives in the directions of periodicity the arrays $ipv(i), kpv(k)$ and $imv(i), kmv(k)$ permit the calculation without introducing any if statement. The indices $jpv(j)$ and $jmv(j)$ are used to discretize the derivatives near the axis and near the wall. Before the time integration starts the coefficients entering in the tridiagonal solvers and in the expression of the radial derivatives are evaluated. The latter are calculated in the subroutine $COETAR$. The saving of CPU is due to the large reduction of multiplications. In this code the tridiagonal matrices of the factorization scheme are inverted in a different way. Instead of a RHS_i, in Eq.(10.9), stored as a three-dimensional matrix, it has been allocated in a one-dimensional array with the index $iadd$ related to the values of the indices of each grid point i, j, k. Some of the coefficients of the forward sweep are evaluated at the beginning and the tridiagonal solvers have been written with the inner loop as long as possible. The order of the loops are different depending on the equation and on the sweep. Since the tridiagonal inversions, in the factorization scheme, are the part of the code which takes more time they are explained in detail. This part is rather different from that for the plane channel; it can be easily adapted and inserted into the channel code, to investigate whether the efficiency increases. The user, however, should remember that the efficiency is related to the computer's architecture.

10.5a Calculation of $\hat{q}_i$ (first step)

The solution of the first step of the fractional step method gives the velocity components $\hat{q}_i$, and it is done in the routine $TSCHEME$. By means of q_i^l at the previous substep, the nonlinear terms for the three velocity components are calculated in the subroutines $HDNL_i$. In these routines also the derivatives of the velocities q_l with $l \neq i$ that appear in Eq.(10.5-7) are evaluated. These routines are simple and do not require any further explanation. The $\hat{q}_i$ components are calculated in the subroutines $INVTU_i$. In each of these routines the viscous terms at the old time step in the right hand side of Eq.(10.9) are evaluated by means of the coefficients of the radial derivatives coming from the routine $COETAR$. The second derivatives in the 1 and 3 directions use the expression for uniform meshes. The right hand side of Eq.(10.9) includes also the pressure gradient and the non-linear terms. The RHS_1 is evaluated in $(n1-1) \times (n2-1) \times (n3-1)$ points and the index of the one-dimensional array is $iadd = i + (j-1)*(n1-1) + (k-1)*(n1-1)*(n2-1)$.

The first implicit sweep is in the direction 1 and is performed by the routine $SOLQ1I$. In this routine a periodic tridiagonal solver (subroutine $TRPV1IJ$) with the inner loop in the j direction is used. To save time in this routine coefficients appear that, at the beginning, have been calculated in the routine $CTPV1IJ$. The price to pay for this saving of time is a greater memory occupancy; in fact the coefficients are different at each of the three substeps and vary in the x_1 and x_2 directions. The structure of the coefficients, in addition, requires a temporary matrix $fi(j, i)$ where $RHS(IADD)$ is transferred. The names of the routines for the tridiagonal inversion are constructed in such a way that the last letter indicates the direction of the inner loop, which is the vectorization index; the index of the sweep is that before the last. The number indicates the velocity component; since q_1 and q_3 have the same coefficients of the matrix the same subroutine is used for both components. The second sweep is in the other periodic direction (direction 3) where the coefficients of the tridiagonal matrices do not depend on the other directions. In this direction the vectorized loop includes the first $(n1 - 1) \times (n2 - 1)$ elements of the array RHS. The tridiagonal solver is indicated by $TRIPKIJ$ where K indicates the sweep index and IJ the vectorization indices. In the name of the subroutine there is no indication of the velocity component since for all of them the coefficients are the same.

The last sweep in the radial direction 2 uses the tridiagonal solver with assigned boundary conditions; even here a fast solution is obtained by vectorization in the direction x_1 and x_3, corresponding to the indices i and k; to have these indices as inner do loop the RHS array is rearranged, before the $SOLQ1J$ routine is called. The routine $BTRJIK$, where B means boundary and, as before, J indicates the sweep direction and IK the vectorization directions, performs the tridiagonal inversion. In $SOLQ1J$ $\Delta\hat{q}_1$ is added to q_1^l at the previous substep and the velocity $\hat{q}_1$ is thus calculated.

A similar procedure furnishes the other velocity components by subroutines with names similar to those used for $\hat{q}_1$. Any subroutine can be easily understood in the code; it should be only pointed out that for the component $\hat{q}_2$ the nonlinear terms and hence $RHS(IADD)$ are evaluated for $2 \geq J \leq n2 - 1$.

10.5b Calculation of q_i^{n+1} *(second step)*

The fractional step requires the scalar Φ projecting the non-solenoidal $\hat{q}_i$ onto a solenoidal velocity field q_i^{n+1}. The right hand side of the elliptic equation for Φ is the divergence of $\hat{q}_i$ that is evaluated in the subroutine $DIVGU$ by the operator $\mathcal{D}$ in Eq.(10.12). The subroutine $PHCALC$ evaluates the scalar Φ by inverting the operator $\mathcal{L}$ by means of two FFTs and by a tridiagonal solver. The reduced wave numbers, that, in the wave number space, are equivalent to the second order centered finite difference operator and the coefficients for the differentiation along x_2 are evaluated only once in the subroutine $PHINI$ and $FFTQUA$. In $FFTQUA$ also are calculated the quantities necessary for the FFT routines.

These finite difference codes use single precision to save memory, since in cylindrical coordinates, the $\mathcal{D}$ and hence the $\mathcal{G}$ operators have the radius in the denominator; each of the terms could be very large near the axis and then the maximum divergence could be not as small as expected. This consideration should be kept in mind when the check is performed whether in the numerical methods the properties of the Navier-Stokes equations are maintained. As an example in the check of total energy conservation, in the inviscid limit, it has been verified that the discretization of the non-linear terms requires a lot of care and that it is necessary to avoid any division by r in the $\mathcal{D}$ and $\mathcal{G}$ operators. However it has not been checked that, by leaving the division by r, a double precision calculation was leading to the same results.

To avoid any iterative scheme, the calculation of Φ requires the expansion in Fourier series and hence the FFT routines. In this code for a more general employ in any computer the NCAR Temperton FFT routines are used. The choice of x_3 as the first direction for the series expansion, means to call the real FFT in the x_3 and the complex FFT in the x_1 direction. By looking at the different ways to solve the elliptic equations, described in chapter 3, the user can speed up the code by taking the more efficient FFT routines available on each computer.

The scalar field Φ permits to evaluate the solenoidal velocity field q_i^{n+1} in the subroutine $UPDVP$ and to evaluate in the subroutine $PRCALC$ the pressure, entering in the $\hat{q}_i$ calculation. In the code there are several routines to check whether the calculations satisfy the desired requirements; for instance, the subroutine $DIVUCK$ evaluates the total and the maximum flow divergence. If the divergence exceeds a certain limit the calculation stops. The calculation stops also when the velocity becomes very large or when the Δt becomes very small. The last condition may occur in simulations at constant CFL. On the other hand the simulations at constant Δt stop if the CFL becomes large.

In the code there are subroutine to print the results, depending on the flow considered.

10.5c Initial conditions

The initial conditions for the dipole are given in the file *pipindip.f* where in the subroutine $DIPOLE$ the vorticity distribution described in chapter 4 is assigned. The streamfunction is evaluated in the subroutine $PSICALC$ by solving the elliptic equation relating ψ to ω. From the streamfunction the velocity components are calculated in the subroutine $INDIPO$.

The initial conditions for the turbulent pipe are given in the file *pipintur.f*; in the subroutine $INVELA$ the Poiseuille laminar profile is calculated. In the routine $INTUR$ a random disturbance with a large amplitude in the central region is added to the parabolic profile; it is reduced five times near the wall. This initial perturbation corresponds to a large initial turbulent kinetic energy; during the transient it is initially reduced and then it increases to fluctuate around a statistical steady state. The random disturbance has been assigned on q_3 and the disturbance on q_1 has been set equal to zero. The radial component is evaluated by the continuity equation. A check has been done to see the influence of the amplitude of the random disturbance; it has been found that, at least on a coarse grid, if the amplitude is small the turbulence is not sustained. If the simulation is a continuation of a previous one, the field is read in $INIREA$.

Even for the pipe the initial conditions at a grid $n1, n3$ can be interpolated or extrapolated from the results on a finer or a coarser simulation. The subroutines for the interpolation are called $INZETA$, $INTHET$, $INTHZE$ if the ratios between the old (indicated by the subscript o) and the new grid (indicated by the subscript n) are respectively $n3_o/n3_n = 2$, $n1_o/n1_n = 1$; $n3_o/n3_n = 1$, $n1_o/n1_n = 2$; $n3_o/n3_n = 2$, $n1_o/n1_n = 2$. The routines for the extrapolations are named $EXZETA$, $EXTHET$, $EXTHZE$ when $n3_n/n3_o = 2$, $n1_n/n1_o = 1$; $n3_n/n3_o = 1$, $n1_n/n1_o = 2$; $n3_n/n3_o = 2$, $n1_n/n1_o = 2$. When the simulation restarts in a pipe with a different radial distribution of points than that by which the restarting file was generated, a spline interpolation is used in the x_2

direction. In this eventuality it is necessary to have the old and the new grid and thus in the data files, in $CODESCH10/DA$, appear the parameters to compute the old and new radial distributions of points. Interpolations or extrapolations are done only for the q_θ and q_z velocities; q_r is obtained from the continuity equation.

10.6 Results

10.6a Check of axis accuracy

In order to investigate the accuracy on the region around the axis Verzicco & Orlandi (1996a) performed simulations of dipoles crossing the axis, by choosing different initial locations for the dipole; here only one case is reported but at a greater Reynolds number. The check was done in a region with the external free-slip boundary at a distance of 5 dipole radii. The same initial conditions but with a solid no-slip wall can be considered as a case with more practical interest. In fact, a vortex couple impinging a solid boundary has been chosen as a simple flow to understand the turbulent wall friction and hence its control. This was done by Choi *et al.* (1994) in their study of optimal control of turbulent drag. For flat boundaries a numerical study of the control of the impact of dipoles with walls has also been recently carried out by Kousmostakos (1997) by means of vortex methods. The study of dipoles impinging cylindrical walls was initiated and not completed by Vega *et al.* (1996) by means of this code; I am suggesting to the user to continue this study which presents very interesting features when the wall has a steady rotation or an oscillatory motion. This is to point out that this code could be used to study drag control within pipes.

Fig.10.2 shows the vorticity distribution of the dipole during the translation across the axis. In this enlarged picture of the region around the axis, oscillations are observed; these disappear when the dipole is far from the center and hence the velocities induced by the dipole around the axis decrease. The origin of these oscillations is also related to the choice of the simulation at constant $CFL = 1.7$. In fact, has been checked that the amplitude of the oscillations decrease by reducing the CFL. The oscillations become even smaller as the number of grid points increases.

10.6b Check of energy conservation

The total energy requirement is important in the simulation of turbulent flows. For 3D flows with three periodic directions and between two free-slip walls it was shown that, for $\nu = 0$, the energy was preserved. For the flow within a pipe the requirement of global energy conservation is difficult to be respected because the grid near the axis is small and there is the possibility that unphysical oscillations, not damped by the viscosity, become so large that the calculation diverges. In the channel it was found that the energy and even the enstrophy remain constant for a large interval of time, as is shown in Fig.9.1. In the pipe independently of the grid and of the Δt it has been found that the solution was diverging. An expression of the non-linear terms that was conserving the energy for a very long time was not found. A preliminary check has shown that the terms responsible for the lack of energy conservation are those which are not differentiated. These terms in the continuous transport equation of the turbulent energy balance each other, and do not balance in a staggered grid because of interpolations. Different ways to perform the

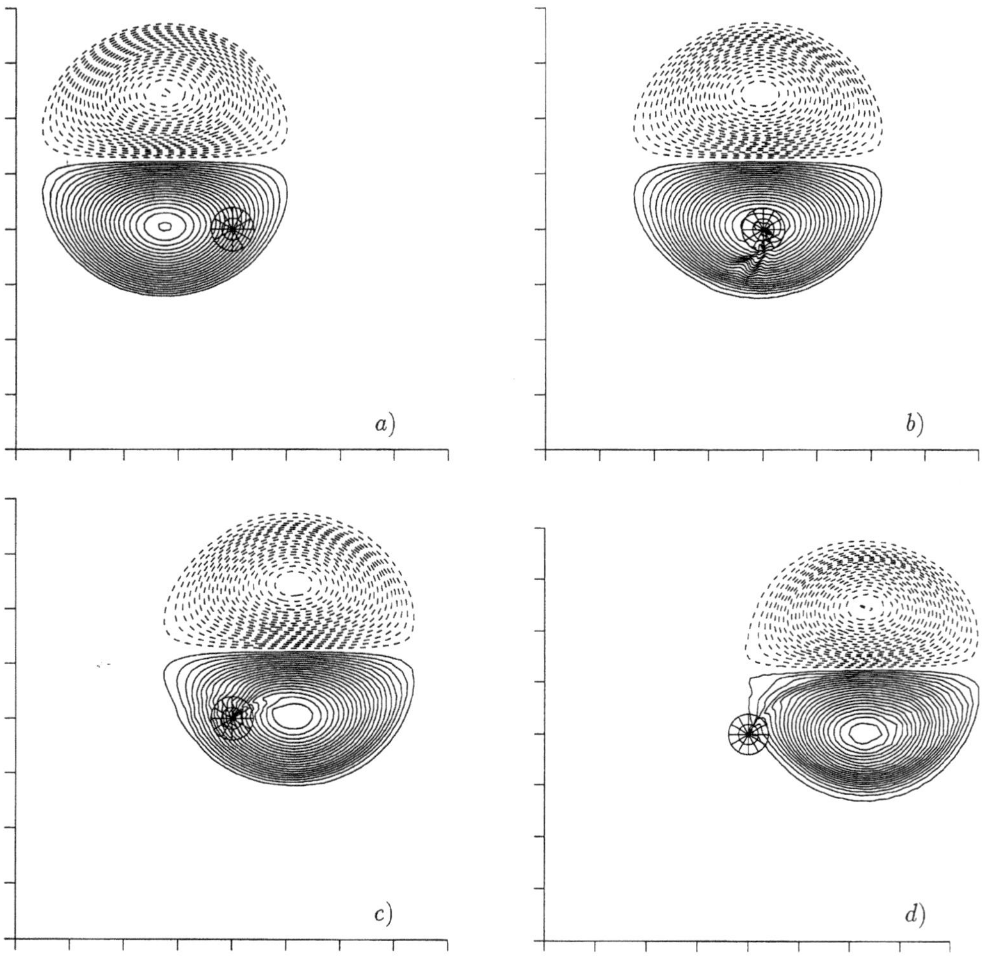

Fig. 10.2 Vorticity contours of a translating dipole around the central region at $Re = 2000$ with a 96×129 grid. The external boundary at $R_F = 5$ has free-slip conditions. a) $t = 0$, b) $t = 0.5$, c) $t = 1.$, d) $t = 1.5$.

interpolations have been tested and each one sooner or later produced a growth of the total kinetic energy. The expression giving the best results is that where, in the q_θ equation, at $i, j + \frac{1}{2}, k + \frac{1}{2}$, q_θ is at its location and q_r is evaluated by the four neighboring points. Furthermore it has been checked that the energy growth did not depend on the time steps by performing simulations at $CFL = 1.7$ and $CFL = 0.5$. The energy conservation check was done on a very coarse grid and in particular on a 24^3 grid. The simulations started with the field that was saved at $t = 200$ in the 48^3 viscous simulation.

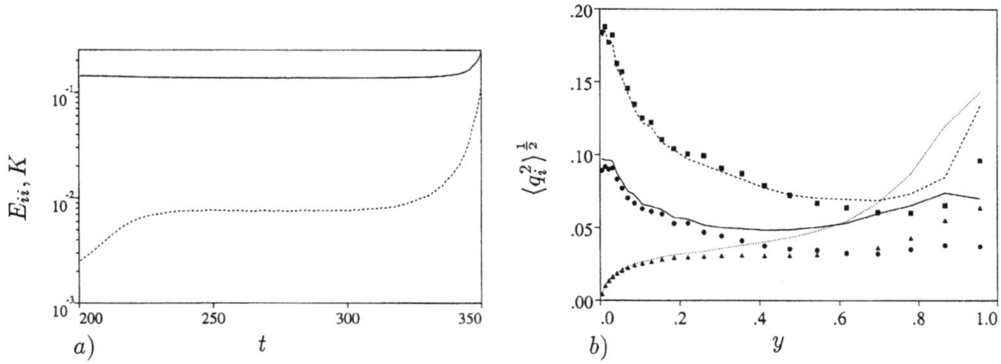

Fig. 10.3 Time evolution of a) —— total energy, ---- turbulent energy b) profiles of normal stress, symbols $t = 300$, lines $t = 340$; —— • v_θ, ······· △ v_r, ---- □ v_z.

Fig.10.3a shows that the 24^3 simulation, for a substantial long time, maintains a constant level of total energy and that even the fluctuating kinetic energy, after a transient, remains constant for a certain time. To investigate the reasons of the solution's divergence and in which region of the pipe it occurs, it is worth to plot in Fig.10.3b the radial profiles of the three *rms* velocity components at $t = 300$ and $t = 340$. Fig.10.3a shows that the energy is constant at $t = 300$; instead, at $t = 340$, it begins to grow exponentially. The radial profiles of *rms* velocities explain that the transient growth of K is related to the change from a no- to a free-slip wall. In fact when the no-slip conditions are removed the fluctuations of the azimuthal and axial velocities increase. As expected the azimuthal fluctuations grow more than the axial ones. Since the conditions for the radial component for a no- or a free-slip wall do not change, the profile of $\langle v_r'^2 \rangle^{\frac{1}{2}}$ in the wall region does not change. These higher fluctuations near the free-slip wall migrate in the pipe and Fig.10.3b shows that at $t = 340$ these become very large near the axis. To understand why there is this growth of the turbulent kinetic energy and to prove that indeed the non-differentiated terms contribute to the loss of energy conservation, in Fig.10.3c the radial profiles of each term of the Euler equations integrated in θ and z are shown. The terms with the derivatives in the x_1 and x_3 directions have a zero contribution and it turns out that the largest are: the term without differentiation, that with the pressure gradient and that with the radial derivative. These terms in time become very large and the imbalance due to the interpolation disrupts the energy conservation.

10.6c Turbulent non-rotating pipe, coarse solution

Orlandi & Fatica (1997) reported in appendix a detailed study of the grid refinement check for the non-rotating pipe flow at $R_\tau \approx 180$ and compared the results with the direct simulation of Eggels *et al.* (1994). The results were obtained by a code slightly different from the version furnished with this book. Here the simulations have been performed with two coarse grids: the first with 48^3 and the second with 64^3 points. The second grid is satisfactory to get reasonable results in investigating the turbulence changes under the

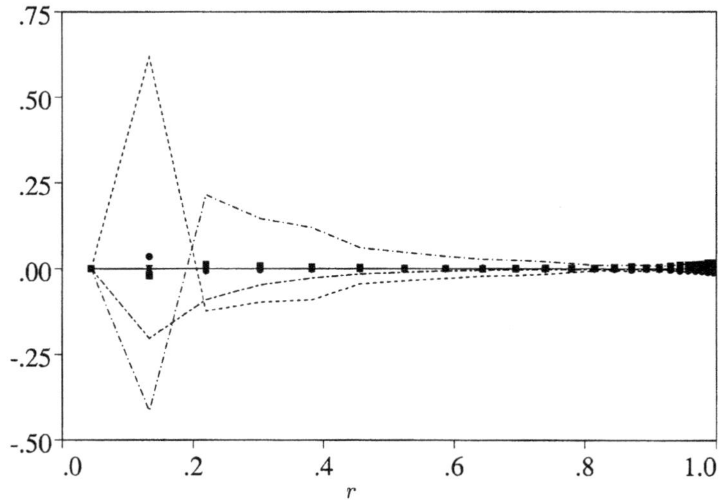

Fig. 10.3c Profiles of the right hand side of the inviscid energy equation lines at $t = 340$ symbols at $t = 300$.

action of external forces. Orlandi & Fatica (1997) analyzed the effects of the steady rotation of the wall. Quadrio (personal communication) did the same for periodic oscillations, and Orlandi (1996) for a magnetic field. In this last study three extra parabolic equations for the magnetic field appeared which imposed a further limitation on the computer memory.

In non-rotating turbulent pipes a length of 5 diameters is sufficient for the longitudinal correlations to go to zero, and this length was fixed in the grid refinement check of Orlandi & Fatica (1997). The Reynolds number is defined by taking, as reference quantities, the radius R, the centerline Poiseuille velocity U_P and the kinematic viscosity ν. $R_\tau \approx 180$ corresponds to $Re = 4900$. The simulations are performed starting from random initial conditions up to a dimensionless time $t = 200$; in this run the statistics were evaluated during the time evolution by starting the averaging at $t = 50$. At $t = 200$ the velocity profile has a reasonable log law and at this point a new simulation for other 210 dimensionless time units was done. The statistics were evaluated in a post-process by 20 fields separated 10 time units from each other. The statistical quantities are indicated by $\langle\rangle$, denoting time and space averages in the two homogeneous directions z and θ. Fig.10.4 shows the velocity profiles for the two grids compared with the resolved profile by Eggels *et al.* (1994) that was checked to reproduce the profiles of laboratory experiments. It is surprising that the results of the 64^3 simulations agree with the refined simulation. The good agreement near the wall is due to the non-uniform grid, which for 48 points allows to locate the first point closer to the wall than for the 96 equidistant points in Eggels *et al.* (1994). The grid clustering near the wall reduces the resolution at the center and this explains why even the results of a 64^3 simulation do not agree, near the center, with those by Eggels *et al.* (1994). The resolution, in wall units, for the 64^3 grid is $\Delta z^+ = 25$ and at $y^+ = 10$

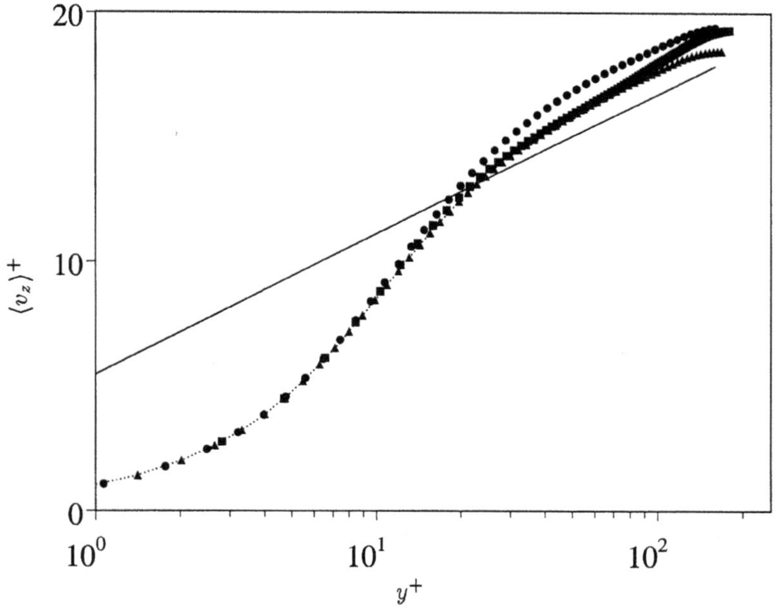

Fig. 10.4 Mean axial velocity profiles in wall coordinates, (—— log law $\frac{1}{0.41} \ln y^+ + 5.5$), ○ 48^3, △ 64^3, □ (Eggels *et al.* (1994).

approximately $\Delta s^+ = (r\Delta\theta)^+ = 15$; this resolution is not small enough to consider the turbulent pipe fully resolved, but it is satisfactory to study the qualitative behavior of the turbulent statistics. Even on the 48^3 grid, that requires a limited amount of memory and CPU time, a turbulent flow not too far from reality is generated. This code, then, can be used by the users as a tool to study, on personal computers, wall turbulent flows.

Orlandi & Fatica (1997) showed that the turbulent stress profiles depend on the resolution more that the mean velocity profile; by physical arguments they explained the resolution dependence. The upward shift of the log law was linked to smaller values of near wall vorticity caused by the reduced resolution. Since these vortices are responsible for the turbulent wall friction, a reduction of their strength produces a reduction of drag. Fig.10.4b-d show indeed, by means of the *rms* profiles in wall units, a decrease of the v_r^+ and v_θ^+ *rms*, and an increase of the v_z^+ *rms*. The increase of the v_z^+ *rms* is related to the reduction of the wall friction. The reduction of the other two components (Fig.10.4c-d) is large since even if u_τ is reduced the *rms* in wall units are smaller than those by Eggels *et al.* (1994). In the present simulations it has not been checked in which direction the refinement is more effective; Orlandi & Fatica (1997) observed that the refinement in θ has a greater influence than that in z. The physical reason is that a better resolution of the azimuthal gradients of ω_z increases the reproduction of the sweep and ejection events and hence leads to greater values of $\langle \omega_\theta \rangle$. These events produce the high and low-speed

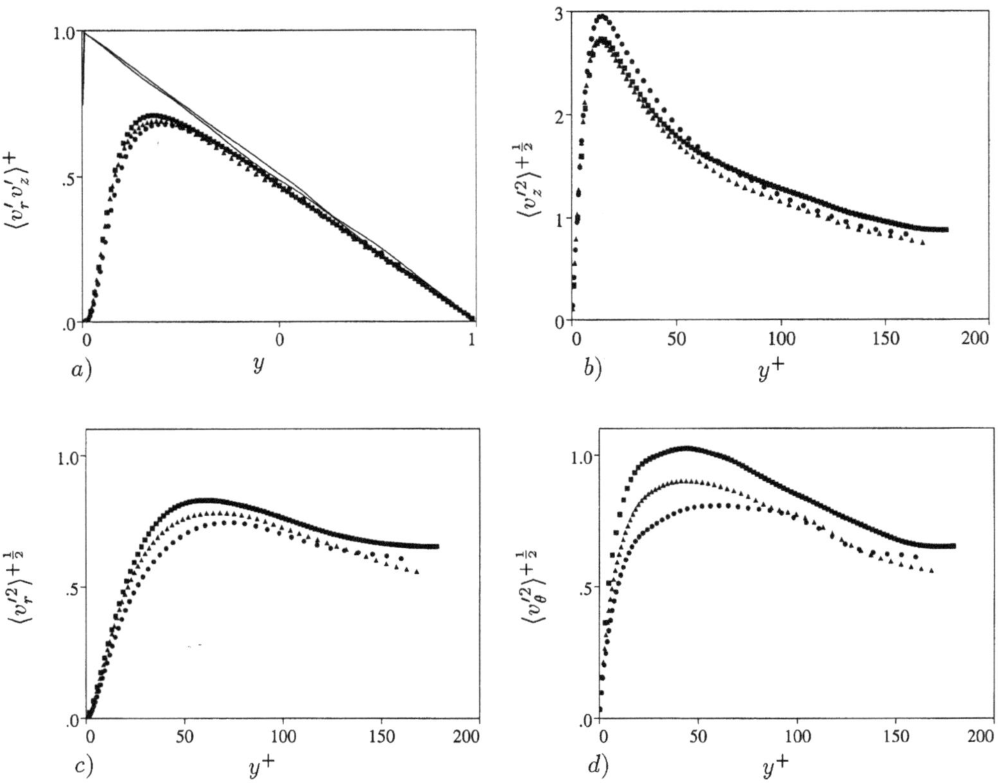

Fig. 10.5 Second order turbulence statistics, a) turbulent stress, b) axial, c) radial, d) azimuthal

streaks, which are responsible of the production of turbulent wall friction and turbulent kinetic energy. The increase in wall units of the streamwise fluctuations and the decrease of the normal fluctuations compensated each other and hence the profile of $\langle v'_r v'_z \rangle^+$ in Fig.10.4a does not largely depend on the resolution.

The profiles of higher order correlations and of the spectra in the non-rotating case are not presented. The post-processing code $PISTAT$ is given in the directory $CODESCH10/STAT$; this code reads the saved files and evaluates a large number of quantities. Orlandi & Fatica (1997) by this code produced the spectra, the correlations and the joint PDF. Orlandi (1997) extended the study to investigate the production, the dissipation rate and the helicity density.

10.6b Turbulent rotating pipe, coarse solution

When the wall of a pipe rotates there is a similarity with a three-dimensional boundary layer, a flow of more practical application. For the latter flow it is difficult to design an experiment with boundary conditions perfectly reproducible in a direct simulation. On the other hand experiments of rotating pipes can be realized and numerical simulations in a one to one correspondence with the geometrical conditions in the experiment can be performed. The detailed knowledge of this simple flow helps to understand the modifications of the flow structures in more practical problems such as the flow on swept wings. To see better the analogy between these two flows the rotating pipe, in a reference frame rotating with the wall, can be thought of as a pipe where the Coriolis force acts as a local azimuthal pressure gradient that affects the orientation, the spacing and the size of the stream-wise vortices. Sendstad & Moin (1992) observed in a direct simulation of a three-dimensional plane channel (that is with a spanwise pressure gradient) drag reduction and explained this phenomenon by the tilting of the stream-wise vortices. In experiments on rotating pipes Nishibori *et al.* (1987) and Reich & Beer (1989) observed that drag as well as turbulence were reduced as the rotation rate was increased. At very high rotation speeds a sort of laminarization characterized by an averaged $\langle v_z \rangle$ similar to the laminar Poiseuille profile was also observed. The changes in the mean and turbulent quantities in these experiments were explained by a general connection with the effects of the centrifugal force, without mentioning any connection to the modifications of the near wall vortical structures. On the other hand changes in wall turbulent flows should strongly depend on the modifications of the near wall vortical structures. Thus, as Sendstadt & Moin (1992), Orlandi & Fatica (1997) visualized the near wall vorticity field in rotating pipes and connected the changes in the profiles of turbulent quantities to the tilting of the wall structures.

The choice of the length of the computational box in direct simulations of wall turbulent flows is an important issue. Eggels *et al.* (1994) verified by axial two point correlations of the fluctuating velocity components that for $N = 0$ a length of the pipe $L_Z = 10R$ is sufficient. When the pipe rotates flow visualizations by Nishibori *et al.* (1987), at high rotation ($N \approx 2$), showed that in the central part of the pipe elongated structures form and that, at higher rotation rates ($N \leq 3$), the variations in the stream-wise direction disappear, in agreement with the Taylor-Proudman theorem (Greenspan, (1980)). These flow visualizations suggested Orlandi & Fatica (1997) to investigate the dependence on L_z by simulations with $L_z = 10, 15$ and 20 times R. They observed that for $L_z = 10$ the steady state was not achieved. On the other hand when the length was increased to $L_z/R = 15$ and to $L_z/R = 20$, by keeping the same Δz, a convergence to the statistical steady state was reached. Furthermore, since the elongated structures in the central region have long time scales, the statistical steady state is reached in a time longer than in the non-rotating case. They also verified that for $N < 2$ a length $L_z/R = 10$ was sufficient.

In the present book the simulations were performed in a longer pipe ($L_z = 30R$) and even the non-rotating case has been simulated with this length. Fig.10.6 shows the velocity profiles, in wall coordinates, for the rotating ($N = 2$) and non-rotating pipe. The profile for the non-rotating flow does not differ from the 64^3 simulation with $l_z/R = 10$ (Fig.10.4), even if the calculation in the longer pipe has a reduced resolution. In fact the simulation was done with 96 points instead of 64 for $L_z/R = 10$. The rotating case has been compared with the results by Orlandi & Fatica (1997) that were obtained by a resolution of $128 \times 96 \times 257$ in the azimuthal, radial and axial directions and with $L_z/R = 15$. In the rotating case there are no reference solution to compare the results. In their study Orlandi & Fatica

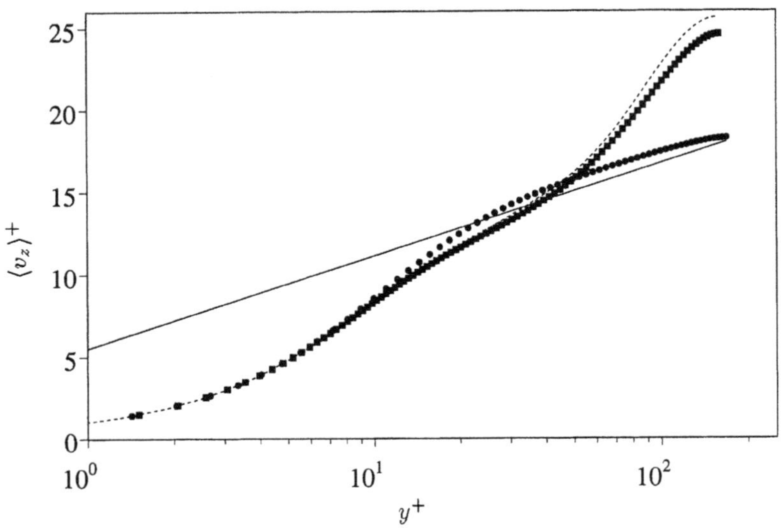

Fig. 10.6 Mean axial velocity profiles in wall coordinates, (—— log law $\frac{1}{0.41}\ln y^+ + 5.5$), (- - - - 64^3 $N = 0$, $L_z/R = 10$), (▵ 64^3 $N = 0$, $L_z/R = 30$), (········ $N = 2$, Orlandi & Fatica, $L_z/R = 15$), (▫ 64^3, $N = 2$, $L_z/R = 30$).

(1997) did not completely investigate the length and the resolution necessary to assert that the data set was validated. In that study they were rather more interested in investigating the reason of drag reduction. From the present simulation, by a coarser grid, the same trend is achieved and in a certain sense it can be considered as a validation of the previous simulations. It is not my aim here to investigate whether the pipe length or the resolution are responsible for the discrepancies; I leave this check to whom is interested. The plots of the mean velocities in dimensionless units in Fig.10.7 show a better agreement than in wall units (Fig.10.6). Fig.10.7 shows also that the profile of the mean tangential velocity $\langle v_\theta \rangle$, in the reference frame rotating with the pipe, agrees well with the more refined results, with insignificant differences near the rotating wall.

The profiles of the normal stresses indicate that the rotation induces large modifications. This investigation was performed by LES (Eggels *et al.* (1994b)) and by DNS (Orlandi & Fatica 1997); both studies found that large effects occur on the stream-wise *rms* velocity; this is expected, since the rotation reduces the production of turbulent energy that is mainly due to $\langle v_z' v_z' \rangle$ near the wal. Orlandi (1997) took the database described by Orlandi & Fatica (1997) and looked in more detail at the modifications of the production and dissipation rate and he linked these modifications to helicity density. The study was done by following a procedure similar to that described in section 9.8.d. The $\langle v_z' v_z' \rangle$ reduction is, furthermore, in accord with the Taylor-Proudman theorem, claiming that a rotating flow reduces the gradients in the direction of the rotation vector. This reduction, however, is not as strong as in homogeneous flows (see section 8.9) because of the turbulence production

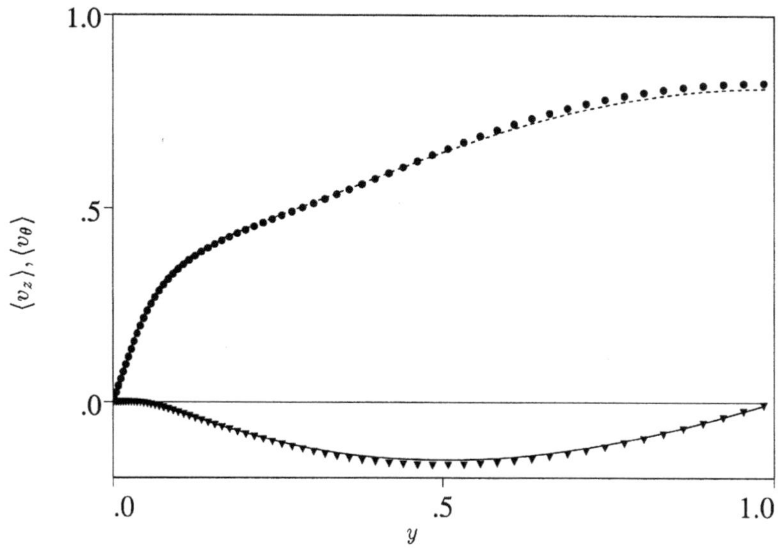

Fig. 10.7 Mean axial velocity profiles in wall coordinates, (---- $\langle v_z \rangle$ $N = 2$ Orlandi & Fatica, $L_z/R = 15$), (▵ $\langle v_z \rangle$ $N = 2$, 64^3 $L_z/R = 30$), (---- $\langle v_\theta \rangle$ $N = 2$ Orlandi & Fatica, $L_z/R = 15$), (▵ $\langle v_\theta \rangle$ $N = 2$, 64^3 $L_z/R = 30$).

near the wall. The *rms* velocity profiles, scaled by the friction velocity show (in Fig.10.8b and Fig.10.8c) a reduction of the stream-wise *rms* and an increase of the azimuthal *rms*. These quantities tend to get the same values when the rotation rate increases. In a certain sense there is an isotropization of turbulence when the pipe rotates. The near wall peaks disappear and a flattening appears. In these figures it can be observed that coarse and fine simulations give a similar trend. In the experiment by Nishibori *et al.* (1987) the measurements of $\langle v_z'^2 \rangle$ at $N \geq 0$ showed a large dependence on the distance from the inlet section; however, even without showing the flattening, the reduction of the wall peak was observed. These authors presented profiles for $Re = 3 \cdot 10^4$ at three stream-wise locations, and the reduction of $\langle v_z'^2 \rangle$ with N in the last station agrees with the present results. Since direct simulations at higher Re are not affordable I am conjecturing that this flattening is a low Reynolds number effect. This is realistic, since even in the channel simulations at $R_\tau = 180$, a strong interaction between the wall and the center occurs. In the rotating pipe Orlandi & Fatica (1997), by vorticity visualizations, observed that the near wall vortical structures increase in size with N and thus the turbulence level increases far from the wall.

Orlandi & Fatica (1997) scaled the normal stresses with the centerline velocity, which largely varies with N, and they found that near the wall the rotation has a reduced influence on $\langle v_\theta'^2 \rangle$ as compared to that on $\langle v_z'^2 \rangle$. Fig.10.8 shows a different behavior than those in wall units. The friction velocity indeed depends less on N. The overall picture, even in this scaling, emphasizes the tendency towards isotropy of the three normal stresses.

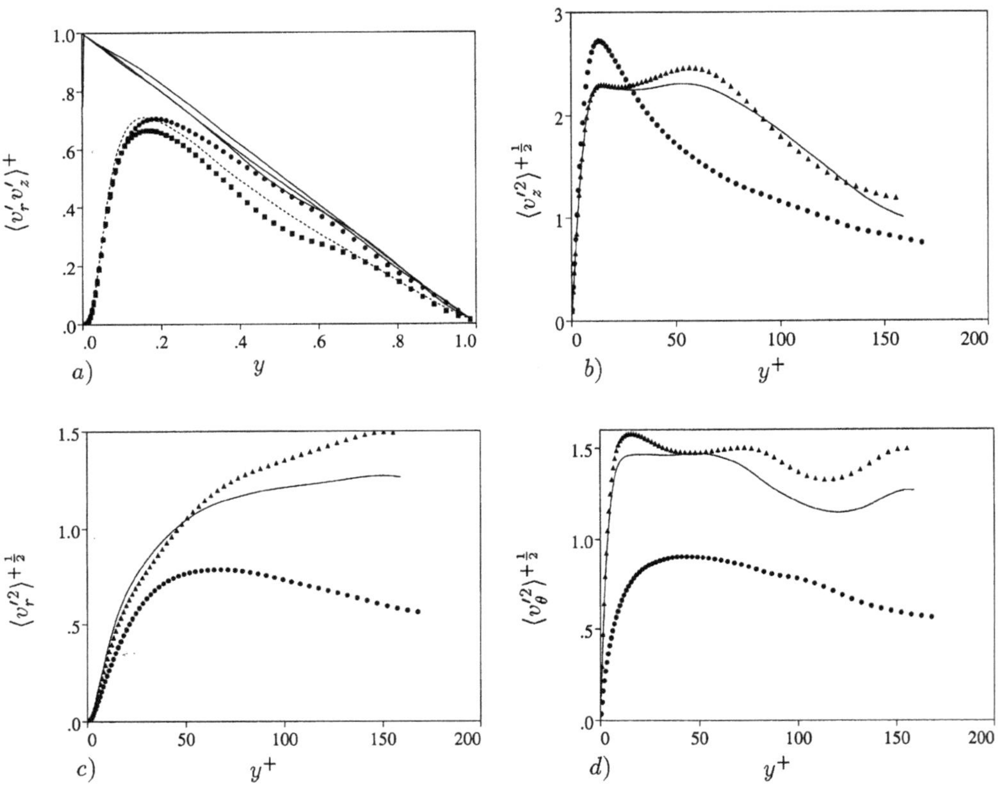

Fig. 10.8 Second order turbulence statistics, a) turbulent stress, b) axial, c) radial, d) azimuthal for the rotating and non-rotating pipe.

In the non-rotating pipe $\langle v_r' v_z' \rangle$ is the only stress; it is reduced when the pipe rotates, as measured by Murakami & Kikuyama (1980) and Kikuyama *et al.* (1983). Fig.10.8a shows that the peak of this stress, in wall units, for $N = 2$, moves towards the wall and that $\langle v_r' v_z' \rangle$ decreases in the central region in accordance with the increase of the viscous stress. Again this trend is predicted by the fine and coarse simulation. The total stress, that is the sum of the viscous and the turbulent stress, shows a satisfactory linear profile; the small oscillations around the straight line are due to the insufficient number of fields saved. In fact, as mentioned before in chapter 9 in the post-processing of turbulent flows, the normal stresses converge faster since they differ in amplitude but not in phase. The turbulent stresses, on the other hand, differ in amplitude and in phase and thus a greater number of samples are requested to get fully converged profiles. The reason for the need of more fields in the rotating case is linked to the low frequency oscillations of the fluctuating velocities. These low frequencies are linked to the formation of large structures when

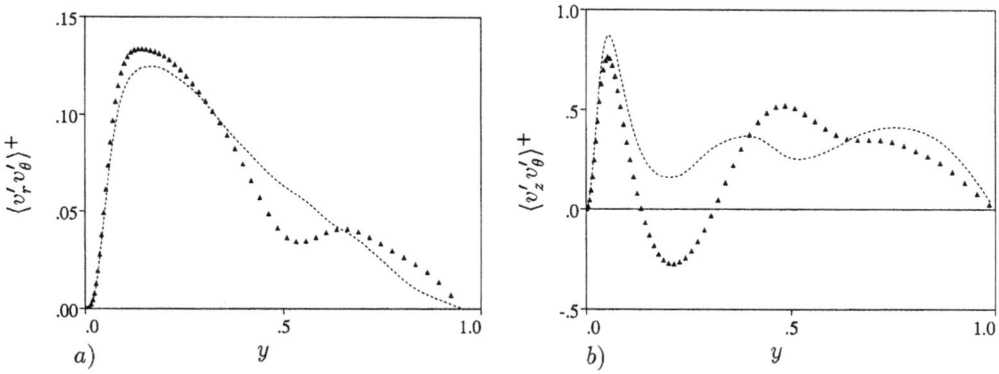

Fig. 10.9 Turbulent Reynolds stress profiles. (lines: fine resolution), (symbols: coarse resolution).

the pipe rotates. Orlandi & Fatica (1997) visualized these large structures by pressure contours. The inverse energy cascade of energy causes the growth of the large scales and solid body rotation produces this effect.

When the pipe rotates the other two Reynolds stresses become comparable to $\langle v_r' v_z' \rangle$ with high values of $\langle v_\theta' v_z' \rangle$ close to the rotating wall as Fig.10.9a-b show. Even these profiles show the same trend between the coarse and the fine grid. The discrepancies in the central region indicate that the rotating pipe requires more resolution in this region because of the greater turbulence produced there. The important features in Fig.10.9q-b consist of a very intense peak of $\langle v_\theta' v_z' \rangle$ near the rotating wall and of the linear profile of $\langle v_r' v_\theta' \rangle$ in the central region. The radial oscillations of $\langle v_z' v_\theta' \rangle$ at $N = 2$ are due to the large scale structures in the central part of the pipe. These scales increase when the resolution decreases and thus the radial oscillations disappear. The intense peak of $\langle v_z' v_\theta' \rangle$ near the wall is related to the tilting of the near wall vortical structures, that increases the amount of correlation between v_z' and v_θ'; Orlandi & Fatica (1997), by the joint PDF, showed that this correlation increases. Fig.10.9b shows that the profile of $\langle v_\theta' v_r' \rangle$, for the fine grid, is approximately linear in the central region of the pipe. The linear behavior theoretically results for the transport equation for $\langle v_\theta \rangle$ if a parabolic profile of $\langle v_\theta \rangle$ is assumed. The parabolic profile was measured and it was observed to be independent of N and Re (Reich & Beer (1989)). Orlandi & Fatica (1997) discussed why in the DNS the parabolic profile has a weaker dependence on N. They, however, did not check the Reynolds independence of the parabolic $\langle v_\theta \rangle$ profile, since they did not have the computational facilities to perform simulations at high Re.

As mentioned before the post-processing code necessary to evaluate other quantities as those considered by Orlandi & Fatica (1997) and by Orlandi (1997), is not described since it does not differ from that used in the channel. The code is given in the directory $CODESCH10/STAT$. A post-processing visualization code has been also written; it reads the pressure and the velocity field. It evaluates the instantaneous vorticity and plots all the quantities in different section in $r - z$, $r - \theta$ and $\theta - z$ planes. The code is given in the directory $CODESCH10/VIS$.

10.7 Conclusions

As a concluding remark, I would like to point out that the rotating pipe has the characteristics which are useful to turbulence modellers. Those interested in LES could check whether subgrid models could predict the large variations of the subgrid dissipation across the pipe. The dynamic model (see chapter 12) should have these capabilities since satisfactory results were obtained by Squires & Piomelli (1993) in simulations of isotropic turbulence subjected to solid body rotation. In addition this subgrid model gave satisfactory predictions in the plane channel. Hence the rotating pipe is a very challenging case to check whether this model could be applied to more complex and realistic situations as those occurring in swirling combustion chambers. The data-base generated by the present code with a satisfactory resolution could be used to improve one-point closure turbulence models. Hirai *et al.* (1988), in fact, proved that the conventional $k - \epsilon$ turbulence model does not work for rotating pipes. The necessity to have reliable one-point turbulence models is mainly dictated by the industries designing combustions chambers where swirl is added to stabilize the flow. From the literature on one-point closure turbulence models it turns out that for such flows the closure does not produce good results. The turbulent rotating pipe is very interesting since a system of one-dimensional partial differential equations appears, and the computational difficulties are reduced.

Finally the rotating pipe is interesting because the rotation is directed in the same direction as the stream-wise vorticity. The same amount of positive and negative ω_z is generated at the wall without rotation; on the contrary, the present simulations show that solid body rotation produces a positive mean stream-wise vorticity near the wall. This occurrence is related to the longer survival of vorticity of the same sign as that of the external rotation. This aspect of vorticity dynamics in the presence of solid body rotation was investigated by Bidokhti & Tritton (1992) in a mixing layer where the external rotation is aligned with the main vorticity and is normal to the rib vortices. The rotating pipe, on the other hand, has the principal vorticity ω_θ orthogonal to the external rotation and the near wall vortices, the analogous of the rib-vortices, aligned with the external rotation. Hence it is a case complementary to the mixing layer. The analysis of the differences or analogies in these two cases could lead to a better comprehension of the vorticity dynamics in the presence of background rotation.

It is interesting, finally, to point out that periodic oscillations can lead to higher drag reduction; this has been measured by Choi (1997) and was also obtained numerically by Quadrio in an unpublished work. The direct simulation could explain why there is a large increase of drag reduction by investigating the modifications of the vortex dynamics produced by the oscillatory motion of the wall.

11 Flows in cylindrical coordinates with two walls

11.1 Introduction

The interaction of axial-symmetric vortices with solid walls necessitates cylindrical coordinates for a more accurate representation. Verzicco & Orlandi (1996a) have shown that a Cartesian coordinate system introduces a perturbation of wave number $n = 4$ of an amplitude is the greater the closer are the boundaries. Orlandi & Verzicco (1993a,1993b) thus used cylindrical coordinates to study the three dimensional interaction of a perturbed vortex ring with a solid wall. They reproduced the experiment by Cerra *et al.* (1972) and explained features difficult to understand by flow visualizations. The analysis of the rate of strain tensor in the principal axis permitted to understand why the secondary ring, generated at the wall, is more unstable than the primary ring. The simulations were performed by a numerical method similar to that described in Chapt.10. A non-uniform coordinate, in the radial direction, clusters the grid points in the regions where the vorticity evolves and locates the external boundary as far as possible. The grid in the direction normal to the wall was uniform and few modifications were made to adapt the code, with periodicity in z, to flows in the presence of walls normal to z. The modifications consisted of an expansion in cos harmonics in the z direction to evaluate the "pressure", and in a few simple changes to impose the no-slip conditions at the walls. The code, in a certain sense, was inefficient, in fact it, was not possible to have sufficient resolution to reproduce the thin vorticity layers forming at the wall when the Reynolds number is high. The study was limited to the minimum Reynolds number at which the three dimensional perturbations are not dumped by viscous effects. At that period I was not aware of the existence of direct solvers for elliptic equations with coefficients variable in two directions.

Recently I found in the literature the *FISHPAK* package, developed by Adams, Swarztrauber and Sweet in 1980 at NCAR, by which an elliptic partial differential equation with non-constant coefficients can be solved directly. This package, very simple to use, permits to use non-uniform grids in the z and in the r direction. The non-uniform grid in the normal direction allows to cluster the grid points near the solid wall and thus, with the same number of computational points, it is possible to solve flows at higher Reynolds numbers. In this chapter the same interaction that was studied by Orlandi & Verzicco (1993a) is repeated for a reduced number of points. The reduction of grid points in the normal direction permits to increase the number of points in the azimuthal direction and to study interactions of vortices with an initial perturbation of greater amplitude.

The code was developed with the intention to simulate numerically flows in a rotating tank, that is widely used to reproduce the vortex dynamics characteristic of geophysical flows. To cite some of the experiments available in literature it is important to mention the experiment of Griffiths & Hopfinger (1986) on the motion of vortices of the same sign and their eventual merging. Van Heijst was one of the principal investigators of interactions of vortical patches of different sign. Flor & van Heijst (1994) investigated the collision of vortex dipoles, and Kloosterziel & van Heijst (1991) the formation of tripolar vortices. In these studies the aim was to be as closer as possible to two-dimensional flows in horizontal planes. The laboratory experiments were performed at low Rossby numbers, that is, at high rotation rates. The large number of experiments done in these conditions are referenced in the review article by Hopfinger & van Heijst (1993) on rotating flows. At a high rotation speed the Ekman layer is very thin and it plays a minor role. Theoretical studies can help to understand flows at low Ro.

More difficult to understand is the vortex dynamics at a low rotation rate, that is, at Rossby numbers of order 1, and to investigate the eventual differences between cyclonic and anticyclonic vortices. These studies, partially, were carried out by Kloosterziel & van Heijst (1991) in their research work on tripole formation. The effects of the Ekman layer were postulated but were not deeply investigated by means of quantitative observations. In fact, such a study requires visualizations on $r-z$ planes that were avoided for their difficulty. Orlandi & Carnevale (1999) performed a study to investigate, in the tripole formation, the differences between cyclonic and anticyclonic vortices. In addition the numerical simulations allow to investigate the importance of the walls by performing simulations with periodicity along the axis. The comparison between the evolution in the two conditions revealed that indeed the Ekman layer was important. In that study Orlandi & Carnevale (1999) argued that the knowledge about the interaction of a rotating wall and a vortex is limited, both in laboratory experiments and in numerical simulations. The code described in this chapter is a useful tool that has not been widely used. This unexplored field leaves space to several scholars to look at the intricate vortex dynamics without spending a large part of their time to write new codes or to develop new methods. A particular benefit can be achieved by those who in the laboratory have a rotating tank and thus can perform simulations and observations simultaneously

11.2 Governing equations and numerical model

The equations governing the motion are Eq.(10.1) and Eqs.(10.2-4), in this case the equation for a passive scalar, has been added for the main reason that we want to investigate the evolution of a quantity similar to the dye that usually is used to visualize the vortex dynamics in a rotating tank. In the simulations, however, the Schmidt number was 1; on the other hand; in the laboratory this number takes values of the order of 10^3, and simulations at this value are onerous and can be performed only if the interest is to understand the dynamics of the very fine scales of a passive scalar. If the interest, instead, is to investigate the differences between vorticity and a passive scalar, simulations at $Sc = 1$ or at $Sc = 10$ are satisfactory. In fact, it is well known, as reported by Hussain & Melander (1991) and confirmed by Verzicco & Orlandi (1992) in a simulation of a round jet, that the scalar is gathered at scales larger than those of the vorticity. Moreover, it was observed that the passive scalar is concentrated in regions of weak vorticity and it is depleted in regions of high vorticity. If these arguments are not taken into consideration, someone could withdraw wrong conclusions on the vortex dynamics from visualizations. The differences between vorticity and markers persist at $Sc = 1$ because vortex stretching produces an intensification of vorticity.

The transport equation for the passive scalar σ is

$$\begin{aligned}\frac{\partial\sigma}{\partial t}+\frac{1}{r^2}\frac{\partial\sigma q_\theta}{\partial\theta}+\frac{1}{r}\frac{\partial\sigma q_r}{\partial r}+\frac{\partial\sigma q_z}{\partial z}+\\ =\frac{1}{ReSc}\left[\frac{1}{r^2}\frac{\partial^2\sigma}{\partial\theta^2}+\frac{1}{r}\left(\frac{\partial}{\partial r}r\frac{\partial}{\partial r}\frac{\sigma}{r}\right)+\frac{\partial^2\sigma}{\partial z^2}\right],\end{aligned} \qquad 11.1$$

This equation together with the continuity and momentum equation are discretized in the cell given in Fig.10.1. The passive scalar σ is located at the center of the cell. In the present case everything was written as in the previous chapter using dimensionless numbers and conservation properties. In this case, however, it has not been checked whether the discretized equations, in a coordinate system with a non uniform grid in the axial direction, maintains the conservation properties. This check is not so important because here the interest is to simulate flows where the viscosity plays a key role.

As in chapter 10 the θ coordinate coincides with x_1, the axial coordinates z and the radius r are indicated, respectively, by x_3 and by x_2. In the axial , as in the radial directions the derivatives are represented by the metric quantity $g_3 = \frac{\delta x_3}{\delta z}$. The subscripts m and c indicate where g_3 is calculated. With these definitions the system of equations for $q_1 = q_\theta$, $q_2 = q_r$, $q_3 = q_z$ and σ becomes:

$$\frac{\partial q_1}{\partial \partial t} + g_{3m}\frac{\delta q_1 q_3}{\delta x_3} + g_{2m}\frac{\delta q_1/r_m q_2}{\delta x_2} + \frac{1}{r_m^2}\frac{\delta q_1 q_1}{\delta x_1} + \frac{q_1 q_2}{r_m^2} + N q_2$$
$$= -\frac{\delta p}{\delta x_1} + \frac{1}{Re}\left[g_{2m}\frac{\delta}{\delta x_2} r_c g_{2c}\frac{\delta q_1/r_m}{\delta x_2} - \frac{q_1}{r_m^2} + \frac{1}{r_m^2}\frac{\delta^2 q_1}{\delta x_1{}^2} + g_{3m}\frac{\delta}{\delta x_3} g_{3c}\frac{\delta q_1}{\delta x_3} + \frac{2}{r_m^2}\frac{\delta q_2}{\delta x_1}\right], \quad 11.2$$

$$\frac{\partial q_2}{\partial \partial t} + g_{2m}\frac{\delta q_2 q_2}{\delta x_2} + g_{2m}\frac{\partial}{\partial x_2}\frac{q_2^2}{r_c} + \frac{\delta}{\delta x_1}\frac{q_1 q_2}{r_m r_c} + g_{3m}\frac{\delta q_2 q_3}{\delta x_3} - \frac{q_1^2}{r_c^2} - N q_1$$
$$= -r_c g_{2c}\frac{\delta p}{\delta x_2} + \frac{1}{Re}\left[g_{2c}\frac{\delta}{\delta x_2} g_{2c} r_c\frac{\delta q_2/r_c}{\delta x_2} - \frac{q_2}{r_c^2} + \frac{1}{r_c^2}\frac{\delta^2 q_2}{\delta x_1{}^2} + g_{3m}\frac{\delta}{\delta x_3} g_{3c}\frac{\delta q_2}{\delta x_3} - \frac{2}{r_c^2}\frac{\delta q_1}{\delta x_1}\right], \quad 11.3$$

$$\frac{\partial q_3}{\partial t} + \frac{g_{2m}}{r_m}\frac{\delta q_2 q_3}{\delta x_2} + \frac{1}{r_m^2}\frac{\delta q_1 q_3}{\delta x_1} + g_{3c}\frac{\delta q_3 q_3}{\delta x_3}$$
$$= -\frac{\delta p}{\delta x_3} - \frac{\overline{\partial P}}{\partial z} + \frac{1}{Re}\left[\frac{g_{2m}}{r_m}\frac{\delta}{\delta x_2} g c r_c\frac{\delta q_3}{\delta x_2} + \frac{1}{r_m^2}\frac{\delta^2 q_3}{\delta x_1{}^2} + + g_{3c}\frac{\delta}{\delta x_3} g_{3m}\frac{\delta q_3}{\delta x_3}\right]. \quad 11.4$$

$$\frac{\partial \sigma}{\partial t} + \frac{g_{2m}}{r_m}\frac{\delta q_2 \sigma}{\delta x_2} + \frac{1}{r_m^2}\frac{\delta q_1 \sigma}{\delta x_1} + g_{3m}\frac{\delta q_3 \sigma}{\delta x_3}$$
$$= \frac{1}{Re}\left[\frac{g_{2m}}{r_m}\frac{\delta}{\delta x_2} g c r_c\frac{\delta \sigma}{\delta x_2} + \frac{1}{r_m^2}\frac{\delta^2 \sigma}{\delta x_1{}^2} + + g_{3m}\frac{\delta}{\delta x_3} g_{3c}\frac{\delta \sigma}{\delta x_3}\right]. \quad 11.5$$

The structure of the numerical method and the sequence of the different steps to solve this system of equation is identical to that described in the previous chapter; the differences between the present code and the code with periodicity in the axial direction follow below.

The code permits free- or no-slip conditions on the $z = 0$ and $z = L_z$ boundaries. The two cases differ in the discretization of the second derivatives in x_3 at the first point close to the boundary. As in the channel, see chapter 9, the second derivative near the walls for a staggered cell with a non uniform grid distribution is given by Eq.(9.6). An index set 0 or 1 permits to impose free- or no-slip conditions independently on the upper ($z = L_z$) and lower ($z = 0$) walls. The free-slip condition on the upper surface reproduces the rotating tank without a cover. However, this condition does not reproduce the real condition of the surface with a parabolic shape; this affects the vortex dynamics as was proved by Carnevale *et al.* (1991b) for the merging of two vortical structures. In that study the geostrophic equations were solved and the deformation of the surface was treated as a topography perturbation. A fully three-dimensional simulation to reproduce the wall deformation requires a large effort that at the moment is not so relevant, since the effects of the solid wall are still not investigated. In addition, in a simulation it is difficult to accurately reproduce the effects of the surface tension generated by the surfactants on a free surface.

The present code reproduces satisfactorily the rotating tank experiments with a small deformation of the free surface, that is, when the aspect ratio between the height of the water layer and the radius of the tank is large. On the other hand, rotating tanks with a solid cover are well reproduced, by clustering the points of the axial grid near both walls, as in the plane turbulent channel.

The fractional step method with the approximate factorization is used, and the same considerations as before applied to the most appropriate sequence of the sweeps brings, in this code, the last sweep to perform in the direction x_3. The first sweep, in the periodic direction, does not require any boundary condition. The intermediate sweep is in the radial direction, where the boundary condition is an impermeable free-slip wall located far enough from the region of vortex dynamics. Then an approximation of the conditions on the intermediate quantities does not disrupt the inner solution. The three sweeps that approximate the inversion of a sparse matrix are:

$$\begin{aligned} (1-\alpha_l A_{i\theta})\Delta\hat{q}_i^{**} &= RHS_i \\ (1-\alpha_l A_{ir})\Delta\hat{q}_i^{*} &= \Delta\hat{q}_i^{**} \\ (1-\alpha_l A_{iz})\Delta\hat{q}_i &= \Delta\hat{q}_i^{*} \end{aligned} \qquad 11.6$$

Since $\hat{q}_i = q_i^{l+1} + O(\Delta t^2)$ the physical boundary conditions for q_i^{l+1} are those for $\hat{q}_i$. This non-solenoidal velocity field is projected into a solenoidal field by the scalar Φ. The elliptic equations for this quantity is obtained as in the previous chapters and it follows that

$$\mathcal{L}\Phi^{l+1} = \frac{1}{\alpha_l \Delta t}\mathcal{D}\hat{\mathbf{q}} = \hat{Q}, \qquad 11.7$$

For this coordinate system the elliptic operator is

$$\mathcal{L}\cdot = \frac{1}{r_m^2}\frac{\delta^2\cdot}{\delta x_1^2} + \frac{g_{2m}}{r_m}\frac{\delta}{\delta x_2} g_{2c} r_c \frac{\delta\cdot}{\delta x_2} + g_{3m}\frac{\delta}{\delta x_3} g_{3c}\frac{\delta\cdot}{\delta x_3}. \qquad 11.8$$

The expansion by Fourier series in θ reduces the band of the matrix. Keeping in mind that Φ is defined at the center of the cell and that the indices $\tilde{i}, \tilde{j}, \tilde{k}$ indicate the position $i+\frac{1}{2}, j+\frac{1}{2}, k+\frac{1}{2}$:

$$\Phi(\tilde{i},\tilde{j},\tilde{k}) = \sum_{-N_1/2}^{-N_1/2+1} \hat{\Phi}_{k_1,\tilde{j},\tilde{k}} e^{\frac{\iota 2\pi k_1 \tilde{i}}{N_1}} \qquad 11.9$$

By substituting this expression and the analogous for $\hat{Q}(\tilde{i},\tilde{j},\tilde{k})$ in Eq.(11.7), for each reduced wave number k_1' defined in Eq.(2.67) the following equation follows

$$\begin{aligned} &AP_3(k)\Phi(k_1,j,k+1) + AC_3(k)\Phi(k_1,j,k) + AM_3(k)\Phi(k_1,j,k-1) + \\ &AP_2(j)\Phi(k_1,j+1,k) + AC_2(j)\Phi(k_1,j,k) + AM_2(j)\Phi(k_1,j-1,k) = \hat{Q}(k_1,j,k) \end{aligned} \qquad 11.10$$

If the subscript c indicates the metric at the grid location and m that at the mid cell, the coefficients are

$$AP_2(j) = \frac{r_c(j+1)g_{2c}(j+1)g_{2m}(j)}{r_m(j)\Delta x_2^2}, \quad AM_2(j) = \frac{r_c(j)g_{2c}(j)g_{2m}(j)}{r_m(j)\Delta x_2^2} \qquad 11.11$$

and

$$AC_2(j) = -AP_2(j) - AM_2(j) - k_1{}'2/r_m(j)^2 \qquad 11.12$$

The coefficients of the matrix in the axial direction are

$$AP_3(k) = \frac{g_{3c}(k+1)g_{3m}(k)}{\Delta x_3^2}, \quad AM_3(k) = \frac{g_{3c}(k)g_{3m}(k)}{\Delta x_3^2} \qquad 11.13$$

with $AC_3(k) = -AP_3(k) - AM_3(k)$. Near the boundary the coefficients change because in Eq.(11.7) the values of q_i^{l+1} at the boundary are not expressed in terms of $\hat{q}_i$ and hence the external values of Φ do not appear. This procedure was encountered in the Channel (see section 9.3). The same values of the coefficients are obtained if it is imposed that the value of Φ at an imaginary point outside the box is equal to that at the interior. This assumption is in principle equal to that made by approximating Φ by a cos transformation. The solution of Eq.(11.9) is obtained by the $FISHPACK$ package. In chapter 2 it was shown that the efficiency of this package is not so high as that based on the use of trigonometric series in two directions. Nevertheless the efficiency is high and solutions by a large number of grid points can be achieved, at reasonable cost. I would like to suggest to the user using a large number of grid points, to investigate whether the results change going from single to double precision. Verzicco (private communication) found a lack of local divergence in single precision that disappears in double precision.

The introduction of the pressure p^l in Eq.(11.2-4) simplifies the velocity boundary conditions but it requires the explicit calculation of the pressure through $p^{l+1} = p^l + \Phi^{l+1} - \frac{\alpha_l \Delta t}{2\mathrm{Re}} \mathcal{L}\Phi^{l+1}$ at the end of each time substep of the third order Runge-Kutta scheme. As usual I would like to remind that one of the most important issues of this scheme is its convenience to perform simulations at constant CFL.

Non-uniform grids in two directions do not require any modification of the axis treatment, thus holds what was written in section 10.4 of chapter 10.

11.3 Code general informations

The code in this chapter was derived from the code used to simulate turbulent flows in a pipe. In that version there were subroutines evaluating the turbulent statistics, which necessitated time and space averages. In the turbulence simulations the goal is to reach a statistical steady state, hence the initial conditions do not play a fundamental role; thus several routines were introduced to start a fine simulation from a coarse solution. In the present code, devoted to vortex dynamics, the results depend on the initial conditions and during the evolution it is important to evaluate averaged vorticities. Then the code is much simpler than that in the previous chapter.

The code consists of several files with names composed of two parts: the initial part of the name is $TR3DNU$, a name indicating that this code was used to perform simulations of tripolar structures in 3D and with non-uniform grids. The last two letters which follow $TR3DNU$ indicate different sections of the code. The code is given in the directory $CODESCH11/TR3D$; a brief description of each file follows

11.3a Description of the file tr3dnuma.f

In this file there is the main program where the input file *tr3dfnu.d* is read. In the input files there are the parameters necessary to assign the initial vorticity distribution, the non-uniformities of the coordinates, the intervals of times when the output files are printed, the time step and the number of grid points. This input file is similar to that for the other codes and the comments in it facilitate the comprehension. The routine $SOLVE$ is the main structure of the code where several subroutines are called and in particular: that generating the mesh $MESHES$, that with the definition of the indices $INDIC$ necessary to express the derivatives, and the routine $PHINI$ that evaluates the coefficients and arrays for the solution of the "pressure" Φ. The code can start from $t = 0$, by assigning the desired velocity field in the routine $INITIA$, or it can start from a previous simulation by reading in $INIREA$ the field written in $CONTWR$. The time evolution is done by calling the routine $TSCHEME$ and certain global quantities are written by calling the routine $OUTH$. The last subroutine is called every *npin* constant time steps ($icfl = 0$) or when the time is a integer multiple of *tpin* in simulations at constant CFL ($icfl = 1$). The full velocity fields are written every *nprint* time steps or when t is a multiple of *tprin*. The field is written in $CONTWR$ and some average quantities is evaluated and written in the routine $OUTP$.

The routine $TSCHEME$ describes the time integration scheme; a do loop accounts for the three substeps of the Runge Kutta scheme. The passive scalar σ, indicated by the word *psc*, is the first quantity to be evaluated since it requires the velocities at the old time step. In this code without any coupling between thermal and velocity field, the advective terms, needed for σ, are calculated in $HDNLPS$; immediately thereafter, the routine $INVTPS$ furnishes the passive scalar at $n+1$. When σ indicates a density perturbation or a temperature there is a coupling with the velocity through the buoyancy and thus the subroutines are called in a different order.

The three subroutines named $HDNLi$ evaluate for each velocity component the non-linear terms H_i in the expression of RHS_i given in Eq.(10.9). These quantities are stored in arrays that successively are used to get Φ. The three routines $HDNLi$ are in the file *tr3dnuhn.f*. The output from the routines $HDNLi$ constitutes partly of the right hand side of Eq.(11.6), that is necessary to compute $\hat{q}_i$ in the $INVTRi$ routines. The divergence of $\hat{q}_i$ is computed in $DIVG$, and the output enters in the routine $PHCALC$ to get Φ. The solenoidal velocities q_i^{n+1} are obtained by $\hat{q}_i$ and Φ in the routine $UPDVP$. The true pressure is calculated in the routine $PRCALC$.

11.3b Description of the file fr3dnuge.f

In this file in the routine $INDIC$ the indices are stored to express the first and second derivatives. These indices eliminate " IF" instructions to express the derivatives near the boundaries. In this file the routine $DIVG$ for the calculation of the divergence of the velocity field $\hat{q}_i$ is given together with a similar routine to check that the field q_i^{n+1} is solenoidal ($DIVGCK$). The divergence is different from zero if an error is made in the evaluation of Φ. It has been found that the maximum divergence is high if in Eq.(11.8) the second derivative in θ is divided by r^2. This occurrence should be expected since the divergence is zero within round-off errors. The division by r^2 makes these terms large near the axis and thus the maximum divergence can be even $O(10^{-1})$. $DIVGCK$ is called as a control together with the *OUTH* routine. The cfl number is evaluated by the routine *CFL*; it is necessary as a check or when one should perform simulations at constant *CFL*.

11.3c Description of the file tr3dnuco.f

The grid spacing in the computational grid $\Delta x_2 = 1/(n2-1)$ and $\Delta x_3 = 1/(n3-1)$, together with the effective grid spacing $\Delta x_1 = 2\pi/(n1-1)$ are calculated in *MESHES*. The coordinate transformations in the radial and axial direction together with the metric quantities are calculated in the routine *CORDIN*. Several kinds of coordinate transformations, described in chapter 2, can be assumed depending on the physical case. In the axial direction there is the option to cluster the grid near the upper and lower boundary, or to cluster near the lower boundary and to stretch near the upper one. Several options can be also used for the radial transformation; in the cases studied here the radial coordinate is stretched near the outer boundary.

11.3d Description of the file tr3dnuintr.f

The routine to assign the initial field for the 3D simulation of the tripole formation is called *INTRIP*. The initial conditions described in section 4.8a.2c for the azimuthal velocity field are assigned, but in 3D ρ is replaced by ρ^*, given by

$$\rho^* = \rho + 0.25\eta \exp(-4(\alpha\rho^\alpha - 2)^2) \qquad 11.14$$

where $\eta(r,\theta,z)$ is a random number that varies between 0 and 1 at each point in the computational domain. The exponential term in Eq. (11.3.1) enforces the perturbation to get its maximum at the radial location where ω_z changes sign. Thus, the perturbation involves part of the core and part of the outer annulus. Since this random perturbation changes in z, it turns out that even an axial perturbation has been imposed. In the code there is also the possibility to add a further axial perturbation, but in the present simulation it has been neglected. The axial velocity q_3 has been set equal to zero.

If in the input file $\alpha = 0$ a Gaussian vortex is considered instead of an isolated vortex. This is an interesting initial condition for the comprehension of the Ekman layer at $Ro = O(1)$, because of its simpler vortex dynamics with respect to that of an isolated vortex. From $q_3 = 0$ and the continuity equation q_2 is derived.

In the routine *INQPR* σ gets a distribution coinciding with the positive ω_z and with $\sigma_{max} = 1$.

11.3e Description of the file tr3dnuinri.f

The initial condition to simulate a perturbed vortex ring impinging a solid wall is described in the subroutine *INRING*. In this case the initial azimuthal vorticity distribution is

$$\omega_\theta(t=0) = \frac{\Gamma}{\pi a^2} e^{-(s/a)^2}, \qquad s^2 \equiv (z - z_0)^2 + (r - r_0)^2, \qquad 11.15$$

where Γ is the circulation of the ring that defines the Reynolds number $Re = \Gamma/\nu$. The azimuthal perturbation has been assigned by replacing r_0 with $r_0^* = r_0 + \epsilon\cos(2\pi n\theta)$. It represents a perturbation of the center of the vortex, that can be reproduced in a laboratory by corrugating the nozzle from which the fluid that generates the ring exits. To account for the fact that the perturbation changes the orientation of the vorticity in the ring, the azimuthal component is multiplied by $\cos(\alpha)$ with $\alpha = atan[-\epsilon \sin(2\pi n\theta)]$. This projection was not introduced by Orlandi & Verzicco (1993a). The disturbance has been assumed to lie on a $r-\theta$ plane hence, with a good approximation, $\omega_z = 0$. ω_r is evaluated by the solenoidal property of the vorticity field. The initial passive scalar is $\sigma = \sqrt{\omega_\theta^2 + \omega_r^2}$, scaled with its maximum value to vary between 0 and 1. From the vorticity field the velocity field is evaluated by the kinematic relationship valid for incompressible flows

$$\nabla^2 \mathbf{v} = -\nabla \times \boldsymbol{\omega} \qquad 11.16$$

This equation has been solved only for the axial component

$$(A_{3\theta} + A_{3r} + A_{3z})q_3 = \frac{1}{r_m}[g_{2m}\frac{\delta r_c \omega_\theta}{\delta x_2} + \frac{1}{r_m^2}\frac{\delta \omega_r}{\delta x_1}] \qquad 11.17$$

To invert the Laplacian in Eq.(11.17) the Fourier expansion in x_1 reduces the equation to a set of 2D elliptic equations for each reduced wave number; each is solved by the $FISCHPACK$ package. The arrays for the FFT and the coefficients for the Laplacian in Eq.(11.17) are slightly different from those for the Φ equation; these have been evaluated in the routine $PHINV$. The velocity is calculated in the routine $PHCALV$, that has the same structure as $PHCAL$ giving Φ.

With the axial velocity and with ω_r the azimuthal velocity is evaluated by the definition of ω_r

$$\omega_r = \frac{\delta q_1}{\delta x_3} - \frac{1}{r_m}\frac{\delta q_3}{\delta x_1} \qquad 11.18$$

The velocity q_2 is evaluated by means of the continuity equation.

11.3f Description of the file tr3dnuhn.f

In this file the four subroutines for the non-linear terms are evaluated together with the linear terms that cannot be implicitly treated. Some of the derivatives require linear interpolations. Approximations are furthermore necessary to evaluate these terms near the axis. The routines for the three velocity components are called $HDNLi$ with $i = 1, 3$ and that for the passive scalar is named $HDNLPS$.

11.3g Description of the file tr3dnutn.f

In this file the part of the code concerning the two steps to get q_i^{n+1} is given. In the right hand side of Eq.(11.6) the viscous terms at the old time appear. These are evaluated in the routines $INVTRi$, and then are added to the combination of the non-linear terms at two previous substeps and to the pressure gradients. The RHS_i matrix is stored as a one-dimensional array with the index function of the indices i, j, k at each grid point. This choice, as has been explained in section 10.5a, leads to a better performance of the tridiagonal solvers. The approximate factorization, necessary to invert the large sparse matrix, is performed by three routines called $SOLQiI$, $SOLQiJ$, $SOLQiK$, where $i = 1, 3$ indicates the velocity component and the last letter indicates the sweep direction. For the passive scalar the name of the routines are obtained by changing "Qi" into "PS". For all the velocity components as well as for the passive scalar the first sweep is along θ, the second along r and the last one along z. In each of these routine the coefficients for the tridiagonal matrices are calculated and then enter into the routines for the tridiagonal solvers. These routines are found in the file *tr3dnutr.f*.

In the routine $UPDVP$ Eq.(10.10) gives q_i^{n+1} and the pressure p, linked to Φ, is calculated in the $PRCALC$ routine.

11.3h Description of the file tr3dnutr.f

The tridiagonal solvers are in this file. $TRIPVMY$ inverts a tridiagonal matrix with periodic conditions. In this routine the inner do loop is in j, for an efficient vectorization in the CRAY computers. $BRTRK$ inverts the tridiagonal matrix in k (the axial direction) and $BRTRJ$ in j; for both routines the two inner loops can be combined to increase the efficiency. The efficiency by one-dimensional arrays depends on the machine and on the grid; other solutions are possible and the effort is worthwhile only if the vortex dynamics requires a large number of computational points.

11.3i Description of the file tr3dnutu.f

In this file the velocity and vorticity field are averaged along the axis to give a first idea of the velocity field during the time evolution. In the routine $VELTUR$ the velocities at the cell center are evaluated from the staggered velocity components; these are averaged and plotted on $r-\theta$ planes. A similar procedure is used for the vorticity field that in the routine $VORC$ is evaluated at the center of the cell, from which the averaged vorticity is calculated. The averaging in z has a physical meaning only if periodic conditions are assumed along the axis; in this code the averages are done only to get a first idea on the vortex dynamics evolution.

11.3j Description of the file tr3dnuio.f

In this file the routines for the output of the results are given. In $PRICOR$ the coordinates are written to be used in flow visualization packages. The flow visualizations of this book were done by the NASA package $TURB3D$, that is the incompressible version derived by $PLOT3D$. The write statements in the code are those requested by the $TURB3D$ package. The writing statements should be changed according to the package at the disposal of each user. However, several packages have now the option to read files written for $PLOT3D$. In the routine $OUTP$ the quantities evaluated in the routines $VELTUR$, $VORC$ and $VORTUR$ are written.

In the routine $OUTH$ the time evolution of the maxima and minima of the velocities, vorticities and passive scalar in the whole field are written. These are quantities that show whether the simulation proceeds in a satisfactory way. The full field that the post-processing code reads is written in the routine $CONTWR$; in this routine the name of the output field is constructed and it indicates the time when it is printed. These fields could also be used as restarting file if there is the necessity of a longer evolution to get a better picture of the complex vortex dynamics.

11.4 Results

11.4a Three-dimensional tripole formation

The tripole formation is an example of vortex dynamics that has attracted the attention of several scholars interested in geophysical flows since on Earth tripolar structures were detected on satellite images. In addition, these structures in simulations of 2D forced turbulence were observed by Legras *et al.* (1988). From these observations Van Heijst and his group were motivated to devote a large effort to the reproduction of these structures in rotating tank experiments. They were successful as shown by their flow visualizations (Klosterziel & van Heijst (1991)). They devoted several studies to the tripole formation describing the differences between cyclonic and anticyclonic experiments. Successively two-dimensional simulations were performed to reproduce these experiments by methods similar to those described in chapter 4. Fully 3D simulations were never done in the past until Orlandi & Carnevale (1999), by the present code, explained features that in the experiments were not clear.

The goal of the 3D simulations was to investigate the effects of the centrifugal instability and of the Ekman layer to explain the different evolutions of isolated vortices with the rotation rate. The investigation through numerical experiments of flows with different boundary conditions permitted to explore these phenomena more carefully than was possible in the laboratory experiments. In the laboratory it is difficult to measure the initial vorticity profile of the vortex relative to the background, and hence to get the exact value of the Rossby number. In some of the cases, the strength and even the horizontal velocity field were measured and reported (Kloosterziel & van Heijst (1991)), but, especially for the centrifugally unstable anticyclones, this information was lacking. In contrast, the initial conditions are exactly controlled in the numerical simulations, and thus it was possible to examine a large range of Rossby numbers.

In the laboratory it is very difficult to measure all three vorticity components; estimates of ω_z, at the surface of the flow, were made in some of the cases from measurements of the velocities of particles dispersed on the surface. These estimates seem inaccurate. No measurement of ω_θ anywhere in the flow was made, which is essential to understanding the effects of centrifugal instability. The original purpose of the laboratory experiments was directed towards the two-dimensional dynamics of vortices. Thus there is very little information on the three dimensional structure of the vortices. It is a great advantage of the numerical simulations to be able to examine all components of the vorticity field at all times during the evolution. Furthermore, in the numerical simulations the bottom boundary condition can change so as to include or exclude the effects of the Ekman layer. Thus simulations were performed with periodic and slip boundary conditions, which exclude the Ekman layer, to test hypotheses about its importance in the experiments without introducing the complications of a bottom layer of different density as in the laboratory. It was important to examine the Ekman effect for a wide range of Rossby numbers, because there was ambiguity about the initial Rossby numbers in many of the laboratory experiments and because only at very low Rossby number there is a good theoretical understanding of the structure of the Ekman layer, while many of the numerical experiments have been performed with Rossby numbers $O(1)$, in which case no guiding theory does exist.

The section relative to the no-slip simulations in Orlandi & Carnevale (1999) is reported here. Also some results for the case of free-slip and periodic conditions are reported for

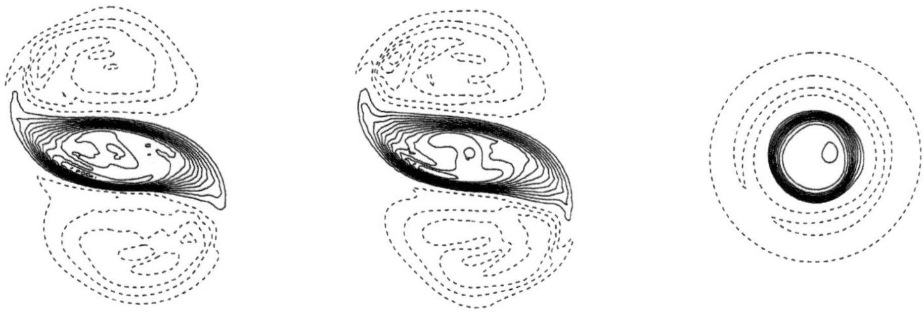

Fig. 11.1 Contours of ω_z on mid-height planes ($\Delta\omega_z = 0.025$) for $N = -1/2$ at $t = 200$ for different boundary conditions along the vertical direction: a) periodic, b) free-slip, c) no slip ($\alpha = 3$).

a clear description of the phenomenon. The results with periodical conditions cannot be obtained by the present code; whoever is interested can produce these results by the code described in chapter 10 by inserting the appropriate initial conditions. However, with free-slip conditions the results are practically identical to those with periodical conditions. The presence of the bottom boundary of the tank causes the formation of the Ekman layer, which significantly enhances the rate of decay of the vortices (cf. Pedlosky (1979)). Also, it could be expected that any inertial wave radiation emanating from the vortex will be damped more rapidly in the presence of a no-slip wall because reflection from such a wall would produce large gradients of vorticity on which viscosity would act effectively (cf. Greenspan (1980)).

As a preliminary look at the effects of the bottom boundary, a series of simulations with $\alpha = 3$ were performed with a free-slip condition on the top. In one set of simulations the bottom boundary was replaced with a free-slip boundary and in another with a no-slip boundary. Figure 11.1 shows a comparison at time $t = 200$ of the vorticity fields for simulations in the case $N = -1$, all starting with the same initial condition but having different boundary conditions. The result in the case where the boundaries on top and bottom are both free-slip (b) is very similar to the case with periodic boundary conditions (a). Fig.11.1 also shows that compared with the periodic case, there is considerably more damping of the field in the no-slip case, and indeed the anticyclonic vortex shown in Fig.11.1c does not reach the tripolar state. It appears that the barotropic instability in this case, and also for all $N < -0.5$, is arrested by the damping effect of the bottom Ekman layer. In contrast, under the same conditions ($|N| = 0.5, \alpha = 3$, with the no-slip boundary), the cyclonic vortex does form a tripole as in the periodic boundary condition case. For the cyclones, even with the bottom Ekman layer, the sequence of predictions from 2-dimensional barotropic instability studies seem to hold approximately. In other words, as the α of the initial condition is increased, a sequence of final structures that goes from stable monopole to stable tripole to unstable tripole is obtained. This is illustrated

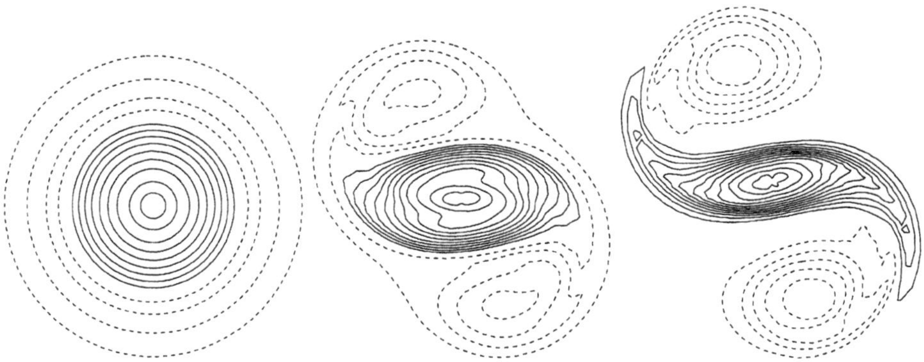

Fig. 11.2 Contours of ω_z on mid-height planes ($\Delta\omega_z = 0.025$) for $N = 1/3$ at $t = 200$ with a no-slip bottom boundary condition a) $\alpha = 2$, b) $\alpha = 3$, c) $\alpha = 3.5$.

in Fig.11.2, by the results at $t = 200$ for $\alpha = 2$, 3, and 3.5 (with $N = 0.33$). Although the two-dimensional inviscid studies predict that the transition between monopole and stable tripole should occur around $\alpha = 1.85$, here in the three-dimensional simulations with viscosity the transition occurs at a somewhat higher α.

In order to understand better the difference between the effect of the bottom boundary in the anticyclonic case and the cyclonic case, a series of simulations with $\alpha = 3.5$ were performed in which the barotropic instability was not completely damped for $N \leq -0.5$. Four representative cases are shown in Fig.11.3.

In Fig.11.3c the formation of an unstable tripole at $N = 0.5$ can be seen. This N is certainly in the centrifugally stable range since for $\alpha = 3.5$ from the analytical theory reported in Orlandi & Carnevale (1999) it results that $N_{cr} = 0.36$. Thus this case is an example of the formation of an unstable cyclonic tripole without the effect of centrifugal instability. In the laboratory experiments discussed above, unstable cyclones usually evolved into stable tripoles; however, Van Heijst *et al.* (1991) showed an example of a cyclone that does become an unstable tripole that undergoes double-dipole instability. As an explanation of the difference between this case of unstable cyclonic tripole formation and the usual stable cyclonic tripole formation, Van Heijst *et al.* (1991) pointed out that the scatter plot of ψ vs. ω in the core of the cyclonic vortex in this case was different from that in the more typical case of the stable tripole. They suggested that it was the anomalous structure of the core itself that led to the unstable tripole formation. In contrast, the example in Fig.11.3c requires nothing else than having initially a sufficiently high value of α for the formation of an unstable tripole. For higher values of N, the damping effect of the Ekman layer becomes very important. At $N = 2$, by $t = 200$ the amplitudes of both the positive and negative vorticity have decreased considerably more than in the lower N cases. A measure of the decay rate of vortices due to the Ekman layer is the Ekman time scale; rewritten in terms of the advective time scale it is

$$t_E/t_a = \frac{D}{\sigma}\left(\frac{2Re}{N}\right)^{1/2}. \qquad 11.19$$

For these experiments, the aspect ratio $\frac{H}{\sigma}$ is π and $Re = 2500$. Thus the final time shown in the figure i.e. $t = 200$ corresponds to $0.64t_e$ for $N = 0.5$ and $1.3t_e$ for $N = 2$. Thus it is natural to see considerably more damping the larger the value of N. Also, since the timescale of the barotropic or two-dimensional instability is controlled by the advective timescale, it is possible for a rapid Ekman decay to arrest the barotropic instability. In the $N = 2$ example (Fig.11.3d), a tripolar structure barely forms by $t = 200$ and although somewhat asymmetrical, it seems to decay as a weak tripole.

Figure 11.3b ($N = -.33$), shows a typical double-dipole instability initiated by the centrifugal instability. In Fig.11.3a the formation of a weak but symmetrical tripole for $N = -2$ is shown. Orlandi & Carnevale (1999) discussed the differences between $N = +2$ and $N = -2$.

The formation of the Ekman layer at the bottom of the vortex sets up a secondary circulation of flow which can be observed in the advection of a passive scalar. Simulations with a passive scalar with $Sc = 1$ were performed; the diffusivity of dye in the laboratory is three orders of magnitude larger than the kinematic viscosity. Even though the present scalar field diffuses more rapidly that the dye in the tank, the important effects are captured qualitatively. In Fig.11.4, the evolution of a passive scale field is shown. The scalar was initially uniformly distributed within the core of the vortex, with a smooth transition to zero concentration in the negative vorticity annulus, reaching zero about one third of the way between the points where the vorticity vanishes and where it is most negative. Comparing Figs.11.4a ($N = -2$) and Fig.11.4b ($N = +2$), it turns out that for the anticyclone/cyclone there is a radial outflow/inflow of fluid in the bottom boundary layer since the contour lines of the scalar, which are initially all vertical, bow out/in at the bottom of the anticyclone/cyclone. Note that in the cyclonic case, the resulting upward flux of passive scalar at the bottom continually raises the level of dye away from the bottom. This phenomenon is also observed in the laboratory experiments when viewed from the side.

Figure 11.4c is shown as a representative case of the evolution of the scalar when there is centrifugal instability. The scalar field reproduces the overturning motion at time $t = 100$ typical of this instability. By time $t = 150$ side cyclones have formed. It appears that passive scalar from the core anticyclone is being transferred through the bottom Ekman layer into the cyclones, which makes sense from the point of view of the simple linear Ekman layer theory for the induced secondary circulation (i.e. axially downward in the anticyclone, radially out form the center of the anticyclone in the Ekman layer, radially in towards the centers of the cyclones in the Ekman layer, and upward into the cyclone.

According to the linear Ekman layer theory (cf. Pedlosky (1979)), the thickness δ_E of the layer should scale as

$$\delta_E = (\nu/\Omega)^{1/2}, \qquad 11.20$$

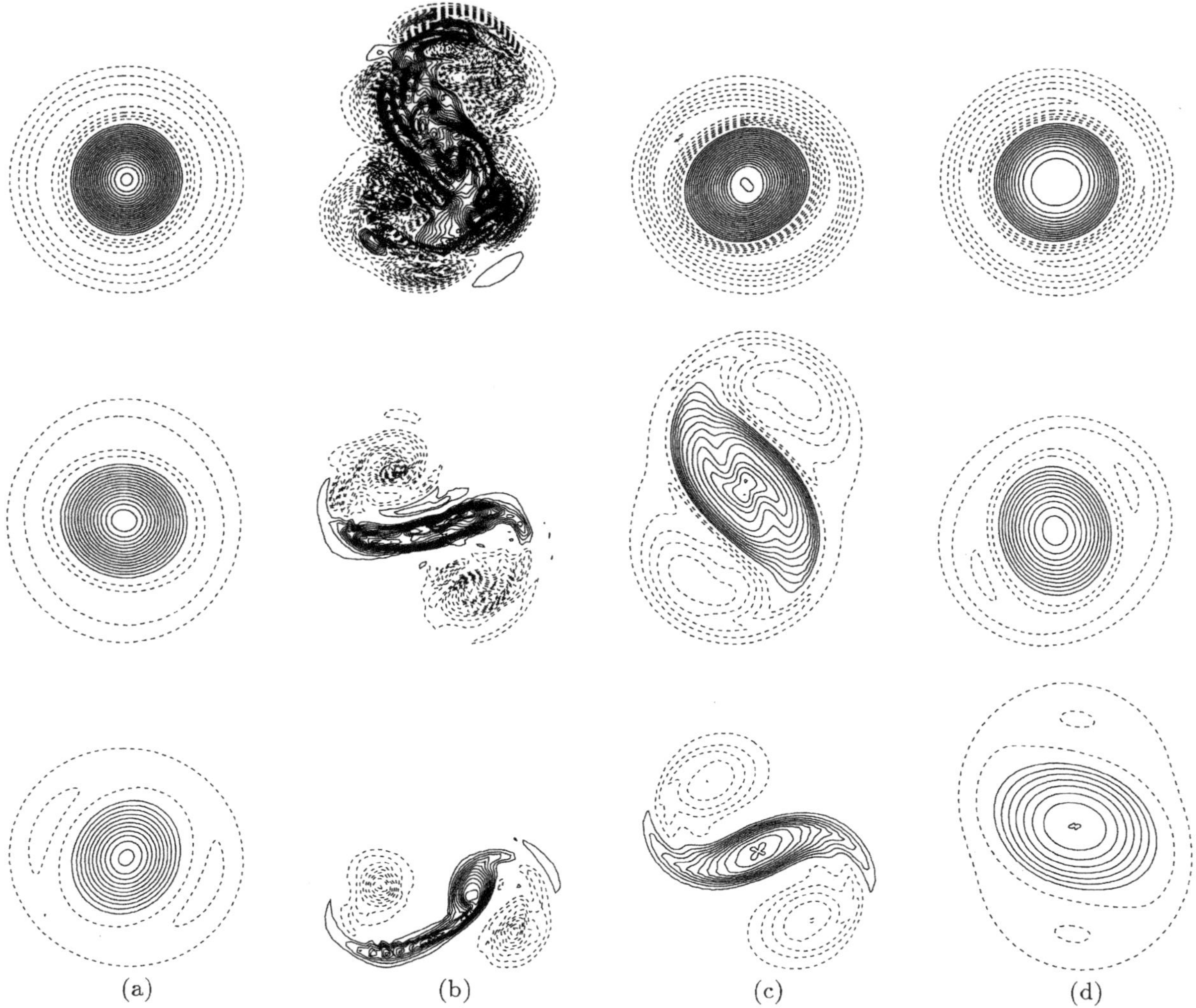

Fig. 11.3 Contours of ω_z on mid-height planes from simulations with the no-slip condition at the bottom boundary. ($\Delta\omega_z = 0.025$) From left to right a) $N = -2$, b) $N = -1/3$, c) $N = 1/2$, d) $N = 2$; from top to bottom $t = 100$, $t = 150$, $t = 200$.

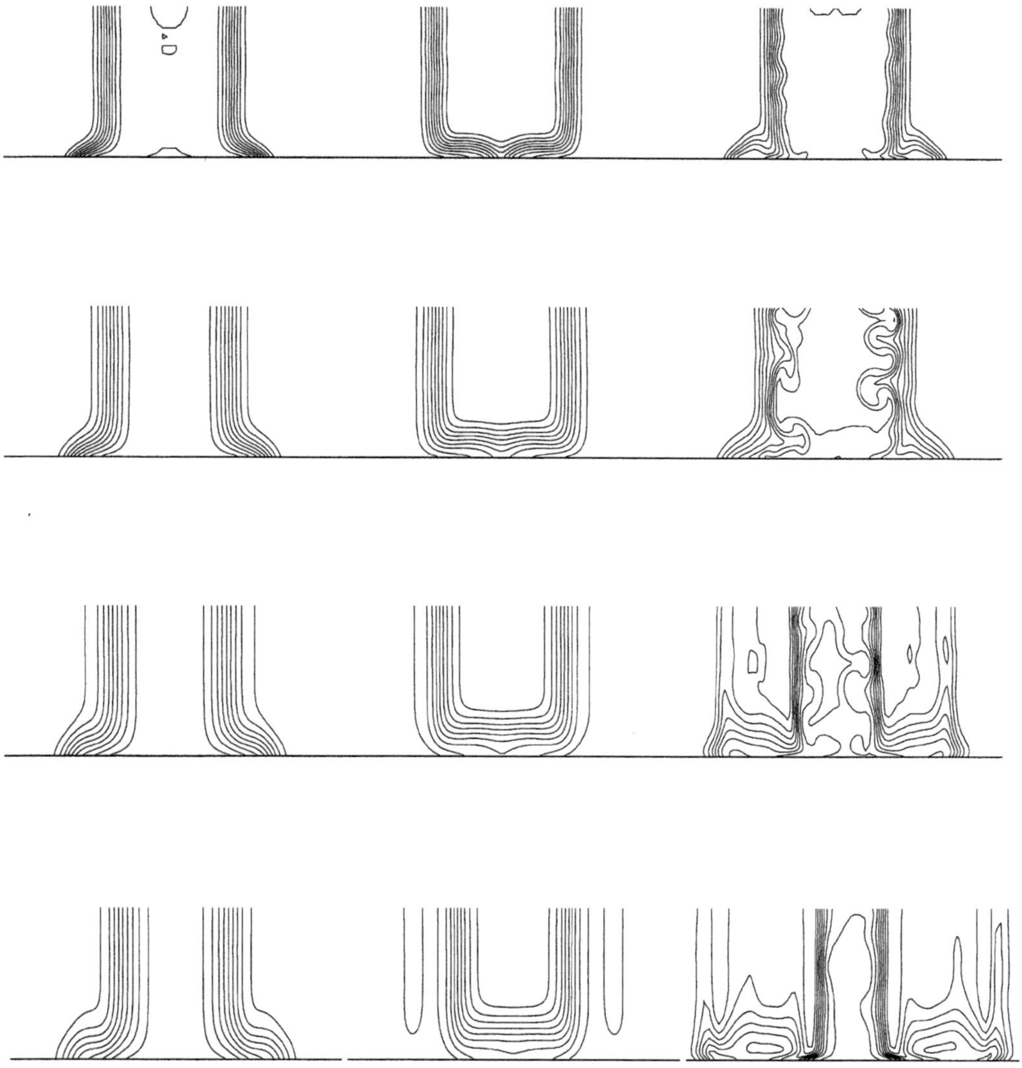

Fig. 11.4 Contours of passive scalar in vertical sections passing through the point where ω_z takes its most negative value. a) $N = +2$, b) $N = -2$, c) $N = -0.5$. From top to bottom $t = 50$, $t = 100$, $t = 150$, $t = 200$.

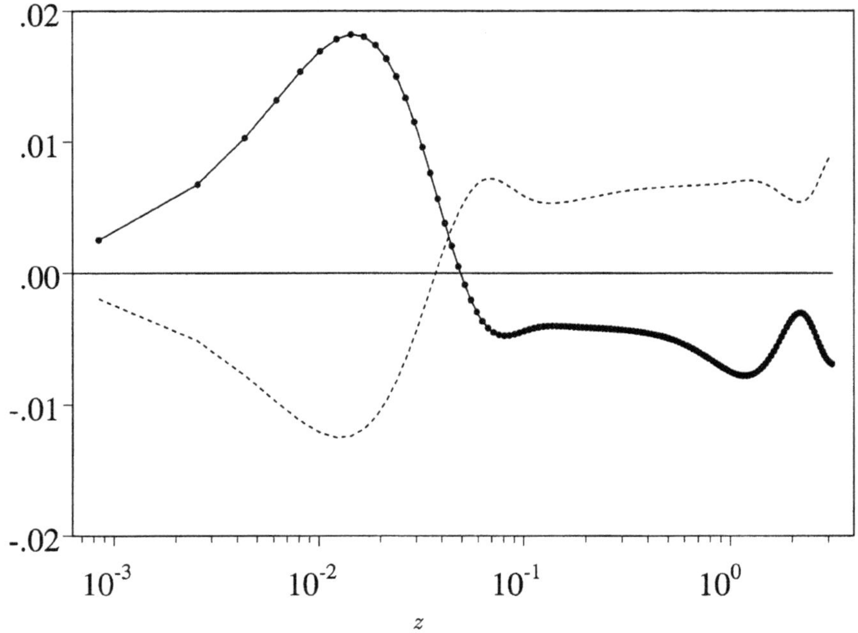

Fig. 11.5 Radial velocity profiles in the z direction for ——— $N = -2$ and - - - - for $N = +2$. The profiles are logarithmic to emphasize the Ekman boundary layer region. The profiles are taken at $r = 0.58$, this position corresponding approximately to the maximum of v_z (see Fig.11.6). On the solid lines the dots indicate the distribution of the computational grid points.

which can alternatively be written in terms the horizontal scale of the vortex as

$$\delta_E = \sigma(\frac{2}{NRe})^{1/2}. \tag{11.21}$$

Thus for the case of $|N| = 2$, with $Re = 2500$, $\delta_E = \sigma/50$. The Ekman layer in the simulations was resolved by a variable mesh spacing with a relatively high density of grid points in the thin Ekman layer. Figure 11.5 shows the radial velocity as a function of z at a radius of $r = 0.58$, which corresponds roughly to the position where v_z takes its maximum value in the cases N=$\pm$2 at time $t = 200$. As predicted by the linear Ekman theory, the radial velocity in the Ekman layer is positive/negative for the anticyclone/cyclone. The radial velocity is maximum in magnitude near $z = \delta_E$. The grid mesh size was illustrated by putting dots at the appropriate radii on the curve for the anticyclonic case. Including the boundary, there are 10 grid points from the wall to the point of maximum radial velocity, and another 18 in the region where the radial velocity decays to the interior value.

The graph for the case of the cyclone is at least qualitatively in accord with the Ekman layer theory. Outside of the Ekman layer it is weakly positive, representing a slow outward flow over a major part of the vortex column.

11.4b Three-dimensional impact of a vortex ring on a wall.

The growth of the azimuthal perturbation during the impact of a vortex ring on a solid wall was depicted by the flow visualizations of dye done in a laboratory by Walker *et al.* (1987). More quantitative data were obtained by Orlandi & Verzicco (1993a) by numerical experiments at $Re_\Gamma = 2985$, a sufficient high Reynolds number that allows the growth of the azimuthal perturbations. In that study two simulations were done, the first on a grid $65 \times 97 \times 128$ in $0 < \theta < 2\pi$, $0 < z < 4r_0$ and $0 < r < r_0$. This grid was not sufficient to resolve the thin vorticity layer of the secondary structure generated at the wall. However, the simulation was useful to see that with a $n = 5$ perturbation only wave numbers which are multiple of $n = 5$ were growing. The coarse simulation showed that in the secondary structure a $n = 10$ wave number was growing faster than the $n = 5$. From this fact a second simulation was performed, in which the number of points in r was increased up to 128 and in θ it was set equal to 48, in a domain $2\pi/5$ which is equivalent to perform a simulation by 240 points in the full domain. This more refined simulation reproduced the vortical interactions that first produce hairpin vortices in the secondary and finally a new vortex ring travelling far from the wall, as was shown in the laboratory experiment by Walker *et al.* (1987). The color plates in Orlandi & Verzicco (1993a) and in particular the surface contours of the passive scalar in their Figure 26 give the same impression of the vortex evolution obtained by photographs of dye in figure 15 in the paper by Walker *et al.* (1987).

For this book the simulations were repeated with the same amplitude of the initial azimuthal vorticity perturbation, but the initial conditions were assigned in a different way. In the previous study Orlandi & Verzicco (1993a) did not project the azimuthal vorticity, multiplying it by $\cos\alpha$; it turned out that the azimuthal perturbation in the previous simulation was smaller. In the previous study, to my knowledge the first numerical simulation of this interesting phenomenon (later on other simulations appeared, Fabris & Liepman (1996), Swearingen (1995)), the aim was directed to investigate whether the simulation reproduced the experimental observations. Thus it was important to give a perturbation that in any case could not reproduce that in the laboratory. Here the perturbation is more realistic and the comparison between the contours plots of ω_θ in the $r - z$ planes in the two simulations show that the initial conditions play a role.

A first simulation has been done with a uniform grid in z and in r with $48 \times 128 \times 128$ points. A second one with the non-uniform grid with $48 \times 96 \times 85$ points and the results were practically the same. In Fig.11.6 the time evolution of ω_θ shows that during the impact there is the generation of a secondary ring which is advected at the center. Later on in time a tertiary ring is generated at the wall that in a certain measure contributes to the vertical motion of the secondary ring. A closer comparison of Fig.11.6 and Fig.25 in Orlandi & Verzicco (1993a) shows that in the present simulation the ejection of the new ring is visible while it was only slightly detectable in the previous one. The evolution of ω_z plays a large role and thus contours of this quantity at $t = 40$ are shown. At this time ω_z reaches its maximum amplitude and hence gives the greatest effect on ω_θ. Vortex stretching is responsible for the growth of ω_z in the region of the secondary ring; this phenomenon begins at $t = 30$ and Fig.11.7a shows that at $t = 40$ ω_z is part of a hairpin vortex with the legs below the primary ring. In the same section contours of ω_r (Fig.11.7b) emphasize that

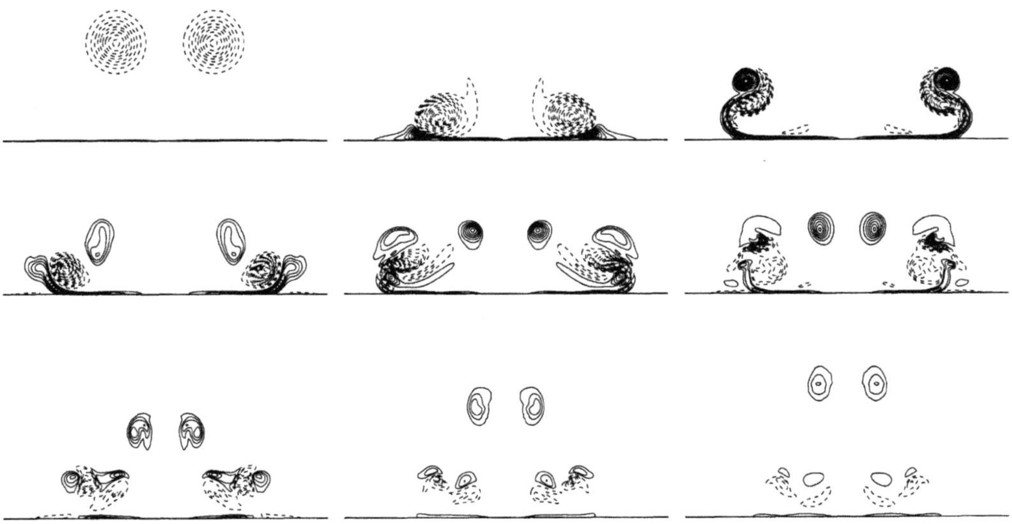

Fig. 11.6 Azimuthal vorticity contour plots at $Re_\Gamma = 2895$: *(a)* $t = 0$, *(b)* $t = 10$, *(c)* $t = 20$, *(d)* $t = 30$, *(e)* $t = 40$, *(f)* $t = 50$, *(g)* $t = 60$, *(h)* $t = 70$, *(i)* $t = 80$. The sequence in time is from top-left to bottom-right and contour increment is ± 027, $\omega(t = 0)_{min} = -1.9$. Initial amplitude perturbation $\epsilon = 0.02$

these new vortex structures are indeed inclined legs of hairpin vortices at 45^0. In the figure below the contours at horizontal planes at two different distances from the wall show the large deformations of the primary vortex and these are produced by the legs of the hairpin vortices. On the external side of the primary vortex there are concentrated regions of ω_z of sign opposite to that at the interior and these deform the tertiary structure. Perhaps an easy and fast comprehension of the gross features of this complex vortex dynamics can be obtained by surface contours of vorticity magnitude in the whole field but these require the storage of very large files. This was done by Orlandi & Verzicco (1993a) to produce their figure 25.

The non-uniform grids permitted to cluster the points in region of highest vorticity activity and to reduce the spacing in regions without vorticity. Then the number of grid points saved can be used to study flows at high Reynolds numbers or flow with large perturbations, that produce smaller and more active vorticity structures. In the present book the same value of the Reynolds number is maintained and the amplitude of the perturbation is changed. A second simulation has been performed by a grid $64 \times 128 \times 128$ and the perturbation amplitude was increased up to 0.1. Fig.11.8 shows that indeed in the early stage of the impact the amount of secondary vorticity produced at the wall is higher and that there is a large deformation in the primary. At $t = 40$, when the secondary patch is well inside the primary, it appears that the secondary ring has a more elongated

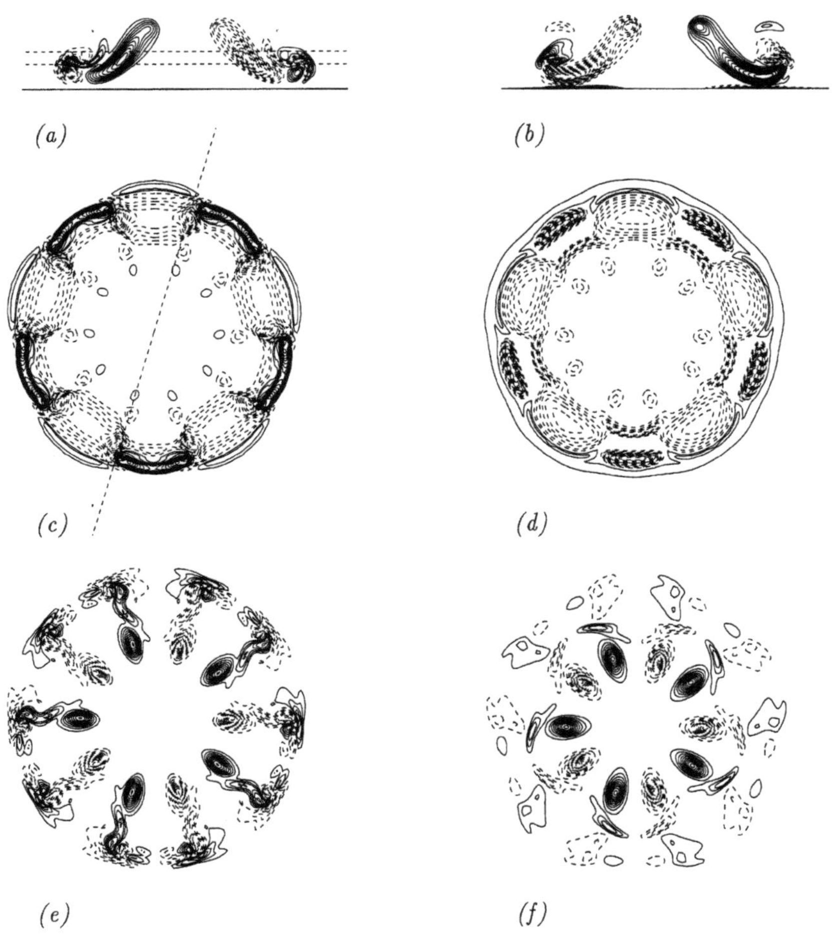

Fig. 11.7 Contour plots of vorticity in $r-z$ planes *(a)* ω_z, *(b)* ω_r; in $r-\theta$ planes *(c)* ω_θ, *(e)* ω_z at $z=0.5$; *(d)* ω_θ, *(f)* ω_z at $z=0.75$. $\Delta\omega=\pm 0.2$ at $t=40$ and $Re_\Gamma=2895$. Initial amplitude perturbation $\epsilon=0.02$

structure than in the previous case and that the head of the hairpin has a smaller size with a greater amount of vorticity. This occurrence causes the secondary ring to travel faster upwards as is shown at $t=80$ in Fig.11.8.

The same view as that in Fig.11.7 with this greater initial perturbation shows the small scales that grow during the interaction. The top pictures in Fig.11.9 show that on the vertical sections there is not anymore one single structure but these are two where ω_r and ω_z grow at the interior of the primary ring. The horizontal sections at the same distance

Fig. 11.8 Azimuthal vorticity contour plots at $Re_\Gamma = 2895$: *(a)* $t = 0$, *(b)* $t = 10$, *(c)* $t = 20$, *(d)* $t = 30$, *(e)* $t = 40$, *(f)* $t = 50$, *(g)* $t = 60$, *(h)* $t = 70$, *(i)* $t = 80$. The sequence in time is from top-left to bottom-right and contour increment is ± 0.2, $\omega(t = 0)_{min} = -1.9$. Initial amplitude perturbation $\epsilon = 0.1$

from the wall, as in Fig.11.7, show that the amplitude of the ω_θ deformation is much higher and that the distortion has generated round structures where this component is concentrated. The secondary ring, that before did not appear at this distance from the wall, now has large deformations and these are localized in thin regions. Fig.11.9d at a greater distance from the wall shows that the length of the inner and outer regions of ω_θ is different; in fact the outer regions of concentrated ω_θ at this distance do not appear any more. Finally the contours of ω_z in the $r-\theta$ planes show that these have been concentrated in regions of much smaller scales.

11.5 Conclusions

In this chapter a numerical method and the related code are described; these permit to study the vortex dynamics of flows with two inhomogeneous directions. The code is of particular utility to study vortical structures with an axis of symmetry. Two examples of interest to whom enjoy the interaction of vorticity with solid boundaries were chosen. The ring impinging a solid wall has been quite largely studied experimentally and numerically, since it has a relevance to the generation of vorticity at walls which is the key structure of wall friction. The ring is a good model, because it is propelled by itself towards the wall and by interacting with it generates vorticity of opposite sign. The instability mechanism of the second ring is different from that of the primary ring. In fact while an isolated ring is perturbed only by the self-induced velocity field, the secondary ring is perturbed by two fields. The vorticity in the secondary is weaker that that in the primary hence it is more unstable that the primary. This was explained by Orlandi & Verzicco (1993a) and there are similarities between their analysis and that used by Blackburn *et al.* (1996) to study the vortex dynamics in a turbulent channel. In the real wall turbulence the near wall

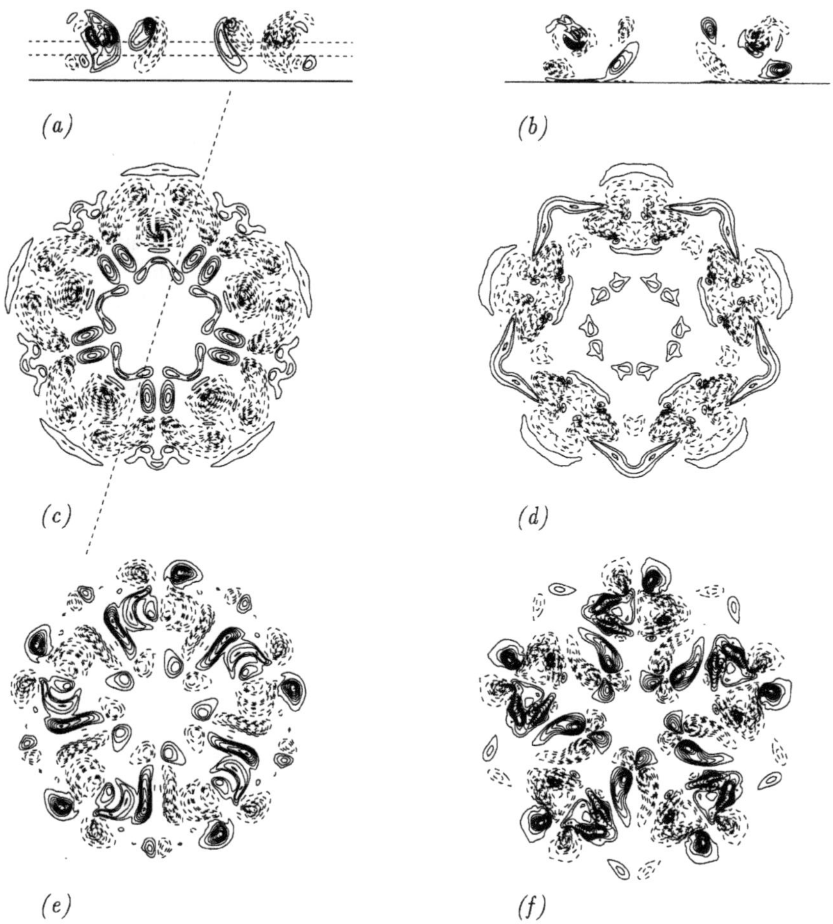

Fig. 11.9 Contour plots of vorticity in $r-z$ planes *(a)* ω_z , *(b)* ω_r; in $r-\theta$ planes *(c)* ω_θ, *(e)* ω_z at $z=0.5$; *(d)* ω_θ,*(f)* ω_z at $z=0.75$. $\Delta\omega=\pm0.2$ at $t=40$ and $Re_\Gamma=2895$. Initial amplitude perturbation $\epsilon=0.1$

region vortical structures are inclined and these do not have any axial symmetry. Thus a more realistic structure closer to those observed in the wall layer could be a vortex ring impacting the wall at an angle. Verzicco & Orlandi (1996b) in their review about the impact of vortical structures with a wall described the differences between the normal and the oblique impact. They resumed that numerical and experimental approaches revealed that in the oblique impact the azimuthal perturbations grow less for effect of the axial motion induced by the pressure gradient along the ring axis. The inclined interaction is more interesting and it can be numerically simulated by a code in Cartesian coordinates similar to that described in chapter 7 for turbulent channel simulations.

The study of interactions between vortical structures and Ekman layers, on the contrary, received less attention and I think that further studies are necessary to understand the effects of this thin wall layer and the evolution of a vortical structures with the axis perpendicular to the wall. Theoretical work was done at very high rotation numbers N (very low Rossby numbers). At intermediate rotation rates the theory does not work and hence laboratory and numerical experiments are necessary. Since there are many parameters playing a role the study should be done by using the two techniques together. The laboratory, by visualizations on $r - \theta$ planes, should give the input to the numerics and indicated the more interesting range of parameter to study. Then the accurate simulations, once validated by the visualizations, will produce a complete understanding of the complex vortex dynamics.

I would like to mention that this code can be modified to study space developing round jets. The coaxial jet previously described in chapter 7 was obtained by a code with an uniform grid along the axis. With the present code similar simulations can be done by changing the outlet conditions at the exit boundary. This is a very easy task that I leave to whom is interested. For space developing jet it is very important to capture the vortex dynamics near the nozzle, when the jet profile at the inlet is laminar, since rollup and pairings near the nozzle are the events controlling the entire flows. This fine resolution is of particular relevance for combustion-related problems where the mixing occurs in the potential region where the instability occurs.

Finally the code could be used to study wall friction control in pipes, especially by slots with suction or blowing. Periodic conditions can not be used at the inlet or outlet of the pipe, and in the present code grid points can be clustered near the slots. To have realistic conditions, two simulations should be performed, one with the code described in chapter 10 to generate the inlet conditions for this code. It is not important to have at the inlet velocity distributions with a good resolution in z. To overcome interpolation errors it is better to leave the same resolution in θ and to maintain the same coordinate transformation in the radial direction.

12 Large eddy simulations

12.1. Introduction

In practical applications high Reynolds numbers flow fields are requested, and the direct simulation cannot be achieved. In the early period of the numerical simulation of the Navier-Stokes equations turbulence models, based on Reynolds averages, were widely used. These one-point closures turbulence models, as was shown in the 1969 Stanford conference, gave reasonable results in boundary-layers with moderate pressure gradients. In the seventies the simulations were extended to complex flows with some success. The ad hoc constants in the models were evaluated by the help of measurements. The reason of the satisfactory results is linked to the physical assumptions introduced in the closure of the high order correlations. Thus, an interesting output of these turbulence models was that with a limited number of constants it was possible to obtain satisfactory results for different flows in a large range of Reynolds numbers. However, the users of these models, those working in the industries, were not satisfied and they were pushing the modellers to find more universal models.

The event confirming the limited utility of one-point closure models was the 1980-81 Stanford conference on complex flows. In this conference experimentalists and computors convened together; from the analysis of post-diction and pre-diction it was found that the models had success in post-diction but not when the experimental data were absent. These were not the results the industries liked because in design it is necessary to explore new geometries or boundary conditions. A further result of the conference was that the worst predictions were obtained for homogeneous flows, where the high order correlations account for the interaction between turbulent scales and that these should be modeled accurately. One example was the prediction of the effects of solid body rotation on the evolution of isotropic turbulence, a flow that cannot be reproduced by one- or two-equation turbulence models.

During the 1980-81 Stanford conference the results of LES for the channel (Moin (1981)) and DNS for homogeneous flows (Rogallo (1981)) were presented. These results were promising and attracted the interest of several conveners. However, it is important to recall that in that period these simulations were feasible only to the scholars who had access to the large computational facilities of research centers as those of NASA, Los Alamos and NCAR in the United States. A historical review of the hopes in that years is given in the article by Chapman (1979). Since then the large growth of computer speed and memory and the contemporary decrease of the costs made those simulations accessible to a large number of scholars.

DNS was described in the previous chapter and it was pointed out that DNS is the most simple turbulence models; it requires only a numerical scheme that does not introduce any numerical viscosity. In the previous chapters it has been proved that finite difference schemes for a staggered array of unknowns satisfy this requirement. Pseudo-spectral methods have been used more often in DNS for their accuracy and conservation properties. These methods should give more reliable results in Large Eddy Simulations, since, as recently pointed out by Ghosal (1996), high order accuracy avoids truncation errors of the same order of the subgrid scale model. This claim by Ghosal (1996) was proved by a theoretical analysis but, to my knowledge, it was not heuristically confirmed by simulations.

In this chapter I would like to prove that, for homogeneous turbulence, the finite difference scheme (described in chapter 8) produces satisfactory results for LES. To investigate the importance and the efficiency of subgrid scale models different models have been tested, starting from the widely used Smagorinsky model to end with the most recent dynamic model. I would like to recall that at the end of the seventies and at the beginning of the eighties LES had a large diffusion, but that, during the late eighties, there was a diminishing interest in LES. This decrease was related to the consideration that, in principle, the subgrid models had the same drawbacks as the one-point closures, since the constant is flow-dependent. This was found by Moin (1979) for the channel and by Mansour (1979) for the mixing layer.

In one of the summer schools at the Center of Turbulence Research the group in LES (Cabot, Germano, Moin and Piomelli) found a very interesting and promising way to evaluate the constant in the Smagorinsky eddy-viscosity expression by using the Germano identity. Germano, in fact in the research proposal submitted to attend the school, proposed to prove, by the data-base of the DNS, that an identity holds between the stresses corresponding to two filter operations. During the discussions within the group the idea came out that this identity could be used to improve the subgrid models by getting the constant from the simulation itself. After that summer, the research on the subgrid models received a new impetus and the number of scholars involved in LES has grown. To account for the success of the LES I would like to point out that just recently commercial codes as the $STAR5$ package have the possibility to introduce s.g.s. models. It is impossible to account for the large amount of research work done in recent years on LES. Those interested to know the improvements done by the Stanford group should read the papers collected in the proceedings of the Summer Programs at the CTR and the Annual Research Briefs.

Other groups around the world were involved in LES and their results are mentioned in the review article on LES by Boris *et al.* (1992) and by Lesieur & Metais (1996). One of the models of the Grenoble group, the structure function model, has been tested in the present chapter. Since finite differences solve the equations in physical space in this book the s.g.s. expressed in physical space are preferred. A different class of s.g.s. models such as the Kraichnan (1976) s.g.s. model are expressed in wave number space (see McComb (1990) pg.130), and are appropriate to pseudo-spectral numerical methods. Since these simulations are not of interest here the large amount of work done by these models is neglected. However, some of the results from these simulations are taken as reference solutions to show that finite differences produce results as good as those obtained with these more accurate numerical methods.

Other groups are claiming that satisfactory results could be obtained by the Very Large Eddy Simulations, that is, by coarse simulations without any model. This philosophy was followed by a group of Japanese researchers (Tsuboi *et al.* (1989)) who performed simulations of practical problems for which the small scales effects are related to some sort of numerical viscosity. This numerical viscosity plays a large role as the grid inhomogeneity increases, and it becomes even larger for a non-orthogonal grid. A discussion on this approach is reported by Reynolds (1989) and by Boris *et al.* (1992). In this case, in fact, it is very difficult to prove that the numerical schemes conserves the total energy. From these considerations it turns out that it should be mandatory in every LES simulation, to perform simulations without any subgrid model to get a quantitative measure of the effects of the numerical scheme used. This procedure perhaps could be avoided when standard numerical schemes such as pseudo-spectral or staggered finite difference schemes are used. On the other hand the check must be performed by commercial codes which introduce approximations when dealing with complicated geometries. In the standard method, however, the check is also useful to investigate whether the major part of bugs, always present in the codes, have been eliminated.

12.2. Filtering

LES is based on unknowns that are representatives of the large scales; these quantities are obtained by a filtering operation.

$$\overline{f}(\mathbf{x}) = \int G(\mathbf{x}, \mathbf{x}')f(\mathbf{x}')d\mathbf{x}' \qquad 12.1$$

I would like to remember that Deardoff (1979) in his LES paper was mentioning that this filtering is the kind of averaging Osborne Reynolds had in mind (Reynolds (1895)) in making a closure for turbulent flows. My opinion is that, therefore it is more appropriate to call LES a Reynolds averaged model. However, since everybody is accustomed to refer to Reynolds averaged models as those with spatial and time averages, it is better to leave the things as these are. The filtering operation applied to the Navier-Stokes operation must satisfy several properties. In this book not all the details of the LES are described, the user interested in the filtering properties can read the papers by Germano (1986a-c). In the early LES, several filters were tested and the Gaussian filter was largely used since it commutes and it has the same expression in wave number as in physical space. In wave number space the simplest filtering operation with a clear physical meaning is the top-hat filter. In fact, it means that in the equations appear only the wave number components at $k_i < k_{ic}$; k_{ic} is the greatest wave number related to the resolution of the LES simulation. To be more explicit $k_{ic} = 1/\Delta x_i$ in a box of length 2π in each of the three directions. The expression of the kernel G is

$$G(x, x') = \begin{cases} 1 & if |x - x'| \leq X_c \\ 0 & otherwise \end{cases} \qquad 12.2$$

I mentioned, the clear physical meaning of the filtering operation in wave number space, since usually the turbulence spectra are the quantities which characterize the flow. In this sense, a large eddy simulation at practical Reynolds numbers should have the filter cut-off at a wave number that is in the inertial range and that is greater than the energy containing range wave number. The part of the spectrum were the energy decays exponentially is thus neglected and this is the part of the spectrum that should be modelled by the s.g.s model. In the Libby (1996) and in the Frisch (1995) books on turbulence the one-dimensional energy spectra at different Reynolds numbers are reported and for different flows (for instance isotropic turbulence, boundary layers, jets, and mixing layers). These spectra, in

Kolmogorov units, show a good collapse of the data at the wave numbers where the energy spectra decay exponentially. The isotropy of the small scales thus was the reason that was driving the hope that LES had the potentiality to give better results than the Reynolds averaged turbulence models. However, in some inhomogeneous flows as those near a wall in a plane channel the size of the energy containing scales and the dissipating scales are not sufficiently separated and then it must be expected that LES presents difficulties. Then a large effort to find satisfactory s.g.s. models is necessary.

The filtering operation in wave number space can be applied even to simulations in physical space; in this case the filter $G(x, x')$ has the same expression as in Eq.(12.2) but in this case $X_c = \Delta x$ is the mesh of the grid. A Fortran code is furnished to understand the filtering procedure. In this code $filter.f$ in the directory $CODESCH12/FILT$ there is the possibility to perform a top-hat filter in spectral and in physical space. The filtering in wave number space is performed in the subroutine $FILSPE$ where the quantity $u(x_1, x_2, x_3)$ enters. After one real and two complex FFT calls (which have been explained in several chapters of this book) the complex quantity $\hat{u}(k_1, k_2, k_3)$ with $0 < k_1 < n_1$, $0 < k_2 < n_2/2 + 1$ and $0 < k_3 < n_3$ is obtained. Where n_i are the numbers of intervals along each direction x_i. If the filter wave number is $k_{ic} = k_i/n_c$, the spectral filter consists of setting

$$\hat{u}(k_1, k_2, k_3) = 0 \begin{cases} for \;\; k_{1c} < k_1 < n_1, 0 < k_2 < n_2/2 + 1, 0 < k_3 < n_3 \\ for \;\; 0 < k_1 < n_1, k_{2c} < k_2 < n_2/2 + 1, 0 < k_3 < n_3 \\ for \;\; 0 < K_1 < n_1, 0 < k_2 < n_2/2 + 1, k_{3c} < k_3 < n_3 \end{cases} \qquad 12.3$$

It is important to remember that, on coding the cutting, it should be taken into account how the wave numbers are stored in the FFT routines.

The top-hat filtering in physical space is done by applying three subsequent one-dimensional filters

$$\overline{f}(i)^n = \frac{1}{N}\Sigma_{l=-(n-1)/2}^{(n-1)/2} f(i+l)2^{n-|l|-1} \qquad with \;\; N = \Sigma_{l=-(n-1)/2}^{(n-1)/2} 2^{n-|l|-1} \qquad 12.4$$

This filter holds for uniform meshes at the moment the analysis is restricted only to equally spaced grids, since the filtering is done in the homogeneous directions.

To understand how the spectrum changes after the filter application the three-dimensional spectrum of a velocity field with random phases and with wave number amplitudes with a $k^{-5/3}$ decay is considered. In Fig.12.1 (solid line) the spectra of the filtered velocity in wave number space lie on the original spectrum. On the other hand, the physical filters of order n_c equal to 2, 4 and 8 show that the filtering operation does not only cut the wave numbers greater than k_c, but also reduces the energy content at scales greater than the filtered scales. From this figure it could be thought that the spectral filter should be used even in finite-difference simulations. On the other hand the results show that it is better to use the filter in physical space in finite difference simulations. An important property of these filters is that the filtered velocity field is solenoidal.

12.3. Equations

The filtered incompressible momentum equations are

$$\frac{\partial \overline{u_i}}{\partial t} + \frac{\partial \overline{u_j u_i}}{\partial x_j} = -\frac{\partial \overline{p}}{\partial x_i} + \frac{1}{Re}\frac{\partial^2 \overline{u_i}}{\partial {x_j}^2} \qquad 12.5$$

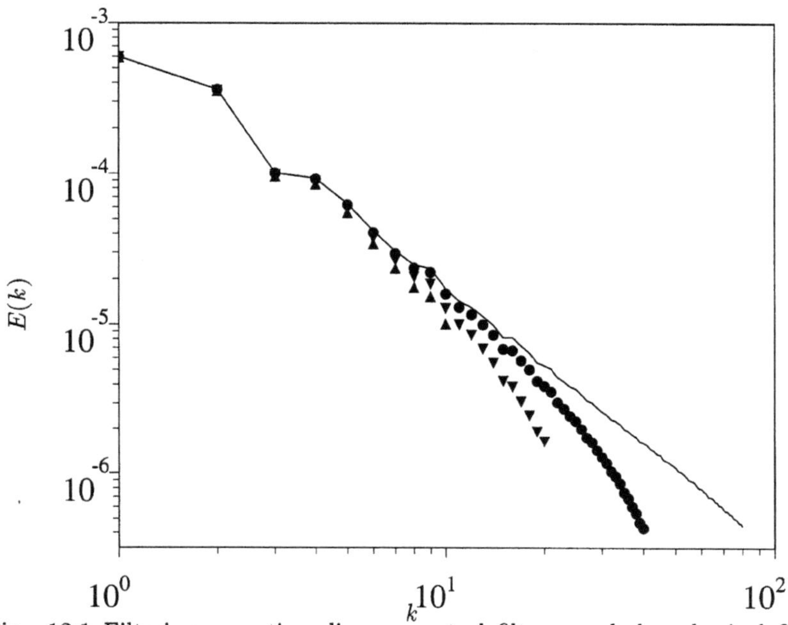

Fig. 12.1 Filtering operation, lines: spectral filter, symbols: physical filter, ▪ $n_c = 2$, $\nabla n_c = 4$, ▵ $n_c = 8$.

In these equations the filtering introduces the tensor $\overline{u_j u_i}$ that should be expressed in the unknowns $\overline{u_i}$. In the early papers on LES this term was expressed by decomposing the field u_i into the filtered $\overline{u_i}$ and the residual u_i' components as

$$u_i = \overline{u_i} + u_i' \qquad 12.6$$

The stress in the advection term becomes

$$\overline{u_j u_i} = \overline{\overline{u_i}\,\overline{u_j}} + R^*_{ij} \qquad 12.7$$

where R^*_{ij} are the residual stresses

$$R^*_{ij} = \overline{u_i'\, u_j'} + \overline{u_i'\, \overline{u_j}} + \overline{u_j'\, \overline{u_i}} \qquad 12.8$$

Since usually the advection terms are expressed by $\frac{\partial \overline{u_i}\,\overline{u_j}}{\partial x_j}$, on the right hand side of Eq.(12.7) is added and subtracted the quantities $\overline{u_i}\,\overline{u_j}$ and it becomes

$$\overline{u_j u_i} = \overline{u_i}\,\overline{u_j} + L_{ij} + C_{ij} + R_{ij} \qquad 12.9$$

where

$$C_{ij} = \overline{\overline{u_i}\, u_j'} + \overline{u_i'\, \overline{u_j}} \qquad , R_{ij} = \overline{u_i' u_j'} \qquad 12.10$$

and

$$L_{ij} = \overline{\overline{u_i}\,\overline{u_j}} - \overline{u_i}\,\overline{u_j} \qquad 12.11$$

are the Leonard stresses. Kwak (1975) performed an analysis of the Leonard stresses and he found that they are second order accurate when a top hat filter in physical space is used. Since in finite differences the scheme is second order accurate, the Leonard stresses can be neglected and the stresses $Q_{ij} = C_{ij} + R_{ij}$ are modelled. If the trace of the tensor Q_{ij} is subtracted the subgrid scale Reynolds stress to be modelled is

$$\tau_{ij} = \overline{u_j u_i} - \overline{u_i}\,\overline{u_j} = Q_{ij} - \frac{1}{3}Q_{kk}\delta_{ij} \qquad 12.12$$

The momentum equations to solve are then

$$\frac{\partial \overline{u_i}}{\partial t} + \frac{\partial \overline{u_i}\,\overline{u_j}}{\partial x_j} = -\frac{\partial P}{\partial x_i} + \frac{1}{Re}\frac{\partial^2 \overline{u_i}}{\partial {x_j}^2} - \frac{\partial \tau_{ij}}{\partial x_j} \qquad 12.13$$

$P = \overline{p} + \frac{1}{3}Q_{kk}$ is not the filtered pressure $\overline{p}$; if the real pressure is requested, this relationship should be used in the post-processing.

12.4. Subgrid models

12.4a Smagorinsky model for the residual stress

As mentioned before the LES is based on the assumption that the small eddies are universal and that their effect is negligible in the total turbulent energy production. Their role is thus limited to give the satisfactory rate of dissipation that is concentrated at small scales. From this consideration it turns out that a fairly crude model should be sufficient at least in LES of homogeneous turbulence at moderate R_λ.

The subgrid model most frequently used is based on the eddy viscosity assumption and it was derived by Smagorinsky (1963) for geophysical flows, that is flows at very large Reynolds numbers. The residual stress is related to the strain rate tensor of the filtered field by

$$\tau_{ij} = 2\nu_T \overline{S}_{ij} \qquad 12.14$$

This is a sort of closure that was suggested by Boussinesq (1870) where ν_T is not a property of the fluid but of the flow. It is interesting to read at Pg.46-48 of the Boussinesq paper that he almost closed the equations as Smagorinsky did almost 100 years later. Boussinesq furnished also a description of the near wall region of a channel not too far from that obtained by the DNS. To determine the eddy viscosity expression Smagorinsky (1963) assumed that in the budget of the subgrid scale kinetic energy the production is balanced by the rate of dissipation. The production $P = -\overline{u_j' u_i'}\,\overline{S}_{ij}$ and the rate of dissipation ϵ by dimensional analysis is proportional to $(\overline{u_i' u_i'})^{3/2}/l$. By supposing that the eddy viscosity is proportional to a characteristic length and velocity of the small scales ($\nu_T = cl(\overline{u_i' u_i'})^{1/2}$), the expression of ν_T is

$$\nu_T = l^2\sqrt{2\overline{S}_{ij}\overline{S}_{ij}} \qquad 12.15$$

The length l is a scale representing the small eddies and it must be related to the filter size, that is to the smallest box of the simulation. Several expressions have been used for l; that widely used is

$$l = C_S(\Delta x_1 \Delta x_2 \Delta x_3)^{1/3} \qquad 12.16$$

If the cutoff is within the inertial sub-range Lilly (1987) related the Smagorinsky constant to the Kolmogorov constant $C_K = 1.4$:

$$C_S \approx \frac{1}{\pi}\left(\frac{3C_K}{2}\right)^{-3/4} \qquad 12.17$$

This expression yields $C_S \approx 0.18$. Different values have been used depending on the flow and on the numerical method. The results are sensible to the changes of this constant as is shown later on.

As mentioned before the Smagorinsky model worked well for isotropic turbulence. When it was applied to inhomogeneous flows and in particular to wall bounded flows the constant was modified, as was done by Moin & Kim (1982) and by Piomelli *et al.* (1987). These deficiencies of the Smagorinsky model promoted the research to find models feasible to practical flows.

12.4b Structure function model

Metais & Lesieur (1992) in the simulation of isotropic turbulence and in the presence of a passive scalar introduced the structure functions model, claiming that the predictions of the compensated spectra were better than those by the Smagorinsky model. See Fig.2 in their review article, where they recalled the basic ideas from which the model is derived. From the consideration that an eddy viscosity in wave number space should have a cusp near the cutoff wave number K_C, they gave the following expression of the eddy viscosity in physical space

$$\nu_T(\mathbf{x}, \Delta x) = \frac{2}{3} C_K^{-3/2} \left[\frac{E_x(K_c)}{K_C}\right]^{1/2} \qquad 12.18$$

They related the local energy spectrum to the local intermittency of turbulence, that is linked to the second-order velocity structure functions. They claimed that the eddy viscosity is thus reduced in those regions where small-scale turbulence was absent. It is not clear from the paper whether they evaluated the longitudinal or the transversal structure functions. The structure function requires a local statistical average

$$F_2(\mathbf{x}, \Delta x) = \langle ||\overline{\mathbf{u}}(\mathbf{x}, t) - \overline{\mathbf{u}}(\mathbf{x} + \mathbf{r}, t)||^2 \rangle_{||\mathbf{r}|| = \Delta x} \qquad 12.19$$

hence they evaluated the structure function at each grid point by averaging the velocity differences between $\mathbf{x}$ and the six closest surrounding points. The expression for the eddy viscosity is

$$\nu_T(\mathbf{x}, \Delta x) = 0.105 \Delta C_K^{-3/2} F_2(\mathbf{x}, \Delta x)^{1/2} \qquad 12.20$$

Since there was an uncertainty how to evaluate the structure functions, in the present scheme based on staggered velocities with the eddy viscosity at the center of the cell, the longitudinal structure functions at the six cells around the cell at $I+\frac{1}{2}, J+\frac{1}{2}, K+\frac{1}{2}$ were evaluated. The average of these six values furnished the structure function at this position. In isotropic turbulence the continuity equation permits to perform the averages only by 12 independent quantities.

The evaluation of the transversal structure functions is neglected to avoid interpolations. Moreover in isotropic turbulence the transversal are linked to the longitudinal structure functions hence the longitudinal is sufficient. Someone argued (see the discussion at the AGARD conference Application of Direct and Large Eddy Simulation to Transition and Turbulence) that the structure function essentially coincides with the Smagorinsky model. In the debate at the conference it turned out that this assessment was partially true and that the model did not produce an improvement of the treating of inhomogeneous flows and in particular in the near wall region of wall-bounded flows.

My opinion is that to account for small scale turbulence it is important to introduce derivatives (Smagorinsky) or differences (structure functions) that are quantities linked to small scales.

12.4c Dynamic subgrid model

To overcome the lacks of the previous models that necessitate corrections of the constant for inhomogeneous flow, the dynamic model received a large success the last few years and, as mentioned, it gave a new impetus to the applications and development of new strategies for LES. In this model the constant C_S is not assigned, but is computed during the simulation from the flow variables. The turbulent viscosity is expressed by the Smagorinsky model, Eq.(12.15), but l^2 is dynamically determined as follows.

If a test filter of width $\widetilde{\Delta}$, larger than the computational filter, is applied to the momentum equations, a residual stress corresponding to this new filter, $T_{ij} = \widetilde{\overline{u_j u_i}} - \widetilde{\overline{u}}_i \widetilde{\overline{u}}_j$, can be defined. This quantity is similar to the quantity Q_{ij} that was modeled by the Smagorinsky model. Introducing the resolved stress

$$\mathcal{L}_{ij} = \widetilde{\overline{u_i}\,\overline{u_j}} - \widetilde{\overline{u_i}}\,\widetilde{\overline{u_j}} \qquad 12.21$$

Germano *et al.* (1991) introduced an identity relating the stresses corresponding to the two filters to the resolved stress

$$\mathcal{L}_{ij} = T_{ij} - \widetilde{\tau_{ij}} \qquad 12.22$$

The right hand side of Eq.(12.22) can be expressed by means of the Smagorinsky model. If $|S| = \sqrt{2S_{ij}S_{ij}}$, it follows

$$\mathcal{L}_{ij} = -\left(2C_S\widetilde{\Delta}^2|\widetilde{\overline{S}}|\widetilde{\overline{S}}_{ij} - 2\Delta^2 C_S\widetilde{|\overline{S}|\overline{S}_{ij}}\right) \qquad 12.23$$

using the questionable hypothesis that C_S does not depend on the filter width, Eq.(12.23) becomes

$$\mathcal{L}_{ij} = -2C_S\left(\widetilde{\Delta}^2|\widetilde{\overline{S}}|\widetilde{\overline{S}}_{ij} - \Delta^2\widetilde{|\overline{S}|\overline{S}_{ij}}\right) = -2C_S\Delta^2\mathcal{M}_{ij} \qquad 12.24$$

In this expression Δ^2 does not depend on the filter and $\mathcal{M}_{ij} = \left(\widetilde{\Delta}^2/\Delta^2|\widetilde{\overline{S}}|\widetilde{\overline{S}}_{ij} - \widetilde{|\overline{S}|\overline{S}_{ij}}\right)$. A contraction permits to evaluate the constant C_S. A least square technique, suggested by Lilly (1992), gives a positive quantity at the denominator, to avoid singular values for the eddy viscosity. The expression for the constants is

$$C_S(\mathbf{x},t) = -\frac{1}{2\Delta^2}\frac{\mathcal{L}_{ij}\mathcal{M}_{ij}}{\mathcal{M}_{ij}\mathcal{M}_{ij}} \qquad 12.25$$

The constant is evaluated at each grid point, and it could be positive or negative. The positive values are linked to energy flowing from large to small scales and the negative to energy going from small to large scales. The possibility of a backward energy transfer, that is when C_S is negative, is the interesting point of the dynamic model. This mechanism occurs in a large part of inhomogeneous flows, for instance near a solid wall where small and large scales have approximately equal sizes, and hence the small scales induce a flow of energy towards the large scales.

Averages in the homogeneous direction eliminate the large part of spatial variations of the constant and leave only the variation in the inhomogeneous direction. For isotropic decaying turbulence the averages are in the three directions and the constant varies with time. Its expression is

$$C_S(t) = -\frac{1}{2\Delta^2}\frac{\langle\mathcal{L}_{ij}\mathcal{M}_{ij}\rangle}{\langle\mathcal{M}_{ij}\mathcal{M}_{ij}\rangle} \qquad 12.26$$

The expression in Eq.(12.25) has also been used in isotropic turbulence and it was found that the constant was positive. This issue is a very interesting property of the contraction, which reproduces the physics of isotropic turbulence, where the energy is flowing from large to small scales. This consideration is linked to the conjecture of Kolmogorov about local interactions, which in wave number space is expressed by triadic interactions among wave numbers of comparable size. Non-locality occurs, but the net transfer is related to local interactions as was found by EDQNM closure (Crocco & Orlandi (1986)) and later on confirmed by Domaradzki & Rogallo (1990) by means of DNS data of isotropic turbulence.

The lack of the dynamic model, from the conceptual point of view, is related to the Smagorinsky model that implies that turbulent stresses and mean rates of strain are aligned. A-priori tests of LES performed by data sets of direct simulations (Clark *et al.* (1979)) and by experimental measurements at high Reynolds numbers (Liu *et al.* (1994)) show that the alignment is poor. Even if there is little correlation between the real and the modelled stresses, Menevau (1994) derived a set of necessary conditions for the modelled stress that make the LES statistics possible. He showed that in isotropic turbulence eddy viscosity closures do follow these conditions satisfactorily well. Models that have the property of high correlation between the real and the modelled stresses are based on the scale similarity hypothesis.

12.4d Similarity scale models

The small scales in a simulation are similar to the smaller scales of the resolved grid that must be modelled. This concept is connected to the constance of the energy transfer in the inertial range, where the cutoff wave number is usually located. This concept was used by Bardina *et al.* (1983) to derive the similarity model

$$\tau_{ij} = -(\overline{u_i}\,\overline{u_j} - \overline{\overline{u_i}}\,\overline{\overline{u_j}}) \qquad 12.27$$

This model has been shown to exhibit considerable correlation with the real stress during a-priori tests; however, when implemented in the simulation, it produced a rate of dissipation not sufficiently high to reproduce the exact energy decay rate. Bardina *et al.* (1983) thus had to add the usual eddy viscosity term to have the right energy decay. Liu *et al.* (1994) performed measurements at the center of a jet to investigate whether the low correlation, from DNS data, was linked to the low Reynolds number; they found that even at high Reynolds numbers the low correlation persists. The similarity model, without any addition of the eddy viscosity, gave more dissipation at low wave numbers and not the right dissipation at high wave numbers.

Mixed dynamic models have also been proposed (Zang & Street (1993), Salvetti & Banerjee (1995)), in which the SGS stress tensor is expressed as a linear combination of an eddy-viscosity part and of the Leonard tensor. One or both of the coefficients of this combination are computed by the dynamic procedure. Mixed models have shown better behavior than eddy-viscosity models in a priori tests and in LES with low order schemes for anisotropic turbulent flows and bounded turbulent flows (Zang & Street (1993), Salvetti *et al.* (1997)).

To account for this insufficient dissipation in the similarity models promising improvements were recently made by Domaradzki & Saiki (1997) and Shah & Ferziger (1996) by modifications in the filtering, respectively in wave number and in physical space. This new filtering produces models that behave satisfactorily well in inhomogeneous flows. These models do not require any constant, and are based on the application of differential operators to quantities evaluated in the simulation. These modifications in the filtering procedure, essentially, do not do anything else than introduce differentiations, which as mentioned, are the driving factors to get the right dissipation at high wave numbers in eddy viscosity models. In the filtering first and second derivatives appear hence the filtering requires the solution of large band matrices that are simplified by approximate factorizations, encountered before in the description of the numerical scheme. As expected the amount of computational time increases.

I am not inserting in this code the similarity models; I prefer instead to show that the eddy viscosity models are satisfactory to describe flows with a passive scalar or flows where density or temperature variations play a substantial role.

12.5 Equations for density-Related problems

In geophysical applications density variations are very important, for example near the coast where a stable stratification occurs. Military interests related to the motion of submarines near the coasts are promoting LES research of stable stratified turbulent flows to predict the evolution of wakes generated by submerged bodies. Chomaz *et al.* (1993) performed an experiment to study this phenomenon. However, even for industrial problems, as those related to the fishing industry, it is important to predict turbulent flows near the coast. Under the most general conditions the momentum equations are linked to an equation representing the motion of density or temperature. The equations for such flows can be adimensionalized in several ways. Batchelor *et al.* (1992) introduced the following dimensionless quantities

$$\theta^* = \frac{\rho' g l_0}{\rho_0 U_0^2}, \qquad t^* = t\frac{U_0}{l_0}, \qquad u^* = u/U_0, \qquad x^* = x/l_0 \tag{12.28}$$

If the density gradient and the gravity are directed along x_3, the momentum equations (neglecting the $*$) become

$$\frac{\partial \overline{u_i}}{\partial t} + \frac{\partial \overline{u_i}\,\overline{u_j}}{\partial x_j} = -\frac{\partial P}{\partial x_i} + \frac{1}{Re}\frac{\partial^2 \overline{u_i}}{\partial {x_j}^2} - I_g \overline{\theta} - \frac{\partial \tau_{ij}}{\partial x_j} \tag{12.29}$$

$I_g = 1$ for buoyant flows; if $\overline{\theta}$ is a passive scalar $I_g = 0$. The equation for the density is

$$\frac{\partial \overline{\theta}}{\partial t} + \frac{\partial \overline{\theta}\overline{u_j}}{\partial x_j} = \frac{1}{ReSc}\frac{\partial^2 \overline{\theta}}{\partial {x_j}^2} - Ri\overline{u_3} - \frac{\partial \overline{u_j \theta}}{\partial x_j} \tag{12.30}$$

Where Re and Sc are the usual Reynolds and Schmidt numbers and $Ri = (\frac{N l_0}{U_0})^2$ is the Richardson number related to the Brunt Waisala frequency $N^2 = \overline{\frac{\partial \rho}{\partial x_3}} g/\rho_0$, where $\overline{\frac{\partial \rho}{\partial x_3}}$ is the mean density stratification. These equations were solved by Batchelor *et al.* (1992) for several conditions with the intent to investigate the scaling laws at low wave numbers in different conditions. The Rogallo (1980) spectral code together with the Kraichnan (1976) eddy viscosity models were used by Batchelor *et al.* (1992); on the contrary, here these equations are solved with the Smagorinsky model together with finite differences. The aim is to show that the present model predicts satisfactory well the inertial ranges for the kinetic and thermal energy.

In presence of the thermal field the numerical scheme, for $Ri = 0$ and with $I_g = 1$, should the conserve the total energy in the discrete space and in the inviscid limit. That is, it should reproduce the transformation of potential energy in kinetic energy and vice-versa. The energy conservation thus enforces, in the staggered velocities layout, definition of the density $\overline{\theta}$ at the same locations as $\overline{u_3}$. The user could verify that a different location of $\overline{\theta}$ (for instance at the center of the cell) disrupts the energy conservation.

12.6 Subgrid model for thermal flows

The thermal field has an effect on the turbulent velocity field in the presence of buoyancy and mean stratification. Since these flows are important in geophysical problems, their large Reynolds number requires LES. In the filtered equations it is necessary to express not only the stress tensor τ_{ij} but also the temperature flux $\overline{u_j \theta}$. The Smagorinsky model was corrected by Lilly (1962) to account for turbulence production due to buoyancy. The turbulent eddy viscosity and the turbulent eddy diffusivity, in dimensionless units, were defined as

$$\nu_T = C_S \overline{\Delta}^2 \sqrt{(2\overline{S}_{ij}\overline{S}_{ij} - \frac{1}{Pr_T}\frac{\partial \overline{\theta}}{\partial x_3})}$$

$$\nu_\theta = \frac{C_S}{Pr_T}\overline{\Delta}^2 \sqrt{(2\overline{S}_{ij}\overline{S}_{ij} - \frac{1}{Pr_T}\frac{\partial \overline{\theta}}{\partial x_3})} \tag{12.31}$$

In several simulations (Wong & Lilly (1994), Siegel & Domaradzki (1994)) these expressions were used by imposing C_S and Pr_T. The critical point was the choice of the turbulent Prandtl number (see Herring *et al.* (1982)).

Thus for thermal flows it will be very useful if both constants (C_S and Pr_T) could be evaluated dynamically . Cabot (1992), Sullivan and Moeng (1992) and Bohnert & Ferziger (1993) extended the dynamic model, in section 12.4c, to these flows, and, because of the interdependence between the eddy viscosity and diffusivity through the Pr_T number, iterative schemes were necessary to calculate C_S and Pr_T. As Wong & Lilly (1994) reported, in certain circumstances it was impossible to reach a convergence. To avoid these occurrences, Wong & Lilly (1994) proposed a simplified version of the dynamic model that does not require the iterative procedure. Instead of deriving the expression of the viscosity by assuming that the rate of dissipation ϵ equilibrates the energy production, they used the scaling arguments of Kolmogorov in his 1942 theory to express the eddy viscosity in terms of a length scale, the grid-filter width $\overline{\Delta}$, and the rate of dissipation ϵ. They got the relation

$$\nu_T = C_L \overline{\Delta}^{4/3} \epsilon^{1/3} \qquad 12.32$$

The constant is indicated by C_L since it is different from the Smagorinsky constant C_S. They assumed also that ϵ is not affected by the filter width, thus ϵ can be included in the constants. The expressions of the eddy viscosity and diffusivity are

$$\nu_T = C_\epsilon \overline{\Delta}^{4/3}, \qquad \nu_\theta = \frac{C_\epsilon}{Pr_T} \overline{\Delta}^{4/3} \qquad 12.33$$

With these expressions the dynamic model is largely simplified. Eq.(12.22) and Eq.(12.23) hold but Eq.(12.24) becomes

$$\mathcal{L}_{ij} = -2C_\epsilon \overline{\Delta}^{4/3} \mathcal{M}_{ij} \qquad 12.34$$

with $\mathcal{M}_{ij} = [1 - \left(\frac{\widetilde{\overline{\Delta}}}{\overline{\Delta}}\right)^{4/3}] \widetilde{\overline{S_{ij}}}$. Due to incompressibility $\overline{S_{kk}} = 0$. To have a positive definite denominator to overcome singularities the contraction of $\mathcal{L}_{ij}$ is done by multiplying Eq.(12.34) by $\widetilde{\overline{S_{ij}}}$. In flows with three homogeneous directions, the constant, in this case the eddy viscosity, is calculated by a volume averaging. The eddy viscosity is constant in space and varies in time; it is

$$\nu_T = \frac{\langle \mathcal{L}_{ij} \widetilde{\overline{S_{ij}}} \rangle}{[1 - \left(\frac{\widetilde{\overline{\Delta}}}{\overline{\Delta}}\right)^{4/3}] \langle \widetilde{\overline{S_{ij}^2}} \rangle} \qquad 12.35$$

The analogous of $\mathcal{L}_{ij}$ for the heat fluxes is

$$\mathcal{R}_{\theta i} = T_{\theta i} - \widetilde{\tau_{\theta i}} = \widetilde{\overline{\theta} \overline{u_j}} - \widetilde{\overline{\theta}} \widetilde{\overline{u_j}} \qquad 12.36$$

The filtered temperature fluxes and the temperature fluxes at the computational level are related to the resolved temperature field by

$$T_{\theta i} = -\frac{C_\epsilon}{Pr_T}\tilde{\overline{\Delta}}^{4/3}\widetilde{\frac{\partial\overline{\theta}}{\partial x_i}}, \qquad \tau_{\theta i} = -\frac{C_\epsilon}{Pr_T}\overline{\Delta}^{4/3}\frac{\partial\overline{\theta}}{\partial x_i} \tag{12.37}$$

By substituting these quantities in Eq.(12.36) it follows

$$\nu_\theta[1 - \left(\frac{\tilde{\overline{\Delta}}}{\overline{\Delta}}\right)^{4/3}]\frac{\partial\tilde{\overline{\theta}}}{\partial x_i} = \mathcal{R}_{\theta i} \tag{12.38}$$

The contraction of the denominator permits to evaluate the eddy diffusivity which after a volume average becomes

$$\nu_\theta = \frac{\langle \mathcal{R}_{\theta i}\widetilde{\frac{\partial\overline{\theta}}{\partial x_i}}\rangle}{[1 - \left(\frac{\tilde{\overline{\Delta}}}{\overline{\Delta}}\right)^{4/3}]\langle\widetilde{(\frac{\partial\overline{\theta}}{\partial x_i})^2}\rangle} \tag{12.39}$$

The interesting feature of this model is that even if the eddy viscosity and diffusivity are constant in space, these are function of the computed quantities and it is not necessary to assign any value to the constants. There are difficulties to simulate flows without any initial velocity; thus during a short transient the Smagorinsky model can be used.

This model as it was shown later on produced results not too encouraging. However, following the ideas lying at the base of this model a slightly different version has been derived with constant eddy viscosity and diffusivity in space but time-dependent. This simplified model (indicated here as SDM) was also tested by Love (1980) for the Burgers equation but was never introduced for the Navier-Stokes equations. This dynamic model, instead of considering the Smagorinsky eddy viscosity and diffusivity with the right hand side varying in space, averages the quantity in the square root of Eq.(12.31). Only positive contributions are considered in averaging. In this term appears the turbulent Prandtl number that is the unknown of the problem. Since Pr_T should not have large variations in time with a good approximation, the value of Pr_T at the previous time step can be assumed. If

$$\overline{\Pi} = \sqrt{\langle(2\overline{S}_{ij}\overline{S}_{ij} - \frac{1}{Pr_T}\frac{\partial\overline{\theta}}{\partial x_3})\rangle} \tag{12.40}$$

the eddy viscosity and diffusivity are

$$\nu_T = C_S\overline{\Delta}^2\overline{\Pi} \qquad \nu_\theta = \frac{C_S}{Pr_T}\overline{\Delta}^2\overline{\Pi} \tag{12.41}$$

which are constant in space. The filtered turbulent stresses in Eq.(12.22) are

$$T_{ij} = -2C_S\tilde{\overline{\Delta}}^2\tilde{\overline{\Pi}}\ \tilde{\overline{S}}_{ij} \qquad \tilde{\tau}_{ij} = -2C_S\overline{\Delta}^2\widetilde{\overline{\Pi}\,\overline{S}_{ij}} \tag{12.42}$$

The expression for $\mathcal{L}_{ij}$ becomes

$$\mathcal{L}_{ij} = 2\nu_T(1 - \frac{\widetilde{\overline{\Delta}}^2\widetilde{\overline{\Pi}}}{\overline{\Delta}^2\overline{\Pi}})\widetilde{\overline{S_{ij}}} \qquad 12.43$$

A contraction similar to that used to get Eq.(12.35), this time by $\widetilde{\overline{S_{ij}}}$, permits to calculate the eddy viscosity.

The temperature fluxes at the two filter levels are, in a similar way,

$$T_{\theta i} = -\frac{C_S}{Pr_T}\widetilde{\overline{\Delta}}^2\widetilde{\frac{\partial\overline{\theta}}{\partial x_i}} \qquad \widetilde{\tau_{\theta i}} = -\frac{C_S}{Pr_T}\overline{\Delta}^2\widetilde{\frac{\partial\overline{\theta}}{\partial x_i}} \qquad 12.44$$

which, when substituted in Eq.(12.36), give

$$\mathcal{R}_{\theta i} = \nu_\theta(1 - \frac{\widetilde{\overline{\Delta}}^2\widetilde{\overline{\Pi}}}{\overline{\Delta}^2\overline{\Pi}})\widetilde{\frac{\partial\overline{\theta}}{\partial x_i}} \qquad 12.45$$

The eddy diffusivity is obtained by a contraction with $\widetilde{\frac{\partial\overline{\theta}}{\partial x_i}}$.

If the results confirm the validity of these simplifications, from the physical point of view, this simplification suggests that in LES of homogeneous decaying flows the eddy viscosity should account for the total turbulent dissipation, and that the fluctuations of the local dissipation are not important. For a better clarification the intermittency of the turbulent dissipation rate should not play a large effect in LES. From the numerical point of view the computational effort is reduced, the extra terms in the right hand side of Eq.(12.13) can be eliminated and the same code for DNS can be used for LES. The only modification consists on adding a routine that evaluates the time variable eddy viscosity and diffusivity. In the code furnished in this book this numerical simplification is not inserted because the same code was used to test the different subgrid models. The user can easily reduce the complexity of the code if he is interested in reducing the computational time.

12.7 Code informations

The code in this chapter is derived from the code described in chapter 8 for DNS of isotropic turbulence. The principal change consists of the introduction of the eddy viscosity in the solution of $\hat{q}_i$. In addition there is the possibility to solve a further equation for σ that can be either a passive scalar or the temperature. The velocity field can be a function of the temperature through the buoyancy term. Different subgrid models can be used. The code $ISOLES$ encloses several files with names of which the first part indicates that the code is solving isotropic flows by LES and the second part indicates different parts of the code. The code is in the directory $CODESCH12/ISOLES$. A brief description of each part follows.

12.7a Description of the file isolesma.f

In this file there is the main program where the input file *isoles.d* is read. In this file there are the parameters necessary to assign the initial energy and temperature spectra and other quantities as the number of points, the dimension of the box, the parameters indicating when the output files should be printed and those related to the time advancing scheme. These parameters are just used in the other codes.

The new parameters are those related to the type of subgrid model ($ics0$), and those indicating the kind of initial spectrum ($imic$). Furthermore there are indices indicating whether the evolution of σ is performed or not. For instance for $istrat = 0$ σ is not calculated. For $istrat = 1$ and $igrad = 0$ σ is a passive scalar, for $istrat = 1$ and $igrad = 1$ there is buoyancy, and if there is stratification the Richardson number is evaluated by $bvais$. The main structure of the code is the routine $SOLVE$ where there are the calls of several subroutines of which we mention in particular: that generating the mesh ($MESHES$), that with the definition of the indices necessary to express the derivatives ($INDIC$), the routine $PHINI$ that evaluates the coefficients for the solution of the "pressure" Φ. The code can start from $t = 0$ by assigning the spectra, or it can start from a previous simulation by reading in $INIREA$ the field written in $CONTWR$. The time evolution is done in the routine $TSCHEME$, and some of the global quantities are written in the routines with names starting with OUT.

The different subgrid models to evaluate the eddy viscosity and diffusivity are called before the routine $TSCHEME$.

12.7b Description of the file isolesnn.f

In this file in the routine $INDIC$ are stored the indices to express the first and second derivatives. The coordinates with uniform meshes, defined in the routine $MESHES$, are calculated in the routine $COORDI$. In this file the divergence of the velocity field is calculated in the routine $DIVG$; a similar routine permits to check whether the field is solenoidal ($DIVGCK$). The CFL number is evaluated by the routine CFL; it is necessary as a check or to perform simulations at constant CFL. In this file there are the routines $TAVER$, $TAVWRI$ and $TAVREA$ that are necessary to perform time averages of certain quantities if a simulation leading to a statistical steady state is performed. In the simulations described here the turbulence evolves in time and hence these routines are not used.

Finally in the routine $UPDVP$ the second part of the fractional step to get the solenoidal velocity field is coded.

12.7c Description of the file isoleshn.f

In this file there are the routines $HDNLi$ evaluating the non-linear terms and the terms related to the subgrid stresses with cross derivatives. In each routine $HDNLi$ has been inserted the term $\frac{\partial}{\partial x_j}\nu_T\frac{\partial \overline{u_i}}{\partial x_i}$ (with $j \neq i$). In the routine $HDNL3$ the buoyancy force appears, since it has been assumed that gravity is parallel to x_3.

12.7d Description of the file isolestn.f

In this file there is the routine $TSCHEME$ where each time substep is performed by calculating first the non-linear terms for the velocity and the thermal fields. The successive call of the routines $INVTRi$ evaluates the non-solenoidal velocity field $\widetilde{\overline{u_i}}$. In these routines part of the subgrid stress $\frac{\partial}{\partial x_i}\nu_T\frac{\partial \overline{u_i}}{\partial x_i}$ is implicit. This choice leads to storage of the coefficients of the tridiagonal matrices of the approximate factorization in two-dimensional instead of in one-dimensional arrays, as in chapter 8. In the file the routine $PRCALC$ evaluates $\overline{p}$ from Φ. In LES the relationship between $\overline{p}$ and Φ is

$$\overline{p}^{l+1} = \overline{p}^{l} + \Phi^{l+1} - \alpha_l \frac{\Delta t}{2} \frac{\partial(\nu + \nu_T)\frac{\partial \Phi}{\partial x_i}}{\partial x_i} \qquad 12.46$$

In this file there is also the routine $INVISC$ to perform a simulation with the Euler equations to check whether the energy is conserved.

12.7e Description of the file isolestnsi.f

This file reports the solution of the σ equation; the non-linear terms are calculated in the routine $HDNL4$ and the transport equation is solved in the routine $INVTR4$. The routine $INVTI4$ permits to investigate whether all the potential energy is transformed in kinetic energy and thus to check whether the total energy is conserved, in the inviscid limit.

12.7f Description of the file isolesintu.f

In this file in the routine $TURBIS$ different choices for the initial energy spectrum can be done depending on the value of *imic*. From the spectrum, as in chapter 8, the velocity field with random phases is calculated. This field is not free divergence-free and hence it is projected onto a solenoidal field by the routine $UPDVP$ that is called in the routine $INITUR$ after a call of $DIVG$ and $PHCAL$. To be sure that there is no mean motion the mean velocities are evaluated and subtracted from the velocity field.

12.7g Description of the file isolesinsi.f

In this file the initial σ field is evaluated by an assigned spectrum by the same procedures used to get the three velocity fields. σ thus has random phases.

12.7h Description of the file isolesph.f

This file evaluates Φ as in chapter 8 for DNS. In fact, there is no difference between the Φ calculation in DNS or LES. In this file the $NCARFFTs$ routines are used.

12.7i Description of the file isolessp.f

In this file the routine $QPHYWA$ permits to transform any variable from the physical to the wave number space. With the output of this routine the 3D energy spectrum is evaluated in the routine $SPECTRE$, where the routine $ENEKCO$ evaluates the 3D energy spectrum for each component.

12.7j Description of the file isolesst.f

In this file the routine $STRPER$ evaluates the six components of the strain tensor s_{ij}, at the center of the cell.

12.7k Description of the file isolestr.f

In this file there are the subroutines relative to the inversion of periodic tridiagonal matrices. These differ from each other depending on the order of the indices of the two-dimensional matrices appearing in the routine $INVTRi$.

12.7l Description of the file isolesio.f

In this file there are the routines concerning the output of the results. The turbulent statistics of any order are calculated in the routine $ENERCA$. The spectra are written in the routine $OUTP$, while in the routine $OUTH$ the total energy, enstrophy, skewness, flatness of the velocity and of the derivatives are written. In the routine $SPEVIS$ the spectrum of ν_T is calculated as it is done for each velocity component.

12.7m Description of the file isolesiosi.f

In this file the quantities relative to σ are calculated and written. In $OUTRHO$ the total rms, skewness and flatness are written. In $OUPRHO$ the σ spectrum is evaluated and written; also here $RHPHWA$ transforms σ from the physical to the wave number space. The spectrum is evaluated in $SPERHO$.

12.7n Subgrid-models

Each subgrid model is evaluated in a different routine that is called by changing the value of the index $ics0$.

With $ics0 = 1$ the dynamic model is calculated by the routine $DYNAMIC$. This model is called only when σ is not calculated.

With $ics0 = 2$ the model (SDM) evaluates ν_T in $DYWOLMO$ and ν_θ in $SMARHO$.

With $ics0 = 3$ the Smagorinsky model gives ν_T in $SMAGO$ and ν_θ in $SMARHO$.

With $ics0 = 4$ $STRFUN$ evaluates ν_T by the structure function model and this holds only without σ.

With $ics0 = 5$ the constant ν_T and ν_θ are calculated in $CONSMA$ and in $COSMRH$.

12.8 Results

12.8a Grid turbulence at intermediate R_λ

One of the best documented flows in grid turbulence, at a sufficiently high R_λ, is that by Compte Bellot & Corrsin (1971). The energy spectra were evaluated at three different downstream positions after the grid. R_λ, evaluated by the Taylor hypothesis, decreased from 71.6 to 60 when the dimensionless time changed from $\frac{tU_0}{M} = 42$ to 171. Since in these simulations, by the procedure described in chapter 8, a velocity field with random phases is assigned at $t = 0$, it is necessary to wait some time to produce the physical structures leading to an exponential decay close to that measured by Compte-Bellot & Corrsin (1971). There are several ways, see e.g. Carati *et al.* (1995), to account for this transient. At this point it is not important to perform a real quantitative comparison with the experimental data, but to see whether the numerical scheme and the different subgrid models predict the correct exponent for the energy decay. To perform the DNS of the Compte Bellot & Corrsin (1971) flow a 256^3 grid is the minimum requirement, as has been shown recently by Mansour & Wray (1994). At this R_λ the inertial range does not have a large extension, hence the LES could be inaccurate since any of the s.g.s. models rely, more or less, on the assumption that the cutoff wave number is within the inertial range. However, since this experiment has been considered in the past as a test case to validate LES, this flow is also computed here. One of the first LES of this flow was performed by Kwak *et al.* (1975) by a fourth order finite difference scheme in a non-staggered grid, and they demonstrated

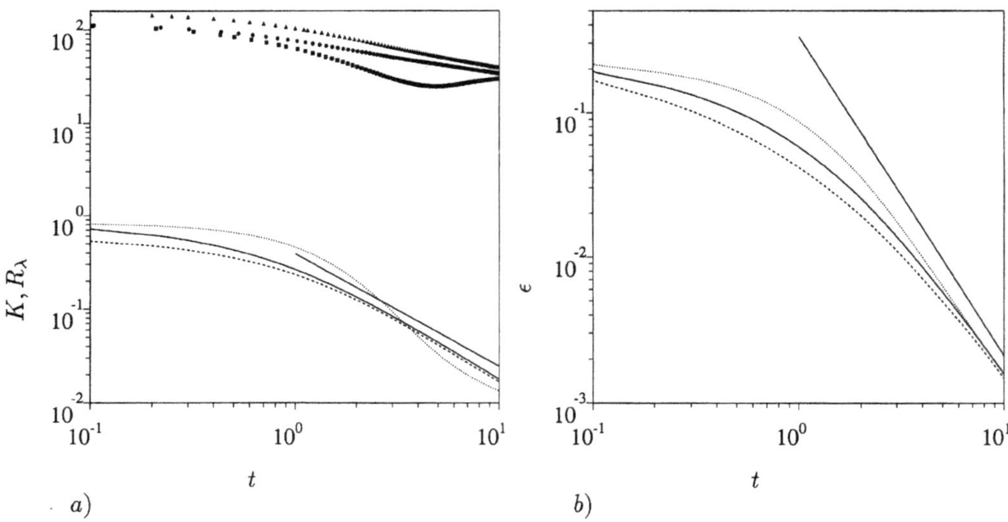

Fig. 12.2 a) Energy (lines) and R_λ (symbols) decay rate grid dependence: —— ○ 48^3, ---- △ 24^3 with Smagorinsky model and $C_S = 0.18$, ········ □ 24^3 no model ; b) effect of C_S on the decay of the energy dissipation rate ϵ: ········ $C_S = 0.13$, —— $C_S = 0.18$, and ---- $C_S = 0.23$.

that LES was an efficient way to get reasonable results. They claimed that the Leonard stress should be evaluated by a fourth order scheme, and that this term can be neglected by second order simulations, since truncation errors partially account for this term. They performed checks on resolution and on the values of the constant and in this chapter their procedure is repeated.

The simulations here have been performed on a 48^3 and a 24^3 grid and Fig.12.2a shows that in the range of $R_\lambda$between 71.6 and 50 of the Compte-Bellot & Corrsin (1971) experiment, the energy decay rate exponent is ≈ 1.2 which is in a good agreement with the experimental value. The values of R_λ for the smaller grid (24^3) are higher than those for the finer grid, in fact the insufficient wave numbers reduce more the dissipation than the energy. The second check has been done to investigate the dependence on the Smagorinsky constant by simulations with the constants set equal to $C_S = 0.13, 0.18$ and 0.23, that is, in the same range that was used in the past. As reported by Lesieur & Metais (1996), the value of C_S was always assumed smaller than the theoretical value $C_S = 0.18$. Fig.12.2b shows that the decay of the dissipation rate is more influenced by the value of the constant. The energy decay is not reported because the three curves do not differ and in the last period the decay exponents are $n \approx 1.2$.

The energy spectra, at three times show (in accordance with the previous considerations) in Fig.12.3a that a small constant reduces, to a smaller degree the amount of energy near the cutoff wave number. This behavior, however, does not give any indication about the dissipation of the numerical scheme. In chapter 8 it was proved that for $\nu = 0$ a k^2 spectrum

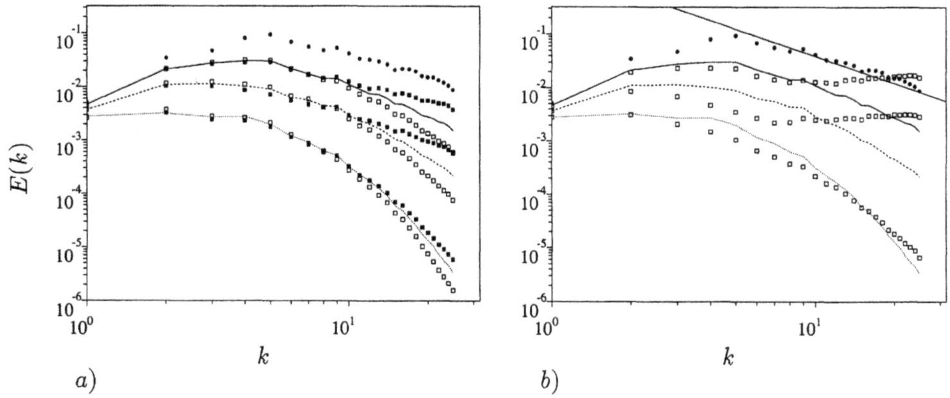

Fig. 12.3 a) Effect of C_S on the energy spectra lines: $C_s = 0.18$, open symbols: $C_S = 0.13$, closed symbols $C_S = 2.3$ b) Comparison between $C_S = 0.18$ and $C_S = 0$. The spectra are evaluated at $t = 0, 1, 3$ and $t = 10$.

was achieved. In this LES a simulation without model ($C_S = 0$) has been performed to show that for $\nu \neq 0$, such that $Re = 5000$, at high wave number there is no relevant dissipation; Fig.12.3b indeed shows that the spectra, from a certain k (depending on the viscous Reynolds number) rise. In the limit of zero viscosity the spectrum tends to a k^2 law. With $C_S = 0$ Fig.12.2a shows that the energy decays, initially with a reduced slope and then, at a certain time, with a more rapid slope; in the last part of the simulation the energy has the correct slope. This final decay has a correct slope because ν is large enough to influence the spectrum at high k.

The Smagorinsky model, the structure function and the dynamic model have been compared for grid turbulence. For the dynamic model it was observed that the filtering procedure, as seen in section 12.2, could be performed in physical or in wave number space. The value of the constant then should depend on the chosen filter. Fig.12.4a shows that the filter in physical space gives a value of the constant coinciding with the theoretical value $C_S = 0.18$. The filtering in wave number space gives a lower value for C_S because of the less dissipative nature of the wave number filtering. The energy spectra at the same time of that in Fig.12.3a shows in Fig12.4b the incorrect dissipation with the filter in wave number space. The comparison between the energy spectra, in Fig.12.4b, with the dynamic model confirms the value of the constant shown in Fig.12.4a. The filter in wave number space predicts in Fig.12.3b spectra with a reduced slope; these agree with those by the Smagorinsky model with $C_S = 0.13$. This consideration leads to the conclusion that when a numerical method is chosen all the differential operation should be congruent with those used to solve the Navier-Stokes equations. The present results agree with those obtained by Carati *et al.* (1995) by a pseudospectral method. It could be asserted, then, that at least for isotropic turbulence at intermediate Reynolds number, Smagorinsky and the dynamic model together with a second order accurate scheme with staggered velocities

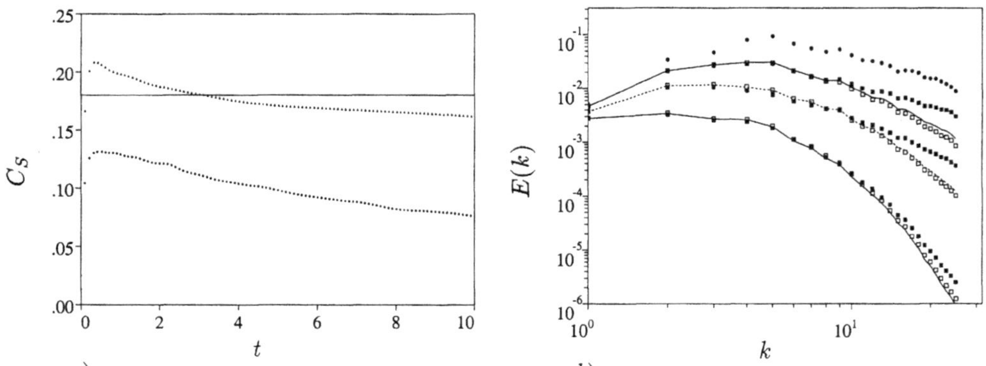

a) *b)*

Fig. 12.4 a) Time history of C_S in the dynamic model obtained with —— physical filter and ---- spectral filter b) Effect of the filtering procedure in the dynamic model on the energy spectra, lines: Smagorinsky model with $C_s = 0.18$, open symbols: spectral filter closed symbols: physical filter.

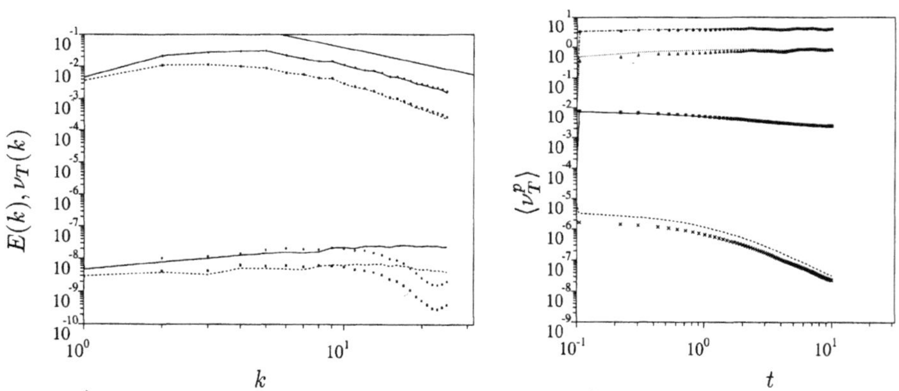

a) *b)*

Fig. 12.5 a) Comparison between Smagorinsky (lines) and structure function (symbols) for the energy and turbulent viscosity spectra b) Time evolution of eddy viscosity statistics, for Smagorinsky (lines), dynamic model (symbols); (—— , ▪) $p = 1$ mean, (---- ,×) $p = 1$ rms, (········ , ▵) $p = 3$ skewness, (—-— , ▪) $p = 4$ flatness.

give very satisfactory results.

The comparison in Fig.12.5a between the spectra from the Smagorinsky and from the structure function model confirms that in isotropic turbulence there are minor differences between the two models, in particular at high wave numbers. These differences perhaps are related to second order longitudinal structure functions that affect the eddy viscosity

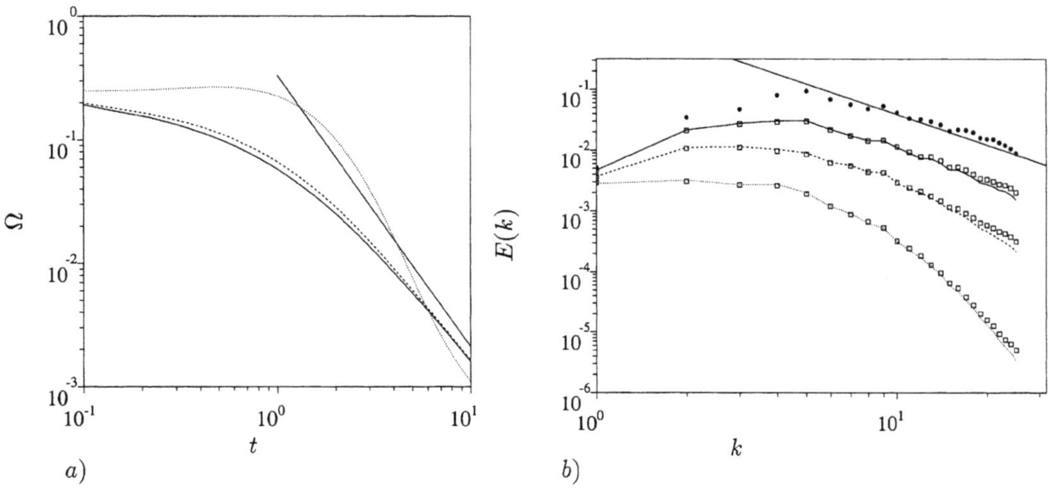

Fig. 12.6 a) Time evolution of the energy dissipation rate for the Smagorinsky model (———), constant eddy viscosity (- - - -) and no-model (········). b) Comparison between the energy spectra at $t = 1, 3$ and 10 with the Smagorinsky model and with a constant eddy viscosity.

at small scales. In Fig.12.5a together with the energy spectra at $t = 1$ and $t = 3$ is also reported the spectrum of the eddy viscosity in wave number space. To evaluate the last quantity, a FFT of $\nu_T(x_1, x_2, x_3)$ gives $\nu_T(k_1, k_2, k_3)$, and then, as done for each velocity component, the eddy viscosity three-dimensional spectrum is evaluated. Since the amplitude of the spectrum is low it can be concluded that the eddy viscosity has a peak at the zero wave number and that its r.m.s. is negligible. The difference between the two models, mainly near the cutoff, does not produce large differences in the spectra, as is shown by $E(k)$ in Fig.12.5a. The flat profile of $\nu_T(k)$ was also found by Siegel & Domaradzki (1994) by the Smagorinsky model in turbulence with stratification and buoyancy. In addition the structure function model, in the present version, gives a viscosity spectrum decreasing near the cutoff instead of increasing as expected. Lesieur & Metais (1996) in fact claimed that the structure function model should reproduce the profile of the Kraichnan (1976) eddy viscosity expression, often used in pseudospectral simulations.

The mean, the *rms*, the skewness and the flatness of the eddy viscosity have been evaluated to investigate whether ν_T is intermittent. This calculation was done for the Smagorinsky and for the dynamic model with physical filtering. Fig.12.5b indeed shows that the *rms* is small and that the eddy viscosity is not intermittent; in fact the flatness varies between 3.6 and 4.2 and moreover the skewness is positive with values around 0.8. Fig.12.5b shows that the statistics of the eddy viscosity evolves in the same way as the Smagorinsky and the dynamic model.

The small *rms* and the relative large mean value for ν_T suggest to investigate whether a

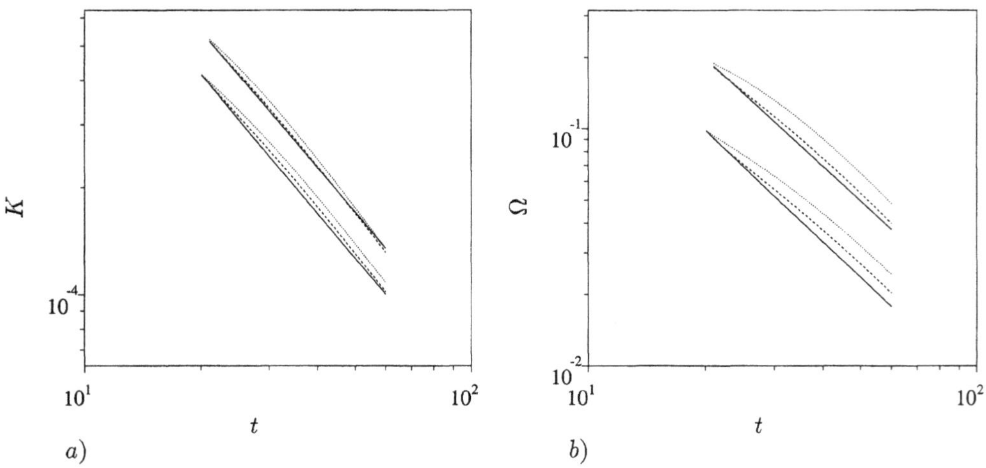

Fig. 12.7 Time evolution of a) energy and b) rate of dissipation for the dynamic model (——), Smagorinsky model (- - - -), constant eddy viscosity (········).

model with a constant eddy viscosity $\nu_T = C_S\epsilon(t)$ gives reasonable results for the decay of isotropic turbulence at this low Reynolds number. Fig.12.6a shows that indeed the constant eddy viscosity model gives a satisfactory decay of the energy dissipation rate. Since this quantity is related to the high wave number spectrum it should be expected that the energy spectra at different instants agree with those predicted by the Smagorinsky model; Fig.12.6b shows that this occurs indeed.

This last point is very important because this model was used by Wong & Lilly (1994) to derive a simple and efficient dynamic subgrid model to be applied in turbulent flows where the thermal field plays an important role. Before we analyze the LES of thermal fields it is worth to see how at much higher R_λ than those of the Compte-Bellot & Corrsin (1971) experiment (that is, when the inertial range extends for several decades) the models work well.

12.8b Isotropic turbulence at high R_λ

As an exercise to investigate the properties of the subgrid models it is important to perform simulations at very high values of R_λ, where a LES operates in the right conditions since the wavenumber cutoff is in the inertial range. The simulation was started with a field with the spectrum of Eq.(8.18) where $K_P = 25$ and $\sigma = 10$ and with random phase; the simulation with the dynamic model was evolving for 20 time units at $Re = 10^6$. The field at $t = 20$ has been assumed as the initial condition for three simulations with different subgrid models. One with the standard Smagorinsky, one with the eddy viscosity constant and proportional to the rate of dissipation and the last one with the dynamic model. In this way the simulations started from a condition with structures which produced an energy decay with a well defined power law. At this Reynolds number, to see whether the inertial

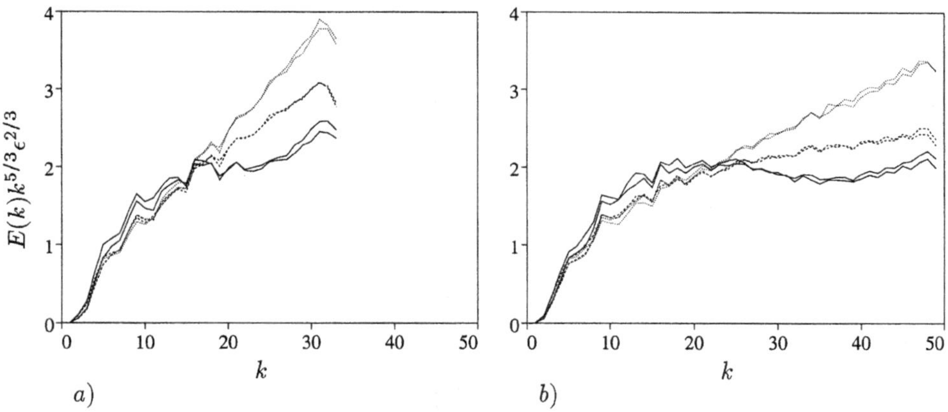

Fig. 12.8 Compensated energy spectra for a) 64^3 and b) 96^3 grid dynamic model (———), Smagorinsky model (- - - -), constant eddy viscosity(········) . Spectra at $t = 40$ and 60 are given.

range is in part predicted the simulations require a larger number of grid points than those in the previous sections. Two simulations were performed, one with a 64^3 and the other with a 96^3 resolution. In Fig. 12.7 the evolution of the total energy and rate of dissipation show that the model with constant ν_T in the first period dissipates less than the other two models. In the last period of the simulation the decay exponents of the turbulent energy for the 96^3 grid are $n = -1.29$, $n = -1.38$, and $n = -1.43$ for the dynamic, the Smagorinsky and for the constant ν_T models respectively.

A better picture of the capability of a subgrid model to produce the inertial range is obtained by plotting the compensate spectrum, defined as $E(k)k^{5/3}\epsilon^{2/3}$. The compensate spectra, at high wave numbers, give also the value of the Kolmogorov constant that recently was measured by Saddoughi & Veeravalli (1994) in an experiment at $R_\lambda \approx 600$, smaller than the present (which is approximately equal to 1800). The compensated spectra have been calculated at $t = 40$ and $t = 60$ and are shown in Fig.12.8 for two different resolutions and with the different subgrid models. Fig.12.8a shows that the 64^3 simulation is insufficient to produce even a very small inertial range since half of the wave numbers are necessary to describe the large scales of turbulence. The last part of the spectrum decays with a slope smaller that $k^{-5/3}$ and thus a value of the Kolmogorov constant can not be evaluated. On the other hand, the finer grid simulations show that the inertial range is predicted only with the dynamic model. The value of the Kolmogorov constant, $C_K \approx 1.85$, is in satisfactory agreement with the experimental values of 1.5. This is a very interesting property of the dynamic model that to my knowledge was never published, but that was investigated by Wray. In a private communication he told me that the reproduction of this property for him was a bug check of its pseudospectral code and of the model. I think, on the other hand, that this property is very important and in addition I am very glad that this result can be obtained by a finite difference code.

From the simulations at intermediate and high Reynolds numbers it turns out that the intermittent character of the eddy viscosity is not important at low Reynolds numbers but also that it is an important ingredient that must be included in a model. In fact Fig.12.8b shows that the Smagorinsky model gives a large improvement of the results obtained by a constant ν_T model. The dynamic model introduces a time variation of the constant that seems to be important in order to produce the right properties of the inertial range. The role played by the intermittency of the eddy viscosity was not largely studied in the past and perhaps it requires more attention.

12.8c Isotropic turbulence with a passive scalar

The evolution of a passive scalar in a turbulent field is very important for applications related to geophysical flows and in combustion. From the fundamental point of view it is interesting to investigate whether the thermal energy spectrum shows a scaling law as that of the velocity field. Lesieur & Rogallo (1989) investigated this flow and they found that, contrary to the theoretical findings, the thermal spectrum does not present an inertial-convective range with a $k^{-5/3}$ law, but instead it presents a k^{-1} range, resulting from the shearing of temperature by velocity gradients that act at wave numbers corresponding to peak energy. Lesieur & Rogallo (1989) performed simulations by means of the Rogallo (1981) pseudo-spectral code with a Kraichnan (1976) eddy viscosity that has a cusp near the cutoff wave number k_c. The same initial velocity and temperature spectrum, imposed by Lesieur & Rogallo (1989) and in similar simulations by Chasnov (1991), were assumed; the energy spectrum is

$$E(k) = \frac{A_n}{k_P}\left(\frac{k}{k_P}\right)^n exp\left[-\frac{n}{2}\left(\frac{k}{k_P}\right)^2\right] \qquad 12.47$$

A_n is chosen so that the root-mean-square density and the velocity are unity at $t = 0$:

$$A_n = \frac{n^{\frac{(n+1)}{2}}}{1 \cdot 3 \cdot \cdot (n-1)} \qquad 12.48$$

where they used $n = 8$.

The parameter n permits to investigate, at very large times, which slope is selected at low wave numbers. To this purpose Chasnov (1995) dedicated a successive study. Chasnov (1991) as well as Lesieur & Rogallo (1989) presented pseudospectral simulations with a sufficiently high resolution (128^3) and they studied the change in time of the spectra. He observed that from k^8 it changes to k^4. A satisfactory inertial range width for $E(k)$ was achieved by a 128^3 simulation; this width reduces in time due to the rate of dissipation. On the other hand $E_T(k)$ shows a $k^{-5/3}$ in the early part of the evolution; in the long run it reaches a k^{-1} slope. Lesieur & Rogallo (1989) performed the simulations at $\nu = 0$ ($Re = \infty$); instead I decided to perform these at $Re = 10^4$.

Here I like to show that simulations with a smaller resolution predict the same behaviors at low and high wave number as those predicted by Lesieur & Rogallo (1989). Fig.12.9 indeed shows that even a 64^3 resolution predicts the slopes of the inertial ranges found by Lesieur & Rogallo (1989). In that paper they furnished only the time step, but it was not clear how long the simulation lasted. The simulation here evolved for 8 dimensionless

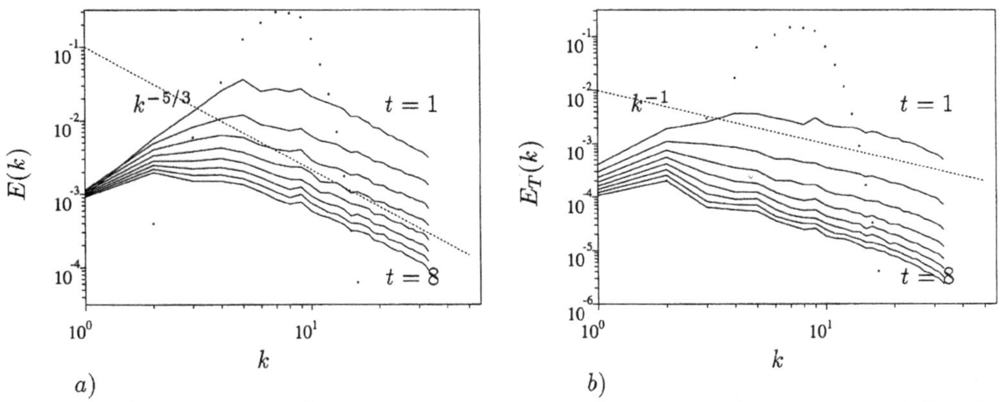

Fig. 12.9 a) Velocity and b) temperature energy spectra evolution between $t = 0$ (▪) and $t = 8$ (——) with the Smagorinsky s.g.s. model and a 64^3 resolution.

time units. The time unit is based on a velocity scale which has a velocity root-mean-square equal to 1 and a length scale such that the peak energy spectrum is at $k = 8$. By comparing the present spectra with those in Fig.1 of Lesieur & Rogallo (1989), it turns out that the present simulation lasts for a longer time. In fact in their figure the energy peak reduces ten times; on the contrary here it reduces more than a hundred times and this difference cannot depend on the different assumption for the Reynolds number. The same occurs for the peak of the temperature spectrum that for them reduces for almost two decades and in Fig.12.9b for four. At this point it is important to point out that here the simulations were done at constant CFL equal to 1.0; thus the final time of 8 units was reached in approximately 270 time steps for the 96^3 simulation and in 220 for the 64^3.

The comparison between Fig.12.9a and Fig.12.10a shows that by the finer grid a substantially large inertial range is achieved and that it lasts for a sufficient long period of time. At low wave numbers the slope is less than k^4 and the same was obtained by Lesieur & Rogallo (1989). In this longer simulation, however, it seems that this law decreases towards the k^2 slope. For the spectrum of the passive temperature the simulation does not produce any $k^{-5/3}$ transitory inertial range but from $t = 1$ up to $t = 8$ a well defined k^{-1} is predicted. The $k^{-5/3}$ appears in the first part of the simulation, between $t = 0$ and $t = 1$, as shown in figure 12.10b. Fig.12.10b and Fig.12.9b show that the temperature spectra at high wave number have a slightly faster decay in the 96^3 simulation than in the 64^3; in both cases the temperature decays faster than the velocity. This behavior at high k was also observed by Lesieur & Rogallo (1989) and it is linked to the value of Pr_T; in their simulation $Pr_T = 0.6$ and in this $Pr_T = 0.5$ were assumed. As a final consideration I would like to draw the attention of the reader to the spectra, that in the pseudospectral simulations show a small energy pile-up at the cutoff k_c which depends on the dealiasing procedure. The present simulations by a conservative finite difference method never present this rise near the cutoff and the spectra seem to decay in better agreement with the physics.

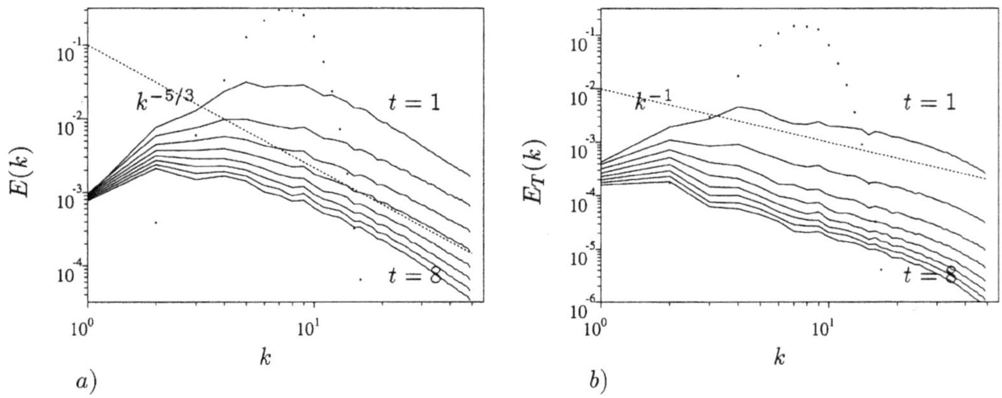

Fig. 12.10 As Fig.12.9 but with a 96^3 resolution.

For the present flow the dependence of temperature spectra from the s.g.s. models is not shown as was previously done. However, it has been observed that the structure function model gives substantially similar results. The conventional dynamic model was not applied since for a passive scalar it is sufficient to assign the turbulent Prandtl number to get the eddy diffusivity, and then the results should not differ largely from those with the Smagorinsky model. I like to show the difference of the spectrum between the Smagorinsky and the constant eddy viscosity because of the utility of the SDM model in more complex flows. Fig.12.11a-b indeed show that a constant eddy viscosity and diffusivity produces spectra that slightly differ from those with the Smagorinsky model at high wave numbers. The kinetic energy spectra show an inertial range with a reduced power law decay. On the other hand, the spectra of the passive scalar agree with those for the Smagorinsky model and in both cases they predict a k^{-1} law.

12.8d Buoyancy-generated turbulence

Simulations of practical problems of relevance in geophysical flows require the correct reproduction of the physical mechanisms generating a turbulent velocity fields by buoyancy forces due to a random distribution of fluid density created by some means. In this flow, as it was assumed by Batchelor *et al.* (1992), the velocity field at $t = 0$ is at rest and a density spectrum distribution with random phases is assigned. The motion develops by action of the gravity and since there is no forcing the mean-square density gradient can only decrease to return to a state of rest. The physics is interesting in the first period of the evolution, when the potential energy is transformed into kinetic energy. As claimed by Batchelor *et al.* (1992) this flow was never studied experimentally and also was never observed in nature. Their LES was the first one and, as they pointed out, this simplified flow is different from the most common cases of "fossil turbulence" (Gibson (1986)) in the ocean where a mean density gradient plays a fundamental role. A simulation close to real conditions in the ocean is presented in the next section. In this section once more the aim is to show that the present numerical method gives satisfactory results. In this case the parameters determining the flow are the initial density spectrum, the two molecular

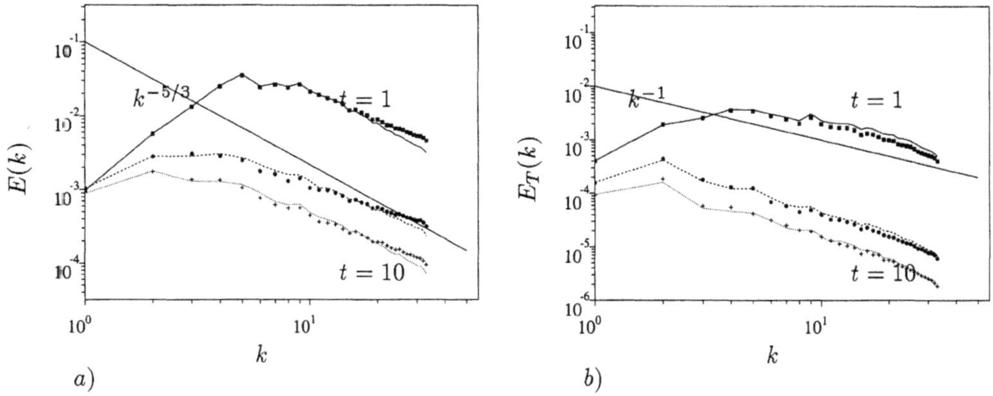

Fig. 12.11 a) Kinetic energy and b) passive scalar spectra at $t = 1, 3$ and 10; lines: Smagorinsky model; symbols: constant eddy viscosity and diffusivity. The resolution is 64^3.

transport coefficients and the gravity.

Batchelor *et al.* (1992) performed several simulations at different Reynolds numbers and even with the linearized set of equations. In their LES they used the Kraichnan (1976) eddy diffusivity model, well suited for pseudospectral simulations. Furthermore they changed the values of the exponent n of the initial spectrum to investigate how the decay exponent, at large time, depends on the large scales. The simulations were done by a 128^3 resolution and they imposed the initial spectrum with the peak at $k_P = 50$. Their simulations are repeated by decreasing the resolution to 64^3 and 96^3 to investigate whether the lower resolutions reproduce the main outcomes of their study. $n = 4$ has been chosen since from their results it turns out that, if the spectrum has $n > 4$, the spectrum, at low k, tends to $n = 4$. The spectra with $n = 2$ maintain the slope in time. They found that the thermal energy decreases in time and reaches around $t = 5$ a power law decay proportional to t^{-q} with an exponent between $q = 2.4$ for $n = 2$ and $q = 2.85$ for $n = 8$. The kinetic energy decays less with q varying from 0.4 for $n = 2$ to 0.857 for $n = 8$. Before we describe the effects of the s.g.s. model and of the grid resolution, even for the 64^3 coarse simulation, the results predict a satisfactorily good power law decay with $q = 0.7$ for the kinetic energy and $q = 3$ for the thermal energy. The Smagorinsky model has been used in a 64^3 and in a 96^3 simulation by setting the kinematic Prandtl number equal to 1.

The spectra dependence on Pr_T has been investigated by performing simulations with $Pr_T = 0.5$ and $Pr_T = 1$; the Reynolds number was $Re = 10^5$. Batchelor *et al.* (1992) assumed $Pr_T = 0.6$ and $Re = \infty$ in their study. The results in Fig.12.12 show that the turbulent Prandtl number largely affects the temperature spectra. At $Pr_T = 1$ the power law at large k is less steep that the k^{-1} law predicted by Batchelor *et al.* (1992). On the other hand the k^{-1} law is achieved in a large range of k when the turbulent Prandtl number is set equal to 0.5. The effect of the Prandtl number has a minor but still important influence on the kinetic energy spectra as Fig.12.12a shows. In fact the inertial range at

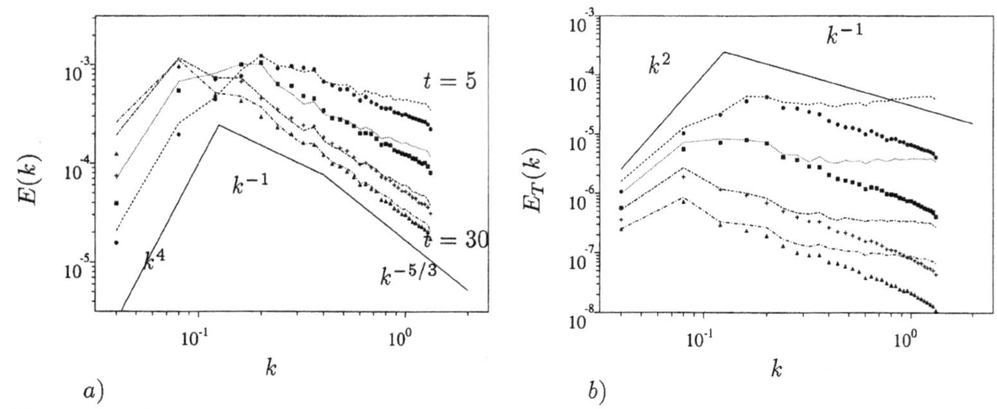

Fig. 12.12 a) Kinetic energy and b) passive scalar spectra at $t = 5, 10, 20$ and 30; lines: Smagorinsky model with $P_{rT} = 1$, symbols: with $P_{rT} = 0.5$. Simulations at $Re = 10^5$ by 64^3 resolution.

$Pr_T = 1$ has a slope that slightly deviates from the $k^{-5/3}$ law at very large t.

From the previous simulations it is clear that the Smagorinsky model for buoyant stratified and non-stratified flows leaves a lot of uncertainty since C_S and P_{rT} should be assigned. To overcome the freedom to impose these values, two simplified models have been described in section 12.4a, where the eddy diffusivity and viscosity are evaluated by the resolved quantities without introducing any ad hoc constants. As mentioned before, these simplified dynamic models have the chances to be satisfactory if in advance has been shown that the Smagorinsky model without the space variability of ν_T and ν_θ is successful. Fig.12.13 shows that the kinetic energy spectra and the temperature spectra with constant ν_T and ν_θ agree in a wide range of k with those obtained when the space variability of the model is maintained. Even the constant eddy diffusivity predicts a quite long k^{-1} inertial range and the agreement with the Smagorinsky model is rather good even at large k. The kinetic energy spectra at large k show a deviation from the $k^{-5/3}$ law, but it reproduces at intermediate k the transition from a k^{-1} to a $k^{-5/3}$ power law. I would like to point out that to quantify the deviation from the $k^{-5/3}$ inertial range it is convenient to plot the compensate spectrum as was done in Fig.12.8. It has not been investigated whether these discrepancies disappear by increasing the resolution, because the aim of this comparison was directed only at checking the opportunity to use the simplified dynamic model.

The results achieved thus suggest to apply the simplified model to calculate the turbulent Prandtl number from the resolved field. The simulations with the simplified dynamic model in section 12.6 have been performed on a 64^3 and 96^3 grid, and the time evolution of the eddy viscosity and the turbulent Prandtl number are given in Fig.12.14. In Fig.12.14a the eddy viscosity for the dynamic model decays as the averaged $\nu_T(x_1, x_2, x_3)$ by the Smagorinsky model; both show the same shift by changing the grid. The turbulent Prandtl number predicted by the dynamic model varies around 0.3, a value close to $Pr_T = 0.4$ found by Moin *et al.* (1991) in a priori tests of different DNS simulations of weakly compressible

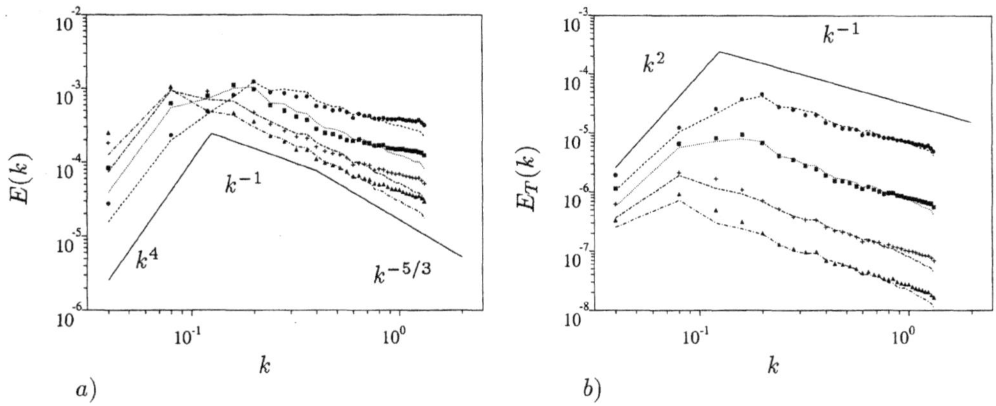

Fig. 12.13 a) Kinetic energy and b) passive scalar spectra at $t = 5, 10, 20$ and 30 (the different times are indicated with the same symbols as Fig.12.12); lines: Smagorinsky model with $Pr_T = 0.5$, symbols: with the space dependence eliminated. Simulations at $Re = 10^5$ by 64^3 resolution.

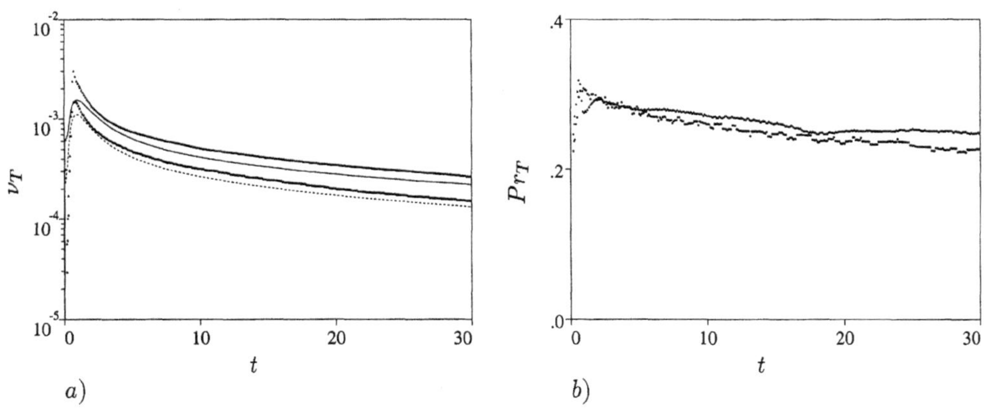

Fig. 12.14 Time history of a) mean eddy viscosity for the dynamic model ▪ 96^3, ▵ 64^3; Smagorinsky model —— 96^3 - - - - 64^3 b) turbulent Prandtl number ▪ 96^3, ▵ 64^3. Simulations at $Re = 10^5$.

and incompressible flows with a passive scalar.

Considering the inability of the simplified models to reproduce the full inertial range, these simplified dynamic models could be used to get an approximate value of the turbulent Prandtl number. The physical LES then can be performed by the Smagorinsky model with the value of Pr_T obtained by the simplified dynamic model. A more sophisticated dynamic model as that described by Moin *et al.* (1991) could be used, but, as previously mentioned,

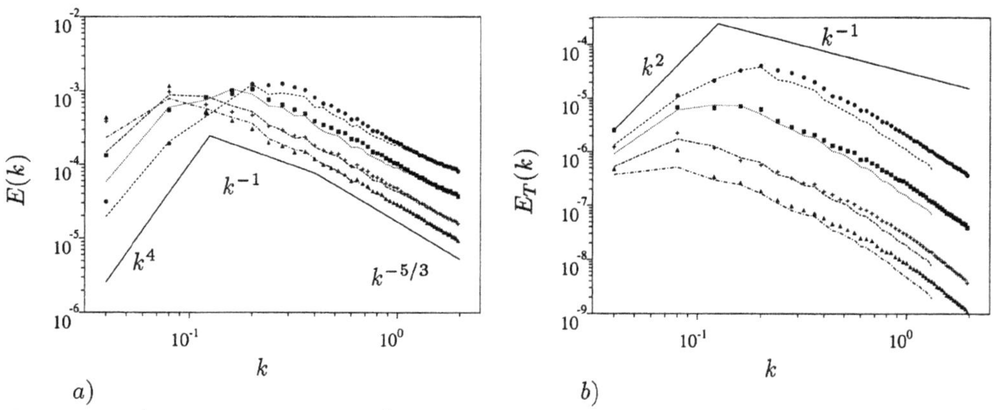

Fig. 12.15 a) Kinetic energy and b) passive scalar spectra at $t = 5, 10, 20$ and 30 with the dynamic simplified model; lines: 64^3, symbols: 96^3. Simulations at $Re = 10^5$

an iterative procedure is requested to get Pr_T. As reported by Cabot (1992) this iterative procedure in some circumstances presents a difficulty in converging. Fig.12.15 shows the kinetic energy and temperature spectra with the simplified dynamic model for the two simulations with different resolutions. The energy spectra show, at large time a $k^{-5/3}$ inertial range, at intermediate k an approximate k^4 law and at small k a short range with k^-1. This trend agrees qualitatively with the results by Batchelor *et al.* (1992). By a finer resolution (128^3) and by a peak energy at different k they achieved longer ranges with k^4 and k^{-1} laws. In agreement with the slight differences in the turbulent Prandtl number shown in Fig.12.14b, Fig.12.15b shows some differences in the temperature spectra. Both simulations show, at high k, a well developed inertial range with a slope close to k^{-3}. It is not clear whether this law agrees with measurements or other simulations but it has been observed that the power law strongly depends on the turbulent Prandtl number.

In literature there are theoretical studies related to temperature spectra at different values of the kinematic Prandtl number. For example Tennekes & Lumley (1972, Pg.286) have shown that, at very low Prandtl, the decay of the temperature spectrum is proportional to $k^{-5/3}$. LES could predict this behavior because at low Pr and high Re the velocity field is truncated and the temperature is fully resolved. Flows at low Pr are not of interest for geophysical flows oriented to meteorology or oceanography. For ocean simulations the kinematic Prandtl number is about 7; at this Pr a simulation confirmed that the LES is not affected by the kinematic Prandtl number.

The turbulent Prandtl number's influence on the spectra, to my knowledge, was never investigated in previous LES; thus I decided to perform this study by the Smagorinsky model by a 64^3 resolution. Simulations at $Pr_T = 1, 0.5$and 0.3 show in Fig.12.16 that the total kinetic and thermal energy, in the final stage, decay with the same exponents. However, to withdraw more satisfactory conclusions the simulations should last longer and I leave this exercise to the reader. On the contrary, in the initial period, there is a large effect of the turbulent Prandtl number. When it decreases the rate of thermal energy

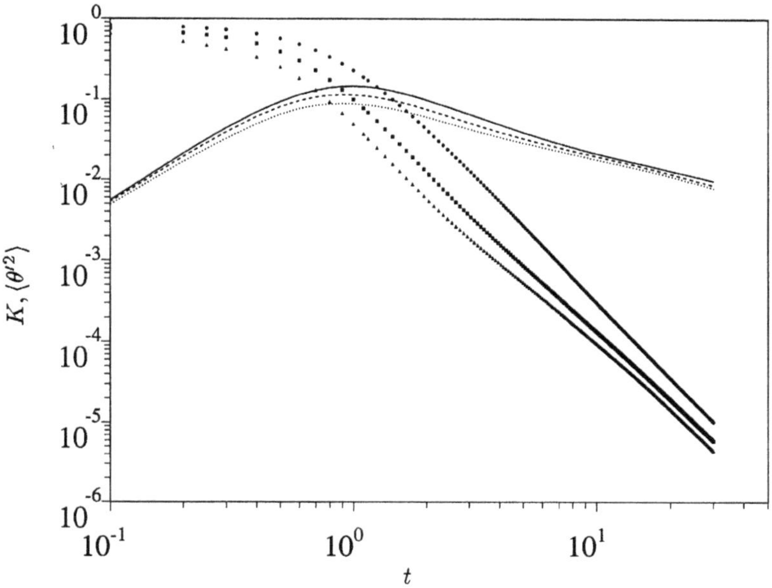

Fig. 12.16 Time history of kinetic energy (lines) and *rms* temperature (symbols) (—— ∘ $Pr_T = 1$, - - - - □ $Pr_T = 0.5$, ········ △ $Pr_T = 0.3$).

dissipation increases and thus the production of kinetic energy is reduced. The transfer of thermal energy to small scales is related to the slope of the thermal energy spectrum at high k. In Fig.12.17 the spectra at $t = 10, 20$ and 30 for the three Prandtl numbers show the fast decay at $Pr_T = 0.3$, with a k^{-3} law. The kinetic energy spectra, with a good approximation shows, after $t = 20$ the achievement of a $k^{-5/3}$, independently from Pr_T. From these figures, it turns out that $Pr_T = 0.5$ produces a thermal energy with a k^{-1} slope, observed in the previous LES by Batchelor *et al.* (1992). As mentioned before, my opinion is that the turbulent Prandtl number should be evaluated from the resolved field. In the simplified dynamic model it varies in time hence it is difficult to have results matching those obtained by the Smagorinsky model if in the latter a value different from that in the simplified dynamic model is used.

Fig.12.18 shows that the two subgrid models predict temperature spectra with the same power spectra at high and low wave numbers. At this 96^3 resolution the k^{-3} power law for the thermal energy spectra is predicted in the last part of the time evolution. The reason for the differences in the thermal spectra is related to the variations, in the dynamic model, of the turbulent Prandtl number, as is shown in Fig.12.14. These variations produce a greater thermal decay than that for a constant Prandtl. Fig.12.18a, on the other hand, shows that both models predict energy spectra with a quite long inertial range.

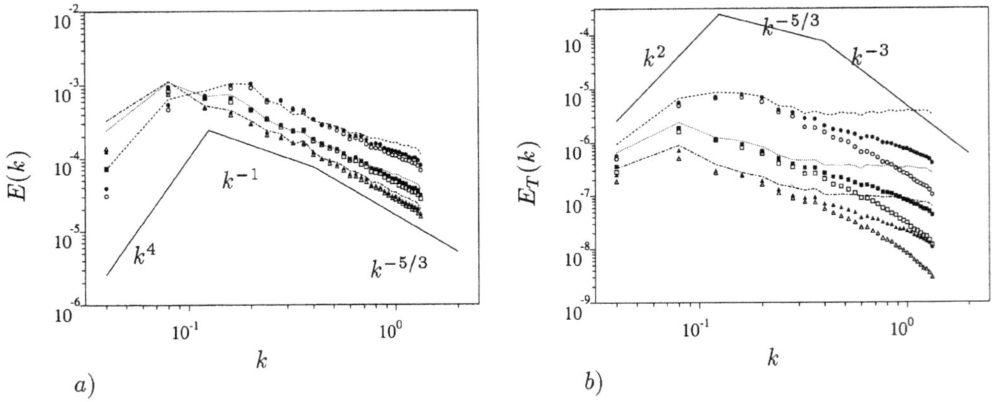

Fig. 12.17 a) Kinetic energy and b) passive scalar spectra at $t = 10, 20$ and 30 for the Smagorinsky model; lines: $Pr_T = 1$, closed symbols: $Pr_T = 0.5$, open symbols: $Pr_T = 0.3$. Simulations at $Re = 10^5$ with 64^3 resolution.

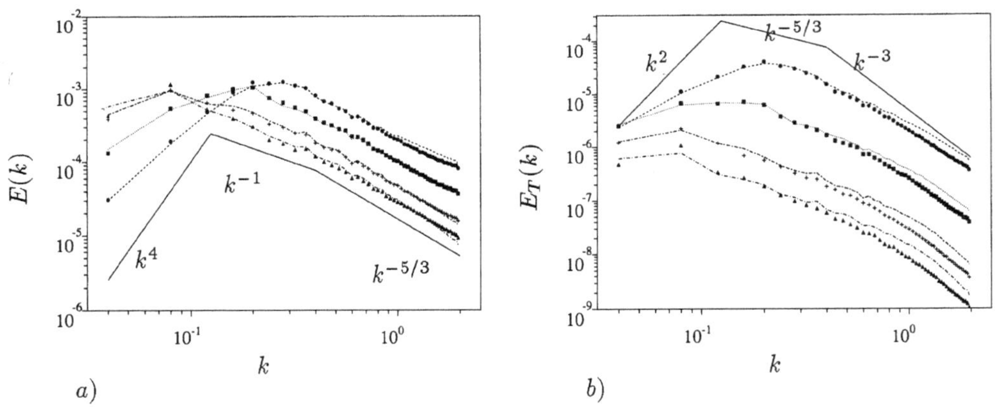

Fig. 12.18 a) Kinetic energy and b) passive scalar spectra at $t = 10, 20$ and 30 with the Smagorinsky model at $Pr_T = 0.3$ (lines), and with the dynamic model (symbols). Simulations at $Re = 10^5$ with 96^3 resolution.

12.8e Turbulence with buoyancy and stratification

The effects of stable stratification on buoyant flows are of large interest to understand the dynamics of geophysical flows and in particular of the sea where the thermocline plays an important role. Thus in this section simulations have been performed taking the reference quantities from the oceans. We have performed a series of simulations with Reynolds number of $Re = 10^6$ and the following bulk Richardson numbers: Ri = 0.03, 0.09, 0.3, 0.76 and 2.3. Recall that Ri is the square of the nondimensional form of the Brunt-Vaisala frequency ($Ri = (NL_0/U_0)^2$). We can try a range of typical length and velocity scales to

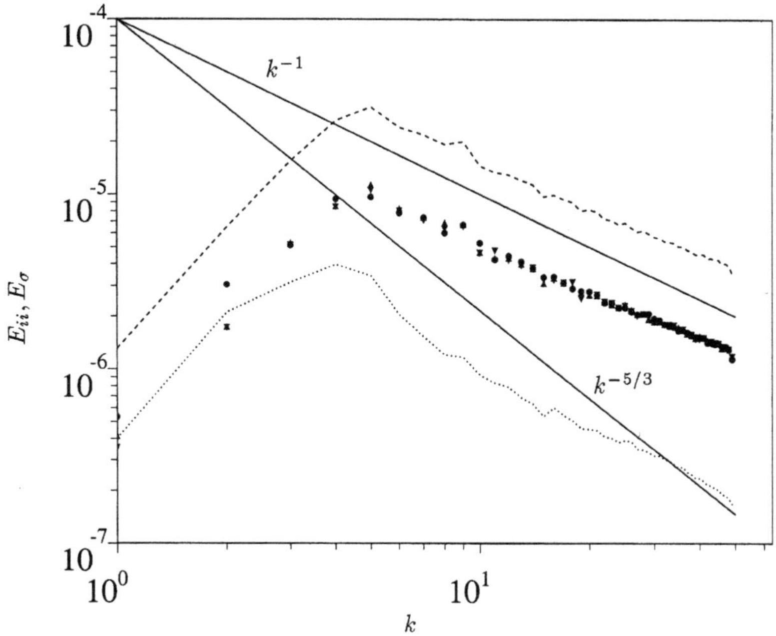

Fig. 12.19 Initial velocity (———) and temperature (- - - -) spectra , ▪ E_{11}, ▫ E_{22}, ▵ E_{33} , ○ thermal energy at $t = 20$.

see how these parameters correspond to oceanographic values. For example, let us take $X = 40$ m to be the length of one side of our computational domain. We nondimensionalize our evolution equations with the length scale $L_0 = X/2\pi$ so that wavenumber 1 corresponds to a wavelength equal to the size of the domain. With this domain size, the smallest resolved scale in simulations with 64^3 grid points will be near the scale where the inertial range typically begins ($\approx 1m$), and for 96^3 the smallest resolved scale will be in the inertial range. For a velocity scale, let us use $U_0 = 15cm/sec$, which is not atypical of horizontal velocities advecting oceanic fine structure. With these values, the dimensional Brunt-Vaisala frequencies corresponding to the Ri that we have used would be $N = 2.4, 4, 7.5, 12, 20$ *cph* respectively. Typical oceanic values of N measured in the thermocline may vary from about 1 to 10 *cph*. The results with the stratification are compared with those without stratification. To have an initial field with realistic vortical structures a simulation with the Smagorinsky model has been performed starting from a spectrum with random distribution of the phases for the velocity and for the passive scalar. The initial spectrum evolved for 20 time units and it gets the profiles shown in Fig.12.19. This transient, necessary to build up the turbulent structures, was simulated by the Smagorinsky model with $Pr_T = 0.55$. At $t = 20$ the velocity and passive scalar spectra do not show the $k^{-5/3}$ inertial range; the spectra span a large range of wave numbers without any pile-up of energy near the cutoff.

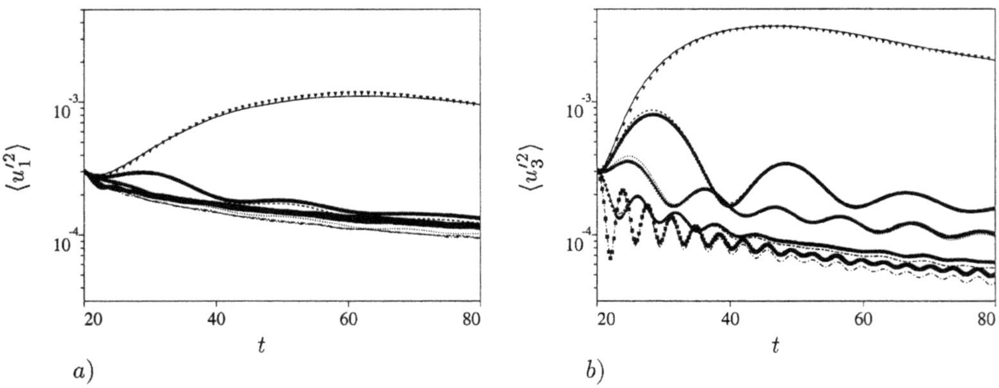

Fig. 12.20 Time history of velocity rms; lines: Smagorinsky, symbols: SDM, (—— , ∇ $Ri = 0.$), (---- , ▪ $Ri = 0.03$), (········ , △ $Ri = 0.09$), (—–— , ◇ $Ri = 0.3$), (—·— , □ $Ri = 2.3$),

Starting from this condition the effect of the stratification are studied by observing the behavior of different quantities. In Fig.12.20 the time history of $\langle u_1'^2 \rangle$ and $\langle u_3'^2 \rangle$ shows that these components increase due to the buoyancy force and, as expected, the component in the direction of the gravity is greater than the other two components. $\langle u_2' \rangle$ behaves as $\langle u_1'^2 \rangle$ and this is the reason why it is not shown. With a stable stratification the velocity components are damped and a modulation in time appears with a frequency the higher the greater is the Brunt Waisala frequency N. In these figures are reported the results obtained by the Smagorinsky model (lines) and by the SDM (symbols) and it turns out that the results do not differ largely. The reason is that the normal stresses are linked to the large scales and the subgrid model acts on the small scales near the cutoff wave number. Even the time evolution of $\langle \sigma'^2 \rangle$ in Fig.12.21a seems to be insensitive to the subgrid model at the different values of Ri.

The history of the density variance in Fig.12.21a shows that without stratification the potential energy decays and, approximately at $t = 60$, it reaches a power law decay t^n with $n = -4.68$. For non-zero values of Ri the potential energy undergoes a damped oscillation. The results are approximately the same for both the Smagorinsky model and the SDM. The oscillation of the potential energy is 90 degrees out of phase with the oscillation of the vertical kinetic energy as shown in Fig.21b where both kinetic and potential energy terms are plotted together (for simplicity, only the results from the Smagorinsky model are shown). This oscillation represents repeated transformation between kinetic and potential energy.

The important feature of the dynamic models, including the SDM, is that they evaluate the values of the coefficients C_S and Pr_T from the flow fields, unlike the Smagorinsky model in which these parameters need to be assigned. Fig.12.22a gives the values of Pr_T obtained for the different stratifications. This shows that the turbulent Prandtl number tends to increase as N increases. The degree of the influence of N changes with the phase

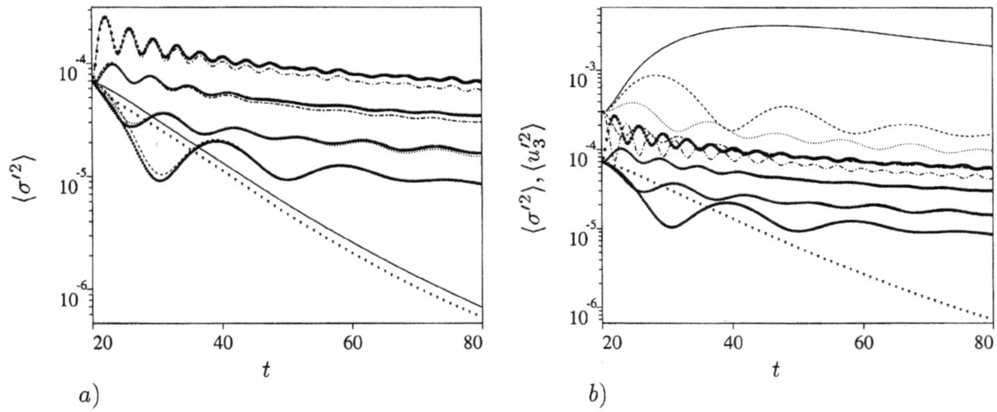

Fig. 12.21 Time history of a) thermal energy b) $\langle u_3'^2 \rangle$ (lines) $\langle \sigma'^2 \rangle$ (symbols) the different Ri have lines and symbols as in the legend of Fig.12.18.

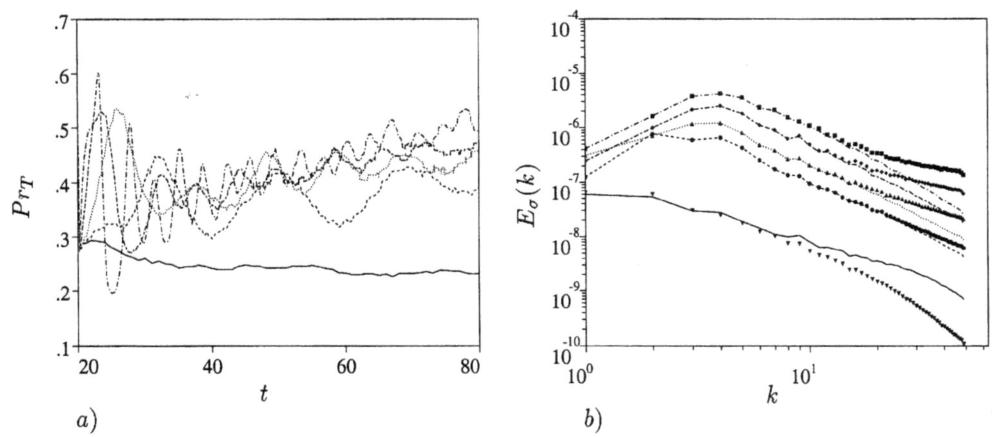

Fig. 12.22 Time history of a) turbulent Prandtl number from the SDM, b) density perturbation spectra at $t = 80$ (the different Ri have lines and symbols as in the legend of Fig.12.18).

of the oscillations. At all times, the largest variation occurs with $Ri = N^2$ in the range 0 to 0.03. For N^2 varying between 0.03 and 2.3, the variation of Pr_T at time t=80 is from about 0.35 to 0.5. These values are smaller than that assumed in the simulations with the Smagorinsky model where we have taken $Pr_T = 0.55$. Note that even the Prandtl number oscillates with a frequency increasing with N. An underestimation of the Prandtl number should produce errors at high wavenumbers. Fig.12.22b shows that the spectra obtained

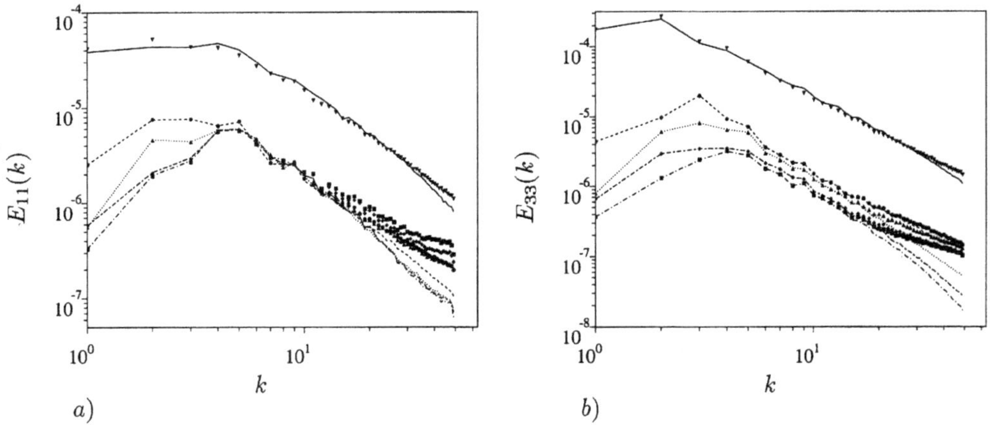

Fig. 12.23 Velocity energy spectra at $t = 80$, a) E_{11}, b) E_{33} (legend as in Fig.12.18).

with the SDM agree well at low wave numbers with those from the Smagorinsky model. At high wavenumbers, the spectra from the Smagorinsky model decay with a slope that is not too far from $-5/3$, while the spectra from the SDM have less steep spectra. The SDM also produces velocity spectra that are too shallow at high wavenumbers as we see in Fig.12.23.

It is remarkable that the SDM can compete well with the Smagorinsky model in reproducing the variance time histories shown in Fig.12.20 and Fig.12.21, while performing so poorly in reproducing the high wavenumber spectra. Clearly the variances are not very sensitive to the behavior of the small scale structures in these flows, at least over the time span of these simulations.

To conclude, the LES simulations in flows with stratification and buoyancy are still not very conclusive since at these high Reynolds numbers the results show a change of slope at high wave numbers that depends on the stratification. To verify that this behavior is indeed true measurements are needed; I am leaving this comparison to the users who are experts in geophysical flows and are interested in the prediction of turbulence in oceans. However, before to draw any conclusion I would like to suggest to the reader to make an effort to introduce into the code the dynamic model as that described by Moin *et al.* (1991) and to see whether the simplifications here used play an important role.

12.9 Conclusions

In this chapter it has been shown that a second order finite difference scheme works satisfactorily well for LES at low and high Reynolds numbers. The reason is related mainly to the fact that the method is energy conserving hence numerical generation or dissipation of energy are eliminated. IN this chapter I was also interested to give a small contribution to the debate about the influence of the numerics on LES. This issue was initiated by Ghosal (1996) where he claimed that second order accurate methods produce unsatisfactory results. From the present results it turns out that at low and high Re the spectra behave quite well. I was not able to find other simulations with the dynamic model at very high Reynolds numbers, even with pseudospectral methods, and I showed that in these flows the spectra show a well defined $k^{-5/3}$ law.

A very interesting outcome of this chapter is that if the interest is in the time evolution of global quantities as energy or dissipation, the simplified model with a constant eddy viscosity works well. At low Reynolds numbers even the spectra are predicted with a very good accuracy; on the contrary at high Reynolds numbers the spectra are poorly predicted. This simplified model could be used to get very rapid results since it can be inserted in a code designed for DNS.

By these simulations the conclusion can be drawn that finite differences have the capability to be used in the simulations of flows of practical interest. To prove this statement I dedicate the next chapter to wall bounded flows. I would like to point out that I pursued this effort by myself and thus I could prove this only for a very limited range of flows. At the CTR this effort was pursued by a much larger group of scholars and they reached more solid conclusions. I would like to suggest to the user to read carefully the large number of papers in the proceedings of the summer schools and on the annual research briefs of the CTR and to get ideas. With the present code my aim was to furnish a tool to test new ideas about more efficient subgrid models that can be easily inserted into this code. For stratified turbulence the goal is more difficult; in fact there the are interactions of waves, large scale structures and fine scale turbulence, as is described by Gibson (1986).

13 Large eddy simulations of wall-bounded flows

13.1. Introduction

In chapter 12 we have shown that LES by finite differences are satisfactory for flows with three periodic directions. Engineering-oriented flows and in particular flows related to aeronautical or mechanical engineering applications have at least a solid boundary. Two homogeneous directions in the simulation of turbulent flows save computational time, in particular in understanding the physics or in checking the physical modelling. In fact, the turbulent statistics can be evaluated by a limited number of flow fields. To investigate the feasibility of LES in wall bounded flows the choice is to consider the flow in a plane channel or in a pipe. The channel is a better choice for the large number of results in literature obtained by different subgrid models and by different numerical methods. Whoever is interested on LES of pipes can use the code described in chapter 10 by modifying the viscous terms as described in this chapter. The equations used by Fatica *et al.* (1994) to perform LES of time developing round jets can be used.

As mentioned, the channel has been simulated since the first LES by Deardoff (1970) on a 32^3 grid, enforcing the use of the wall function. Later on Moin *et al.* (1978) in his Ph.D thesis again a 32^3 simulation, but without the wall function. From that simulation initiated the large effort, still active, to get solutions at high Reynolds numbers. Piomelli introduced the different subgrid models, starting from the Smagorinsky subgrid model with the van Driest dumping in a pseudospectral code. He resumed his thesis in a paper with Moin and Ferziger (Piomelli *et al.* (1987)) where he also made a-priori checks of the models by using the DNS by Kim *et al.* (1987). Later on, at one of the CTR summer schools, Piomelli in a group lead by Moin together with Cabot and Germano contributed to the development of the dynamic model that has been largely used in recent years (Germano *et al.* (1991)). Piomelli (1990), finally, has the merit to have shown that this new model is efficient in transitional flows, due to the fact that the model is inactive during the transition from a laminar to a turbulent regime. He recently summarized all his efforts in two articles (Piomelli (1998a) and Sarghini & Piomelli (1998b)). Piomelli was always using a pseudospectral numerical scheme; in the mean time, the group at the CTR was performing simulations also with finite differences. This research work can be found in several papers in the Annual Research Briefs of the CTR and in the proceeding of the summer schools (Lund (1991), Cabot (1993), Ghosal *et al.* (1992), Moin *et al.* (1994)). The group lead by Lesieur at Grenoble furnished a minor contribution to the LES of a plane channel by using the structure functions model. Their work is mentioned in the review article by Lesieur & Metais (1996) and in the book of Lesieur (1990).

In this chapter the code based on the finite difference scheme is described and the results obtained by different subgrid models are compared. Even for the channel the simplified model with the eddy viscosity constant along $x - z$ planes and varying with y is applied but the constant is not dynamically evaluated. For the channel the dynamic model can be coded easily and the results depend on the numerics, as is shown, by comparing the present, Cabot's (1993) and Piomelli's (1993) results.

13.2. Equations

The filtered incompressible momentum equations for wall bounded flows are those of

Eq.(12.5) valid for the three periodic directions. The same considerations about the filtering that should be consistent with the numerics suggest to filter in physical space when filtering at the test level is done in the dynamic model. The filtering is applied only in the homogeneous directions. The closure for the residual stress, in Eq.(12.14) by means of the symbols of chapter 12, leads to the discretized equations for LES

$$\frac{\partial \overline{u_i}}{\partial t} + g_j \frac{\delta \overline{u_i}\, \overline{u_j}}{\delta x_j} = -g_i \frac{\delta P}{\delta x_i} + g_j \frac{\delta g_j}{\delta x_j}\left[\frac{\widetilde{\nu_T}}{2}\left(\frac{\delta \overline{u_i}}{\delta x_j} + \frac{\delta \overline{u_j}}{\delta x_i}\right)\right] \qquad 13.1$$

where $\widetilde{\nu_T} = (\frac{1}{Re} + \nu_T)$. In this chapter the detailed description of the code in the directory $CODESCH13/CHLE$ is not given, since the structure of the code is similar to that in the previous chapters. The subroutines are commented during the short description of the numerical method in this chapter.

The differences between the LES and the DNS equations rely on the appearance of the cross terms derivatives in the right hand side of the momentum equations. This consideration slightly modifies the numerical procedure, which consists of the evaluation of the cross terms in the routines $HDNLi$ (in the file *chaleshn.f*) where the non-linear terms are calculated. This is a consequence of the explicit treatment of the cross terms in the time advancing scheme. Further minor corrections are necessary in *chalestr.f* for the inversion of the tridiagonal matrices caused by the necessity of the code vectorization. The structure of the code does not differ from that of the channel described in chapter 9; the files *chalesin.f*, *chalesma.f*, *chaleoun.f*, *chalestn.f*, *chalestu.f* and *chalesphnc.f* are almost identical to those in $CODESCH9$.

It is more important to describe briefly the differences in the subgrid models between wall bounded flows and flows with three homogeneous directions. The eddy viscosity expression for the Smagorinsky model is given in Eq.(12.5), but in wall-bounded flows the length scale l varies with the distance from the wall for the variations of $\Delta\hat{x}_2$ in Eq.(12.16). In the early LES of the channel it turned out that the Smagorinsky model required a further correction to account for the physical observation that the turbulent scales, approaching the wall, reduce in size. Piomelli *et al.* (1987) reported the different expressions used by Moin & Kim (1982), by Schuman (1975) and by themselves for the Van Driest damping factor. Moin & Kim (1982), by neglecting the wall units in the damping, used the expression

$$l = C_S(1 - (y/\delta)^2)(\Delta x_1 \Delta x_2 \Delta x_3)^{1/3} \qquad 13.2$$

In the present simulation $\delta/h = 0.1$ has been assumed. The Smagorinsky eddy viscosity is calculated in the subroutine $SMAGO$ in the file *chalessma.f*. The same damping function has been introduced in the subroutine $COSMA$ calculating the eddy viscosity with the rate of dissipation averaged on $x_1 - x_3$.

The dynamic model in wall bounded flows requires the definition of the test filter. It has been assumed that $\widetilde{\Delta x_i} = 2\Delta x_i$ and that the filter is applied only in the homogeneous directions with the indices $i = 1, 3$. The coefficients in the subroutine $FILTER$, in the file *chalesdyn.f* representing a top hat filter in physical space, are different from those in the routine used in chapter 12 because of the two-dimensionsonal filters. The constant in wall bounded flows depends on x_2 and the expression is

$$C_S(x_2, t) = -\frac{1}{2\Delta(x_2)^2} \frac{\langle \mathcal{L}_{ij} \mathcal{M}_{ij} \rangle}{\langle \mathcal{M}_{ij} \mathcal{M}_{ij} \rangle} \qquad 13.3$$

with $\Delta(x_2) = (\Delta x_1 \widehat{\Delta x_2} \Delta x_3)^{\frac{1}{3}}$. The average $\langle\rangle$ eliminates the points with negative constants that could cause numerical instabilities. However, if at certain locations (and not always at the same grid point) $C_S(x_1, x_2, x_3)$, obtained without any average on the right hand side of Eq.(13.3), becomes negative, this accounts for an energy back transfer. This was the mechanism, absent in the Smagorinsky model, that necessitated the Van Driest correction near the wall. In fact, near the wall small scales contribute to the energy production and these account for the sweeps events. These small scales are not those that withdraw energy as in isotropic turbulence. With the expression in Eq.(12.3) this physical mechanism is averaged and the values of the constant is reduced at each plane and does not become negative at any point on this plane. There are several studies performed by different scholars at the CTR devoted to investigate the best ways to get the Smagorinsky constant for isotropic and wall bounded flows; the relative papers can be found in the CTR publications. Some of these methods such as that suggested by Ghosal *et al.* (1992), require more computational power to solve an integro-differential equation which does not largely improve the results.

The dynamic model has been coded as was done for isotropic turbulence, thus the subroutine $DYNAMIC$ in the file *chalesdyn.f* do not substantially differ from that. In this as well in the subroutine $SMAGO$ the deformation tensor $\overline{S}_{ij}$ is evaluated at each grid point in the routine $STRAI$ in the file *chalesdyn.f*, and is contracted to get $|\overline{S}|$. The filtering algorithm is applied to $\overline{S}_{ij}|\overline{S}|$, which is stored in the six matrices mij in the routine $DYNAMIC$. Due to inefficient programming work, the dynamic model necessitates a higher number of extra matrices with respect to the other models. Since in this code several subgrid models are tested, even if not necessary, the six matrices sij have been introduced, which in addition facilitate the comprehension of the routine $DYNAMIC$. Those who have computers with a limited memory should use their skill to reduce the memory occupancy.

13.3. Results

As was done in the past here the LES for the channel have also been performed at different Reynolds number and have been compared with experimental and numerical results. In the first simulation performed by Deardoff (1970) the Reynolds number was substantially high and thus the wall layer, the more active part of the channel, was neglected and modeled as in one-point closure by the wall function. The first grid point was within the logarithmic layer. Moin & Kim (1982) considered the near wall region because they had the opportunity to use the ILIAC IV, a special machine that permitted to double the number of points used by Deardoff in each direction. Moreover, since they used Chebitchev transformations in the normal direction they located some computational points in the viscous and buffer region and thus they captured the wall region. Ferziger (1987), in his review of LES, mentioned the relevance of the wall structures in LES calculations. Different from DNS, where the structures have to be fully resolved, in LES they should only be roughly described.

The DNS of Kim *et al.* (1987) at $R_\tau = 180$ permitted the comparison of the LES with validate results and thus, at this low Re, it was not necessary to consider experimental data. The highest Re DNS, to my knowledge, was performed by Kim and the data were reported by Kim & Antonia (1995). This simulation at $Re_\tau = 395$, corresponding to $Re = U_P h/\nu = 13000$ as in the experiment by Hussain & Reynolds (1972), was used by Piomelli (1993), Cabot (1992) and Lamballais *et al.* (1996) to compare the results of LES, and it will be also used here. Piomelli (1993) and Cabot (1992) performed a LES at $Re_\tau = 1000$, corresponding to $Re = 30000$ for which measurements of Wei & Willmarth (1989) exist. Thus these two LES simulations permit to investigate the Reynolds number dependence of the different subgrid models. The simulation at $R_\tau = 180$ has not been done and it is left as an exercise to the reader; the simulations could be very fast since a very coarse grid can be used, for instance $32 \times 48 \times 32$. The simulations, here, have been performed in a domain with a 2π streamwise length and a spanwise dimension equal to 0.75π. The simulations used different grids; at the small Re the grid was $48 \times 64 \times 48$, at the highest Re it was $64 \times 96 \times 64$. The grid was uniform in the x_1 and x_3 directions and the resolution in the spanwise direction (x_1) was less than 50 wall units. Piomelli (1998a) reported that this is the maximum resolution possible to get reasonable results. This statement was confirmed by the present code; in fact for an error on assigning the data, a simulation was performed by a greater Δx_1 and the results did not show any log region. In the normal direction the coordinate transformation in Eq.(2.20) with a stretching parameter $\alpha = 4.3$ located the first grid point at approximately $y^+ = 1.25$. The LES are coarse, hence it is not necessary to start the simulations from a restarting file; in each case the simulations started from a parabolic velocity profile with a random perturbation and evolved for 150 dimensionless time units without any model. At $t = 150$ the field was used as the initial condition for simulations with different subgrid models. These simulations lasted for 150 time units and the statistics were evaluated starting at $t = 180$ for 120 time units. The statistics were calculated by considering fields separated by 5 time units for $R_\tau = 400$ and 10 units for $R_\tau = 1000$.

13.3.a Results at low Reynolds number

As the good sense suggests a simulation was done without any model (labeled as NOM) to see whether any numerical viscosity could overcome the subgrid model. The damped Smagorinsky model as used by Moin & Kim (1982) and with the modification to assume the eddy viscosity constant on $x_1 - x_3$ planes was also employed. In chapter 12 the Smagorinsky constant was linked to the Kolmogorov constant and theoretically its value was $C_S = 0.18$. In previous channel simulations $C_S = 0.10$ was assumed. The simulations have been performed for these two values of C_S to check the dependency on this constant. For the classical Smagorinsky model the simulations are labeled $SM18$ and $SM10$ respectively for $C_S = 0.18$ and $C_S = 0.1$. $CS18$ and $CS10$ indicate simulations using the model with a constant eddy viscosity on horizontal planes with $C_S = 0.18$ and $C_S = 0.1$. DYN indicates simulations with the dynamic model where the constant is obtained by averaging in time and in $x_1 - x_3$ planes.

The results show the large influence of the values of the constants in the Smagorinsky models. These results furthermore differ from those with the constant calculated dynamically. Fig. 13.1 indeed shows that the values of the constant affect in a large measure the $\langle u_3 \rangle^+$ profile at $R_\tau = 400$. The coincidence of the profiles by $SM10$ and by $CS10$ shows that, even in inhomogeneous flows, the key factor in LES is to withdraw the correct amount of energy by the subgrid model at each distance from the wall. The results up to

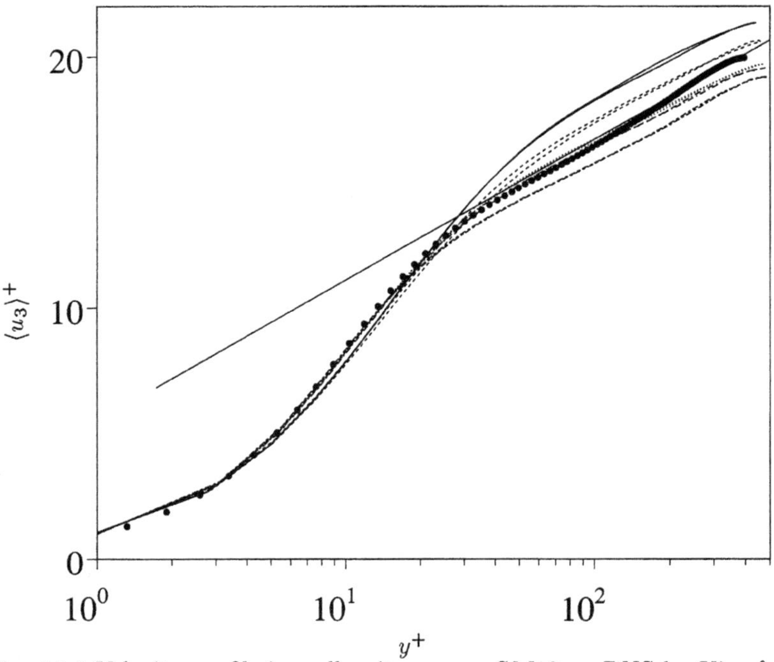

Fig. 13.1 Velocity profile in wall units: —— $SM18$, ▪ DNS by Kim & Antonia (1993), —— $U^+ = \frac{1}{0.41} \log y^+ + 5.5$, - - - - $CS18$, ········ $SM10$, —·— $CS10$, — — — NOM.

$y^+ = 150$ are in a good agreement with those by DNS. The present simulations show some undulation around the *log* law; this could be attributed to the limited number of points in the normal direction. Piomelli (1993) was using 97 points; here the same number of 64 points of Cabot (1994) were used, and undulations were shown also in his results. When C_S is increased the amount of subgrid dissipation increases and the strength of the near wall vortical structures is reduced and at their turn the fluctuating velocities are reduced. This occurrence can be visualized in horizontal planes, at a distance of 12 wall units from the wall, and vertical planes, at mid channel, by contours of ω_2' and u_3'. Fig.13.2 shows the differences amongst NOM, $SM18$ and $SM10$. NOM, using this non-dissipating finite difference method, produces eddies with high vorticity that shift the *log* profile below that of the DNS. On the other hand, the higher dissipation, in $SM18$ produces larger and weaker vortices that shifts the law of the wall upwards, as in drag reducing flows. $SM10$ produces the right amount of dissipation and thus the streaks have the desired strength and spacing.

The profiles of the normal and tangential stresses in wall units are compared with those obtained by DNS in Fig.13.3. For some of the normal stresses the agreement is very poor and then, to emphasize the differences in the wall region, the profiles have been plotted in semi-log scales. Even if $SM10$ and $CS10$ predict the best agreement with the DNS results

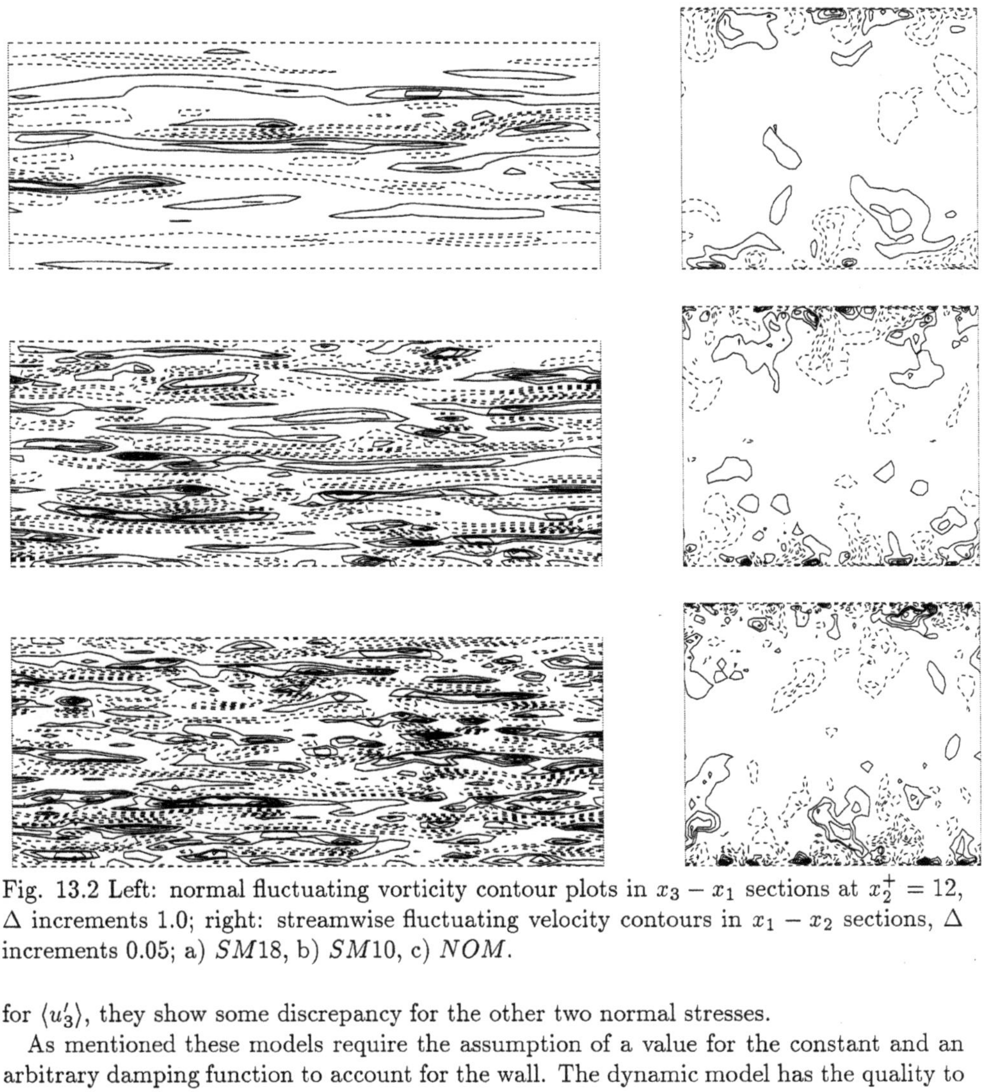

Fig. 13.2 Left: normal fluctuating vorticity contour plots in $x_3 - x_1$ sections at $x_2^+ = 12$, Δ increments 1.0; right: streamwise fluctuating velocity contours in $x_1 - x_2$ sections, Δ increments 0.05; a) $SM18$, b) $SM10$, c) NOM.

for $\langle u_3' \rangle$, they show some discrepancy for the other two normal stresses.

As mentioned these models require the assumption of a value for the constant and an arbitrary damping function to account for the wall. The dynamic model has the quality to relate the constant to the resolved field and then it could be an universal model; however, the model requires other assumptions such as the filtering and the averaging procedure in the evaluation of the constants. Cabot (1994) investigated the dynamic model and a possible improvement with an equation for the subgrid kinetic energy, by inserting the models in a code similar to the present one. Cabot (1994) found in both cases a very good agreement with the DNS results; see his Fig.3. Before we analyze the velocity and turbulent profiles of the different models it is interesting to see how the eddy viscosity behaves in each model. Fig.13.4 indeed shows that in the dynamic model the eddy viscosity gets

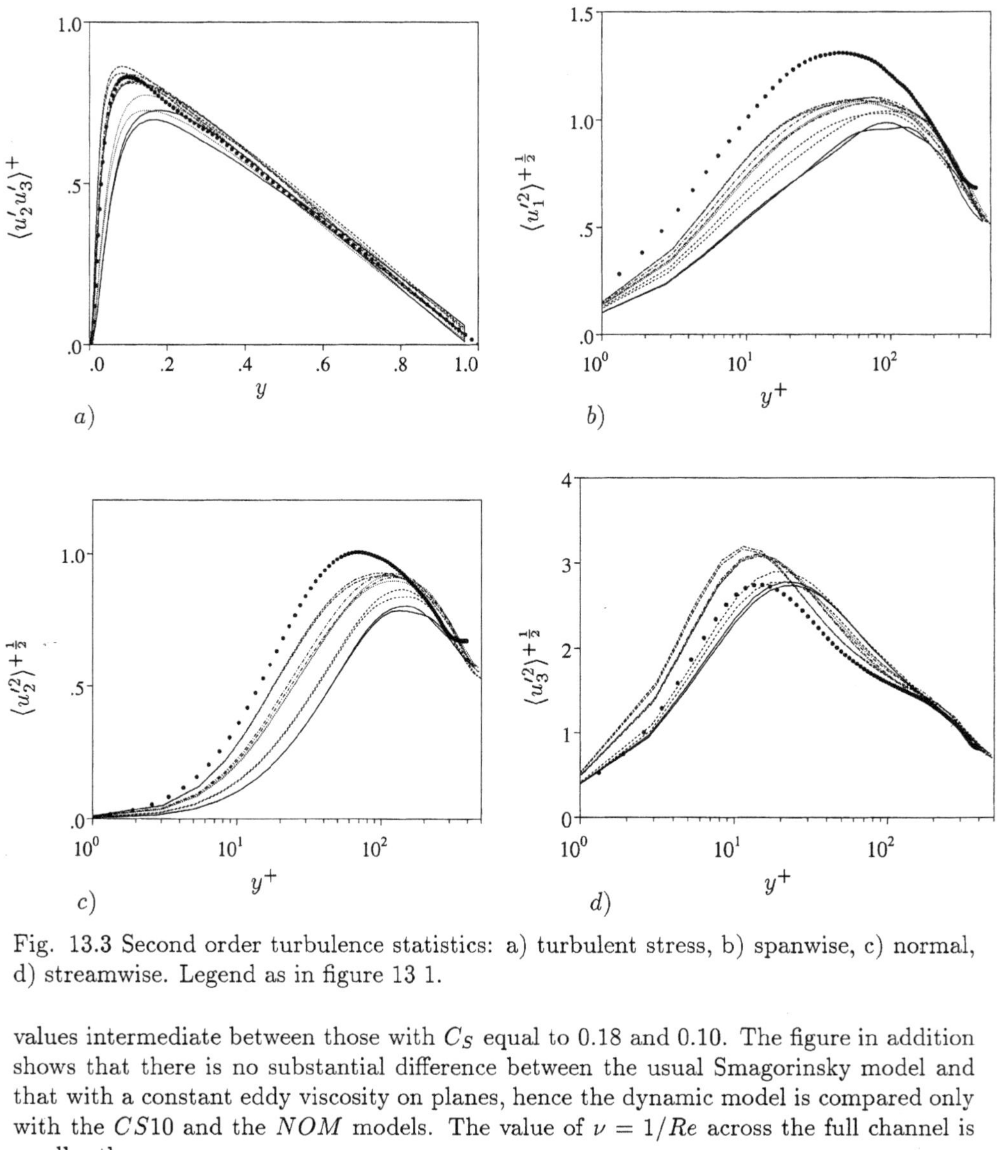

Fig. 13.3 Second order turbulence statistics: a) turbulent stress, b) spanwise, c) normal, d) streamwise. Legend as in figure 13 1.

values intermediate between those with C_S equal to 0.18 and 0.10. The figure in addition shows that there is no substantial difference between the usual Smagorinsky model and that with a constant eddy viscosity on planes, hence the dynamic model is compared only with the $CS10$ and the NOM models. The value of $\nu = 1/Re$ across the full channel is smaller than ν_T.

Fig.13.5 shows that the higher values of ν_T for DYN model with respect to those for $CS10$ shift upwards the log law. The dynamic model at this Reynolds number is too dissipative. The influence of filtering or of the different possibilities to evaluate the constant have not been investigated. Sarghini & Piomelli (1998) found at a higher Re the same behavior even by different filtering procedures in the pseudospectral code. The

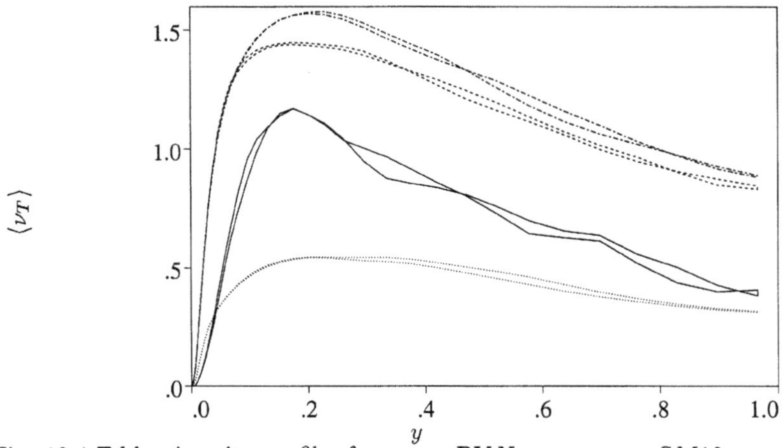

Fig. 13.4 Eddy viscosity profiles for ——— DYN, - - - - $SM18$, —-— $CS18$, ········ $SM10$, —·— $CS10$.

profiles of the normal and tangential stress for the dynamic model do not differ too much from those obtained by $CS10$ and are not shown.

A comparison of the present results with those by Cabot (1994), (his figure 3) shows that in the present case the dynamic model is too dissipative. Perhaps this could be due to the different grids used in the streamwise and spanwise directions. To my experience the effects of resolution in LES are large, and this point still has not been deeply investigated. By purpose the same grid of Cabot (1994) was not used; thus who is interested could use this code to investigate the reasons of the discrepancies with Cabot's simulations. The simulation by Lambalais *et al.* (1996) at the same Re was done on a much finer grid and this is the reason why the results in Fig.XII-6 at Pg.393 of Lesieur's (1992) book agree well with those obtained by DNS.

As a first comment on these simulations I would like to affirm that the present very coarse LES produces reasonable results from an engineering point of view.

13.3.a Results at high Reynolds numbers

To investigate how the present code behaves when the Reynolds number increases, simulations at $Re = 30000$ (approximately 2.5 times the previous one) have been performed. In this simulation the number of grid points was increased 1.5 times, but the box size as well as the clustering in the normal coordinate were the same as in the previous simulation. The simulation started from the Poiseuille profiles and evolved for 150 dimensionless units without any model. From this field the different models have been checked as it was done for $R_\tau = 400$; however, at this R_τ, the simulations with $C_S = 0.18$ were not done any longer. The Reynolds number achieved by this simulation ($R_\tau = 1020$) is close to that in Sarghini & Piomelli (1998). Fig.13.6 shows that when the Reynolds number increases the results by NOM do not differ largely from those by $CS10$ and $SM10$, close to the log law. In this figure the experimental data by Wei & Willmarth (1989) are not plotted since

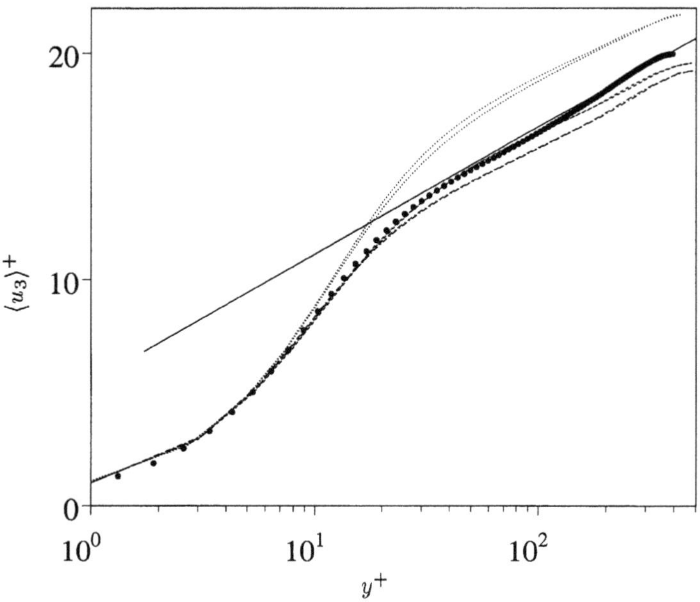

Fig. 13.5 Velocity profile in wall units: ▪ DNS by Kim & Antonia (1993), —— $U^+ = \frac{1}{0.41}\log y^+ + 5.5$, - - - - $CS10$, ········ DYN, —·— NOM, at $R_\tau = 395$.

these coincide with the log law. The dynamic model is shifted upwards the same amount of that found by Sarghini & Piomelli (1997); even though they, by the pseudospectral code, found that without model the log law was shifted downwards. The present $\langle u_3 \rangle^+$ from the dynamic model differs largely from that in figure 5 of Cabot (1994) by a similar code. The reason of this difference should be investigated and possible reasons are the grid, the averaging procedure, and the filtering. From Cabot's paper it was not clear which filter he was using. The upward shifting of the log law by DYN is related to $u_\tau = 0.0297$ which was smaller than that for the other models: for $CS10$ $u_\tau = 0.034$ and for NOM $u_\tau = 0.0344$. This difference again is explained by the eddy viscosity profiles in Fig.13.7 that show a greater value in the wall region. This high viscosity reduces the strength of the near wall vortical structures and as a consequence the friction is reduced.

The comparison between the normal stresses for the different models considered is a further check to see how the results depend on the models. For a better analysis of the results the original figures by Sarghini & Piomelli (1998) and by Cabot (1994) are reported. The plots of Sarghini & Piomelli (1998) show the stress profiles in the near wall region. In the present and in the Cabot (1994) simulation, a semi-log plot has been used to emphasize the wall region and in the mean time to have an idea of the behavior in the full channel, a semi-log plot has been used. Fig.13.9 reveals that the present $\langle u_1'^2 \rangle$ agrees better with the results by Sarghini & Piomelli (1998) by locating the maximum of this stress in the

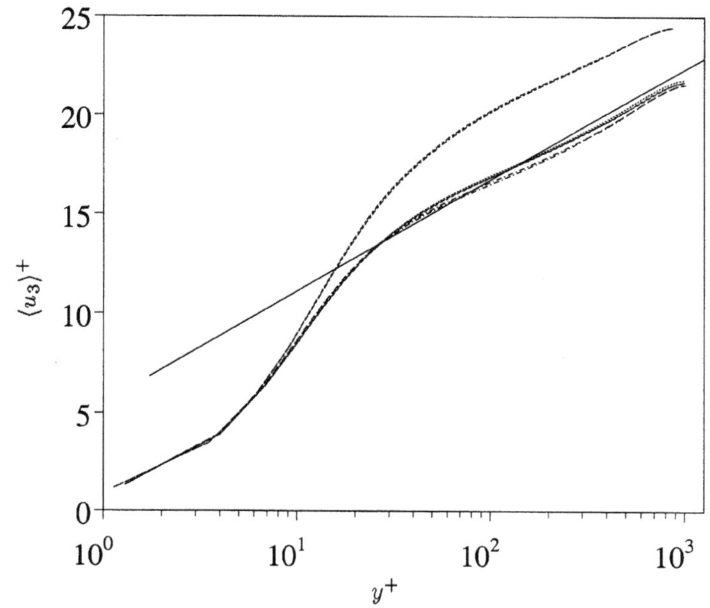

Fig. 13.6 Velocity profile in wall units: —— $U^+ = \frac{1}{0.41} \log y^+ + 5.5$, ---- $CS10$, ... DYN, —·— NOM, at $R_\tau = 1020$.

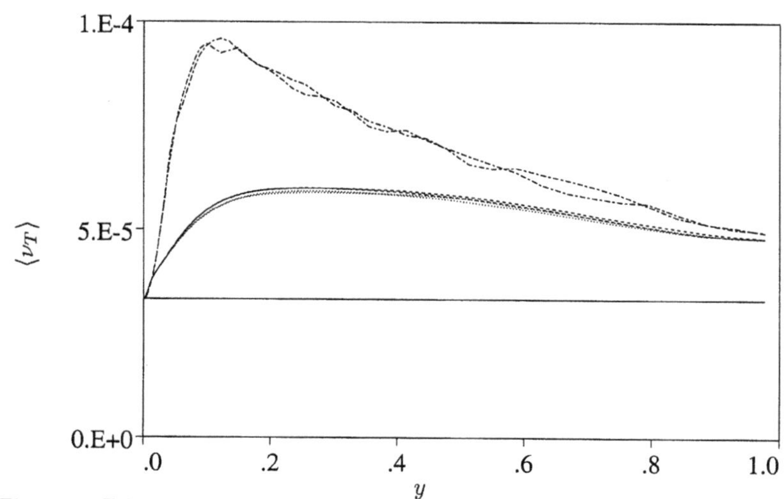

Fig. 13.7 Eddy viscosity profiles for: —— $CS10$, ---- DYN, at $R_\tau = 1020$

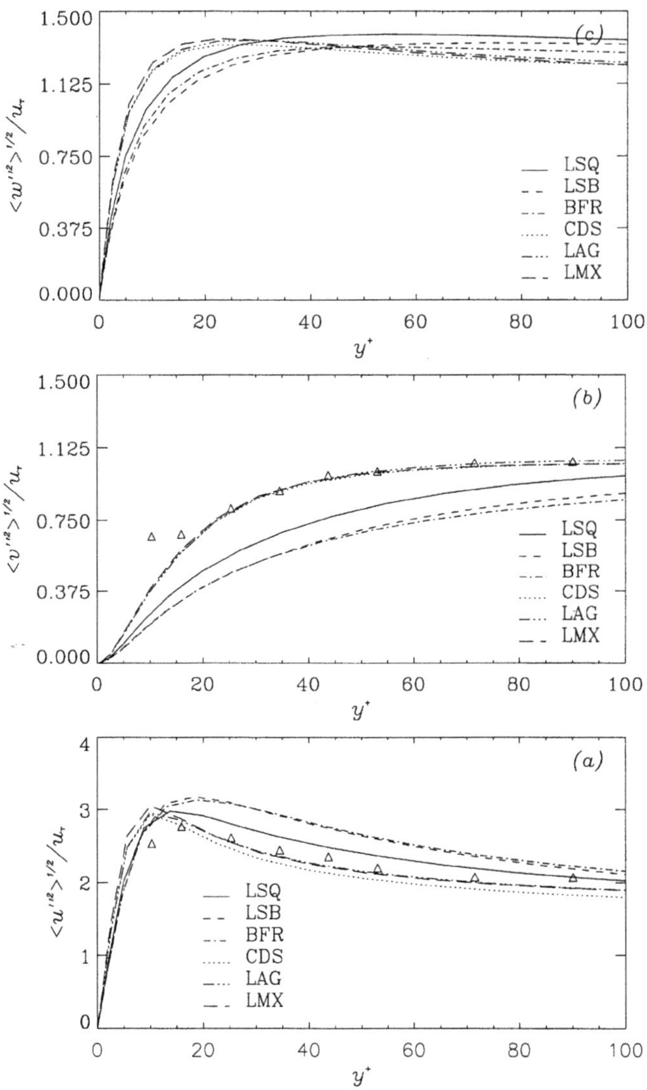

Fig. 13.8 Second order turbulence statistic from Sarghini & Piomelli (1998): a) spanwise, b) normal, c) streamwise .

NOM case closer to the wall than for the *DYN* model. The present values are smaller than those by Sarghini & Piomelli (1998) and slightly higher than those by Cabot (1994). The differences found by Sarghini & Piomelli (1998) between the *DYN* and *NOM* model

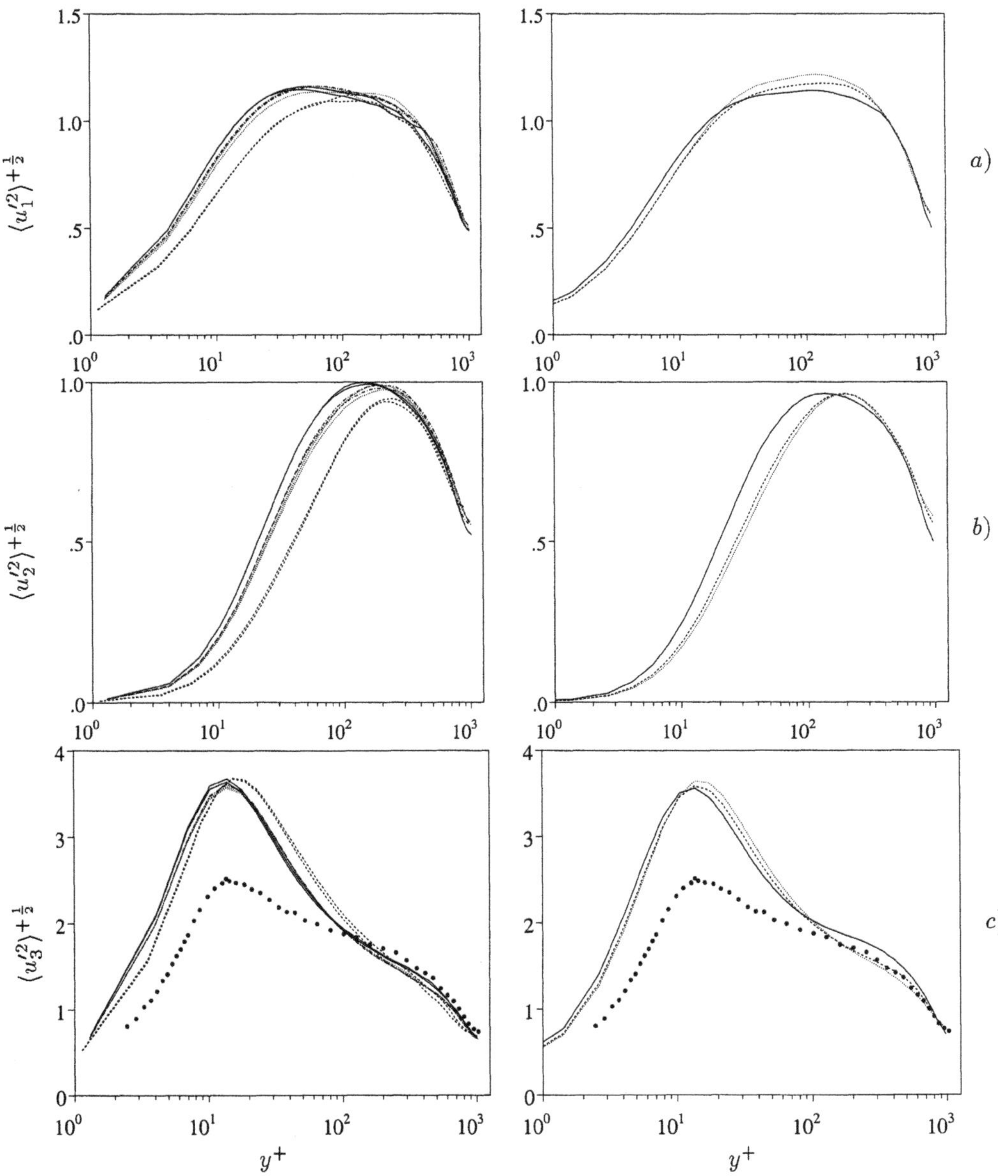

Fig. 13.9 Second order turbulence statistics: a) spanwise, b) normal, c) streamwise. Left present: ········ DYN, ---- $CS10$, —— NOM; right Cabot (1994): ········ DYN, ---- $CS10$, —— NOM; symbols Wei & Willmarth (1989).

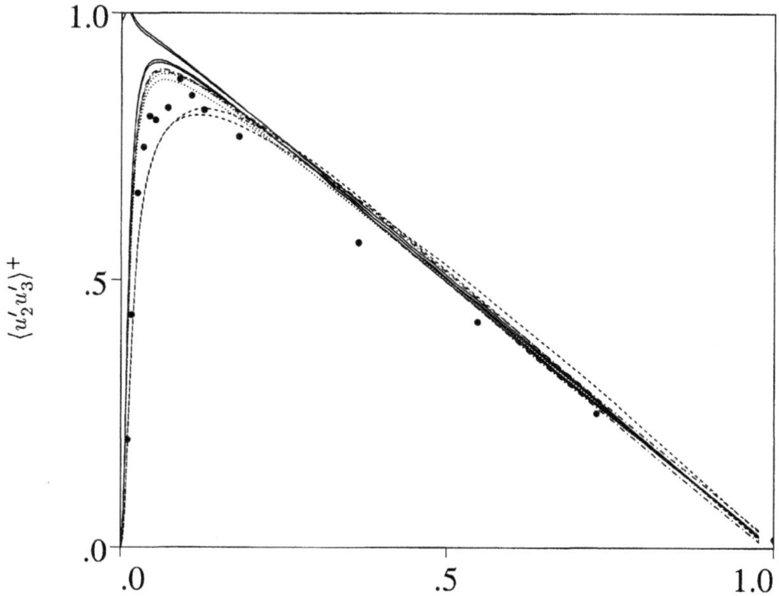

Fig. 13.10 Reynolds stress profile: —— DYN, ---- $CS10$, ········ NOM, symbols Wei & Willmarth (1989).

are much larger than they are here and with a good probability are linked to the numerical scheme. For $\langle u_2'^2 \rangle^{+\frac{1}{2}}$ it turns out that there is some difference between NOM and $CS10$ and since the eddy viscosity does not act implicitly in the NOM, the peak closer to the wall indicates more active wall structures. The peak values $\langle u_1'^2 \rangle^{+\frac{1}{2}}$ for the previous stress are intermediate between the results of Sarghini & Piomelli (1998) and those of Cabot (1994). The comparison with the experimental results by Wei & Willmarth (1989) in Fig.13.10 indeed shows that the $CS10$ model works well at this high Reynolds number. $\langle u_3'^2 \rangle^{+\frac{1}{2}}$ in the present case shows in Fig.13.9 that (as found by Cabot) it is independent from the model and that the peak values agree in both simulations. The pseudospectral simulations by Sarghini & Piomelli (1998) in this case predict lower peak values which are in a better agreement with the experimental values of Wei & Willmarth (1989). The tangential stress in Fig13.10 confirms the greater viscosity near the wall of the DYN model and a comparison with the measurement of Wei & Willmarth (1989) shows that the $CS10$ model reproduces the experimental results very well.

13.4 Conclusions

In this chapter the numerical method and the relative code to investigate the features of different subgrid models in the LES of a plane channel have been described. The results show that the Smagorinsky model with an eddy viscosity constant on horizontal planes gives very satisfactory results, but it requires wall damping and a value for the Smagorinsky constant. The dynamic model on the other hand gives results that agree substantially well with previous results obtained by different numerics and by different filtering procedures, but which do not reproduce well the experimental results especially at high Reynolds numbers. This model, in principle, is very challenging because it does not require a priori assumptions. This is confirmed by the large amount of papers that appeared in the last years with different modifications and improvements. The aim of this chapter was not devoted to investigate which is the best subgrid model available. The main aim, as for homogeneous flows, is to furnish to whom is interested a tool to make improvements on the different subgrid models. It has been shown, together with the results in chapter 10, that a staggered finite difference scheme is accurate enough to get satisfactory results for DNS and LES in wall-bounded flows. As found by Cabot, the subgrid models produce large effects at low Reynolds number and these decrease by increasing the Reynolds number. At the moment it is difficult to give a reasonable explanation why at very high Re without any model the results are so good. To understand this fact different resolutions should be tested and this is a very pleasant job for a graduate student. By investigating the influence of the resolution perhaps more can be learned about the link, that exists between numerics and subgrid models in LES. To this purpose several authors are devoting a large part of their time and to my knowledge several of these issues can be find in the publications of the CTR at Stanford.

References

ADRIAN, R.J. 1991 Particle-Imaging Techniques for Experimental Fluid Mechanics. *Annual Review of Fluid Mechanics.* **23,** 261–304.

AKSELVOLL, K. & MOIN, P. 1996 An efficient method for temporal integration of Navier–Stokes equations in confined axisymmetric geometries. *J. Comp. Phys.* **125,** 454–463.

AKSELVOLL, K.& MOIN, P. 1996 Large Eddy Simulation of Turbulent Confined Coannular Jets . *J. Fluid Mech.* **315,** 387-411.

ANDERSON, D.A., TANNEHIL, J.C. & PLETCHER, R.H. 1994 Computational fluid mechanics and heat transfer. *Hemisphere New York.* ,

ANDREOPOULOS, J.& AGUI, J.H. 1996 Wall-vorticity flux dynamics in a two-dimensional turbulent boundary layer. *J. Fluid Mech.* **309,** 45–84.

ANSELMET, F., GAGNE, Y., HOPFINGER, E. J. & ANTONIA, R. A. 1984 High order velocity structure functions in turbulent shear flows . *J. Fluid Mech.* **140,** 63.

ANTONIA, R.A., ZHU,Y. & SOKOLOV, M. 1995 Effect of concentrated wall suction on a turbulent boundary layer. *Phys. Fluids.* **10,** 2465–2474.

ARAKAWA, A. 1966 Computational design for long term numerical integration of the equations of fluid motion: Two-dimensional incompressible flow. Part I. . *J. Comp. Phys.* **1,** 119.

ASHURST, W.T., KERSTEIN, A.R., KERR, R.M. & GIBSON, C.H. 1987 Alignment of vorticity and scalar gradient with strain in simulated Navier Stokes turbulence. *Phys. Fluids.* **30,** 3243–3253.

AUERBACH, D. 1991 Stirring properties of vortex rings. *Phys. Fluids A.* **3,** 1351–1355.

BARDINA J., FERZIGER J.H. & REYNOLDS, W.C. 1983 Improved turbulence models based on large eddy simulation of homogeneous, incompressible, turbulent flows. *Stanford Univ., Dept. Mech. Eng. Thermosciences Division Report TF-19.*

BARDINA, J., FERZIGER, J.H. & ROGALLO, R.S. 1985 Effect of rotation on isotropic turbulence: computation and modelling. *J. Fluid Mech.* **154,** 321–336.

BASDEVANT, C.,& SADOURNY, R. 1984 Numerical simulation of incompressible flow: the Mask method. *Laboratoire de Metereologie Dynamique Ecole Normale Superieure, Paris (Unpublished).* ,

BASDEVANT, C., DEVILLE, M., HALDENWANG,P., LACROIX, J.M. , QUAZZANI,J. PETRET, R., ORLANDI P.& PATERA, A.T. 1986 Spectral and Finite Difference Solutions of the Burgers Equation. *Computers and Fluids.* **14,** 23–41,

BATCHELOR, G.K. & TOWNSED, A.A. 1949 The nature of turbulent motion at large wave numbers. *Proc. R. Soc. Lon. A.* **199,** 238–255.

BATCHELOR, G.K. 1953 The theory of homogeneous turbulence . *Cambridge University Press, Cambridge.*

BATCHELOR, G.K. 1970 An introduction to Fluid Dynamics . *Cambridge University Press, Cambridge.*

BATCHELOR, G.K., CANUTO, V.M., & CHASNOV, J.R. 1992 Homogeneous buoyancy generated turbulence. *J. Fluid Mech.* **235,** 349–378.

BEAM, R.M. & WARMING, R.F. 1976 An implicit finite-difference algorithm for hyperbolic systems in conservation-law form. *J. Comp. Phys.* **22,** 87–110.

BEAUDAN, P. & MOIN, P. 1994 Numerical experiments on the flow past a circular cylinder at sub-critical Reynolds number. *Report No TF-62 Dept. Mech.Eng., Stanford University, Stanford California.*

BECHERT, D.W., BRUSE, M., HAGE, W., HOEVEN, J.G.T. & HOPPE, G. 1997 Experiments on drag reducing surfaces and their optimization with an adjustable geometry. *J. Fluid Mech.* **338,** 59–87.

BENZI, R., CILIBERTO, S., TRIPICCIONE, R., BAUDET, C., MASSAIOLI, F. & SUCCI, S. 1993 Extended self-similarity in turbulent flows. *Phys. Rev. E.* **48,** R29–R32.

BERNAL, L.P., HIRSA A., KWON J.T., & WILLMARTH, W.W. 1989 On the interaction of vortex rings and pairs with a free surface for varying amounts of surface active agent. *Phys. Fluids.* **A12,** 2001–2004.

BIDOKHTI, A.A. & TRITTON, D.J. 1992 The structure of a turbulent free shear layer in a rotating fluid. *J. Fluid Mech.* **241,** 469–502.

BLACKBURN, H.M., MANSOUR, N.N. & CANTWELL, B.J. 1996 Topology of fine-scale motion in turbulent channel flow. *J. Fluid Mech.* **310,** 269-292.

BOHNERT, M. & FERZIGER, J.H. 1993 The dynamic subgrid scale model in large-eddy simulation of stratified Ekman layer. *Proceedings of the 2nd International symposium on Engineering Turbulence Modelling and Measurements, Florence, Italy.*

BORATAV, O.N. & PELZ, R.B. 1997 Structures and structure functions in the inertial range of turbulence. *Phys. Fluids.* **9,** 1400–1415.

BORIS, J.P., GRINSTEIN, F.F., ORAN, E.S. & KOLBE, R.L. 1992 New insight into large eddy simulation. *Fluid Dynamics Research.* **10,** 199–228.

BOUSSINESQ, M. 1870 Essai sur la theorie des eaux courantes. *Memoires a l'academie des sciences de l'institut national de France, Tome XXIII N°.*1.

BRILEY W.R. & MCDONALD, H. 1975 Solution of three-dimensional compressible Navier-Stokes equations by an implicit technique. *Proceedings of the fourth international conference on numerical methods in Fluid Dynamics. Lecture Notes in Physics.* **Vol 35,** Springer Verlag, NewYork.

BRISCOLINI, M. & SANTANGELO, P. 1989 Development of the mask method for incompressible unsteady flows. *J. of Comp. Phys.* **84,** 57–75.

BRISCOLINI, M. & SANTANGELO, P. 1994 The non-Gaussian statistics of the velocity field in low-resolution large-eddy simulations of homogeneous turbulence. *J. Fluid Mech.* **270,** 199–217.

BROWN, G.L. & ROSHKO, A. 1974 On density effects and large structure in turbulent mixing layers. *J. Fluid Mech.* **64,** 775–816.

BROWN, G.L. & LOPEZ, J.M. 1990 Axisymmetric vortex breakdown. Part.2 Physical mechanism. *J. Fluid Mech.* **221,** 553–576.

BUELL, J.C. & CATTON, I. 1983 The effect of wall conduction on the stability of a fluid in a right circular cylinder heated from below. *J. of Heat Tran.* **105,** 255–260.

BUELL, J.C. 1989 Incompressible spatially-developing free-shear flows. *Annual Research Brief 1989, Center for Turbulence Research, Stanford.* , 195–213.

BUELL, J.C., & HUERRE, P. 1988 Inflow/Outflow Boundary Conditions and Global Dynamics of Spatial Mixing Layers. *Center for Turbulence Research, Proceedings of the Summer Program, 1988.* , 19–28.

CABOT, W. 1992 Large eddy simulations of time-dependent and buoyancy-driven channel flows. *Center for Turbulence Research, Annual Research Briefs-1992.* , 45–60.

CABOT, W. 1993 Dynamic localization and second-order subgrid-scale models in large eddy simulations of channel flow. *Center for Turbulence Research, Annual Research Briefs-1993.* , 129–144.

CABOT, W. 1994 Local dynamic subgrid-scale models in channel flows. *Center for Turbulence Research, Annual Research Briefs-1994.* , 143–160.

CAMBON, C., MANSOUR, N.N. & GODEFERED, F.S. 1997 Energy transfer in rotating turbulence. *J. Fluid Mech.* **337,** 303–332

CAMUSSI, R., BARBAGALLO, D., GUJ, G. & STELLA, F. 1996 Transverse and longitudinal scaling laws in non-homogeneous low *Re* turbulence. *Phys. Fluids.* **8,** 1181–1191.

CANTWELL, B.J. 1981 Organized motion in turbulent flow. *Annual Review Fluid Mech.* **13,** 457–515.

CANUTO, C., HUSSAINI, M.Y., QUARTERONI, A. & ZANG, T.A. 1987 Spectral Methods in Fluid Mechanics. *Springer-Verlag, New York.*

CARATI D., GHOSAL, S & MOIN, P. 1995 On the representation of backscatter in dynamic localization models. *Phys. Fluids.* **3,** 606–6161.

CARATI, D., WRAY, A. & CABOT, W. 1996 Ensemble averaged dynamic modeling. *Proceedings of the 1996 Summer Program, Center for Turbulence Research Stanford Ca.* , 237–248.

CARNEVALE, G.F., KLOOSTERZIEL, R. & VAN HEIJST, G.J. 1991a Propagation of barotropic vortices over topography in a rotating tank. *J. Fluid Mech.* **233,** 119-139.

CARNEVALE, G.F., CAVAZZA, P., ORLANDI, P. & PURINI, R. 1991b Anomalous vortex merger in laboratory experiments. *Phys. Fluids A.* **5,Part. 2,** 1411–1415.

CARNEVALE, G.F., MCWILLIAMS, J.C., POMEAU Y., WEISS J.B., & YOUNG W.R. 1991c Evolution of vortex statistics in two-dimensional turbulence. *Phys. Rev. Letters.* **64,** 2735-2737.

CARNEVALE, G.F. & KLOOSTERZIEL, R.C. 1994 Emergence and evolution of triangular vortices. *J. Fluid Mech.* **259,** 305–331

CARNEVALE, G.F.,BRISCOLINI, M., KLOOSTERZIEL, R.C. & VALLIS, G.K. 1997 three-dimensionally perturbed vortex tubes in a rotating flow. *J. Fluid Mech.* **341,** 127-163.

CARNEVALE, G.F., FUENTES, O.V. & ORLANDI, P. 1997 Dipole rebound from a coast. *J.Fluid Mech.* **351,** 75–103.

CARNEVALE, G.F., SMITH, S.G.L., CRISCIANI, F. PURINI, R. & SERRAVALL, R. 1998 Bifurcation of a Coastal Current at an Escarpment. *in print on J. Phys. Oceanography.*

CARTON, X.J., FLIERL, G.R. & POLVANI, L.M. 1989 The generation of tripoles from unstable axisymmetric isolated vortex structures. *Europhys. Lett.* **9,** 339–344.

CAVAZZA, P. 1992 Dinamica di strutture di vorticita' per lo studio di flussi geofisici. *Tesi di Dottorato di Ricerca di Meccanica Applicata, Universita' di Roma "La Sapienza".*

CAVAZZA, P., HEIJST, G.J.F. VAN & ORLANDI, P. 1992 The stability of vortex dipoles. *Proceedings of the 11th Australasian Fluid Mech. Conf., Hobart, Australia.* , 67.

CERRA, A.W. & SMITH, C.R. 1983 Experimental observation of vortex ring interaction with the fluid adjacent to a surface. *Lehigh University* Rep, **FM-4**

CHAPMAN, D.R. 1979 Computational aerodynamics development and outlook. *AIAA Journal.* **17,** 193–1313.

CHASNOV, J.R. 1991 Simulation of the Kolmogorov inertial subrange using an improved subgrid model. *Phys. Fluids A.* **3,** 188–200.

CHASNOV, J.R. 1995 Similarity states of passive scalar transport in buoyancy generated turbulence. *Phys. Fluids A.* **7,** 1498–1506.

CHOI, H., MOIN, P. & KIM, J. 1993 Direct simulation of turbulent flow over riblets. *J. Fluid Mech.* **255,** 503–539.

CHOI, H. & MOIN, P. 1994 Effects of the computational time step on numerical solutions of turbulent flow. *J. of Comp. Phys.* **113,** 1–4.

CHOI, H., MOIN, P. & KIM, J. 1994 Active turbulence control for drag reduction in wall-bounded flows. *J. Fluid Mech.* **262,** 75–110.

CHOI, K.S. & GRAHAM, M. 1998 Drag reduction of turbulent pipe flows by circular-wall oscillations. *Phys. Fluids.* **1,** 7–9.

CHOMAZ, J.M., BONNETON, P., & HOPFINGER, E.J 1993 The structure of the near wake of a sphere moving horizontally in a stratified fluid. *J. Fluid Mech.* **254,** 1–21.

CHU, D.C. & KARNIADAKIS, G.E. 1993 A direct numerical simulation of laminar and turbulent flow over riblet-mounted surfaces. *J. Fluid Mech.* **250,** 1–42.

CLARK, R.A., FERZIGER, J.H. & REYNOLDS, W.C. 1979 Evaluation of subgrid models using an accurately simulated turbulent flow. *J. Fluid Mech.* **91,** 1–16.

COMTE-BELLOT,G. 1965 Ecoulement turbulent entre deux parois paralleles. *Publ. Sci. Techn. Min. Air. No.419 Paris.*

COMTE-BELLOT, G. & CORRSIN, S. 1971 Simple Eulerian time correlations of full- and narrow-band velocity signals in grid generated isotropic turbulence. *J. Fluid Mech.* **48,** 273–337.

COOK, A.W. & RILEY, J.J. 1996 Direct numerical simulation of a turbulent reactive plume on a parallel computer. *J. of Comp. Phys.* **129,** 263–283.

COOK, A.W. & RILEY, J.J. 1998 Subgrid-scale modeling for turbulent reacting flows. *Combustion and Flame.* **112,** 593–606.

CORCOS, G.M. & SHERMAN, F.S 1984 The mixing layer: deterministic models of a turbulent flow. Part 1. Introduction and the two-dimensional flow. *J. Fluid Mech.* **139,** 29–66.

CORCOS, G.M. & LIN, S.J 1984 The mixing layer: deterministic models of a turbulent flow. Part 2. The origin of the three-dimensional motion. *J. Fluid Mech.* **139,** 67-96.

CROCCO, L. & ORLANDI, P. 1986 A Transformation of the Triade Integral for Isotropic Turbulence. *J. Fluid Mech.* **16,** 405–424.

CROW, S.C. 1970 Stability theory for a pair of trailing vortices. *AIAA J..* **8,** 2172–2179.

COUDER,Y. & BASDEVANT, C. 1986 Experimental and numerical studies of vortex couples in two-dimensional flows. *J. Fluid Mech.* **173,** 225–251.

CUNSOLO, D. & ORLANDI, P. 1978 Accuracy in Non Orthogonal Grid Reference Systems. *Numerical Methods in Laminar and Turbulent Flow, Swansea, U.K., July 1978.*

DIDDEN, N. 1979 On the formation of vortex rings: rolling-up and production of circulation. *ZAMP.* **30,** 101–116.

DAHM, W.J.A., FRIELER C.E., & TRYGGVASON, G. 1992 Vortex structure and dynamics in the near field of a coaxial jet. *J. Fluid Mech.* **241,** 371–402.

DEARDOFF, J.W. 1970 A numerical study of three-dimensional turbulent channel flow at large Reynolds numbers. *J. Fluid Mech.* **41,** 453–480.

DOMARADZKI, A.J. & ROGALLO, R.S. 1990 Local energy transfer and non-local interactions in homogeneous isotropic turbulence. *Phys. Fluids.* **3,** 413–426.

DOMARADZKI, A.J. & SAIKI, E.M. 1997 A subgrid-scale model based on the estimation of unresolved scales. *Phys. Fluids.* **7,** 2148–2164.

DOMENICHINI, F. & PEDRIZZETTI, G. 1998 Impulsivell started flow separation in wavy-walled tubes. *J. Fluid Mech.* **359,** 1–22.

DOUADY, S., COUDER, Y. & BRACHET, M.E. 1991 Direct observation of intermittency of intense vorticity filaments in turbulence. *Phys. Rev. Lett..* **67,** 983–986.

DRAZIN, P.G. & REID, W.H. 1981 Hydrodynamic Stability. *Cambridge University Press, Cambridge.*

DRIEST , E.R. VAN 1956 On the turbulent flow near a wall. *J. Aeronautical Sci.* **23,** 1007–1017.

DURBIN, P.A. 1991 Near-wall turbulence closure modelling without "damping functions". *Theoret. Comp. Fluid Dynamics.* **3,** 1–13.

DURST, F., JOVANOVIC', J. & SENDER, J. 1995 LDA measurements in thew near-wall region of a turbulent pipe flow. *J. Fluid Mech.* **295,** 305–335.

DYKE, M. VAN 1982 An album of fluid motion. *The Parabolic Press, Satnford Ca.*

ECKELMANN, H. 1974 The structure of the viscous sublayer and the adjacent wall region in a turbulent channel flow. *J. Fluid. Mech.* **65,** 439–459.

EGGELS, J.G.M., UNGER, F., WEISS, M.H., WESTERWEEL, J., ADRIAN, R.J., FRIEDRICH, R. & NIEUWSTADT, F.T.M. 1994a Fully developed turbulent pipe flow: a comparison between direct numerical simulation and experiment. *J. Fluid Mech.* **268,** 175–209.

EGGELS, J.G.M., BOERSMA, B.J. & NIEUWSTADT, F.T.M. 1994b Direct and large eddy simulations of turbulent flow in an axially rotating pipe. *Submitted to J.Fluid Mech.*

EISENGA, A.H.M., VERZICCO, R., & HEIJST, G.J.F. VAN 1998 Dynamics of a vortex ring moving perpendicular to a background rotation. *J. Fluid Mech.* **354,** 69–100.

ETLING, D. & FERNANDO, H.J.S. 1994 On the influence of background rotation on turbulent jets. *Recent research advances in the fluid mechanics of turbulent jets and plumes (P.A. Davies and M.J. Valente Neves, Eds.), Kluwer, Dordrecht.* , 401–412.

FADLUN, E.V. 1998 Flussi in geometrie complesse con forze di massa su griglie Cartesiane:simulazioni numeriche e validazione sperimentale. *Thesis of master degree Universita' di Roma "La Sapienza".*

FABRIS, D. & LIEPMANN, D. 1996 Quantitative experimental and numerical investigation of a vortex ring impinging on a wall. *Phys. Fluids.* **10,** 2640–2649

FATICA, M., ORLANDI, P. & VERZICCO, R. 1994 Direct and Large Eddy Simulation of Round Jet. *Direct and Large Eddy Simulation I, selected papers from the First ERCOFTAC Workshop on Direct and Large Eddy Simulation, edited by Voke, Kleiser and Chollet, Kluwer.*

FERZIGER, J.H. 1987 Simulation of incompressible turbulent flows. *J. of Comput. Phys.* **69,** 1–48.

FERZIGER, J.H., & PERIC, M. 1996 Computational Methods for Fluid Dynamics . *Springer-Verlag, New York.*

FLETCHER, R. 1987 Computational Techniques for Fluid Dynamics, Vol I and II. *Springer-Verlag, New York.*

FLOR, J.B. & HEIJST, G.J.F. VAN 1994 An experimental study of dipolar vortices in a stratified fluid. *J. Fluid Mech.* **279,** 101–133.

FRISCH, U. 1995 Turbulence. *Cambridge University Press, Cambridge.*

GENT, P.B. & MCWILLIAMS, J.C. 1986 Barotropic and baroclinic instabilities of axisymmetric vortices in a quasi-geostrophic model. In . *Mesoscale/Synoptic Coherent Structures in Geophysical Turbulence (eds. J.C.J. Nihoul & B.M. Jamart).* , 225-244.

GERMANO, M. 1986a Differential filters for the large eddy numerical simulation of turbulent flows. *Phys. Fluids.* **6,** 1755–1757.

GERMANO, M. 1986b Differential filters of elliptic type . *Phys. Fluids.* **6,** 1757–1758.

GERMANO, M. 1986c A proposal for redefinition of the turbulent stresses in the filterd Navier-Stokes equations. *Phys. Fluids.* **7,** 2323–2324.

GERMANO, M., PIOMELLI, U., MOIN, P. & CABOT, W.H. 1991 A dynamic subgrid-scale eddy viscosity model. *Phys. Fluid A.* **3,** 1760–1765.

GHARIB, M., RAMBOD, E., & SHARIFF, K. 1998 A universal time scale for vortex ring formation. *J. Fluid Mech.* **360,** 121–140.

GHOSAL, S., LUND, T.S. & MOIN, P. 1992 A local dynamic model for large eddy simulation. *Center for Turbulence Research, Annual Research Briefs-1992.* , 3.

GHOSAL, S. 1996 An analysis of numerical errors in large-eddy simulations of turbulence. *J. Comp. Phys.* **125,** 187–206.

GIBSON, C.H. 1986 Internal waves, fossil turbulence and composite ocean microstructure spectra. *J. Fluid Mech.* **168,** 89–117.

GLAZER, A. 1988 The formation of vortex rings. *Phys. Fluids A.* **31,** 3523–3542.

GLEDZER, E., VILLERMAUX, E., KAHALERRAS, H. & GAGNE,Y. 1996 On the log-Poisson statistics of the energy dissipation field and related problems of developed turbulence. *Phys. Fluids.* **8,** 3367–3378.

GOGINENI, S. & SHIH, C. 1997 Experimental investigation of the unsteady structure of a transitional plane wall jet. *Exp. in Fluids.* **23,** 121–129.

GOLDSTEIN, D., HANDLER, R. & SIROVICH, L. 1993 Modeling a no-slip flow boundary with an external force field. *J. Comp. Phys.* **105,** 354–366.

GRAMMELVEDT, A. 1969 A survey of finite-difference schemes for the primitive equations for the barotropic fluid. *Monthly Weather Review.* **97/5,** 384–406.

GREENSPAN, H.P. 1980 The theory of rotating fluids. *Cambridge University Press, Cambridge.*

GRIFFITS, R.W. & HOPFINGER, E. 1986 Experiments with baroclinic vortex pairs in a rotating fluid. *J. Fluid Mech.* **173,** 501–518.

HEIJST, G.J.F. VAN, KLOOSTERZIEL, R.C. & WILLIAMS. C.W.M. 1991 Laboratory experiments on the Tripolar vortex in a rotating fluid. *J. Fluid Mech.* **225,** 301–331.

HEIJST, G.J.F. VAN & FLOR, J.B. 1989 Dipole formation and collision in a stratified fluid. *Nature.* **340,** 212–215.

HERRING, J.R., SCHERTZER, D. LESIEUR, M., NEWMAN, G.R., CHOLLET, J.P. & LARCHEVEQUE, M. 1982 A comparative assessment of spectral closures as applied to passive scalar diffusion. *J. Fluid Mech.* **124,** 411–437.

HINZE, J.O. 1975 Turbulence. *Mc-Graw-Hill, New York.*

HIRAI, S., TAKAGI, T. & MATSUMOTO, M. 1988 Prediction of the laminarization phenomena in an axially rotating pipe flow. *Journal of Fluids Engineering.* **110,** 424–430.

HIRSCH, C. 1991 Numerical computation of internal and external flows, Vol I and Vol II. *John Wiley & sons, New York*

HO, C.M. & TAI, Y.C. 1998 Micro-electro-mechanical-systems (MEMS) and fluid flows. *Annual Review of Fluid Mechanics.* **30,** 579–612

HOMA, J., LUCAS, M. & ROCKWELL. D. 1988 Interaction of impulsively generated vortex pairs with bodies . *J. Fluid Mech.* **197,** 571–594.

HOPFINGER, E.J., BROWAND,K.F. & GAGNE Y. 1982 Turbulence and waves in a rotating tank. *J. Fluid Mech.* **125,** 505–534.

HOPFINGER, E.J. & HEIJST, G.J.F. VAN 1993 vortices in Rotating Fluids. *Annual Review of Fluid Mechanics.* **25,** 241–289.

HORIUTI, K. 1987 Comparison of conservative and rotational forms in large-eddy simulation of turbulent channel flow. *J. of Comp. Phys.* **71,** 343–370.

HUSSAIN, A.K.M.F. 1986 Coherent structures and turbulence. *J. Fluid Mech.* **173,** 303-356.

HUSSAIN, A.K.M.F. & REYNOLDS, W.C. 1972 The mechanics of an organized wave in turbulent shear flows. Part. 2 . Experimental results.. *J. Fluid. Mech.* **54,** 241–261.

HUSSAIN, A.K.M.F. & MELANDER, M.V. 1991 Understanding turbulence via vortex dynamics . *Studies in Turbulence. Spinger-Verlag.* , 157-178.

JIMÉNEZ, J. & MOIN, P. 1991 The minimal flow unit in near wall turbulence. *J. Fluid. Mech.* **225,** 213–241.

JIMÉNEZ, J. 1992 Wall friction and the wall structure in near wall turbulence. *Proc. 11th. Australasian Fluid Mech. Conf., Hobart, Dec. 1992.*

JIMÉNEZ, J., WRAY, A.A., SAFFMANN, P.G. & ROGALLO, R.S. 1993 The structure of intense vortices in isotropic turbulence. *J. Fluid. Mech.* **255,** 65–90.

JIMÉNEZ, J., MOFFATT, H.K. & VASCO, C. 1996 The structure of vortices in freely decaying two-dimensional turbulence. *J. Fluid. Mech.* **313,** 209–222.

JUNG, W.J., MANGIAVACCHI, N. & AKHAVAN, R. 1992 Suppression of turbulence in wall-bounded flows by high frequency spanwise oscillations. *Phys. Fluids A.* **8,** 1065–1067.

KARNIADAKIS, G.E. & TRIANTAFYLLOU, G.S. 1992 Three-dimensional dynamics and transition to turbulence in the wake of bluff objects. *J. Fluid Mech.* **238,** 1–30.

KASAGI, N., TOMITA, Y. & KURODA, A. 1992 Direct numerical simulation of passive scalar field in a turbulent channel flow. *Transaction of the ASME, J. of Fluids Engineering.* **114,** 598–606.

KERR, R.M. 1985 Higher-order derivative correlations and alignment of small-scale structures in isotropic numerical turbulence. *J. Fluid Mech.* **153,** 31–58.

KIKUYAMA, K., MURAKAMI, M. & NISHIBORI, K. 1983 Development of three dimensional turbulent boundary layer in an axially rotating pipe. *Journal of Fluids Engineering.* **105,** 154–160.

KIM, H.T. , KLINE, S.J. & REYNOLDS, W.C. 1971 The production of turbulence near a smooth wall in a turbulent boundary layer. *J. Fluid Mech.* **50,** 133–160.

KIM, J. & MOIN, P. 1985 Application of a fractional-step method to incompressible Navier-Stokes equations. *J. Comp. Phys..* **59,** 308–323.

KIM, J., MOIN, P. & MOSER, R. 1987 Turbulence statistics in fully developed channel flow at low Reynolds number. *J. Fluid. Mech.* **177,** 133–166.

KIM, J. 1989 On the structure of pressure fluctuations in simulated turbulent channel flow. *J. Fluid Mech.* **205,** 421–451.

KIM, J. & MOIN, P. 1989 Transport of passive scalars in a turbulent channel flow. *Turbulent shear flows 6, Springer Verlag, Berlin.* , 85–96

KIM, J. & HUSSAIN, F. 1993 Propagation velocity of perturbations in turbulent channel flow. *Phys. Fluids.* **3,** 695–706

KIM, J. & ANTONIA, R. A. 1993 Isotropy of the small-scales of turbulence at small Reynolds number. *J. Fluid Mech.* **251,** 219–238.

KLINE, S.J., REYNOLDS, W.C., SCHRAUB, F.A. & RUSTANDLER, P.W. 1967 The structure of turbulent boundary layers. *J. Fluid Mech.* **30,** 741–773.

KLINE, S.J., MORKOVIN, M.V., SOVRAN, G. & COCKRELL, D.J. 1968 Computation of Turbulent boundary layers- 1968 AFOSR-IFP Stanford Conference. *Stanford University, Ca.*

KLINE, S.J., CANTWELL, B.J. & LILLEY, G.M. 1982 198-81 AFOSR-HTTM Stanford Conference on Complex Turbulent Flows: Comparison of Computation and Experiments. *Stanford University, Ca.*

KLOOSTERZIEL, R.C. 1990 Barotropic vortices in a rotating fluid. *Ph.D. thesis, University of Utrecht.*

KLOOSTERZIEL, R.C. & HEIJST, G.J.F. VAN 1991 An experimental study of unstable barotropic vortices in a rotating fluid. *J. Fluid Mech.* **223,** 1–24.

KLOOSTERZIEL, R.C. & HEIJST, G.J.F. VAN 1992 The evolution of stable barotropic vortices in a rotating free-surface fluid. *J. Fluid Mech.* **239,** 607–629

KOUMOUTSAKOS, P. 1997a Active control of turbulent channel flow. *Annual Research Briefs.* , 289–297.

KOUMOUTSAKOS, P. 1997b Active control of vortex-wall interactions. *Phys. Fluids.* **12,** 3808–3816.

KORCZAC, K.Z. & PATERA, A.T. 1986 An isoparametric spectral element method for the solution of the Navier-Stokes equations in complex geometries. *J. Comput. Physics.* **62,** 361–382.

KRAICHNAN, R.H. 1976 Eddy viscosity in two and three dimensions. *J. Atmos. Sci..* **33,** 1521–1536

KRAVCHENKO, A.G. & MOIN, P. 1998 B-Spline methods and zonal grids for numerical simulations of turbulent flows. *Report TF-73, Dept. of Mech. Engineering, Stanford University.*

KWAK, D., REYNOLDS, W.C. & FERZIGER, J.H. 1975 Three-dimensional time dependent computation of turbulent flow. *Report No. Tf-5 Thermoscience division Dept. of Mech. Eng. Stanford Ca.*

LAMB H. 1932 Hydrodynamics. *Cambridge University Press, Cambridge.*

LAMBALLAIS, E., LESIEUR, M. & METAIS, O. 1996 Effects of spanwise rotation on the vorticity stretching in transitional and turbulent channel flow. *International Journal of Heat and Fluid Flow* . **17,** 324–332.

LAUFER, J. 1954 The structure of turbulence in fully developed pipe flow. *NACA Report 1174.*

LAUNDER, B.E.,REECE, G.J. & RODI, W. 1975 Progress in the development of a Reynolds stress turbulence closure. *J. Fluid Mech.* **67,** 537–566.

LE, H., MOIN, P. & KIM, J. 1997 Direct numerical simulation of turbulent flow over a backward-facing step. *J. Fluid Mech.* **330,** 349-374.

LEE, C., KIM, J. & CHOI, H. 1998 Suboptimal control of turbulent channel flow for drag reduction. *J. Fluid. Mech.* **358,** 245–258.

LEGRAS, B, SANTANGELO, P. & BENZI, R. 1988 High resolution numerical experiments for forced two-dimensional turbulence. *Europhys. Lett.* **5,** 37–42.

LEITH, C.E. 1984 Minimum enstrophy vortices. *Phys. Fluids.* **7,** 1388-1395.

LEONARD, B.P. 1984 Third-order upwinding as a rational basis for computational fluid dynamics . *Computational Techniques & Applications J.Noye & C. Fletcher (Editors) Elsevier Science Publisher.*

LEONARD, A. & WRAY, A. 1975 A new numerical method for the simulation of three-dimensional flow in a pipe. *Fourth Inter. Conf. on Num. Meth. in Fluid Dyn. (Colorado). Lecture Notes in Physics.* **135,** 335.

LESIEUR, M. & ROGALLO, R. 1989 Large-eddy simulation of passive scalar diffusion in isotropicturbulence. *Phys. Fluids.* **1,** 718–722.

LESIEUR, M. 1990 Turbulence in fluids. *2nd Edition, Kluwer, Dordrecht.*

LESIEUR, M., STAQUET, C., LE ROY, P. & COMTE, P. 1988 The mixing layer and its coherence examined from the point of two-dimensional turbulence. *J. Fluid Mech.* **192,** 511–534.

LESIEUR, M. & METAIS, O. 1996 Large eddy simulation of turbulence. *Annual Review of Fluid Mechanics.* **28,** 45–81.

LIBBY, P.A. 1996 Introduction to Turbulence. *Taylor rm & Francis London.*

LEWEKE, T. & WILLIAMSON, C.H.K. 1998 Cooperative elliptic instability of a vortex pair. *J. Fluid Mech.* **360,** 85–119

LEVI-CIVITA, T. 1977 The absolute Differential Calculus. Dover Publications, *New York.*

LILLY, D.K. 1962 On the numerical simulation of buoyant convection . *Tellus.* **15,** 148–162.

LILLY, D.K. 1965 On the computational stability of numerical solutions of time-dipendent non-linear geophisical fluid dynamics problems. *Monthly Weather Review.* **93/1,** 11–26.

LILLY, D.K. 1992 A proposed modification of the Germano subgrid scale closure method. *Phys. Fluid A.* **4,** 633–635.

LIM, T.T., NICKELS, T.B. & CHONG, M.S. 1991 A note on the cause of rebound in the head–on collision of a vortex ring with a wall. *Exp. in Fluids.* **12,** 41–48.

LIU, S., MENEVEAU, C. & KATZ, J. 1994 On the properties of similarity subgrid-scale models as deduced from mesaurements in a turbulent jet. *J. Fluid Mech.* **275,** 83–119.

LOMBARDI, P., DE ANGELIS, V. & BANERJEE, S. 1997 Direct numerical simulation of near-interface turbulence in coupled gas-liquid flow. *Phys. Fluids.* **6,** 2429–2442.

LOPEZ, J.M. 1990 axisymmetric vortex breakdown. Part.1 confined swirling flow. *J. Fluid Mech.* **221,** 533–552.

LOULOU, P. 1996 Direct simulation of incompressible pipe flow using a B-spline spectral method. *Thesis of Department of Aeronautics and Astronautics* $N^0 SUDAAR683$, *Stanford University.*

LOVE, M.D. 1980 Subgrid modelling studies with Burgers' equation. *J. Fluid Mech.* **100,** 87–110.

LUCHIK, T.S. & TIEDERMAN, W.G. 1988 Turbulent structure in low-concentration drag-reducing channel flows. *J. Fluid Mech.* **190,** 241–263.

LUGT, H.J. 1983 Vortex flows in nature and technology. *John Wiley & sons, New York*

LUGT, H.J. & HORING, S. 1993 The oblique rise of a viscous vortex ring toward a deformable free surface. *Meccanica.* **29,** 313–330.

LUGT, H.J. 1996 Introduction to Vortex Theory . Vortex Flow, Potomac, *Potomac MD.*

LUMLEY, J. & BLOSSEY, P. 1998 Control of turbulence. *Annual Review of Fluid Mechanics.* **30,** 311–328.

LUND, T.S. 1991 On dynamic models for large eddy simulation. *Center for Turbulence Research, Annual Research Briefs.* **177–190.**

MAAS, L.R.M. 1993 Nonlinear and free-surface effects on the spin-down of barotropic vortices. *J. Fluid Mech.* **246,** 117–141.

MANSOUR, N.N., REYNOLDS, W.C. & FERZIGER, J.H. 1978 Large eddy simulation of incompressible channel flow. *Report* N_oTF-12, *Mechanical Engineering Department, Stanford University.*

MANSOUR, N.N. , KIM, J. & MOIN, P. 1988 Reynolds-stress and dissipation-rate budgets in a turbulent channel flow. *J. Fluid Mech.* **194,** 15–44.

MANSOUR, N.N. & WRAY, A.A. 1994 Decay of isotropic turbulence at low Reynolds number. *Phys. of Fluids* . **6,** 808–814.

MCCOMB, W.D. 1990 The Physics of fluid turbulence. *Oxford University Press, Oxford.*

MELANDER, M.V. & HUSSAIN, A.K.M.F. 1990 Topological aspects of vortex reconnection. *Topological Fluid Mechanics,* Moffat & Tsinober Eds., 485–499.

MELLOR, G.L. & HERRING, H.J. 1973 A survey of mean turbulent field closure models. *AIAA J..* **5,** 590–599.

MENEVEAU, C. 1994 Statistics of turbulence subgrid-scale stresses: Necessary conditions and experimental tests. *Phys. Fluids A.* **2,** 815–833.

METAIS, O. & LESIEUR, M. 1992 Spectral large-eddy simulation of isotropic and stably stratified turbulence. *J. Fluid Mech.* **239,** 157–194.

MOFFATT, H. K. 1988 Generalized vortex rings with and without swirl. *Fluid Dyn. Res.* **3**, 22-30.

MOHD-YOUSOF, J. 1997 Combined immersed-boundary B-spline methods for simulations of flow in complex geometries. *Annual Research Briefs 1997 Center for Turbulence Research.*

MOIN, P., REYNOLDS, W.C. & FERZIGER, J.H. 1978 Large eddy simulation of incompressible channel flow. *Report N°TF – 12, Mechanical Engineering Department, Stanford University.*

MOIN, P. 1981 Numerical simulation of wall bounded turbulent flows. *1980-81 AFOSR-HTTM Stanford Conference on Complex Turbulent Flows: Comparison of Computation and Experiments, September 1981.*

MOIN, P. & KIM, J.H. 1982 Numerical investigation of turbulent channel flow. *J. Fluid Mech.* **118,** 341–377.

MOIN,P. & KIM, J.H. 1985 The structure of the vorticity field in turbulent channel flow.Part.1 Analysis of instantaneous fields and statistical correlations. *J. Fluid Mech.* **155,** 441–464.

MOIN, P. ,SQUIRES, K., CABOT, W. & LEE, S. 1991 A dynamic subgrid-scale model for compressible turbulence and scalar transport. *Phys. Fluids A.* **11,** 2746–2757.

MOIN, P., CARATI, D., LUND, T.S., GHOSAL, S. & AKSELVOLL, K. 1994 Developments and applications of dynamic models for large eddy simulation of complex flow. *Presented at 74th Fluid Dynamics AGARD Panel of Chania, Crete 18-21 April 1994*

MOIN, P. & MHAHESH, K. 1998 Direct numerical simulations: A tool in Turbulence research . *Annual Review of Fluid Mechanics.* **30,** 539–578.

MONIN, A.S. & YAGLOM, A.M. 1971 Statistical Fluid Mechanics: Mechanics of Turbulence. *MIT Press, Cambridge,Ma.*

MONTGOMERY, D., MATTHAEUS, W.H., STRIBLING, W.T., MARTINEZ, D. & OUGHTON, S. 1992 Relaxation in two dimensions and the sinh-Poisson equation . *Phys. Fluids A.* **4,** 3

MONTGOMERY, D., SHAN, X. & MATTHAEUS, W.H. 1993 Navier-Stokes relaxation to sinh Poisson states at finite reynolds numbers. *Phys. Fluids A.* **5,** 2207–2216.

MOREAU, R. 1983 Two-dimensional turbulence. *Journal de Mecanique Theorique et Applique.* **5.**

MOSER, R.D. & ROGERS, M.M., 1993 The three-dimensional evolution of a plane mixing layer: Pairing and transition to turbulence. *J. Fluid Mech.* **247,** 275–320

MURAKAMI, M. & KIKUYAMA, K. 1980 Turbulent flow in axially rotating pipes. *Journal of Fluids Engineering.* **102,** 97–103.

NA, Y. & MOIN, P. 1998 Direct numerical simulations of separated turbulent boundary layers. *J. Fluid Mech.* **370,** 175–201.

NARIMOUSA, S., MAXWORTHY, T. & SPEDDING, G.R. 1991 Experiments on the structure and dynamics of forced quasi-two-dimensional turbulence. *J. Fluid Mech.* **223,**

113–133.

NIEDERSCHULTE, M.A., ADRIAN, R.J. & HANRATTY, T.J. 1990 Measurements of turbulent flows in a channel at low Reynolds number. *Exp. in Fluids..* **9,** 222–230.

NIELSEN, A.H. & RASMUSSEN, J.J. 1997 Formation and temporal evolution of the Lamb dipole. *Phys. Fluids.* **4,** 982–991.

NISHIBORI, K., KIKUYAMA, K., & MURAKAMI, M. 1987 Laminarization of turbulent flow in the inlet region of an axially rotating pipe. *JSME International Journal.* **30,** 255–262.

NITSCHE, M. & KRASNY, R. 1994 A numerical study of vortex ring formation at the edge of a circular tube. *J. Fluid Mech.* **276,** 139–161.

NORRIS, H.L. & REYNOLDS, W.C. 1975 Turbulent channel flow with a moving wavy boundary. *Technical Report TF-7, Mechanical Engineering Department, Stanford University, Stanford Ca.*

ORLANDI, P. & BRISCOLINI, M. 1983 Direct Simulation of Burgulence. *CTAC-83, Sydney, Australia, 1983.*

ORLANDI, P. 1987 Vorticity Velocity Method for High Reynolds Numbers. *Computers and Fluids.* **15,** 137–149.

ORLANDI, P. 1989 Numerical solution of 3-D flows periodic in one direction and with complex geometries in 2-D. *Annual Research Briefs Center for Turbulence Research-1989.* **215–230.**

ORLANDI, P. 1990 Vortex dipole rebound from a wall. *Phys. Fluids A.* **8,** 1429–1436.

ORLANDI, P. & HEIJST, G.J.F. VAN 1992 Numerical simulation of tripolar vortices in 2D flow. *Fluid Dyn. Res.* **9,** 179–206.

ORLANDI, P. 1993 Vortex dipole impinging circular cylinders. *Phys. Fluids A.* **9,** 2196–2206.

ORLANDI, P. & JIMÉNEZ, J. 1993 On the generation on turbulent wall turbulence. *Phys. Fluids A.* **2,** 634–641.

ORLANDI, P. & VERZICCO, R. 1993a Vortex ring impinging on a wall: axisymmetric and three-dimensional simulations. *J. Fluid Mech.* **256,** 615–645.

ORLANDI, P. & VERZICCO, R. 1993b Vortex rings interacting with walls, analogies with wall turbulence structures. *Proc. Near-Wall Turbulent Flows (*So, Speziale, and Launder Eds.*)* 497–506.

ORLANDI, P., VERZICCO, R. & HEIJST, G.J.F. VAN 1993 Stability of vortex dipoles. *Proceedings of the "Modeling of Oceanic Vortices" congress, Amsterdam 1993.*

ORLANDI, P. & VERZICCO, R. 1994 Identification of zones in a free evolving vortex ring. *Proceedings of the IUTAM Congress, Poitier 1992, App. Sci. Res.* **53,** 387–399.

ORLANDI, P. 1995 A tentative approach to the direct simulation of drag reduction by polymers . *Journal of Non-newtonian Fluid Mechanics.* **60,** 277–301.

ORLANDI, P. 1996 Drag reduction in turbulent MHD pipe flows. *Proceedings of the 1996 Summer Program, Center for Turbulence Research.*

ORLANDI, P. 1997 Helicity fluctuations in rotating and non-rotating pipes. *Phys. Fluids A.* **7,** 2045–2056.

ORLANDI, P. & FATICA, M. 1997 Direct simulations of a turbulent pipe rotating along the axis. *J. Fluid Mech.* **343,** 43–72.

ORLANDI, P. ,CARNEVALE, G.F., LELE,S., & SHARIFF, K. 1998a DNS study of stability of trailing-vortices. *Proceedings of the 1998 Summer Program, Center for Turbulence Research.*

ORLANDI, P., EGERMANN, P. & HOPFINGER, E. 1998b Vortex rings moving in a stratified medium. *Phys. Fluids.* **10,** 2819-2827.

ORLANDI, P. & CARNEVALE, G.F. 1999 The three-dimensional tripole formation, effects of the walls. *J. Fluid Mech.* **381,** 239–269.

ORLANSKI, I. 1972 A simple boundary condition for unbounded hyperbolic flows. *J. Comp. Phys.* **21,** 251–281.

ORSZAG, S.A. 1971 Numerical simulation of incompressible flows within simple boundaries: accuracy. *J. Fluid. Mech.* **49,** 75–112.

ORSZAG, S.A. & PATERA, A.T. 1980 Subcritical transition to turbulence in plane channel flows. *Physical Review Letters.* **45,** 989-993.

PANTON, R.L. 1984 Incompressible flow. *John Wiley & sons, New York*

PATEL, V.C., RODI, W. & SCHEUERER, G. 1985 Turbulence models for near-wall and low-Reynolds number flow: A Review. *AIAA Journal.* **23,** 1308–1319.

PAULEY, L.L., MOIN, P. & REYNOLDS, W.C. 1992 The structure of two-dimensional separation. *J. Fluid Mech.* **220,** 397–441.

PEARSON, B. R. & ANTONIA, R. A. 1997 Velocity structure functions in a turbulent plane jet. *Eleventh Symposium on Turbulent Shear Flows, Grenoble.*

PEDRIZZETTI, G. 1996 Unsteady tube flow over an expansion. *J. Fluid Mech.* **310,** 89–111.

PEDLOSKY, J. 1986 Geophysical Fluid Dynamics. *Springer-Verlag, New York.*

PELTZ, R.B., YAKHOT, V., ORSZAG, S.A., SHTILMAN, L. & LEVICH, E. 1985 Velocity vorticity patterns in turbulent flows. *Phy. Rev. Lett.* **54,** 2505–2508.

PERIDIER, V.J. & WALKER, J.D.A. 1989 Vortex-induced boundary-layer separation. *Report FM-13 Department of Mechanical Engineering and Mechanics, Lehigh University, Bethlehem.*

PEROT, B. & MOIN, P. 1995 Shear-free turbulent boundary layers. Part 1 Physical insights into near-wall turbulence. *J. Fluid Mech.* **295,** 199–227.

PEROT, B. & MOIN, P. 1995 Shear-free turbulent boundary layers. Part 2 New concepts for Reynolds stress modelling of inhomogeneous flows. *J. Fluid Mech.* **295,** 229–245.

PESKIN, C.S. 1972 Flow pattern around heart valves:a numerical method. *J. Comp. Phys.* **10,** 252–268.

PEYRET, R. & TAYLOR, T.D. 1983 Computational Methods for Fluid Flow. *Springer-Verlag, New York.*

PHILIPS, N.A. 1959 An example of non-linear computational instability . *Atmosphere and the Sea in Motion, Rossby Memorial Volume, New York, Rockfeller Institute Press.* **501-504.**

PIOMELLI, U., FERZIGER, J.H. & MOIN P. 1987 Models for large eddy simulations of turbulent channel flows including transpiration. *Stanford Univ., Dept. Mech. Eng. Thermosciences Division Report TF-32.*

PIOMELLI, U., BALINT, J.L. & WALLACE, J.M. 1989 On the validity of the Taylor hypothesis for wall-bounded flows. *Phys. Fluids.* **3,** 609–611.

PIOMELLI, U., ZANG, T.A., SPEZIALE, C.G. & HUSSAINI, M.Y. 1990 On the large-eddy simulation of transitional wall-bounded flows. *Phys. Fluids A.* **2,** 257–265.

PIOMELLI, U. 1993 High Reynolds number calculations using the dynamic subgrid-scale stress model. *Phys. Fluids A.* **5,** 1484–1490.

PIOMELLI, U. 1998a Large-eddy simulation: present state and future directions. *AIAA paper 98-0534.*

PRATESI, C., MICHELAGNOLI, S. & PEDRIZZETTI, G. 1995 La chirurgia della carotide extracranica: considerazioni di tecnica alla luce di valutazioni fluidodinamiche. *Abstracts XVII Congresso della Societa' Italiana di Chirurgia Cardiaca e Vascolare. Archivio di chirurgia toracica e cardiovascolare, edizioni L. Pozzi.* **17, n.1,** 167.

PRESS, W.H., TEUKOLSKY, S.A., VETTERLING, W.T. & FLANNERY, B.P. 1993 Numerical recipes in Fortran. The art of scientific computing(II Ediction). *Cambridge University Press, Cambridge.*

RAMBOD, E., & GHARIB, M. 1995 Gallery of fluid motion. *Phys. Fluids A .* **7,** 5.

RAI, M.M. & MOIN, P. 1991 Direct simulations of turbulent flow using finite-difference schemes. *J. Comp. Phys.* **96,** 15–53.

RAO, K.H., NARASIMHA, R. & BADRI NARAYAMA, M.A. 1971 The bursting phenomenon in a turbulent boundary layer. *J. Fluid Mech.* **48,** 339–352.

REICH, G. & BEER, H. 1989 Fluid flow and heat transfer in axially rotating pipe 1. Effect of rotation on turbulent pipe flow. *Int. J. Heat Mass Transfer.* **32,** 551–561.

REYNOLDS, O. 1895 On the dynamic theory of incompressible viscous fluids and the determination of the criterion. *Phil. Trans. Roy. Soc. Lond.* **535–577.**

REYNOLDS, W.C. 1989 The potential and limitations of Direct and Large Eddy Simulations. *Whither Turbulence (Edited by Lumley), Springer Verlag, Berlin.*

ROACHE, P.J. 1972 Computational Fluid Dynamics. *Hermosa Albuquerque, New Mexico*

ROBINSON, S.K. 1991 Coherent motions in the turbulent boundary layer . *Annual Review of Fluid Mechanics.* **601–639.**

ROGALLO, R.S. 1981 Numerical experiments in homogeneous turbulence. *NASA Tech. Mem. 81315 and 1980–81 AFOSR-HTTM Stanford Conference on Complex Turbulent Flows: Comparison of Computation and Experiments, September 1981.*

ROGERS, M.M. & MOSER, R.D. 1992 The three-dimensional evolution of a plane mixing layer: The Kelvin-Helmholtz rollup. *J. Fluid Mech.* **243,** 183–226.

ROGERS, M.M. & MOIN, P. 1987 Helicity fluctuation in incompressible turbulent flows. *Phys. Fluids.* **30,** 2662–2671.

ROSENFELD, M., KWAK, D. & VINOKUR, M. 1991 A fractional step solution method for the unsteady incompressible Navier-Stokes equations in generalized coordinate systems. *J. Comp. Phys.* **94,** 102–137.

SADDOUGHI, S.G. & VEERAVALLI, S.V. 1994 Local isotropy in turbulent boundary layers at high Reynolds number. *J. Fluid Mech.* **268,** 333–372.

SAIKI, E.M. & BIRINGEN, S. 1996 Numerical simulation of a cylinder in uniform flow - application of a virtual boundary method. *J. Comp. Phys.* **123,** 450–465.

SALVETTI, M.V., ORLANDI, P. & VERZICCO, R. 1995 Direct simulations of transitional axisymmetric coaxial jets. *AIAA J.* **34,** 736–743.

SALVETTI, M.V. & BANERJEE, S. 1995 A priori tests of a new dynamic subgrid-scale model for finite-difference large-eddy simulations. *Phys. Fluids.* **7,** 2831-2847.

SALVETTI, M.V, ZANG, Y., STREET, R.L. & BANERJEE, S. 1997 Large-eddy simulation of free-surface decaying turbulence with dynamic subgrid-scale models. *Phys. Fluids.* **9,** 2405-2419.

SANTANGELO, P., BENZI R. & LEGRAS B. 1989 The generation of vortices in high-resolution, two-dimensional decaying turbulence and the influence of initial conditions on the breaking of self-similarity. *Phys. Fluids A.* **6,** 1027–1034.

SARGHINI, S. & PIOMELLI U. 1998 Scale-similar models for large-eddy simulations. *AIAA paper 98-0538.*

SARPKAYA, T. 1983 Trailing vortices in homogeneous and density-stratified media. *J. Fluid Mech.* **136,** 85–109.

SARPKAYA, T. 1994 Interaction of Vorticity, Free-surface and Surfactants. *Annual Review Fluid Mechanics.* **28,** 83–128.

SENDSTAD, O. & MOIN, P. 1991 On the mechanism of 3-D turbulent boundary layers. *8th Symposium on turbulent shear flows, Munich 20-3.*

SERRAVALL, R., ZAVALA-SANSÓN, L. & HEIJST, G.J.F. VAN 1998 Experiments on the bifurcation of a coastal current in presence of a topographic slope. *EGS98 General European Geophysical Society, General Assembly, 20-24 April, 1998, Vortex dynamics section.*

SHAH, K.B. & FERZIGER, J.H. 1996 New SGS model and its application to channel flow. *Annual Research Briefs Center for Turbulence Research-1996.* , 73–90.

SHARIFF, K., VERZICCO, R. & ORLANDI, P. 1994 A numerical study of three dimensional vortex ring instabilities: viscous corrections and early non-linear stage. *J. Fluid Mech.* **279,** 351–374.

SHE, Z.S., JACKSON, E. & ORSZAG, S.A. 1990 Structure and dynamics of homogeneous turbulence, models and simulations. *Proc. R. Soc. Lond. A.* **434,** 101–124.

SHE, Z.S. & ORSZAG, S.A. 1991 Physical model of intermittency in turbulence: Inertial-range non-Gaussian statistics. *Physical Review Letters.* **66,** 1701–1704.

SHRENIVASAN ,K.R. & ANTONIA, R.A. 1997 The phenomenology of small-scale turbulence. *Annual Review Fluid Mechanics.* **29,** 435–472.

SCHUMANN, U. 1975 Subgrid model for finite difference simulation of turbulent flows in plane channels and annuli. *J. Comp. Phys.* **18,** 376–404.

SIEGEL, D.A. & DOMARADZKI, J.A. 1994 Large-Eddy simulation of decaying stably stratified turbulence. *J. Phys. Ocean.* **24,** 2353–2386.

SMAGORINSKY, J. 1963 General circulation experiments with the primitive equations. I. The basic experiment. *Mon. Weather Rev.* **91,** 99–163.

SONG, M., KACHMAN, N., KWON, J.T., BERNAL, L.P. & TRYGGVASON, G. 1991 Vortex ring interaction with a free surface. *Proc. 18th Symp. Naval Hydrod.* Nat. Aca. Press, Washington D.C.-*1991.* , 479–489.

SONG, M., BERNAL, L.P. & TRYGGVASON, G. 1992 Head–on collision of a large vortex ring with a free surface. *Phys. Fluids A.* **4,** 1457–1466.

SPALART, P.R. 1988 Direct simulation of turbulent boundary layers up to $Re_\theta = 1410$. *J. Fluid Mech.* **187,** 61–98.

SPALART, P.R. & WATMUFF, J.H. 1993 Experimental and numerical study of a turbulent boundary layer with pressure gradients. *J. Fluid Mech.* **249,** 337–371.

SPALART, P.R. 1996 On the motion of laminar wing wakes in a stratified fluid . *J. Fluid Mech.* **327,** 139–160.

SPALART, P.R. 1998 Airplane trailing vortices. *Annual Review of Fluid Mechanics.* **30,** 107–138.

SQUIRES, K.D. & PIOMELLI, U. 1993 Large-eddy simulation of rotating turbulence using the dynamic model. *9th Symposium on turbulent shear flows, Kyoto 17-3.*

SQUIRES, K.D., CHASNOV, J.R., MANSOUR, N.N. & CAMBON, C. 1993 Investigation of the asymptotic state of rotating turbulence using large eddy simulation. *Annual Research Briefs-1993.* , 157–170.

STANAWAY, S.K., SHARIFF, K. & HUSSAIN, A.K.M.F. 1988 Head-on collision of viscous vortex rings. *Proceedings of CTR Summer Program 1988.* , 287–310.

STANAWAY, S.K., CANTWELL, B.J. & SPALART, P.R. 1988 A numerical study of viscous vortex rings using a spectral method. *Ph.D. thesis, NASA,* **TM 101041**.

STERN, M.E. 1975 Minimal properties of planetary eddies. *J. Mar. Res.* **33,** 1–13.

STERN, M.E. & AUSTIN, J. 1995 Entrainment of shelf water by a bifurcating continental boundary current. *J. Phys. Ocean.* **25,** 3118-3131.

SWEARINGEN, J.D. & BLACKWELDER, R.F. 1987 The growth and breakdown of streamwise vortices in the presence of a wall. *J. Fluid Mech.* **182,** 255–290.

SWEARINGEN, J.D., CROUCH, J.D. & HANDLER, R.A. 1995 Dynamics and stability of a vortex ring impacting a solid boundary. *J. Fluid Mech.* **297,** 1–28.

TAFTI, D. 1996 Comparison of some upwind-biased high-order formulations with a second-order central-difference scheme for time integration of the incompressible Navier Stokes equations. *Computers & Fluids.* **25**, 647–665.

TARDU, S. 1997 Control of drag by time-space periodical suction and blowing . *Int.Symp. Turbulent Shear FlowsGrenoble.*

TAVOULARIS, S., BENNET, J.C. & CORRSIN, S. 1978 Velocity-derivative skewness in small Reynolds number, nearly isotropic turbulence. *J. Fluid Mech.* **88**, 63–69.

TENNEKES, H. & LUMLEY, J. 1972 A firts course in Turbulence. *MIT Press, Cambridge, Ma.*

THOMPSON, J.F. , WARSI, Z.U.A. & MASTIN, C.W. 1985 Numerical grid generation Foundations and Applications . *North-Holland, New York.*

TIEDERMAN, W.G., LUCHIK, T.S. & BOGARD, D.G. 1985 Wall-layer structure and drag reduction. *J. Fluid Mech.* **156**, 419–437.

TOWNSEND, A.A. 1956 The structure of turbulenct shear flow. *Cambridge University press, Cambridge.*

TSINOBER, A., EGGELS, J.G.M., & NIEUWSTADT, F.T.M. 1995 On alignments and small scale structures in turbulent pipe flow. *Fluid Dynamics Research.* **16**, 297–310.

TRAUGOTT, S.C. 1958 Influence of solid body rotation on screen produced turbulence. *NACA Technical note 4135.*

TSUBOI, K., TAMURA, T. & KUWAHARA, K. 1989 Numerical study of vortex induced vibration of a circular cylinder in high Reynolds number flow. *AIAA 89-0294, Reno, Nevada.*

TURKINGTON, B. 1989 Vortex rings with swirl: axisymmetric solutions of the Euler equations with nonzero helicity. *SIAM J. Math. Anal.* **20**, 57–73.

UZKAN, T. & REYNOLDS, W.C. 1967 A shear-free turbulent boundary layer. *J. Fluid Mech.* **28**, 803–821.

VASILENKO, V. M., LYUBIMTSEV, M. M. & OZMIDOV, R. V. 1975 Fluctuations of the turbulent energy dissipation rate and of the higher-order structure functions of the velocity field in the ocean. *Iz. Atmos. Ocean. Phys.* **11**, 926–932.

VEGA, M., VERZICCO, R. & ORLANDI,P. 1996 The impact of dipolar vortices on curvesd walls. *Fourth Year Project of Aeronautical Engineering Dept. of Aeronautics, Imperial College of Science , Technology and Medicine, London, Unpublished Manuscript.*

VERZICCO, R. & ORLANDI P. 1993 Direct simulations of the transitional regime of a circular jet. *Phys. Fluids A.* **6**, 751–759.

VERZICCO, R. & ORLANDI, P. 1995 Mixedness in the formation of a vortex ring. *Phys. of Fluids.* **7**, 151–153.

VERZICCO, R., FLÓR,J.B., HEIJST G.J.F. VAN & ORLANDI, P. 1995 Numerical and experimental analysis of the interaction between a dipole and a circular cylinder. *Exp. in Fluids.* **18**, 153–163.

VERZICCO, R., JIMÉNEZ, J. & ORLANDI, P. 1995 On steady columnar vortices under local compression. *J. Fluid Mech.* **299,** 367–388.

VERZICCO, R. & ORLANDI, P. 1996a A finite-difference scheme for the three-dimensional incompressible flows in cylindrical coordinates. *J. Comp. Phys.* **123,** 402–414 .

VERZICCO, R. & ORLANDI, P. 1996b Wall/vortex ring interactions. *App. Mech. Review, ASME.* **49,** 447-461.

VERZICCO, R., ORLANDI, P., EISENGA, A.H.M. HEIJST, G.J.F. VAN & CARNEVALE, G.F. 1996 Dynamics of a vortex ring in a rotating fluid. *J. Fluid Mech.* **317,** 215–239.

VERZICCO, R., LALLI, F. & CAMPANA, E. 1997 Dynamics of baroclinic vortices in a rotating stratified fluid: a numerical study. *Phys. Fluids.* **9,** 419–432

VERZICCO, R. & CAMUSSI, R. 1997 Transitional regimes of low–Prandtl number thermal convection in a cylindrical cell. *Phys. Fluids.* **9,** 1287–1295.

VIECELI, J.A. 1969 A method for including arbitrary external boundaries in the MAC incompressible fluid computing technique . *J. Comput. Phys.* **10,** 543–563.

VINCENT, A. & MENEGUZZI, M. 1991 The spatial structure and statistical properties of homogeneous turbulence. *J. Fluid Mech.* **204,** 1–25.

VINOKUR, M. 1989 An analysis of finite-difference and finite-volume formulations of Conservation Laws. *J. Comp. Phys.* **81,** 1-52.

VREMAN, A.W; SANDHAM, N.D. & LUO, K.H. 1996 Compressible mixing layer growth rate and turbulence characteristics. *J. Fluid Mech.* **320,** 235–258.

WALLACE, J.M. & FOSS, J.F. 1995 The measurements of vorticity in turbulent flows. *Annual Review of Fluid Mechanics.* **27,** 469–514.

WALEFFE, F. 1990 On the three-dimensional instability of strained vortices. *Phys. Fluids A.* **2,** 76–80.

WALKER, M. 1964 The Schwarz-Christoffel transformation and its applications; a simple exposition (formerly titled: Conjugate functions for engineers). *Dover Publications, New York.*

WALKER, J.D.A., SMITH, C.R., CERRA, A.W.,& DOLIGASKI, J.L. 1987 The impact of a vortex ring on a wall. *J. Fluid Mech.* **181,** 99–140.

WEI, T. & WILLMARTH, W.W. 1989 Reynolds number effects on the structures of turbulent channel flow. *J. Fluid Mech.* **204,** 57–95.

WELCH, J.E. & HARLOW, F.H. 1965 The MAC method. *Los Alamos Scientific Laboratory Report LA 3425.*

WIDNALL, S.E., BLISS, D.B. & TSAI, C.-Y. 1974 The instability of short waves on a vortex ring. *J. Fluid Mech.* **66,** 35–47.

WIGELAND, R.A. & NAGIB, H.B. 1978 Grid generated turbulence with and without rotation about the streamwise direction. *IIT Fluids and Heat Transfer Rep.R78-1.*

WILCOX, D.C. 1993 Turbulence Modelling for CFD. *DCW Industries, Inc. La Canada, California.*

WONG, V.C. & LILLY, D.K. 1994 A comparison of two dynamic subgrid closure methods for turbulent thermal convection. *Phys. Fluids A.* **2,** 1016–1022.

WOODS, L.C. 1954 A note on the numerical solution of fourth order differential equations. *Aero. Q.* **5,** 176–184.

WRAY, A.A. 1987 Very low storage time-advancement schemes. *Internal Report, NASA Ames Research Center, Moffett Field, California.*

YANENKO, N.N. 1971 The method of fractional sreps . *Springer-Verlag, New York.*

ZANG, Y, STREET, R.L., KOSEFF, J.R. 1993 A dynamic mixed subgrid-scale model to turbulent recirculating flows. *Phys. Fluids A.* **5,** 3186–3196.

ZHANG, Y., GHANDI, A., TOMBOULIDES, A.G., & ORSZAG, S.A. 1994 Simulation of pipe flow. *AGARD Conference Proceedings* $N^0$551 *Applications of Direct and Large Eddy Simulations to transition and turbulence.* , 17-1,17-9.

ZHOU, T., ANTONIA, R. A., ZHU, Y., ORLANDI, P. &ESPOSITO, P.G. 1998 Performance of a transverse vorticity probe in a turbulent channel flow. *Exp. in Fluids.* **24,** 510–517.

Diskette information

The diskette content is in a tar file that has been also compressed by the gzip instruction. To have the content after the diskette has been copied in the computer with the tar instruction,(e.g. tar -xvf/dev/fd0) the file "diskette.tar.gz" is obtained. It should be uncompressed by gunzip and finally by the tar instruction the different directories are created with their content.

As it is easily comprehensible in every directory with name starting with CODESCH there are the codes related to the different chapters of the book. In each of this directory there are several other subdirectories.

In the subdirectories with names starting with POST there are the eventual postprocessing codes.

In the subdirectories with name VIS there are the eventual codes to perform visualizations of the results. The files written by these visualization codes are in a format compatible with the TURB3D or PLOT3D packages. These packages where developed several years ago in the NASA centers. Now other graphics packages reading files written for PLOT3D are available.

In the subdirectories DAT there are the file with the data necessary to perform the simulations.

In the other subdirectories with names related to the subject of the chapter there are the source of the codes. The make files permit to compile the codes, in IBM or Digital computers. Those with name ending with "mak" are for IBM, instead those with names ending with "makdig" are for digital.

As it is well known the codes are continuously modified and this happened also in this case, due to the reason that I decided to insert more comments and to eliminate some part that could be unclear to the users. Thus I check that the compilation was without any error on the Digital compiler, and since my experience is that this compiler is more restrict that that on the IBM, the compilation should be o.k. even on the IBM computers. As the user can notice on each of the subdirectories with the code there is a subsubdirectory named *prova*, where the data files in the subdirectories DAT have been copied. I checked that the compiled files in a Digital computer started correctly, it means that the data files are compatible with the read statements in the codes. Since some of the results in the book where obtained with code slightly differents, there is the possibility that the results obtained with the version of the codes, I am distributing with the book, will be different. I did not have the time to check this occurrence and I leave this task to the users.

I will be glad to help anybody that could have difficulties to understand the codes or to use them by answering by e-mail to their questions; so could you please write to me to the following address:

orlandi@navier.ing.uniroma1.it

In the case the user will find any bug could he be so kind to advise me and this will be of help to other users.

Plates Section

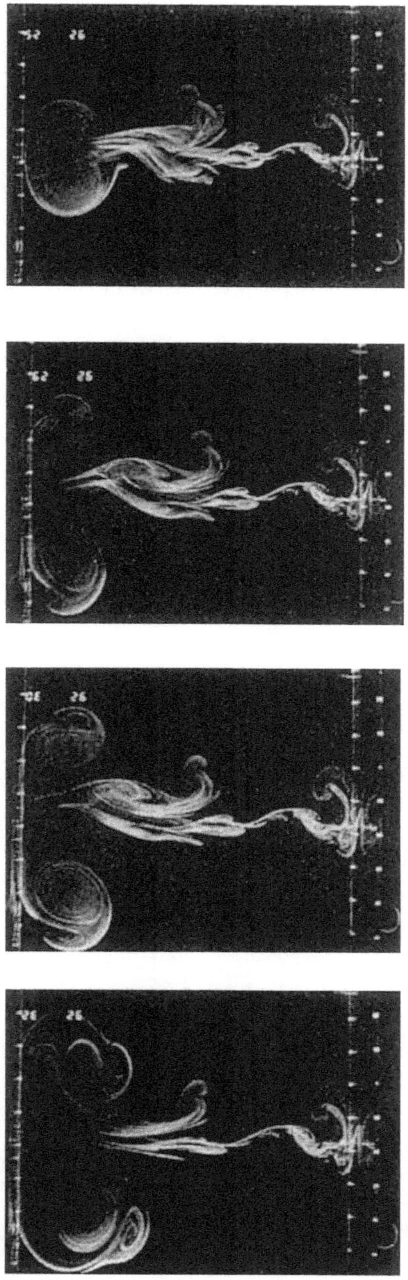

PLATE 1. Vortex dipole rebound from a solid wall, pictures by Van Heijst & Flor from Orlandi (1990) *Courtesy of Physics of Fluids.*

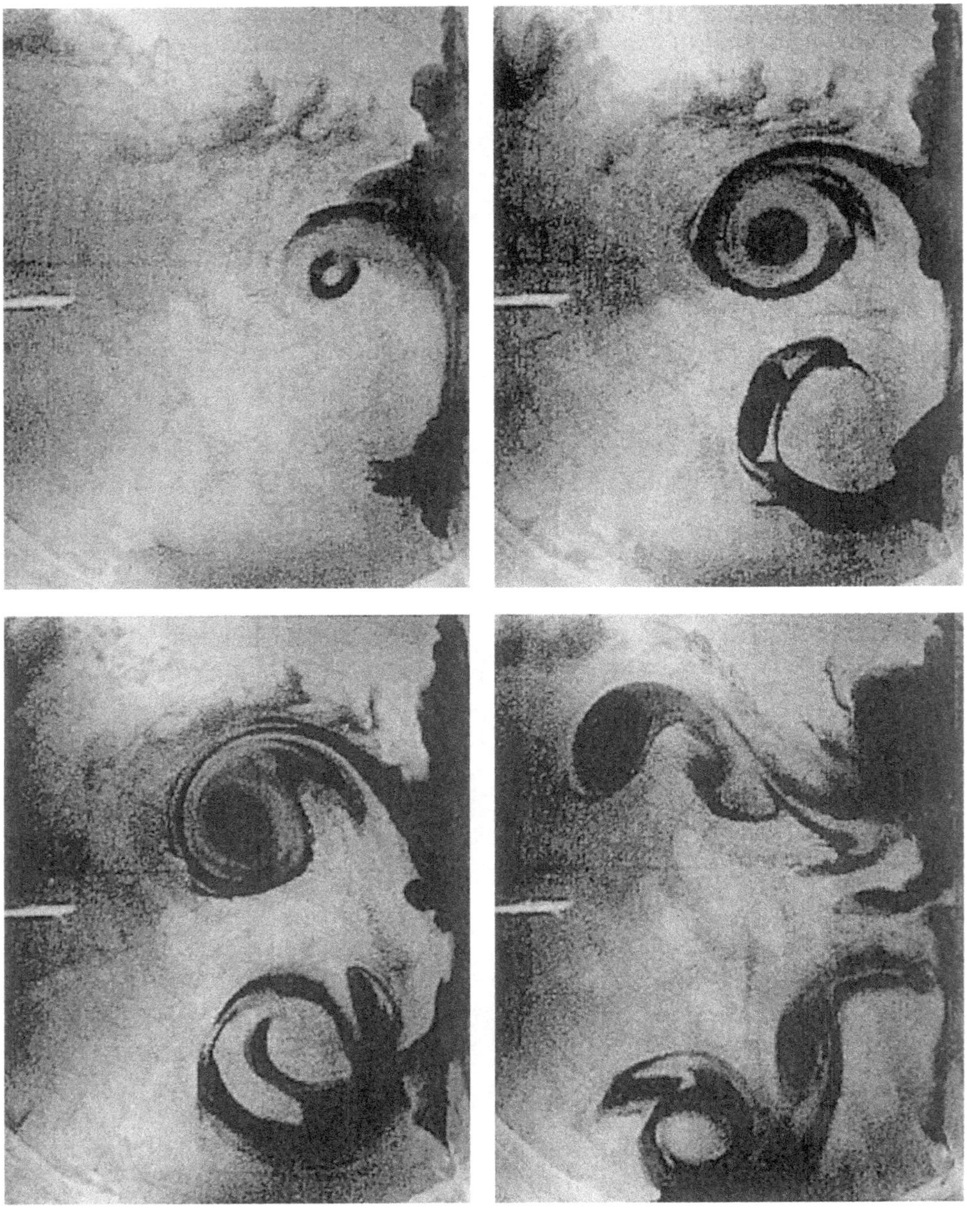

PLATE 2. Rebound of a modon in a β plane from a free-slip wall, from Carnevale *et al.* (1997) *Courtesy of J. Fluid Mech. Cambridge University Press* .

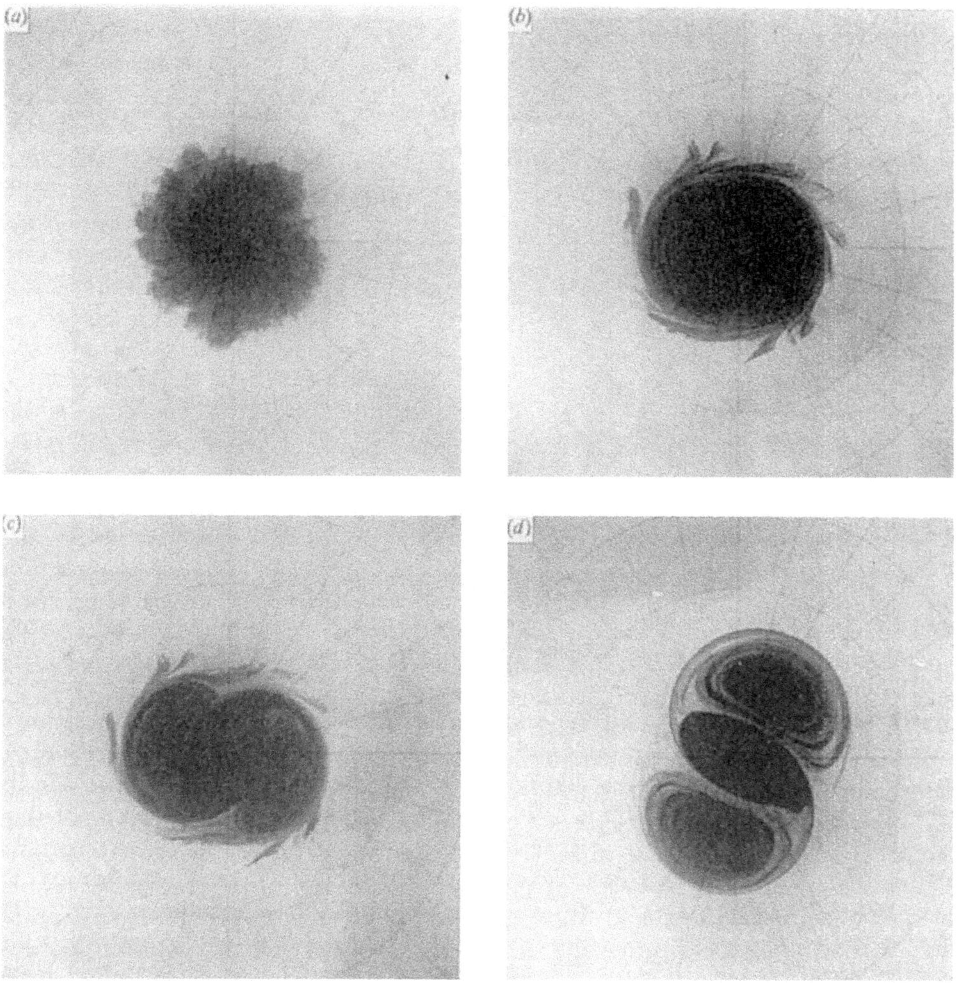

PLATE 3. Tripole formation, from Kloosterziel & Van Heijst (1991) *Courtesy of J. Fluid Mech. Cambridge University Press* .

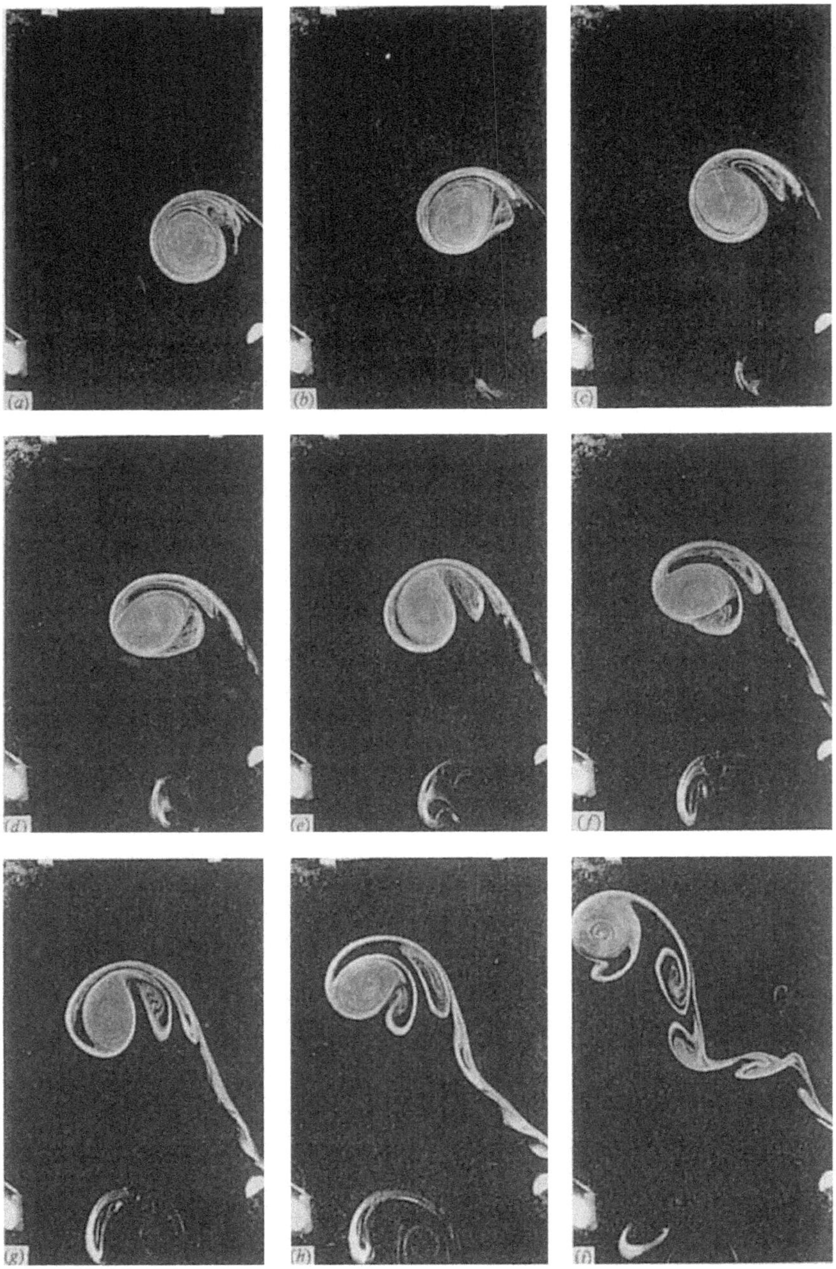

PLATE 4. Motion of a stir vortex in a β plane, from Carnevale *et al.* (1991a) *Courtesy of J. Fluid Mech. Cambridge University Press* .

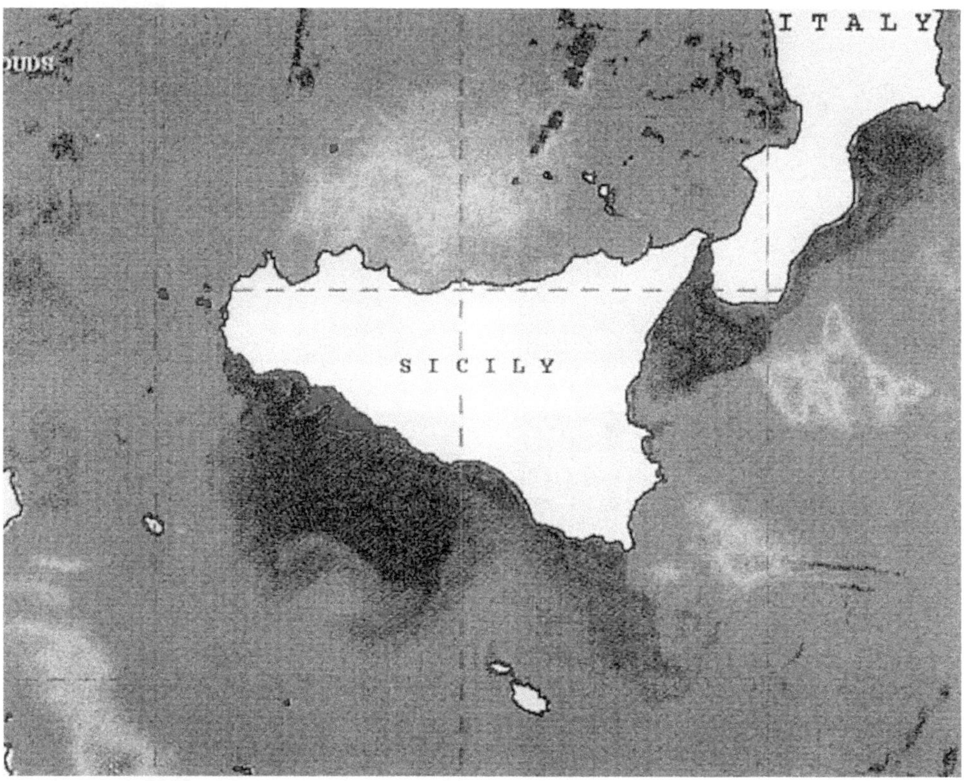

PLATE 5. Picture of isotermal at 26^0C obtained august 25th of (1996) by NOAA14 satellite, from Serravall Ph.D Thesis "Dinamica di una corrente costiera in presenza di un gradino topografico" (1998). *Courtesy of SACLANT Center in La Spezia Italy* .

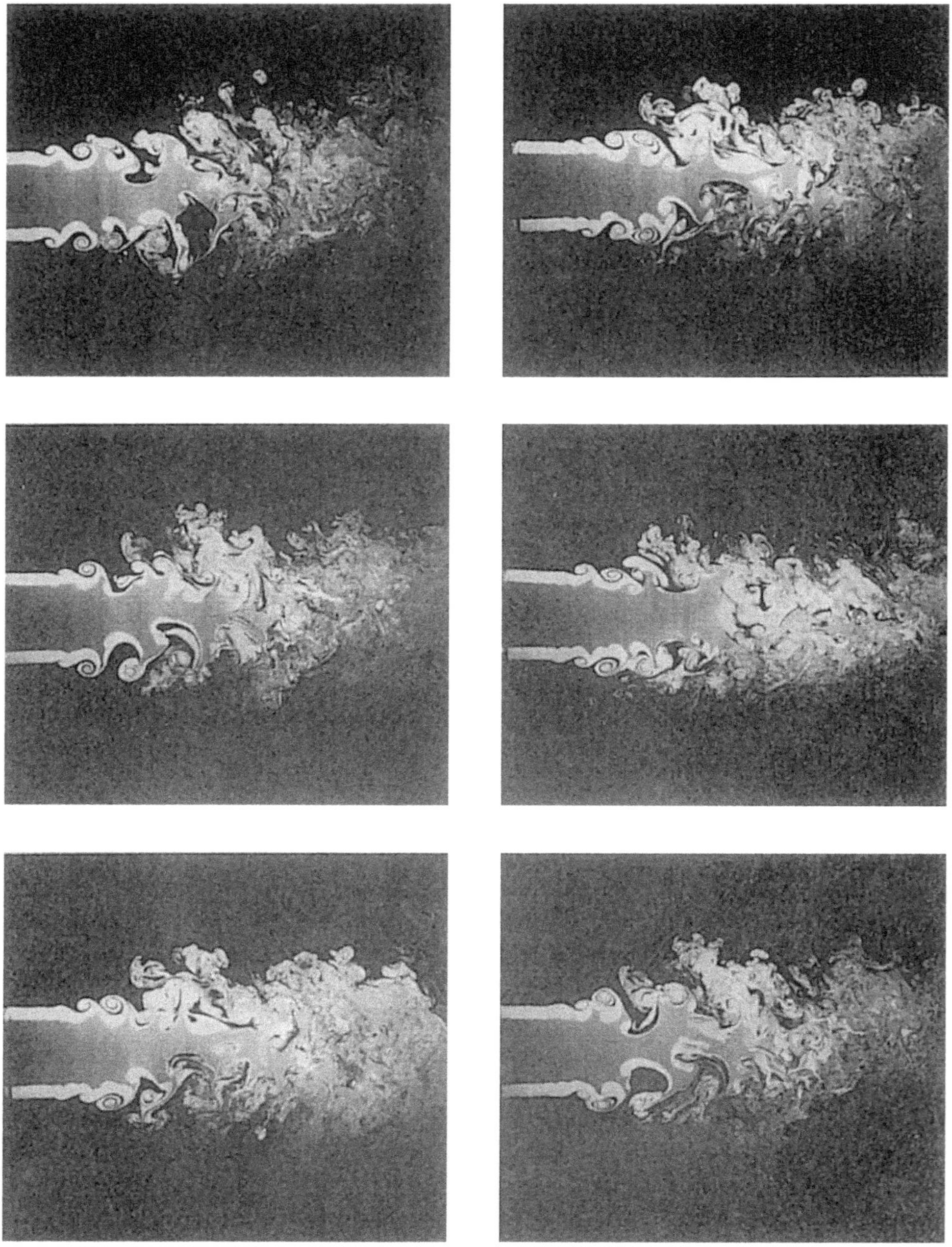

PLATE 6. Pictures of a coaxial jet, from Dahm *et al.* (1992) *Courtesy of J. Fluid Mech. Cambridge University Press* .

Mechanics

FLUID MECHANICS AND ITS APPLICATIONS

Series Editor: R. Moreau

Aims and Scope of the Series

The purpose of this series is to focus on subjects in which fluid mechanics plays a fundamental role. As well as the more traditional applications of aeronautics, hydraulics, heat and mass transfer etc., books will be published dealing with topics which are currently in a state of rapid development, such as turbulence, suspensions and multiphase fluids, super and hypersonic flows and numerical modelling techniques. It is a widely held view that it is the interdisciplinary subjects that will receive intense scientific attention, bringing them to the forefront of technological advancement. Fluids have the ability to transport matter and its properties as well as transmit force, therefore fluid mechanics is a subject that is particularly open to cross fertilisation with other sciences and disciplines of engineering. The subject of fluid mechanics will be highly relevant in domains such as chemical, metallurgical, biological and ecological engineering. This series is particularly open to such new multidisciplinary domains.

1. M. Lesieur: *Turbulence in Fluids.* 2nd rev. ed., 1990 ISBN 0-7923-0645-7
2. O. Métais and M. Lesieur (eds.): *Turbulence and Coherent Structures.* 1991 ISBN 0-7923-0646-5
3. R. Moreau: *Magnetohydrodynamics.* 1990 ISBN 0-7923-0937-5
4. E. Coustols (ed.): *Turbulence Control by Passive Means.* 1990 ISBN 0-7923-1020-9
5. A.A. Borissov (ed.): *Dynamic Structure of Detonation in Gaseous and Dispersed Media.* 1991 ISBN 0-7923-1340-2
6. K.-S. Choi (ed.): *Recent Developments in Turbulence Management.* 1991 ISBN 0-7923-1477-8
7. E.P. Evans and B. Coulbeck (eds.): *Pipeline Systems.* 1992 ISBN 0-7923-1668-1
8. B. Nau (ed.): *Fluid Sealing.* 1992 ISBN 0-7923-1669-X
9. T.K.S. Murthy (ed.): *Computational Methods in Hypersonic Aerodynamics.* 1992 ISBN 0-7923-1673-8
10. R. King (ed.): *Fluid Mechanics of Mixing.* Modelling, Operations and Experimental Techniques. 1992 ISBN 0-7923-1720-3
11. Z. Han and X. Yin: *Shock Dynamics.* 1993 ISBN 0-7923-1746-7
12. L. Svarovsky and M.T. Thew (eds.): *Hydroclones.* Analysis and Applications. 1992 ISBN 0-7923-1876-5
13. A. Lichtarowicz (ed.): *Jet Cutting Technology.* 1992 ISBN 0-7923-1979-6
14. F.T.M. Nieuwstadt (ed.): *Flow Visualization and Image Analysis.* 1993 ISBN 0-7923-1994-X
15. A.J. Saul (ed.): *Floods and Flood Management.* 1992 ISBN 0-7923-2078-6
16. D.E. Ashpis, T.B. Gatski and R. Hirsh (eds.): *Instabilities and Turbulence in Engineering Flows.* 1993 ISBN 0-7923-2161-8
17. R.S. Azad: *The Atmospheric Boundary Layer for Engineers.* 1993 ISBN 0-7923-2187-1
18. F.T.M. Nieuwstadt (ed.): *Advances in Turbulence IV.* 1993 ISBN 0-7923-2282-7
19. K.K. Prasad (ed.): *Further Developments in Turbulence Management.* 1993 ISBN 0-7923-2291-6
20. Y.A. Tatarchenko: *Shaped Crystal Growth.* 1993 ISBN 0-7923-2419-6

Kluwer Academic Publishers – Dordrecht / Boston / London

Mechanics

FLUID MECHANICS AND ITS APPLICATIONS

Series Editor: R. Moreau

21. J.P. Bonnet and M.N. Glauser (eds.): *Eddy Structure Identification in Free Turbulent Shear Flows.* 1993 ISBN 0-7923-2449-8
22. R.S. Srivastava: *Interaction of Shock Waves.* 1994 ISBN 0-7923-2920-1
23. J.R. Blake, J.M. Boulton-Stone and N.H. Thomas (eds.): *Bubble Dynamics and Interface Phenomena.* 1994 ISBN 0-7923-3008-0
24. R. Benzi (ed.): *Advances in Turbulence V.* 1995 ISBN 0-7923-3032-3
25. B.I. Rabinovich, V.G. Lebedev and A.I. Mytarev: *Vortex Processes and Solid Body Dynamics.* The Dynamic Problems of Spacecrafts and Magnetic Levitation Systems. 1994 ISBN 0-7923-3092-7
26. P.R. Voke, L. Kleiser and J.-P. Chollet (eds.): *Direct and Large-Eddy Simulation I.* Selected papers from the First ERCOFTAC Workshop on Direct and Large-Eddy Simulation. 1994 ISBN 0-7923-3106-0
27. J.A. Sparenberg: *Hydrodynamic Propulsion and its Optimization.* Analytic Theory. 1995 ISBN 0-7923-3201-6
28. J.F. Dijksman and G.D.C. Kuiken (eds.): *IUTAM Symposium on Numerical Simulation of Non-Isothermal Flow of Viscoelastic Liquids.* Proceedings of an IUTAM Symposium held in Kerkrade, The Netherlands. 1995 ISBN 0-7923-3262-8
29. B.M. Boubnov and G.S. Golitsyn: *Convection in Rotating Fluids.* 1995 ISBN 0-7923-3371-3
30. S.I. Green (ed.): *Fluid Vortices.* 1995 ISBN 0-7923-3376-4
31. S. Morioka and L. van Wijngaarden (eds.): *IUTAM Symposium on Waves in Liquid/Gas and Liquid/Vapour Two-Phase Systems.* 1995 ISBN 0-7923-3424-8
32. A. Gyr and H.-W. Bewersdorff: *Drag Reduction of Turbulent Flows by Additives.* 1995 ISBN 0-7923-3485-X
33. Y.P. Golovachov: *Numerical Simulation of Viscous Shock Layer Flows.* 1995 ISBN 0-7923-3626-7
34. J. Grue, B. Gjevik and J.E. Weber (eds.): *Waves and Nonlinear Processes in Hydrodynamics.* 1996 ISBN 0-7923-4031-0
35. P.W. Duck and P. Hall (eds.): *IUTAM Symposium on Nonlinear Instability and Transition in Three-Dimensional Boundary Layers.* 1996 ISBN 0-7923-4079-5
36. S. Gavrilakis, L. Machiels and P.A. Monkewitz (eds.): *Advances in Turbulence VI.* Proceedings of the 6th European Turbulence Conference. 1996 ISBN 0-7923-4132-5
37. K. Gersten (ed.): *IUTAM Symposium on Asymptotic Methods for Turbulent Shear Flows at High Reynolds Numbers.* Proceedings of the IUTAM Symposium held in Bochum, Germany. 1996 ISBN 0-7923-4138-4
38. J. Verhás: *Thermodynamics and Rheology.* 1997 ISBN 0-7923-4251-8
39. M. Champion and B. Deshaies (eds.): *IUTAM Symposium on Combustion in Supersonic Flows.* Proceedings of the IUTAM Symposium held in Poitiers, France. 1997 ISBN 0-7923-4313-1
40. M. Lesieur: *Turbulence in Fluids.* Third Revised and Enlarged Edition. 1997 ISBN 0-7923-4415-4; Pb: 0-7923-4416-2

Kluwer Academic Publishers – Dordrecht / Boston / London

Mechanics

FLUID MECHANICS AND ITS APPLICATIONS

Series Editor: R. Moreau

41. L. Fulachier, J.L. Lumley and F. Anselmet (eds.): *IUTAM Symposium on Variable Density Low-Speed Turbulent Flows.* Proceedings of the IUTAM Symposium held in Marseille, France. 1997 ISBN 0-7923-4602-5
42. B.K. Shivamoggi: *Nonlinear Dynamics and Chaotic Phenomena.* An Introduction. 1997 ISBN 0-7923-4772-2
43. H. Ramkissoon, *IUTAM Symposium on Lubricated Transport of Viscous Materials.* Proceedings of the IUTAM Symposium held in Tobago, West Indies. 1998 ISBN 0-7923-4897-4
44. E. Krause and K. Gersten, *IUTAM Symposium on Dynamics of Slender Vortices.* Proceedings of the IUTAM Symposium held in Aachen, Germany. 1998 ISBN 0-7923-5041-3
45. A. Biesheuvel and G.J.F. van Heyst (eds.): *In Fascination of Fluid Dynamics.* A Symposium in honour of Leen van Wijngaarden. 1998 ISBN 0-7923-5078-2
46. U. Frisch (ed.): *Advances in Turbulence VII.* Proceedings of the Seventh European Turbulence Conference, held in Saint-Jean Cap Ferrat, 30 June–3 July 1998. 1998 ISBN 0-7923-5115-0
47. E.F. Toro and J.F. Clarke: *Numerical Methods for Wave Propagation.* Selected Contributions from the Workshop held in Manchester, UK. 1998 ISBN 0-7923-5125-8
48. A. Yoshizawa: *Hydrodynamic and Magnetohydrodynamic Turbulent Flows.* Modelling and Statistical Theory. 1998 ISBN 0-7923-5225-4
49. T.L. Geers (ed.): *IUTAM Symposium on Computational Methods for Unbounded Domains.* 1998 ISBN 0-7923-5266-1
50. Z. Zapryanov and S. Tabakova: *Dynamics of Bubbles, Drops and Rigid Particles.* 1999 ISBN 0-7923-5347-1
51. A. Alemany, Ph. Marty and J.P. Thibault (eds.): *Transfer Phenomena in Magnetohydrodynamic and Electroconducting Flows.* 1999 ISBN 0-7923-5532-6
52. J.N. Sørensen, E.J. Hopfinger and N. Aubry (eds.): *IUTAM Symposium on Simulation and Identification of Organized Structures in Flows.* 1999 ISBN 0-7923-5603-9
53. G.E.A. Meier and P.R. Viswanath (eds.): *IUTAM Symposium on Mechanics of Passive and Active Flow Control.* 1999 ISBN 0-7923-5928-3
54. D. Knight and L. Sakell (eds.): *Recent Advances in DNS and LES.* 1999 ISBN 0-7923-6004-4
55. P. Orlandi: *Fluid Flow Phenomena.* A Numerical Toolkit. 2000 ISBN 0-7923-6095-8

Kluwer Academic Publishers – Dordrecht / Boston / London

ERCOFTAC SERIES

1. A. Gyr and F.-S. Rys (eds.): *Diffusion and Transport of Pollutants in Atmospheric Mesoscale Flow Fields.* 1995 ISBN 0-7923-3260-1
2. M. Hallbäck, D.S. Henningson, A.V. Johansson and P.H. Alfredsson (eds.): *Turbulence and Transition Modelling.* Lecture Notes from the ERCOFTAC/IUTAM Summerschool held in Stockholm. 1996 ISBN 0-7923-4060-4
3. P. Wesseling (ed.): *High Performance Computing in Fluid Dynamics.* Proceedings of the Summerschool held in Delft, The Netherlands. 1996 ISBN 0-7923-4063-9
4. Th. Dracos (ed.): *Three-Dimensional Velocity and Vorticity Measuring and Image Analysis Techniques.* Lecture Notes from the Short Course held in Z~rich, Switzerland. 1996 ISBN 0-7923-4256-9
5. J.-P. Chollet, P.R. Voke and L. Kleiser (eds.): *Direct and Large-Eddy Simulation II.* Proceedings of the ERCOFTAC Workshop held in Grenoble, France. 1997 ISBN 0-7923-4687-4

KLUWER ACADEMIC PUBLISHERS – DORDRECHT / BOSTON / LONDON

GPSR Compliance
The European Union's (EU) General Product Safety Regulation (GPSR) is a set of rules that requires consumer products to be safe and our obligations to ensure this.

If you have any concerns about our products, you can contact us on

ProductSafety@springernature.com

In case Publisher is established outside the EU, the EU authorized representative is:

Springer Nature Customer Service Center GmbH
Europaplatz 3
69115 Heidelberg, Germany

www.ingramcontent.com/pod-product-compliance
Ingram Content Group UK Ltd.
Pitfield, Milton Keynes, MK11 3LW, UK
UKHW021315130926
15127UKWH00002B/18
9780792360957